T0185937

Green Energy and Technology

Climate change, environmental impact and the limited natural resources urge scientific research and novel technical solutions. The monograph series Green Energy and Technology serves as a publishing platform for scientific and technological approaches to "green"—i.e. environmentally friendly and sustainable—technologies. While a focus lies on energy and power supply, it also covers "green" solutions in industrial engineering and engineering design. Green Energy and Technology addresses researchers, advanced students, technical consultants as well as decision makers in industries and politics. Hence, the level of presentation spans from instructional to highly technical.

Indexed in Scopus.

More information about this series at http://www.springer.com/series/8059

Adriano Bisello · Daniele Vettorato ·
Håvard Haarstad · Judith Borsboom-van Beurden
Editors

Smart and Sustainable Planning for Cities and Regions

Results of SSPCR 2019

 Springer

Editors
Adriano Bisello
EURAC Research
Bolzano/Bozen, Italy

Daniele Vettorato
EURAC Research
Bolzano/Bozen, Italy

Håvard Haarstad
Centre for Climate and Energy
Transformation and Department
of Geography
University of Bergen
Bergen, Norway

Judith Borsboom-van Beurden
Locality
Driebergen-Rijsenburg, The Netherlands

ISSN 1865-3529   ISSN 1865-3537   (electronic)
Green Energy and Technology
ISBN 978-3-030-57334-8   ISBN 978-3-030-57332-4   (eBook)
https://doi.org/10.1007/978-3-030-57332-4

This Springer imprint is published by the registered company Springer Nature Switzerland AG
The registered company address is: Gewerbestrasse 11, 6330 Cham, Switzerland

SSPCR 2019 Committees

Project Manager and Scientific Committee Coordinator

Adriano Bisello—Eurac Research

Scientific Committee Members

Adriano Bisello—Eurac Research
Daniele Vettorato—Eurac Research, ISOCARP Institute
Michael Nippa—Free University of Bolzano
Dodo zu Knyphausen–Aufseß—TU Berlin
Giuliano Marella—University of Padua
Håvard Haarstad—University of Bergen
Francesco Calabrò—Mediterranea University Reggio Calabria
Lukas Kranzl—TU Wien
Luca Mora—Edinburgh Napier University
Valentina Antoniucci—University of Padua
Giuseppina Cassalia—Mediterranea University Reggio Calabria
Paola Clerici Maestosi—ENEA, Smart Energy Division
Håkan Perslow—RISE Research Institutes of Sweden
Grazia M. Fiore—Eurisy
Elisa Ravazzoli—Eurac Research
Jessica Balest—Eurac Research
Valentina D'Alonzo—Eurac Research
Nives Della Valle—Eurac Research
Sonia Gantioler—Eurac Research
Antonio Novelli—Eurac Research
Simon Pezzutto—Eurac Research
Alyona Zubaryeva—Eurac Research

Organizing Committee

Maria Pruss—Eurac Research, Meeting Management
Camilla Piz—Eurac Research, Meeting Management
Eliana Begal—Eurac Research
Alessandra Barbieri—Eurac Research

Contacts

www.sspcr.eurac.edu

Preface

Third Time's a Charm: Highlights from SSPCR 2019

Several years have passed since we brainstormed the idea to organize an international conference on innovative, smart and sustainable urban and regional planning approaches at Eurac Research in Bolzano.

When we decided to propose the SSPCR conference in autumn 2015, it was a leap in the dark. We did not know whether anyone would actually answer the call for papers (because the conference was competing with a high number of already well known international events) and travel to Bolzano/Bozen (which for sure is a lovely place, but not exactly easy to reach!) to take part in the first edition of *Smart and Sustainable Planning for Cities and Regions*. There was some skepticism, but there was also a lot of excitement. In the end, we decided to organize it around the concept of "debate" with two full days of engaging discussion, exchange of opinions, open arenas, and inspiring keynote speeches.

Although from the beginning international and sparkling, thanks to the involvement of ISOCARP—the International Society of City and Regional Planners —SSPCR 2015 was still a small scale conference, with just some dozens of participants, but the feedback from delegates and presenters encouraged us to proceed further in this direction. Moreover, teaming up with the international publisher Springer for the publication of the conference results in the valuable book series "Green Energy and Technology" ensured adequate standards and broad dissemination of scientific results.

So, it was quickly decided to invest in SSPCR, enlarging and improving the format. In the spring of 2017, at the Eurac Research headquarter, we held the second edition of the conference, SSPCR 2017, under the motto "inspiring the transition of urban areas towards smarter and more sustainable places to live". The conference experienced quite some growth, more than doubling the number of presentations and participants. We started to team with symposium "New Metropolitan Perspectives," of high reputation at the national level; at the same time, the presence of international speakers and representatives from global institutions such as the OECD—the Organization for Economic Co-operation and Development—ensured the necessary attractiveness and diversity of the related scientific ecosystem.

In parallel, with the evolution of the SSPCR, also the "smartness" concept evolved. Researchers, practitioners, and decision-makers have experienced the diversity, contradictions, and evolution of the "smart city," "smart region," and "smart specialization" concepts, the criticism emerging about purely technological approaches, and the need to recombine these with the previous concept of "sustainability," leaving more space to the "co-creation" of innovative solutions. In particular, mobilizing urban planning conceptual frames for open, integrated, and common solutions, defining ICT smart city governance investments and driving sustainability transitions for carbon-neutral cities and contributing to the overall resilience of our settlements and achievement of the Sustainable Development Goals (SDGs). Thus, SSPCR 2019, the third edition of the conference, friendly named the "winter edition," aimed to provide orientation to delegates within the giant galaxy of smart and sustainable planning, supporting the transposition of research into practice and visionary approaches into groundbreaking planning policies and tools. This objective has been transposed into various thematic tracks, each investigating specific topics and recurrent keywords. Around 150 research results, including poster exhibitions, originating in 38 countries were presented at the SSPCR 2019.

Delegates have been inspired by the keynote speech of Håvard Haarstad (University of Bergen) addressing the question of how to catalyze sustainable transformation in places and cities and Claudia Baranzelli (European Commission —Joint Research Center) about the challenges influencing the future of cities in Europe and beyond, with a focus on the interoperability of data.

For the first time ever, the SSPCR 2019 was also complemented by special sessions or side events organized and promoted by relevant national or European networks and partnerships, which decided to come to Bolzano/Bozen to have one of their recurrent meetings (e.g., EERA JPSC; SMARTENCITY project, HAPPEN Programme), providing training (HOTMAPS project), an update on funding opportunities and integrated approaches offered by the forthcoming Horizon Europe program (UERA, JPI, EIP—SCC), as well as national strategies (ENEA, EIP—SCC).

By summing this all up, the publication "*Smart and Sustainable Planning for Cities and Regions: Results of SSPCR 2019*" will provide the reader with a comprehensive overview of recent and original work in the field and expand shared knowledge among researchers, professionals, decision-makers, and civil society.

A selection of the top contributions is offered within this volume, while other research outputs are distributed free of cost in a separate volume available under an Open Access license.

In Part One, the smart climate–energy transition at the urban and regional level is addressed from various perspectives. Teremranova and Mutule face the question of how the stakeholders, such as energy policy-makers, governance, utilities, and municipalities, can choose a model that would promote maximum efficiency by implementation in the energy sector of the city. A similar purpose drives the research done by Rodrigues et al.; those looking at European Smart City projects (MAtchUP and MAKING-CITY) propose a city-level evaluation framework to support cities in their strategic-planning process to progress toward sustainability and smartness. Becchio et al. propose a new evaluation framework at the district level that combines energy, environmental and social impacts in the COmposIte Modeling Assessment (COSIMA) model for ranking alternative retrofit scenarios for existing districts, pointing out the relevance of co-benefits in the assessment phase. Aboulnaga, Sala, and Trombadore review smart and green strategies that mobilize stakeholders in introducing adaptive natural-based green solutions to ensure city resilience, including urban farming and green facades. Cetara et al. focus on climate change in the Alpine region by implementing a mixed analysis model (qualitative and quantitative), providing a standard procedure aimed at identifying suitable adaptation actions for sub-regional geographical units in a mountain environment. Guastella et al., using downscaled population scenarios, derive some projections of electricity demand in European cities in 2050, while investigating the correlation of consumption to land-use and other relevant variables. Costanzo and Baldissara go even further into detail at the building level, investigating how a pilot project combining data from energy-performance certificates with bottom-up information on building renovation may effectively support the decision making at different territorial scales. Mahmoud and Ragab present the result of a modeling activity supporting the evidence that effective architectural features such as courtyards can mitigate heat stress in hot arid regions. An appropriate configuration of the courtyard leads to significant improvement in the thermal performance of the building and directly influences the behavior of its users and the functionality of the space. Borsboom-van Beurden and Costa report on a special session organized by UERA and EIP-SCC, which discussed how integrated planning and implementation can help to boost the transition to smart, low-carbon cities, how good examples of such an integrated approach can foster wider replication of smart and energy efficient solutions in cities across Europe, and which research and innovation gaps persist.

Part Two investigates how the availability of urban (big) data, spatial data, and ICT tools poses great challenges and opportunities for information retrieval and knowledge discovery. Bukovszki et al., by analyzing the impact of the ICT-accessorized planning process, outline the necessity of a new professional profile bridging urban planning knowledge with specific data-science skills. Aligned with that, Samperio-Valdivieso et al. report on a Web-based application called City Indicators Visualization and Information System (CIVIS) that supports city planners when handling and interpreting data, solving automatically the complex processes of data processing and storage, selecting the indicators and defining the city-related values. Schweigkofler et al. offer insights on how the research project "OPENIoT4SmartCities," developed and tested in the city of Merano (Italy), an operating tool that supports defining a smart city roadmap for strategic decision making. Abastante, Gaballo, and La Riccia introduce a multi-methodological framework to analyze urban walkability by analyzing qualitative and quantitative attributes in an open-source Geographic Information Systems. What's more, Gorrini et al. present a data-driven approach developed within the DIAMOND project based on spatial data and data analytics for assessing the level of accessibility for female passengers of the railway network in Catalunya (Spain).

Part Three addresses how new value propositions emerge at this time of urban innovation ecosystems and sharing economy. Andreucci and Croci, by providing an overview of selected international initiatives, showcase how districts and cities are advancing progress of the circular economy concept in practice by implementing "nexus" initiatives in redevelopment projects. Mangialardo and Micelli draw insights from their recent research comparing the approach of the demolition and reconstruction of obsolete buildings to the reuse of existing assets, thanks to a model that holds together spatial and economic variables. De Franco tackles the issue of private abandoned buildings in urban areas under a neo-institutional approach, discussing multiple causes and potential solutions to the problem. Coscia and Rubino discuss how to integrate the social-impact perspective into a redevelopment project of the system of the historic farmhouses in Volpiano (Italy), focusing on youths not in education, employment or training. Casalicchio et al. explore the new dimension of the renewable-energy community, by simulating various business models to be offered to prosumers of multi-family housing buildings in which the economic return of every participant in the energy community is evaluated. Antoniucci et al., by reviewing contingent valuation studies applied to the assessment of environmental goods and municipal waste management, define a set of statistically significant variables influencing the feasibility of waste treatment at the municipal level and enhancing the recycling of materials. Kim and Wu investigate how the usual linear take-make-waste approach of the fashion industry in New York City (USA) may shift to a circular model that minimizes nonrenewable resource consumption and landfilling of textile waste, introducing the Internet of Things concept.

In Part IV, new strategies to dissolve categorical and social borders are discussed, aiming to develop integrated territorial approaches, from smart cities to smart regions. Svanda and Hirschler discuss how the establishment of governance structures in urban regions in Austria will help convert steering and coordination areas into functional areas. In this way, it will be possible to solve spatial challenges jointly, to bundle resources, and to raise the willingness to cooperate among stakeholders. Aragona, by recalling the 2007 EU Charter of Leipzig—which calls for integrated planning strategies between rural and urban, small, medium, large and metropolitan areas—suggests how to solve potential conflicts between urban areas and rural areas, stressing the potential of this latter in delivering well-being and a high quality of life. Xu and Zhang analyze the case study of Yanhe Village in Hubei Province (China) as an effective example of expert knowledge combined with local knowledge, thus delivering the precondition to establish smart and sustainable regions. Dabis concludes this section focusing on the need to maintain linguistic and cultural diversity in society as a crucial part of the human rights-based approach to the achievement of the SDGs.

In Part Five, the authors explore new environments and approaches that foster civic engagement and collaborative action toward an innovative governance practice. Lissandrello analyzes the interplay between citizens and data through the case study of the Mobility Urban Values project, in which planning theory is combined with theories of societal change under a critical pragmatism. Galassi, Petríková, and Scacchi discuss how to remove the barriers to digital participation for community development and propose good-practice examples for facilitating the process of adopting and integrating digital technologies within such settings. Monardo and Massari argue that social innovation is strictly path-dependent, enabled by "opportunity windows." In this regard, they analyze the experiences of Innovation Centers established in Bologna (Italy) and Boston (USA), which, operating as "intermediate places," can recover direct relationships between the various stakeholders in the urban arena, providing a perspective on long-term change. Andreucci investigates the concept of a smart creative city. Looking at the Lyon metropolis (France) case study, it becomes clear how creativity, art, and culture represent strategic assets in economic development in the urban regeneration process. Massa reports on National Research and Innovation Programmes related to the urban topic, stressing how alignment is the strategic approach taken by the Member States to modify their strategies, priorities, or activities as a consequence of the adoption of joint research priorities in the context of Joint Programming. Santoro and Motta Zanin discuss the differences between a traditional water risk-management tool and an innovative approach in a problem-structuring method that considers stakeholders' risk perception, socioeconomic dynamics, interaction, previous experience, values, and cultural factors. Rasheed, based upon the experience of recent floods in Kerala (India), develops an approach-based framework for decision-makers and physical planners to understand and reduce vulnerability to disasters and ensure community participation. Benda focuses on how sustainable urbanization and SDG 11 are strictly related to the promotion of human rights and the improved meaning of security, which involves respect for fundamental rights.

In Part Six, the energy poverty issue is discussed from various points of view, providing a multi-disciplinary framework. Some of the contributions have been discussed within the special session organized by OIPE, the Italian Observatory on Energy Poverty. Núñez-Peiró et al. referring to the city of Madrid (Spain) address the so-called feminization of energy poverty, urging a revision of the existing studies from a gender perspective to foster its inclusion within energy-poverty-alleviation policies. Martini questions the effectiveness of incentive measures, such as the Ecobonus in Italy, in supporting energy-poverty eradication, while Fanghella and Della Valle examine the behavioral process underlying tenants' usage of in-home display in smart buildings retrofitted within the SINFONIA project in Bolzano (Italy), with a special focus on cognitive biases. Santangelo, Tondelli, and Yan point out how building energy renovation may be a key opportunity for rolling-out a comprehensive urban regeneration strategy against energy poverty but only if the human factor (i.e., the behavior of occupants) is properly considered. D'Alpaos and Bragolusi provide a methodological framework to identify cost-effective and cost-optimal strategies of intervention that match technological advancements and knowledge in energy retrofitting, with both social and environmental needs and end-user behavior.

At the end of the volume, Part Seven discusses new approaches to rural–urban relationships for better territorial development. Ravazzoli et al. provide an overview of the topic and synthesize the discussions occurring during the special session they chaired. Lengyel and Friedrich present a seven-step integrated methodology for unraveling and modeling urban processes at the regional scale, applied to the Ruhr region (Germany). Grilli and Curtis performed a latent-class analysis to reveal the attitude of Irish citizens living in major cities toward using green spaces in view of dedicated policy design. Calcatinge argues that cultural landscape-policy research should include a more technological approach, by including topics like open-source technologies to design effective future policies.

Additional thematic contributions may be found in the "Open Access" volume *"Smart and Sustainable Planning for Cities and Regions: Results of SSPCR 2019— Open Access Contributions"*.

Concluding, we would like to thank all the people actively involved in the organization of the SSPCR 2019 conference, in particular the Eurac's Meeting Management team for their priceless technical assistance, the Scientific Committee, the keynote speakers and sessions' moderators, and all who actively contributed to the conference through their initiatives, posters, and presentations. A special thanks to our supporting partners for providing financial support to the event and contributing to communication and dissemination activities. Our gratitude to Judith Borsboom-van Beurden (Locality—EU) and Håvard Haarstad (University of Bergen) for joining the editorial board of this volume and to the anonymous reviewers who donated their precious time and competence in reading, assessing, and commenting on the submitted manuscripts, providing a valuable contribution to ensure high-level quality (despite the COVID-19 outbreak and personal, as well as work-related, difficulties). Looking forward to welcoming you at the next SSPCR "Summer Edition" (the only one missing so far!).

Bolzano, Italy   Adriano Bisello
                                                                                     Daniele Vettorato

About This Book

Investigating the potential of urban planning to make cities and regions more sustainable in a smart way: This was the purpose of the 3rd international conference "Smart and Sustainable Planning for Cities and Regions," held in December 2019 in Bolzano/Bozen, Italy.

This book offers a selection of research papers and case studies presented at the conference and exploring the concept of smart and sustainable planning, including top contributions from academics, policy-makers, consultants, and other professionals.

Innovation processes such as co-design and co-creation help establish collaborations that engage with stakeholders in a trustworthy and transparent environment while answering the need for new value propositions.

The importance of an integrated, holistic approach is widely recognized to break down silos in local government, in particular, when aimed at achieving a better integration of climate–energy planning. Despite the ongoing urbanization and polarization processes, new synergies between urban and rural areas emerge, linking development opportunities to intrinsic cultural, natural, and man-made landscape values. The increasing availability of big, real-time urban data and advanced ICT facilitates frequent assessment and continuous monitoring of performances, while allowing fine-tuning as needed. This is valid not only for individual projects but also on a wider scale. In addition, and circling back to the first point, (big) urban data and ICT can be of enormous help in facilitating engagement and co-creation by raising awareness and by providing insight into the local consequences of specific plans. However, this potential is not yet fully exploited in standard processes and procedures, which can therefore lack the agility and flexibility to keep up with the pulse of the city and dynamics of society.

The aim of this book is thus to provide a multi-disciplinary outlook based on experience to orient the reader in the giant galaxy of smart and sustainable planning, support the transposition of research into practice, scale up visionary approaches, and design groundbreaking planning policies and tools.

Highlights

- Offers empirical and theoretical insights into planning for smart and sustainable cities and regions
- Combines multi-disciplinary approaches, giving new suggestions to researchers, policy, and decision-makers
- Delivers a grounded perspective on contemporary challenges for smartness, a circular economy, climate-neutrality, and overall sustainability through a wealth of local and regional case studies from Europe and beyond
- Constitutes an excellent overview of up-to-date tools, models, and methods for implementing and scaling up smart city solutions

Contents

About the Editors

Adriano Bisello is a senior researcher in the Urban and Regional Energy Systems research group at Eurac in Bolzano, Italy. Coordinator of the SSPCR conference series, he manages international European co-funded projects in the field of smart cities and visionary urban solutions. His research focuses on the assessment of the multiple benefits of energy transition and urban transformation processes.

Daniele Vettorato is the coordinator of the Urban and Regional Energy Systems research group at Eurac in Bolzano, Italy. He has been the Vice-President of ISOCARP (International Society of City and Regional Planners) since 2017.

Håvard Haarstad is professor of human geography and Director for the Centre for Climate and Energy Transformation, University of Bergen, Norway. His work addresses urban sustainability and the broader processes of climate and energy transformation.

Judith Borsboom-van Beurden is director of Locality, a research and innovation management consultancy. Her main research interest is how integrated planning and implementation of smart and climate-neutral solutions can help to decarbonize cities. She coordinates the Action Clusters in the European Innovation Partnership (EIP) on Smart Cities and Communities and chairs the Urban Europe Research Alliance (UERA).

Shaping the Climate and Energy Transition: Clean Energy and Robust Systems for All

Smart Approach to Management of Energy Resources in Smart Cities: Evaluation of Models and Methods

Jana Teremranova and Anna Mutule

Abstract A smart city represents a complex approach to the development of all systems, directions and spheres aimed at providing comfortable living conditions for citizens and qualitative management and reliable infrastructure for the city. The paper answers questions about how to estimate the ability of the implementation approach within a smart city, and how the stakeholders, such as energy policy makers, governance, utilities and municipalities, can choose a model that would promote maximum efficiency by the implementation in the energy sector of the city. The paper has three main sections, the first of which provides a literature review with a classification of different approaches to the implementation of a smart city depending on the scope and tasks. The main emphasis is placed on the analysis of the energy component of smart city decisions. The second section proposes a series of criteria for evaluating the existing approaches to a smart city. The ability to flexibly react to the changes in consumption, unpredicted situations, changes in the needs, sustainability in generation and the supply of electricity and heat, as well as considering alterations in the legislative base, has been taken into account when analysing smart city infrastructure. Finally, the third section presents a decision-making scheme related to choosing the development methodology in a smart city that enables accounting of multifaceted nature of urban environment development, as well as describing a scheme of the block-integrated approach to the methodology of smart city development. In the authors' opinion, it can have the largest potential for flexible, sustainable and stable development and will produce the greatest qualitative impact on smart city development. As a result, convolving the approaches to the development of a smart city according to various criteria makes possible highlighting the most optimal decision-making process.

Keywords Assessment of approaches · Platformization · Smart city · Smart energy

J. Teremranova (✉)
Riga Technical University, Riga, Latvia
e-mail: jana.teremranova@rtu.lv

A. Mutule
Institute of Physical Energetics, Riga, Latvia
e-mail: amutule@edi.lv

1   Introduction

At the moment, there is a wide variety of definitions and approaches to resolving issues within the so-called smart city. Comfortable living conditions (Jiménez et al. 2016; Bornholdt et al. 2019), high-quality and reliable management (De Filippi et al. 2019; Suwa 2020) are the key characteristics of a city that can call itself "smart", "intelligent". Over time, not only the definitions of a smart city have changed, but also have the ways to implement this sphere of development (Lim et al. 2019). Nevertheless, the multitude of approaches to solving smart city tasks is not only an advantage; it also brings difficulties in choosing the most appropriate model or solution.

In addition, it should be noted that the task of smart city modelling and evaluating is so complex that it requires a breakdown by application areas, namely, energy, economics, education, etc., in order to find a clear and unambiguous solution. There is a lack and ambiguity in assessing models of the urban environment as a whole in existing empirical literature on the energy sector of a smart city. Furthermore, assessment approaches are usually used for local purposes, limited by separate initiatives. This has served as an incentive for research with the development of a system of criteria for assessing the energy sector of a smart city. What can contribute to improving the feasibility of projects within the framework of a smart city? What should be taken into account when choosing a model that can bring the greatest results at the lowest cost? These problems have been solved by creating a spectrum of criteria for evaluating approaches to the energy aspect of a smart city, followed by the proposed decision-making scheme for choosing the best development methodology.

2   Description of Existing Smart City Models: Literature Review

There are many general models that use an integrated approach to transforming a city into a smart city. The present paper focuses on one of the various aspects, namely, energy. Undoubtedly, each specific model should be chosen in accordance with the tasks facing the city. Nowadays, a plentitude of approaches, methods and algorithms are available for solving tasks related to ensuring the sustainable operation of the energy sector of a city. Nevertheless, a shortage of systematization and assessment can be observed that are aimed at ensuring assistance to the stakeholders such as municipality and energy city planners that could help choose an optimal approach to implementing a plan of urban development and making changes towards a smart energy city. To achieve this goal, a table taken from Mutule et al. (2018) was further elaborated to create Table 1 in order to discuss existing approaches to the development of a smart city.

In order to encompass changes in a specific sector or part of an urban structure, local solutions to various energy issues are acceptable and sufficient (Calvillo

Table 1 Regimentation of the approaches to smart city development

Smart city approach	Description	Sources
Employment of optimization or automation to single sectors of development	Smart solutions aimed at solving the energetic problems of individual areas of the city. Additionally, the KPI sets, standardization and innovative IT-based approaches are used to create a suitable solution	Calvillo et al. (2016), Mattoni et al. (2015), EU Smart Cities Information System
Smartainability	This approach uses qualitative and quantitative indicators of technology assessment for intelligent solutions that are designed to improve energy efficiency and environmental sustainability in the city; it is more focused on integrated intelligent mobile platforms	Girardi and Temporelli (2017), Bounazef and Crutzen (2018), Ambrogi et al. (2016), Giordano et al. (2012)
New city planning	Planning and implementation of new smart districts or a city with pre-laid smart energy infrastructure (e.g. the use of 100% renewable energy sources for energy consumption and heating/cooling of buildings and the use of electric vehicles only, use of sensors, etc.) for further development and expansion. The approach considers state-of-the-art technologies and requirements for the level of comfort of residents and the preservation of the environment	BSI (2014), Farag (2019)
Smart city infrastructure architecture model (SCIAM)	Multi-level holistic approach to energy in a smart city; it uses separation of the energy infrastructure of the city into layers, levels and zones, considering their interactions	Uslar and Enge (2015), Bawany and Shamsi (2015)
Development of smart energy city (SEC)	Smart energy is presented as the most important and necessary aspect of the successful and sustainable development of a smart city	Mosannenzadeh et al. (2017), Nielsen et al. (2013), García-Fuentes et al. (2017), Papastamatiou et al. (2017)

(continued)

Table 1 (continued)

Smart city approach	Description	Sources
Energy hubs, multi-energy systems	Development and operation of a smart city through the creation of the so-called energy hubs aimed at the flexible integration of the diversity of the city's energy resources for the most efficient, cost-effective and stable resource management	Dall'Anese et al. (2017), Mancarella et al. (2016), Weisi and Ping (2014), Yu et al. (2014), Rayati et al. (2015), Si et al. (2018)
Blockchains	Considers blockchain technology application in a smart city context and in energy aspect as a focused task, by using for energy supply operations, measuring the amount of electricity consumed, billing for consumed resources and making payments	Marsal-Llacuna (2017), Hwanga et al. (2017), Li (2018), Kotobi and Sartipi (2018), Qian et al. (2018), Lazaroiu and Roscia (2018a), Guan et al. (2018), Xie et al. (2019), Shen and Pena-Mora (2018), Lazaroiu and Roscia (2018b)
Platformization	Combining information resources on energy generation, transmission, distribution and use in a smart city, which usually are not connected to each other on a unified platform; therefore, it makes it possible to simplify and clarify the procedure for monitoring and managing energy resources for both citizens and the administration of the city	Aguilera et al. (2017), Smart Cities Councel; International Electrotechnical Commission (2014), Bollier (2006), Elering; Energinet; Elhub; APCS; Anttiroiko (2016), Thornton (2016), Anttiroiko (2015), Kaulio (2010)
Frugal social smart city	A new concept for smart city proposed for Casablanca, Morocco. It is based on a global bottom-up multidisciplinary approach that relies on the informational and functional cost-effective integration of various urban complex systems such as energy, transport, health and governance	Hayar and Betis (2017), Ramesh (2015)

et al. 2016; Mattoni et al. 2015; EU Smart Cities Information System). However, the expansion of the area of influence, aimed to create integrated progress towards a smart city, has led to the formation of a special area called smartainability (Girardi and Temporelli 2017) which originated in Italy. One of the main priorities for the development of smart energy city highlights mobility as having a strong impact on the energy sector of the city as a whole (Bounazef and Crutzen 2018). Ambrogi et al. (2016), apart from the mobility, also include energy and telecommunication to the model under study, which helps to expand the view on smart city energy problems and solutions. The topics of methodology and the smartainability strategy were developed by Giordano et al. (2012). Along with more traditional approaches to changes in the direction of a smart city, the articles (BSI 2014; Arab News 2019) present an approach implementing modelling of an entire district or city, which makes possible, despite the high cost of such projects, integrating the advanced innovative energy developments into the model. Another way to approach the modelling of a smart city as a holistic structure is presented in the Smart City Infrastructure Architecture Model (SCIAM) (Uslar and Enge 2015; Bawany and Shamsi 2015). The main idea is to integrate the city information systems into the developed multi-level architecture in order to effectively provide public services to citizens. Investigating the concept of a smart energy city (Mosannenzadeh et al. 2017; Nielsen et al. 2013; García-Fuentes et al. 2017) primarily aims at optimizing urban energy systems and improving the quality of life of citizens, thus clarifying the relationship between a smart energy city, a smart city and a sustainable city, and offers a set of solutions and technologies in the energy field. Cooperation of multiple stakeholders and the integration of urban energy areas included in specific energy goals enable devising the most effective energy solutions to promote the sustainable development of smart cities (Papastamatiou et al. 2017). A different view on the solution of the flexibility and sustainability of the city energy structure is presented in research (Dall'Anese et al. 2017), where the authors propose integrating the so-called energy hubs into the city energy structure. In turn, Mancarella et al. (2016), Weisi and Ping (2014), Yu et al. (2014), Rayati et al. (2015) and Si et al. (2018) suggest applying a multi-energy management framework, thus making it possible for the city to achieve a new level of development of flexibility, sustainability, security and efficient use of energy resources through their optimal combination and solving critical issues. The same issues have been given the opportunity to be resolved with the help of the blockchain technology, which has gained popularity in recent years: from a literature review (Shen and Pena-Mora 2018) and the possibilities of the blockchain technology in an urban environment (Marsal-Llacuna 2017; Li 2018; Smart Cities Council), architecture models for the interaction of energy prosumers (Hwanga et al. 2017) to issues of effective communication in an urban environment using blockchain technology (Kotobi and Sartipi 2018; Xie et al. 2019) and the application of the blockchain technology and smart metres to help prosumers in energy production and consumption management (Lazaroiu and Roscia 2018b). In this case, it is also necessary to take into account possible negative aspects of this technology that may be encountered (Kiran et al. 2019).

To develop effective approaches to handling big data, both from city services and the individual data of city residents, Aguilera et al. (2017) use the idea of a city-wide platform architecture, while the Smart Cities Council report presents the advantages and benefits of open data, which are an integral part of the urban infrastructure platformization that is revealed by International Electrotechnical Commission 2014 in terms of collaboration, integration and interoperability. The user-centric city model focuses on citizens who are "co-designers, co-producers and co-learners" (Bollier 2006) with the government, while the Aguilera et al. (2017) study developing this topic also emphasizes the interaction of smart city management and the activity of its citizens and provides recommendations for state policy. In recent years, there have been an increasing number of examples regarding the practical application of existing platforms (Elering; Energinet; Elhub; APCS), combining various city infrastructures under one "roof" and considering a number of theoretical and practical issues for using platformization to improve the citizens' life in a smart city (Anttiroiko 2016, 2015; Thornton 2016; Kaulio 2010).

Another area that has deserved attention in recent years has made the frugal social smart city (Hayar and Betis 2017; Ramesh 2015) its main goal, making changes in the urban environment affordable and feasible with small means, which makes possible extending the experience of changes to a wide range of cities with small financial opportunities, owing to competent administration and increased activity of citizens. However, it is less focused on the use of new advanced technologies and thereby limits the range of possibilities for solving total tasks.

A wide variety of approaches to solving the problems of a smart city provides, on the one hand, the opportunity of applying an individual approach to a specific task for specific conditions of the urban environment and the present stage of development, and on the other hand, it creates a difficult situation for those who are forced to make the final decision on the use of tools that are most relevant to the needs of the city. For a clear and balanced choice, it is necessary to classify approaches with the subsequent methodology, allowing an optimal choice of a tool.

3 Methodology

To understand the choice of the appropriate model for the implementation of the planned development of the city, many aspects should be taken into account. To promote an understanding of the strengths and weaknesses of the approaches (see Table 1), a variety of criteria has been proposed that enables a comprehensive assessment of each of the approaches. The evaluation is provided for each criterion based on the frequency of occurrence in the literature and the importance of a particular criterion given by experts.

Flexibility is one of the main parameters that meets the modern requirements of a smart city, according to the literature review provided in Table 1. It manifests itself heterogeneously in the selected approaches, prevailing as much as possible in new city planning, energy hubs, blockchains, platformization and frugal social smart city. A vast majority of sources indicate the presence of flexibility in the mentioned approaches, while in the description of single sector optimization, smart energy city and SCIAM, the flexibility to respond to changes in needs is mentioned to a lesser extent, though it is present in these tools.

Transparency, as a tool for data openness, accessibility, and data-based operations, is part of a smart city system and is presented moderately among the selected sources. The platformization-based approach differs in the role of transparency that is put at the forefront, being one of the cornerstones of this method.

The economic affordability of the selected model ranges from low (with significant financial investments requiring a long payback time) in new city planning to high (requiring relatively low costs) at a separately selected optimization point, as well as in platformization and frugal social smart city, which originally incorporated the idea of low financial investments.

Attracting citizens to active participation (see Table 1) is becoming an increasingly valuable and necessary resource for the effective operation of a smart city. Many authors of publications point out the need for the active participation of citizens in urban management. The positions occupying the last three columns in Table 1 gained the maximum number of references about the possibility and desirability of participation of citizens in the formation of the urban environment. Accessibility of standardization and unification was mentioned in the literature as a significant and necessary criterion for the quality and sustainable development of the city.

The use of ICT has firmly entered the concept of a smart city, and without it, the development of a smart city is impossible. In most models (see Fig. 1), the readiness and necessity of transformations are noted, taking into account innovative information technologies. Only the frugal social smart city model, due to its specificity and initially set goal, focuses to a great extent on the human resources and to a lesser extent on computerization and technology.

The top–bottom and bottom-top criteria in the context of the approach to solving smart city issues can be considered in relation to the specific tasks set for the city and can serve as an additional support for clarifying the choice of the model. To a great extent, it depends on the institution initiating the transformation in the city: utilities, government, municipalities, electricity generation or distribution companies, energy efficiency projects and initiatives, etc. The consideration of these criteria also becomes important when applying some innovative technologies (e.g. blockchain, which involves the foundations of interaction on the basis of partnership and an equal contribution of the resources of the participants of the chain to the process).

To coordinate urban changes with the requirements of the EU, UN and others, for example, on energy efficiency or reducing CO_2 emissions in the atmosphere, many authors highlight an environment-friendly criterion. Its application will enable the city to successfully claim high positions in the ranking of smart cities in the future. To evaluate the approach from the prospects for further development, it is worth taking into account the criterion "opportunities for further development after the implementation of the model" (e.g. when planning a smart city, prospects for development are already considered in advance, as well as opportunities are envisaged

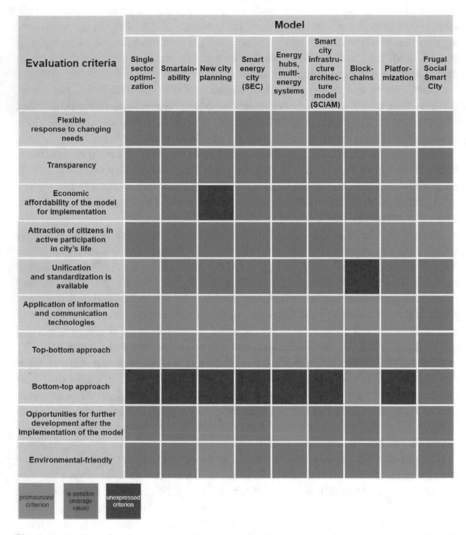

Fig. 1 Evaluation of various models of smart city development according to the criteria selected

for embedding the latest technologies that may appear after the implementation of the project of a new smart city).

The evaluation of all the criteria mentioned is illustrated in Fig. 1.

A general analysis of the evaluation criteria of the approaches demonstrates that platformization has a strong position in almost all criteria, which means that this approach can be highly recommended for the implementation of almost any activity within the framework of the development of a smart city. Platformization will allow naturally integrating the additional data, as well as the requirements for new methods of processing and presenting information into the existing structure. The blockchain approach also confidently leads in the number of functions shown, being a progressive and promising factor in the successful development of the city, where standardization and unification will follow the technology as it develops. Economic affordability (availability of the necessary financial resources) of the model for implementation certainly does not occupy the last place in the list of criteria, without which it is impossible to reasonably and responsibly plan any changes in the urban infrastructure. Although planning of a new city with the latest technologies and opportunities for environmentally friendly and active participation of residents is very attractive, the economic factor is the strongest negative factor for the wide implementation of this approach. In contrast, single sector optimization, platformization and frugal social smart city are examples of cost-effective spending, which are suitable for almost all cities, irrespective of the initial conditions and amount of resources. The optimal, from the point of view of the authors, model of development of a modern city is considered in the next section.

4 Findings

It would be improper to consider a certain model of the development of a smart city without considering the needs of the city, to which it should be applicable. The available resources, legislation, the degree of resident activity and the existing and proposed development plans should be considered, as well as the previous urban initiatives must be taken into account. Only after fulfilling the just-mentioned tasks, does it becomes possible to set an achievable city development goal, which implies the selection of an optimal methodology (see Fig. 2) that meets the requirements of all stakeholders. The coordination of interests of all stakeholders of the city will contribute to concretization of the necessary requirements and confidence that all opinions are maximally considered in an integrated approach.

The development of the urban environment is an ongoing cycle of changes that is based on the same algorithm each time, but with the use of an individual approach (methodology) for an urgent task, in order to advance to a smart city.

5 Discussion

Based on the analysis of evaluation criteria of approaches to the development of a smart city (see Fig. 1) and elaborated decision-making scheme (see Fig. 2), the authors conclude that the block-integrated approach, combining platformization, blockchain and one of the specific industry approaches (energy, transport, health care, economics, etc.), can have the greatest potential to flexible, sustainable and stable development and will enable the greatest possible quantum leap towards the development of a smart city (see Fig. 3). Such a developmental method presupposes the necessary transformation of the industry (using specific knowledge gained, indices and standards, attracting high-level professionals), enables accessibility, convenience

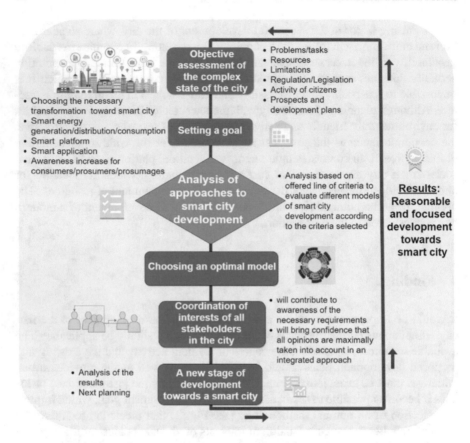

Fig. 2 Decision-making scheme for choosing the developmental methodology in a smart city

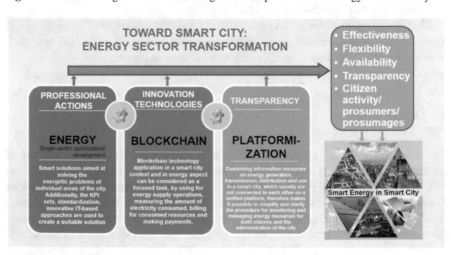

Fig. 3 Block-integrated approach to the developmental methodology of a smart city

and transparency of the data approach through an open platform and enables implementing plans with the help of modern reliable and fast blockchain technology as an effective platform for transactions within the network combining equal partners. A successful experience introducing blockchain technology into the energy industry can be illustrated by initiatives in Chile on a governmental level (CNE 2018), as well as in France, Italy, Germany, Japan and other countries where large energy companies or government organizations are already transferring energy data to the blockchain. They are interested in developing this concept from the conceptual level to a practical result, sharing the views of many world-class experts that blockchain is the advanced technology of the past decade and can become part of everyday life in the next few years. Potential options for using blockchain in the energy sector include not only the possibility of improving security, but also the use of peer-to-peer (P2P) distribution networks, customer billing (tracking of paid work and costs and billing and payment control) and renewable energy certificates.

The developed energy infrastructure of the city intends to use the advanced technologies in the field of energy, e.g. a smart approach, the organization of sustainable and flexible energy supply in the context of the country's specifics, where much attention is paid to training programs for residents on energy efficiency. In this case, the block-integrated approach will serve the replication of experience in a competitive environment and address the needs for flexible, sustainable and transparent city, e.g.

- the use of big data in the design of the platform: the necessary and sufficient amount of data, correct use of data, perception of different types of processed data by users, the ease of adding new data and the ease of handling them;
- combination on one platform of all available energy sources in the city and the ability to add new sources;
- the impact of the blockchain technology on the consumer market and the activity of residents, the accumulation of successful experience of its use, understanding of areas where blockchain can be used to the greatest benefit, etc.

6 Conclusions

The proposed decision-making scheme for selecting a smart city developmental methodology will consider all the results achieved by the analysis of opportunities and resources of the region, as well as will contribute to the accumulation and replication of best practices. This is the key to a successful, qualitative and long-term development. Serving as a guideline for optimal model selection, the evaluation criteria for a model of smart city development help administrators, urban planners and stakeholders. The most viable, efficient and stable smart city model must be flexible and adaptive to the resources and current needs of the city. Such a reliable model, along with a qualitative assessment using the multifunctional criteria proposed in the paper, can satisfy the needs of residents and participants in the urban infrastructure at any request and affordability level, and the positive experience can be disseminated and replicated to other cities and areas.

Acknowledgements This work has been supported by a project of the Latvian Council of Science: Management and Operation of an Intelligent Power System (I-POWER) (No. Izp-2018/1-0066) and the Latvian State Education Development Agency (VIAA), project "An ICT Platform for Sustainable Energy Ecosystem in Smart Cities" (ITCity), ID: ELAC2015/T10-0643.

References

Aguilera U, Peña O, Belmonte O, del Pina DL (2017) Services focused on civil data for smarter cities. Future Gener Comput Syst 76:234–247

Ambrogi R, Nigris M, Girardi PP, Temporelli A (2016) Contributions from research projects on the Italian power system: Accountability of sustainable energy projects. In: 2016 AEIT international annual conference (AEIT)

Anttiroiko A-V (2015) Smart cities: building platforms for innovative local economic restructuring. In: Rodríguez-Bolívar MP (ed) Transforming city for successful smart cities. Springer International Publishing, Cham, pp 23–41

Anttiroiko A-V (2016) City-as-a-platform: the rise of participatory innovation platforms in finnish cities. Sustainability 8:922

APCS. https://www.apcs.at/en. Accessed 20 Dec 2019

Arab News (2019) https://www.arabnews.com/node/1436436/business-economy. Accessed 10 Oct 2019

Bawany NZ, Shamsi AJ (2015) Smart city architecture: vision and challenges. Int J Adv Comput Sci Appl 6(11)

Bollier D (2006) The city of platform: how digital networks are changing urban life and governance. The Aspen Institute, Washington, DC, USA. https://csreports.aspeninstitute.org/documents/CityAsPlatform.pdf. Accessed 20 Dec 2019

Bornholdt H, Jost D, Kisters P, Schmidt TC, Fischer M (2019) SANE: smart networks for urban citizen participation. In: 26th international conference on telecommunications, ICT 2019. 8798771, pp 496–500

Bounazef D, Crutzen N (2018) Smartainability and mobility strategy: the case of Belgian local governments. In: The 6th international conference innovation management, entrepreneurship and sustainability (IMES 2018), Prague

BSI (2014) PD 8101 Smart city planning guidelines. https://www.bsigroup.com/en-GB/smart-cities/Smart-Cities-Standards-and-Publication/PD-8101-smart-cities-planning-guidelines/. Accessed 20 Dec 2019

Calvillo CF, Sánchez-Miralles A, Villar J (2016) Energy management and planning in smart cities. Renew Sustain Energy Rev 55:273–287

CNE, https://www.cne.cl/prensa/prensa-2018/04-abril-2018/ministra-jimenez-lanza-tecnologia-blockchain-en-datos-del-sector-energetico/. Accessed 27 Jan 2020

Dall'Anese E, Mancarella P, Monti A (2017) Unlocking flexibility integrated optimization and control of multi energy systems. IEEE Power Energy Maga 15(1):43–52

De Filippi F, Coscia C, Guido R (2019) From smart-cities to smart-communities: How can we evaluate the impacts of innovation and inclusive processes in urban context? Int J E-Planning Res 8(2):24–44

Elering. https://elering.ee/. Accessed 20 Dec 2019

Elhub. https://elhub.no/en/elhub. Accessed 20 Dec 2019

Energinet. https://energinet.dk/El/DataHub#Dokumenter. Accessed 20 Dec 2019

EU Smart Cities Information System. The making of a smart city: best practices across Europe. https://www.smartcities-infosystem.eu/sites/default/files/document/the_making_of_a_s mart_city_-_best_practices_across_europe.pdf. Accessed 10 Oct 2019

Farag AA (2019) The story of NEOM city: opportunities and challenges. New cities and community extensions in Egypt and the Middle East

García-Fuentes MA, Quijano A, de Torre C, García R, Compere P, Degard C, Tomé I (2017) European cities characterization as basis towards the replication of a smart and sustainable urban regeneration model. Energy Proc 111:836–845

Giordano V, Onyeji I, Fulli G, Sanchez Jimenez M, Filiou C (2012) Guidelines for conducting a cost-benefit analysis of Smart Grid projects. JRC, 2012 [online]. www.jrc.europa.eu. Accessed 10 Oct 2019

Girardi P, Temporelli A (2017) Smartainability: a methodology for assessing the sustainability of the smart city. Energy Proc 111:810–816

Guan Z, Si G, Zhang X, Wu L, Guizani N, Du X, Ma Y (2018) Privacy-preserving and efficient aggregation based on blockchain for power grid communications in smart communities. IEEE Commun Mag 56(7):82–88

Hayar A, Betis G (2017) Frugal social sustainable collaborative smart city Casablanca paving the way towards building new concept for "Future Smart Cities: By and for All". In: SENSET2017

Hwanga J, Choia M, Leea T, Jeona S, Kima S, Parka S, Park S (2017) Energy prosumer business model using blockchain system to ensure transparency and safety. Energy Proc 141:194–198

International Electrotechnical Commission (2014) Orchestrating infrastructure for sustainable Smart Cities. White Paper. Switzerland. https://www.iec.ch/whitepaper/pdf/iecWP-smartcities-LR-en.pdf. Accessed 20 Dec 2019

Jiménez CE, Falcone F, Solanas A, Puyosa H, Zoughbi S, González F (2016) Smart government: opportunities and challenges in smart cities development (Book Chapter). In: Civil and environmental engineering: concepts, methodologies, tools, and applications, vol 3, 2016, pp 1454–1472. IEEE Computer Society E-Government STC, Switzerland

Kaulio MA (2010) Customer, consumer and user involvement in product development: a framework and a review of selected methods. Total Qual Manag pp 141–149

Kiran S, Niranjan P, Pakala S (2019) A study on the limitations of blockchain and the usage of consensus models in blockchain technology. Int J Res

Kotobi K, Sartipi M (2018) Efficient and secure communications in smart cities using edge, caching, and blockchain. In: 2018 IEEE international smart cities conference (ISC2)

Lazaroiu C, Roscia M (2018a) RESCoin to improve prosumer side management into smart city. In: 7th international conference on renewable energy research and applications (ICRERA)

Lazaroiu GC, Roscia M (2018b) Blockchain and smart metering towards sustainable prosumers. In: 2018 international symposium on power electronics, electrical drives, automation and motion (SPEEDAM)

Li S (2018) Application of blockchain technology in smart city infrastructure. In: 2018 IEEE international conference on smart internet of things (SmartIoT)

Lim Y, Edelenbos J, Gianoli A (2019) Identifying the results of smart city development: findings from systematic literature review. Cities, Volume 95, Article number 102397

Mancarella P, Andersson G, Pecas-Lopes JA, Bell KRW (2016) Modelling of integrated multi-energy systems: drivers, requirements, and opportunities. In: Power systems computation conference (PSCC).

Marsal-Llacuna M-L (2017) Future living framework: is blockchain the next enabling network? Technol Forecasting Soc Change 128:226–234

Mattoni B, Gugliermetti F, Bisegna F (2015) A multilevel method to assess and design the renovation and integration of Smart Cities. Sustain Cities Soc 15:105–119

Mosannenzadeh F, Bisello A, Vaccaro R, D'Alonzo V, Hunter GW, Vettorato D (2017) Smart energy city development: a story told by urban planners. Cities 64:54–65. https://doi.org/10.1016/j.cit ies.2017.02.001. Accessed 20 Dec 2019

Mutule A, Teremranova J, Antoškovs Ņ (2018) Smart city through a flexible approach to smart energy. Latvian J Phys Techn Sci 55(1):3–14

Nielsen PS, Ben Amer-Allem S, Halsnæs K (2013) Definition of smart energy city and state of the art of 6 transform cities using key performance indicators. Deliverable 1.2, Technical University of Denmark

Papastamatiou I, Marinakis V, Doukas H, Psarras J (2017) A decision support framework for smart cities energy assessment and optimization. Energy Proc 111(2017):800–809

Qian Y, Liu Z, Yang J, Wang Q (2018) A method of exchanging data in smart city by blockchain. In: 2018 IEEE 20th international conference on high performance computing and communications

Ramesh G (2015) Frugal smart solutions for pan city development. https://udaipursmartcity.in/wp-content/uploads/2015/10/Frugal_Smart_Solutions-IIMB_Delhi.pdf. Accessed 20 Dec 2019

Rayati M, Sheikhi A, Ranjbar AM (2015) Applying reinforcement learning method to optimize an Energy Hub operation in the smart grid. In: 2015 IEEE power & energy society innovative smart grid technologies conference (ISGT)

Saba D, Sahli Y, Berbaoui B, Maouedj R (2020) Towards smart cities: challenges, components, and architectures. Stud Comput Intell 846:249–286

Shen C, Pena-Mora F (2018) Blockchain for cities - a systematic literature review. IEEE Access, pp 1–1

Si F, Wang J, Han Y, Zhao Q, Han P, Li Y (2018) Cost-efficient multi-energy management with flexible complementarity strategy for energy internet. Appl Energy 231:803–815

Smart Cities Councel. Smart cities data guide. Advice, best practices and tools for creating a data-driven city. https://smartcitiescouncil.com/resources/smart-cities-open-data-guide. Accessed 20 Dec 2019

Suwa A (2020) Local government and technological innovation: lessons from a case study of "Yokohama Smart City Project". In: Advances in 21st century human settlements, pp 387–403

Ha T, Zhang Y, Huang J, Thang VV (2016) Energy Hub modeling for minimal energy usage cost in residential areas. In: 2016 IEEE international conference on power and renewable energy (ICPRE)

Thornton B (2016) City-as-a-platform: applying platform thinking to cities. https://www.platfo rmed.info/city-as-a-platform-applying-platform-thinking-to-cities/. Accessed 11 Nov 2018

Uslar M, Enge D (2015) Towards generic domain reference designation: how to learn from smart grid interoperability. In: Conference D-A-Ch Energieinformatik 2015, Karlsruhe https://www.researchgate.net/publication/282292571_Towards_Generic_Domain_Refere nce_Designation_How_to_Learn_from_Smart_Grid_Interoperability. Accessed 6 Jan 2020

Weisi F, Ping P (2014) A discussion on smart city management based on meta-synthesis method. Manage Sci Eng 8(1):68–72

Xie J, Tang H, Huang T, Yu FR, Xie R, Liu J, Liu Y (2019) A survey of blockchain technology applied to smart cities: research issues and challenges. In: IEEE communications surveys & tutorials, pp 1–1

Yu D, Lian B, Dunn R, Le S (2014) Using control methods to model energy hub systems. In: 2014 49th International Universities Power Engineering Conference (UPEC)

City-Level Evaluation: Categories, Application Fields and Indicators for Advanced Planning Processes for Urban Transformation

Carla Rodríguez, Cecilia Sanz-Montalvillo, Estefanía Vallejo, and Ana Quijano

Abstract Advanced urban-planning processes need methods and analysis tools to guide cities toward sustainable urban planning. This provides a realistic and specific urban strategy for each city. Leveraging current developments in various European Smart City projects (MAtchUP and MAKING-CITY), a city-level evaluation framework is proposed to support cities in their strategic-planning process. This process goes through the definition of a methodology to help municipalities in the progress of cities toward sustainability and smartness. Then, through the city-level evaluation, city needs and challenges are identified in order to help municipalities when prioritizing a city strategy. It also aims to evaluate the current status of the city to know to what extent the city is sustainable and smart, as well as to monitor the progress of the city to become sustainable and smart. This evaluation also provides information to the municipality and sets the methodology included in this paper to benchmark and compare multiple aspects within and between cities. This indicator-based methodology enables assessing specific characteristics of the city, diagnosing challenges or discovering patterns through reliable metrics, but also allows comparing various aspects of cities thanks to the normalization and weighting of indicators and the calculation of indexes. Based on the outcome of the literature review, the structure of the city-evaluation framework is performed under the concept of sustainable development, establishing a framework for comparison and evaluation. This framework works as a baseline for all cities in the context of defining their targets, needs, priority areas or action lines. The definition of this methodology has been a learning process based on the study of several methods of normalization, weighting and aggregation, such as the ones described by the *Joint Research Centre* (JRC) from the European

C. Rodríguez (✉) · C. Sanz-Montalvillo · E. Vallejo · A. Quijano
Fundación CARTIF, Parque Tecnológico de Boecillo, Boecillo, Valladolid, Spain
e-mail: caraln@cartif.es

C. Sanz-Montalvillo
e-mail: cecsan@cartif.es

E. Vallejo
e-mail: estval@cartif.es

A. Quijano
e-mail: anaqui@cartif.es

© The Author(s), under exclusive license to Springer Nature Switzerland AG 2021
A. Bisello et al. (eds.), *Smart and Sustainable Planning for Cities and Regions*,
Green Energy and Technology, https://doi.org/10.1007/978-3-030-57332-4_2

Commission through the *Competence Centre on Composite Indicators and Score-boards* (COIN), to build the indexes by category and application field. The sensitivity analysis stage would be the last step to examine the suitability of the proposed evaluation methodology, from the selection of indicators to aggregation techniques. As a result, city's needs, priorities and progress will be easily identified thanks to the both numerical and graphical results, from which to establish strategic objectives and priority actions in the planning processes for urban transformation.

Keywords Smart cities · City diagnosis · Sustainability · Indicators · Indexes

1   Introduction

To evaluate and characterize city performance and whenever possible to benchmark this performance with other relevant references, it is important to use a recognized Evaluation framework.

MAtchUP[1] and MAKING-CITY[2] are both Smart City projects of the HORIZON2020 program supported by the European Commission. These projects involve cities in which the methodology depicted in this paper is being developed. Valencia (Spain), Dresden (Germany) and Antalya (Turkey) are the lighthouse cities of MAtchUP project and Herzliya (Israel), Ostend (Belgium), Skopje (FYROM) and Kerava (Finland) are the followers. The MAKING-CITY project has already developed this methodology in the lighthouse cities of Groningen (Netherlands) and Oulu (Finland) and in Bassano del Grappa (Italy), León (Spain), Kadıköy (Turkey), Trenčín (Slovakia), Vidin (Bulgaria) and Lublin (Poland), as the follower cities.

Within these projects, this framework for city characterization is based on previous experiences in Smart City projects and initiatives supported by the EC, such as SCIS, or CITYkeys (Bosch et al. 2016) and ESPRESSO. The framework methodology includes a number of indicators (e.g., final energy consumption, modal split or unemployment rate) for city characterization in various fields.

The methodological approach gives continuity to previous developments in Smart City projects, such as R2CITIES with the holistic design methodology based on the application of a multi-criteria decision-making approach (García-Fuentes et al. 2018), the CITyFiED methodology to enable the replicability of energy-efficient retrofitting of districts (Vallejo et al. 2019) and the Urban Regeneration Model carried out in the REMOURBAN project (Vallejo et al. 2018).

The evaluation framework of the MAtchUP project is based on the concept of sustainable development and on indicators classified under the three dimensions that comprise the term sustainability (environmental responsibility, economic development and social well-being) and the combination of them (bearable, equitable, viable and sustainable). Additionally, four fields or pillars have been defined to cover the technical solutions to be implemented in the MAtchUP (energy efficiency,

---

[1]MAtchUP EC project website: https://www.matchup-project.eu/.
[2]MAKING-CITY EC project website: https://makingcity.eu/.

mobility and transport, ICT infrastructures and urban platform, and citizens and society). On the other hand, each pillar has been split in several domains to focus the characterization on specific topics.

MAKING-CITY project has refined the MAtchUP methodology and presents data in all its cities and the urban dimension of a set of key performance indicators featuring energy and environment, mobility, governance and society and citizens.

The city diagnosis is a key phase for the deployment of an urban energy-transition methodology because, by an accurate diagnosis, the priority action lines can be identified.

There are other similar studies whose purpose was to make a ranking of cities in term of a sustainable index or environmental impact (Meijering et al. 2014), such as the Green City Index (Siemens AG and EUI 2012) or its European Version (Siemens AG and EUI 2009; Venkatesh 2014), or the Smart City Index (IMD, SCO and SUTD 2019). The main difference between these indexes and the ones developed within the present methodology is that the first ones have the aim of establishing rankings and comparing cities among them, while the focus on the methodology for these projects is put into the study of the strengths and weaknesses of the cities and their future development, knowing their needs and priorities, and establishing a comparative framework to obtain a baseline.

Cities are characterized according to some areas through a methodology that normalizes and prioritizes the city-level indicators to achieve city indexes that characterize and reflects its strengths and weaknesses. Indicators are a useful diagnosis tool to know the starting point of the cities revealing those areas in which it is performing well and those in which a significant improvement is needed.

Visualizing information plays a key role in creating and sharing knowledge. The final indexes obtained through the methodology are presented in a visually attractive way, in a series of diagrams. Thus, a city's needs and priorities are easily identified thanks to those spider radars that visualize data.

2 City-Level Evaluation Methodology

The methodology attempts to characterize the cities through the city-level indicators. The final purpose of this method is to obtain indexes to make a proper analysis of the cities, showing the areas of the city that are stronger and that need improvement through more attention. It also serves to fix future medium- and long-term goals and objectives.

The main steps of the development of the methodology are the following (depicted also in Fig. 1):

- Calculation of the city-level indicators, where objective results are obtained.
- Normalization of the city-level indicators, where results are scored on a 0–10 scale.

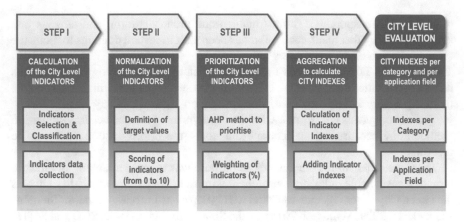

Fig. 1 Proposed methodology for city-level evaluation

- Prioritization of the categories, application fields and city-level indicators (through the AHP method), where the cities reflect their priorities, targets and concerns.
- Aggregation to calculate city indexes, where final indexes are obtained both per category and per application field, to achieve a complete city diagnosis.

This methodology is also based on the 10 Step Guide from the European Commission (OECD 2008) to assess the construction of composite indicators.

Although this methodology was first developed in the MAtchUP project, due to the long list of indicators in the project, cities have encountered several difficulties in calculating all of them, so the process is still on-going. On the other hand, in the MAKING-CITY project, the proposed methodology in MAtchUP was largely adopted and, thanks to the fewer number of indicators, the work has been completed, and results were obtained for all cities. That is why the following steps and results mainly focus on this last project.

2.1 Step I: Calculation of the City-Level Indicators

The first step is to define and calculate the city-level indicators by each city.

First, city-level indicators are selected from various sources, such as CITYkeys (Bosch et al. 2017) and ESPRESSO projects, SCIS and ISO 37120:2014 *Sustainable development of communities–Indicators for city services and quality of life*. Those indicators have been grouped into thematic areas, to develop a deep diagnosis of the city and not take into account only a single aspect.

In the MAtchUP project, the information collected from the cities was a set of variables, from which the indicators were calculated. These are divided into four pillars (energy efficiency, mobility, ICT infrastructures, and citizens), with different domains within them, for a total of almost 200 indicators.

The MAKING-CITY project has 20 indicators, grouped into ten application fields, and these, in turn, into four categories each. Thus are established three levels at which to make diagnoses. The indicators have been calculated using cities available data, according to their definition and formula provided (see Table 1).

Table 1 MAKING-CITY list of categories, application fields and indicators

Category	Application field	Indicator	Description
Energy and environment	City energy profile	Final energy consumption per capita	Annual final energy consumption for all uses and forms of energy
		Primary energy consumption per capita	Annual primary energy consumed in the city
		Primary energy sources	Shares of fuel types used for energy generation
		Buildings connected to DH (District Heating)-Network	% of buildings connected to DH-Network in the city
	GHG (Greenhouse Gas) Emissions	GHG emissions per capita	CO_2 emissions generated over a calendar year
	Waste management	Recycling rate	% of city's solid waste that is recycled
Mobility	City mobility profile	Modal split	Shares of different modes of transport
		Fuel mix in mobility	Shares of fuel types used for transport
		Energy use for transportation	Final energy consumption of the transport sector
		Access to public transport	Share of population with access to a public transport stop within 500 m
		Public infrastructure for low-carbon mobility	Length of lanes in the city for cycling mobility per 100,000 inhabitants
Governance	Economy	Unemployment rate	% of the labor force unemployed
		Gross Domestic Product (GDP)	City's GDP per capita
	Initiatives and Strategies of the Public Administration	Smart City factor in a city development strategy	Inclusion and Level of Detail (LoD) of Smart cities strategies in the urban strategic plans of the city

(continued)

Table 1 (continued)

Category	Application field	Indicator	Description
	Public ICT/data	Quality of open data	The extent to which the quality of open data produced by the city increases
Society and citizens	Affordable housing	Development of housing prices	Development of average price for buying an apartment in the city
		Housing cost overburden rate	Percentage of population living in households where total housing costs represent more than 40% of income
	Citizen engagement	Citizen engagement to climate conscious actions	Appreciation of the benefits of city actions, satisfaction, people happiness
		Encouraging a healthy lifestyle	The extent to which policy efforts are taken to encourage a healthy lifestyle
	Urban structure	Inhabitants in dense areas	% of population living in dense areas

2.2 Step II: Normalization of the City-Level Indicators

The normalization approach consists of assigning scores to the values obtained for the indicators. For that purpose, in the MAKING-CITY project, a target value is defined for each of the indicators. This target or "ideal" value acts as a baseline and comes from the European targets, or it is selected based on data from multiple cities and seems appropriate.

This target value received a score of 10, and this will be the score for all cities that reach or exceed it (which means improve that value). That target value defined is the minimum or maximum value to calculate the normalized indicator. The opposite value for each indicator (maximum or minimum), which would be the "worst" value, is calculated using the standard deviation of the cities values for each indicator. This value will be the 0 score. Formulas for the worst value calculation are the following:

Max value = Average value of the cities + (2.5 × Standard deviation of the cities values)
Min value = Average value of the cities − (2.5 × Standard deviation of the cities values)

where the average value of the indicators in the cases of the study (eight cities in this case), the standard deviation of those values and the coefficient of 2.5 are considered. This coefficient is recommended by the EC guide (OECD 2008) to normalize a small size sample like this one. It should be noted that outliers were detected before the normalization step using graphical methods since the number of cities and indicators was manageable. For studies with a larger sample size, the coefficient 2.5 can be used to detect outliers.

Next, Table 2 depicts the baseline values established (target and worst values) for

Table 2 MAKING-CITY indicators normalization methodology

Indicator	Unit	Target	Worst	Source of the Target value
Final energy consumption	MWh/cap	16.93	34.84	CoM (*Covenant of Mayors*) (Kona et al. 2016)
Primary energy consumption	MWh/cap	33.73	48.04	EEA (*European Environment Agency*) (EEA 2019)
Primary energy sources (renewables and biofuels)	%	20%	0%	Eurostat (2018a)
	MWh/cap	2.29	0.00	CoM (Kona et al. 2016)
Buildings connected to the District Heating-network	% of buildings/city	Values in Table 3		Euroheat (2015)
GHG emissions per capita	Tons of CO_2/cap	4.11	9.76	CoM (Kona et al. 2016)
Recycling rate	% of tons	50%	0%	EEA (2019)
Modal split[a](non-car mobility)	%	74%	21%	Best value of project cities
Fuel mix in mobility (e-mobility)	%	10.0%	0.0%	Eurostat (2018b)
Energy use of transport	MWh/cap	5.16	11.08	CoM (Kona et al. 2016)
Access to public transport	% of people	100%	64%	Best value of project cities
Public infrastructure for low-carbon mobility (cycling)	Km/100,000 people	217.00	0.00	Average value of project cities
Unemployment rate	% of active population	5.00%	18.64%	Sustainable Development Goal 8 (SDGs 2019)
Gross Domestic Product	€/cap	22,095	0	Average value of project cities
Smart City factor in a city development strategy	Likert scale	5	1	Qualitative indicator
Quality of open data	Likert scale	5	1	Qualitative indicator

(continued)

Table 2 (continued)

Indicator	Unit	Target	Worst	Source of the Target value
Development of housing prices	% of change	2.50%	− 14.24% / 17.28%	Web site: The balance[b](Amadeo 2019)
Housing cost overburden rate	%	6.60%	20.17%	Best value of project cities
Citizen engagement to climate conscious actions	Likert scale	5	1	Qualitative indicator
Encouraging a healthy lifestyle	Likert scale	5	1	Qualitative indicator
Inhabitants in dense areas	% of people	78.1%	10.6%	Average value of project cities

[a]For *modal split*, in the normalization only the use of non-car (private motor vehicles) has been taken into account, i.e., walking, cycling and public transport
[b]Target based on the ideal GDP growth, which resembles the growth that housing should consequently have

each indicator, calculated with the formulas above. The source or way in which the target value has been obtained is included as well.

The cities have not been clustered because they are all European, and European targets do not vary according to the circumstances of each city. However, a clustering for the indicator *Building connected to the DH-network* has been carried out since, for this specific indicator, climate influences on the greater or lesser need to implement a district-heating (DH) system.

For this classification, the heating degree days of each city have been taken into account to group the cities. A heating degree-day is a "measurement to quantify the demand for energy needed to heat a building; it is the number of degrees that a day's average temperature is below 18 °C, which is the temperature below which buildings need to be heated".[3] Therefore, a percentage is assigned as minimum buildings connected to the district-heating network for each group of cities as a target value (see Table 3).

And then, the score of the indicator, which is normalized indicator, is calculated using the min–max method, obtaining a punctuation between 0 and 10. The formula is the following:

$$\text{Score of the indicator} = \left(\frac{x - \min \text{value}}{\max \text{value} - \min \text{value}} \right) \times 10$$

---

[3]Definition from https://www.investopedia.com/terms/h/heatingdegreeday.asp.

where x is the value of the indicator, min and max values are the target and worst values defined (or vice versa, depending on the indicator), and multiplied by ten to have the scored in a 0–10 scale.

2.3 Step III: Prioritization of the Categories, Application Fields and City-Level Indicators

Prioritization consists of assigning various weights to every indicator or group of indicators. In this way, cities set weights to indicate the mutual importance of different KPIs and/or application fields. Thus, a city will prioritize and target particular areas of action, as it is very unlikely that a city has the capacity to target many areas at the same time in the short term. The decision on what areas or topics to target first is a complex exercise that should include some initial perspective of economic, environmental, technical or social benefits, combined with citizens' opinions and acceptance.

There are several methods to accomplish weighting process. The one selected for this methodology is the Analytic Hierarchy Process (AHP), which is a multi-criteria decision-making approach to solve complex decision problems developed by Thomas L. Saaty (Saaty 1980, 1987, 2008; Triantaphyllou and Mann 1995).

The AHP process consists of a pair-wise comparison between each element, using as reference the intensity of importance defined in the following Table 4.

Table 3 MAKING-CITY heating degree-days classification and target values regarding DH-network indicator

Classification of cities according to the heating degree-days		Building connected to the DH-network	
Climate	Heating degree-days	Target value (%)	Worst value (%)
Cold climate	>3000	50	0
Temperate climate	2000–3000	15	0
Warm climate	<2000	2	0

Table 4 Scale of relative importance of the AHP (Saaty 2008)

Intensity	Definition	Comments
1	Equal importance	Element A is just as important as element B
3	Moderate importance	Experience and judgement slightly favor element A over B
5	Essential importance	Experience and judgement strongly favor element A over B
7	Very strong importance	Element A is much more important than element B
9	Extreme importance	The greater importance of element A over B is beyond doubt
2, 4, 6, 8	Intermediate values	

With this system, the weights of the prioritization can be obtained in various ways: by all categories, by all indicators, or by indicators within their correspondent categories. It is also a helpful decision-making tool that can show the priority order among the areas of action.

This pair-wise comparison is intended to be done by the municipalities through groups of experts in each of the categories to be compared: groups of experts in energy, mobility and society for those three categories, as well as a heterogeneous group of the municipality for the rest of categories.

2.3.1   Example of Application of the AHP

The pair-wise comparison of the following example, which is the mobility category of the MAKING-CITY indicators, is carried out as shown in the following Table 5.

The results of the pair-wise comparison are represented in the comparison matrix to obtain the weights of each element (indicator), as can be seen in Table 6.

Weights are calculated with the sum of each row and dividing it by the total sum of the matrix: $n = 55.68$.

Table 5 Example of AHP application to the mobility category (pair-wise comparisons)

A		B	More important	Scale (1–9)
Modal split	Compared with	Fuel mix in mobility	A	3
		Energy use for transportation	B	4
		Access to public transport	A	2
		Public infrastructure for low-carbon mobility	A	7
Fuel mix in mobility		Energy use for transportation	B	5
		Access to public transport	A	3
		Public infrastructure for low-carbon mobility	A	6
Energy use for transportation		Access to public transport	A	7
		Public infrastructure for low-carbon mobility	A	9
Access to public transport		Public infrastructure for low-carbon mobility	A	2

Table 6 Example of mobility indicators comparison matrix and weights

	Modal split	Fuel mix in mobility	Energy use for transport	Access to public transport	Public infrast. for low-carbon mobility	Weight (%)
Modal split	1	3	1/4	2	7	21
Fuel mix in mobility	1/3	1	1/5	3	6	15
Energy use for transport	4	5	1	7	9	53
Access to public transport	1/2	1/3	1/7	1	2	7
Public infrastructure for low-carbon mobility	1/7	1/6	1/9	1/2	1	4
SUM	5.98	9.50	1.70	13.50	25.00	

Table 7 Random index (RI) for the different m matrix (Saaty 2008)

m	2	3	4	5	6	7	8	9	10
RI	0	0.58	0.90	1.12	1.24	1.32	1.41	1.45	1.51

With the weights calculated, the consistency of the matrix needs to be checked. The comparison matrix is considered to be adequately consistent if the corresponding consistency ratio (CR) is less than 10%. The CR is calculated from the consistency index (CI), whose formulas are the following:

$$CR = \frac{CI}{RI}; \quad CI = \frac{\lambda \max - m}{m - 1}$$

where RI is the random index given for a matrix (see Table 7), λ max is the eigenvalue of the matrix and m the matrix size.

For the example of the mobility category, CI = 0.08; RI = 1.12; and then, CR = 0.07, which means that the matrix is consistent.

2.4 Step IV: Aggregation to Calculate City Indexes

Aggregation is the last step to calculate the city indexes. Indexes are constructed by linear aggregation. After the normalization step, all indicators are scored on the same scale. It is also convenient for this diagnostic analysis that some of them can be compensated by other better rated ones that fall within the same category or application field (since they are grouped together because of the interrelation they have). With the prioritization and obtaining of the weights, the city needs, targets and priorities will be taken into account.

The aggregation is carried out in two ways, and thus, two types of city-level evaluation are obtained: one more general with four indexes of the categories, and another more detailed which includes the ten indexes of the application fields. The formulas for those Indexes are detailed below:

$$\text{CATEGORY Index} = \text{Indicator normalized} \times \text{Weight of indicator per category}$$

$$\text{APPLICATION FIELD Index} = \text{Indicator normalized} \times \text{Weight of indicator per application field}$$

3   Results for the Cities of the MAKING-CITY Project

Along the development of the MAKING-CITY project, on its deliverable D1.2—*City Diagnosis: analysis of existing city plans*, this methodology has been applied in the eight cities. Figures 2, 3, 4 and 5 shows the application in lighthouse cities: Groningen and Oulu.

Next, Figs. 6, 7, 8, 9, 10 and 11 show the results for the follower cities: Bassano del Grappa, León, Kadıköy, Trenčín, Vidin and Lublin.

The results of the application of the method have been analyzed particularly for each city, and conclusions have been drawn for the current reality of the cities. They

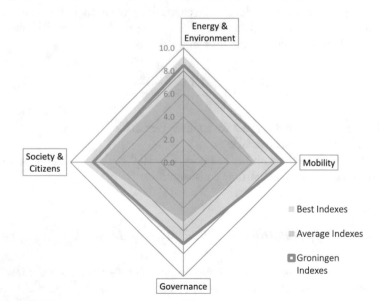

Fig. 2 Indexes per category of Groningen

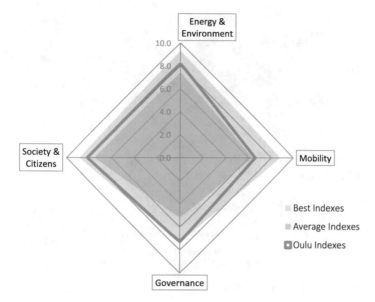

Fig. 3 Indexes per category of Oulu

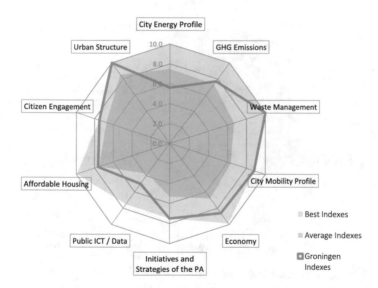

Fig. 4 Indexes per application field of Groningen

are expressed both numerically and graphically: The latter are represented together with the best and average values obtained by the cities to put it into context or have some baseline indexes.

From these analyses, several points can be drawn in common for the cities in terms of the diagnosis based on the city-level indicators. First, both lighthouse cities

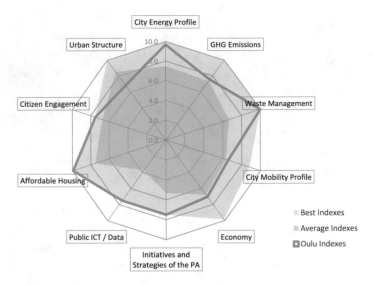

Fig. 5 Indexes per application field of Oulu

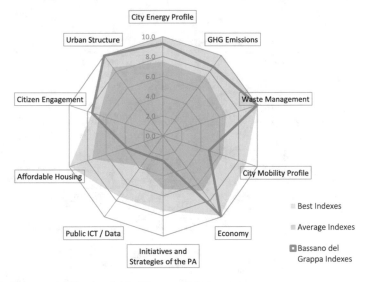

Fig. 6 Indexes per Application field of Bassano del Grappa

of MAKING-CITY are in general terms those that obtain better indexes in the various categories and application fields. As for the categories indexes, lighthouse cities are the only cities above the average value in all of them.

Comparing figures on application field indexes, lighthouse cities have good values, while in the follower cities, a variety of indexes can be found, clearly identifying some areas of action that deserve more attention or have more needs and others in

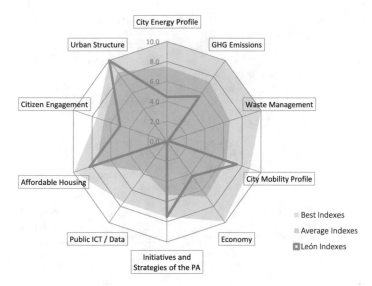

Fig. 7 Indexes per application field of León

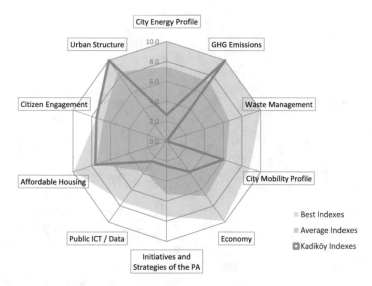

Fig. 8 Indexes per application field of Kadıköy

which it is doing well. This will be very useful for urban-transformation processes, and cities can benefit from this evaluation to identify strengths and weaknesses and consequently set priorities for the selection of actions to be implemented to transform the city into sustainability.

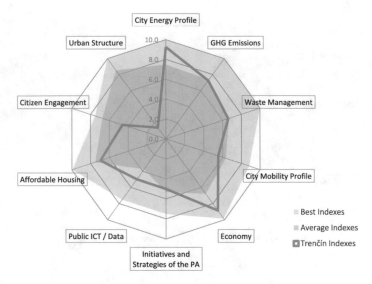

Fig. 9 Indexes per application field of Trenčín

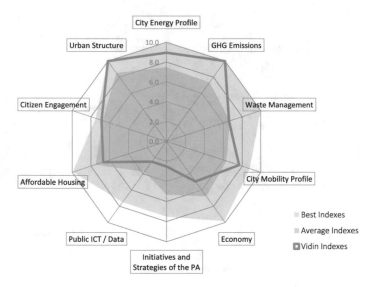

Fig. 10 Indexes per application field of Vidin

4 Discussion and Conclusion

The methodology proposed, based on the normalization and prioritization of indicators and its classification, can help decision-makers in the cities to identify priority areas and needs.

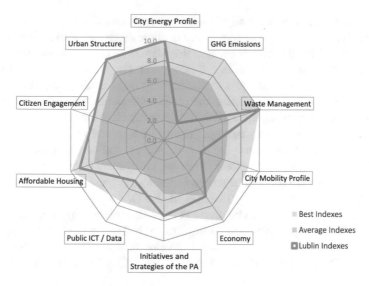

Fig. 11 Indexes per application field of Lublin

It is a four-step methodology, starting with the calculation of the indicators, then normalizing them using a method that takes into account the European targets for those indicators. After that, weights are assigned to indicators and application fields, through the AHP, establishing their priorities through pair-wise comparisons; and finally, the normalized indicators with their corresponding weights are aggregated. Thus, indexes are built compiling individual indicators into a single index through the application of mathematical models. As a result, the cities can rate the sustainability of the urban development of the cities.

The main obstacles encountered in this study rest on the work of the cities to find the necessary data for indicators or variables. This occurs mainly due to the lack of available data in the city or to their privacy.

Several differences have also been found between how to carry out the methodology in these two projects. On the one hand, in the case of MAtchUP, almost 200 indicators were established, grouped into 35 domains, which are within the four pillars. This makes the evaluation and prioritization be done by domains within the pillars, and not by indicators, which gives a more complete and objective vision of each aspect of the city, since within each one a multitude of aspects are taken into account. The negative part of this way to proceed has been that cities have great difficulties in finding and calculating these indicators. Thus, the final number of indicators within each domain will have to be reduced to the common calculated indicators to be normalized (based on the comparison of the data of the cities for each indicator).

On the other hand, in the MAKING-CITY case, there were only 20 indicators divided into ten application fields and four categories. The limited number of indicators makes it possible to calculate them for all cities so the same parameters are evaluated in the same way for all of them. Nevertheless, it also makes some application fields contain a single indicator, and in some categories, there are aspects that have not been taken into account because they cannot be contained with such a small number of indicators. Because of this, it is important to put in the context of the city the results obtained in the analysis so that nothing important is passed up.

This evaluation framework helps the cities to evaluate their current status and improvements that are being achieved during their transformation process. Thus, once the characterization of the city through the city-level indicators is available, it could be analyzed in comparison with the city objectives, and benchmarking indicators with other cities whenever possible. However, cities seem to be more interested in using the indicators to set own targets and monitor progress, rather than comparing to each other and benchmarking (Bosch et al. 2017).

The results in the cities will be used to establish strategic objectives and lines of action, as well as to prioritize actions in terms of budget, in the planning processes for urban transformation, such as in the development of City Vision, Sustainable Energy and Climate Action Plan (SECAP) or Smart City Strategic Plans.

Acknowledgements Authors acknowledge the financial support to the MAKING-CITY project of the Horizon 2020 program supported by the European Commission under contract No. 824418. All information related to MAKING-CITY project is available in website https://makingcity.eu/.

Authors acknowledge the financial support to the MAtchUP project of the Horizon 2020 program supported by the European Commission under contract No. 774477. All information related to MAtchUP project is available in website: https://www.matchup-project.eu/.

References

Amadeo K (2019) What Is the Ideal GDP Growth Rate? How Fast Should the Economy Grow? The balance online: https://www.thebalance.com/what-is-the-ideal-gdp-growth-rate-3306017. Accessed 27 Jan 2020

Bosch P, Jongeneel S, Neumann HM, Branislav I, Huovila A (2016) Deliverable 3.3: Recommendations for a smart city index. *CITYkeys EU Project.* Available online: https://www.citykeys-project.eu/citykeys/resources/general/download/CITYkeys-D3-3-Recommendations-for-the-Smart-City-Index-WSWE-AJENM8. Accessed 27 Jan 2020

Bosch P, Jongeneel S, Rovers V, Neumann HM, Airaksinen M, Huovila A (2017) CITYkeys indicators for smart city projects and smart cities. CITYkeys EU Project. Available online: https://citykeys-project.eu/citykeys/resources/general/download/CITYkeys-D1-4-Indicators-for-smart-city-projects-and-smart-cities-WSWE-AJENUD. Accessed 27 Jan 2020

Euroheat Statistics overview. 2015 Survey. TOP District Heating and Cooling Indicators 2013. Online: https://www.euroheat.org/wp-content/uploads/2016/03/2015-Country-by-country-Statistics-Overview.pdf. Accessed 27 Jan 2020

European Environment Agency (EEA) (2019) Trends and projections in Europe 2019: Tracking progress towards Europe's climate and energy targets. *Publications of the Office of the European Union,* Luxembourg. https://www.eea.europa.eu/publications/trends-and-projections-in-europe-1. Accessed 27 Jan 2020

Eurostat (2018a) Renewable energy in the EU: Share of renewables in energy consumption in the EU reached 17% in 2016. *Eurostat newsrelease.* Available online: https://ec.europa.eu/eurostat/documents/2995521/8612324/8-25012018-AP-EN.pdf/9d28caef-1961-4dd1-a901-af18f121fb2d. Accessed 27 Jan 2020

Eurostat (2018b) Share of transport fuel from renewable energy sources. *Eurostat newsrelease.* Available online: https://ec.europa.eu/eurostat/web/products-eurostat-news/-/DDN-20180312-1. Accessed 27 Jan 2020

García-Fuentes MÁ, García-Pajares R, Sanz-Montalvillo C, Meiss A (2018) Novel design support methodology based on a multi-criteria decision analysis approach for energy efficient district retrofitting projects. Energies 11(9):1–19

IMD World Competitiveness Center, Smart City Observatory (SCO) and Singapore University of Technology and Design (SUTD) (2019) *Smart City Index.* Available online: https://www.imd.org/research-knowledge/reports/imd-smart-city-index-2019/. Accessed 27 Jan 2020

Kona A, Bertoldi P, Melica G, Rivas S, Zancanella P, Serrenho T, Iancu A, Janssens-Maenhout G (2016) Covenant of Mayors: Monitoring Indicators. Publications of the Office of the European Union, Luxembourg. Available online: https://publications.jrc.ec.europa.eu/repository/bitstream/JRC97924/jrc97924%20_%20com%20monitoring%20indicators_online.pdf. Accessed 27 Jan 2020

Meijering JV, Kern K, Tobi H (2014) Identifying the methodological characteristics of European green city rankings. Ecol Ind 43:132–142

OECD (2008) Handbook on constructing composite indicators: methodology and user guide. OECD publications. Available online: https://composite-indicators.jrc.ec.europa.eu/?q=10-step-guide/. Accessed 27 Jan 2020

Saaty RW (1987) The analytic hierarchy process—what it is and how it is used. Math Modell 9(3–5):161–176. Available online: https://core.ac.uk/download/pdf/82000104.pdf. Accessed 28 Jan 2020

Saaty TL (1980) The analytic hierarchy process. McGraw Hill, New York

Saaty TL (2008) Decision making with the analytic hierarchy process. Int J Services Sci 1(1):83–98. Available online: https://www.rafikulislam.com/uploads/resourses/197245512559a37aadea6d.pdf. Accessed 28 Jan 2020

Siemens AG and The Economist Intelligence Unit (EUI) (2009) European Green City Index: Assessing the environmental impact of Europe's major cities. Available online: https://www.prensa.siemens.biz/phocadownload/estudios/greencityeurope.pdf. Accessed 27 Jan 2020

Siemens AG and The Economist Intelligence Unit (EUI) (2012) The Green City Index: A summary of the Green City Index research series. Available online: https://apps.espon.eu/etms/index.php/this-big-city/qr/534-siemens-green-cities-index. Accessed 27 Jan 2020

Sustainable Development Goals Knowledge Platform (SDGs) Progress of Goal 8 in 2019: https://sustainabledevelopment.un.org/sdg8. Accessed 27 Jan 2020

Triantaphyllou E, Mann SH (1995) Using the analytic hierarchy process for decision making in engineering applications: some challenges. Int J Ind Eng: Appl Pract 2(1):35–44. Available online: https://www.researchgate.net/publication/241416054_Using_the_analytic_hierarchy_process_for_decision_making_in_engineering_applications_Some_challenges. Accessed 28 Jan 2020

Vallejo E, Criado C, Arriabalaga E, Massa G, Vasallo A (2019) The CITyFiED Methodology for city renovation at district level. Build Manage 3(2):23–33

Vallejo E, de Torre C, García-Fuentes MÁ (2018) Urban regeneration model. Info package 2 of REMOURBAN EU Project. Available online: https://www.remourban.eu/technical-insights/infopacks/urban-regeneration-model.kl. Accessed 27 Jan 2020

Venkatesh G (2014) A critique of the European City Index. J Environ Plann Manage 57(3):317–328. Available online: https://www.tandfonline.com/doi/abs/https://doi.org/10.1080/09640568.2012.741520. Accessed 28 Jan 2020

Proposal for an Integrated Approach to Support Urban Sustainability: The COSIMA Method Applied to Eco-Districts

Cristina Becchio, Marta Carla Bottero, Stefano Paolo Corgnati, Federico Dell'Anna, Giulia Pederiva, and Giulia Vergerio

Abstract Cities represent the places with the greatest environmental and energy impacts in the world. Their transformation through a sustainable key would make possible reducing the pressures registered in these areas. According to the Sustainable Development Goals, attention has shifted more and more to the creation of sustainable and safe communities, characterized by low energy-consuming buildings due to smart heating and cooling systems, and sustainable transport solutions based on the use of private electric and hybrid vehicles. Besides the energy and environmental impacts, actions to tackle climate change provide the opportunity to create collateral benefits that can potentially generate economic and social improvement for the whole community. The co-benefits inclusion in the decision analysis is crucial to remove barriers and reveal the real potential of renovation projects at the urban/district scale. Following the guidelines of the European Commission, the tool used when evaluating public projects and policies is the Cost–Benefit Analysis (CBA). One of the main limitations of the CBA method is the estimation of all positive and negative externalities in monetary value that can lead to imprecise assessment. To overcome this obstacle,

C. Becchio · S. P. Corgnati · G. Vergerio
Department of Energy, Politecnico di Torino, Corso Duca degli Abruzzi 24, 10129 Turin, Italy
e-mail: cristina.becchio@polito.it

S. P. Corgnati
e-mail: stefano.corgnati@polito.it

G. Vergerio
e-mail: giulia.vergerio@polito.it

M. C. Bottero (✉) · F. Dell'Anna
Interuniversity Department of Regional and Urban Studies and Planning, Politecnico di Torino, Corso Duca degli Abruzzi 24, 10125 Turin, Italy
e-mail: marta.bottero@polito.it

F. Dell'Anna
e-mail: federico.dellanna@polito.it

G. Pederiva
Politecnico di Torino, Corso Duca degli Abruzzi 24, 10129 Turin, Italy
e-mail: giulia.vergerio@polito.it

a growing scientific literature on the application of Multi-Criteria Decision Analysis (MCDA) to assess the sustainability of investment at district scale is emerging. In this study, we propose a new assessment framework based on the COmpoSIte Modeling Assessment (COSIMA) to address the multidimensionality that characterizes the redevelopment process of eco-districts considering energy, environmental, economic and social evaluation criteria. The COSIMA method enables considering both the tangible and intangible aspects of the problem and the opinion of the various stakeholders involved in the decision-making process, which are crucial aspects in urban transformations.

Keywords Decision-making · Co-benefit · Urban renovation · SDG 11 · Energy transition

1 Introduction

Nowadays, cities occupy about 2% of the Earth's surface, but they are responsible for 70% of the global primary energy consumption. About 50% of the world's population lives there, and it is estimated that it can reach 75% in 2050. Therefore, given their high concentrations of people, services and consumption, cities play an essential role in the process of energy conversion towards a sustainable society. This process, closely linked at climate change mitigation, represents one of the most significant challenges of the twenty-first century, as recognized on Paris Agreement. Cities are seen as the starting point for achieving the objectives set by the UNFCC in the context of the Sustainable Development Goals (SDGs) (UNFCCC 2015; UNFCCC 2017). In particular, SDG11 aims to create inclusive, safe, resilient, and sustainable cities by stressing the need to take concrete steps to promote a process of transformation towards a green vision of urban areas.

Moreover, in the transition to a more sustainable future, the role of cities is increasingly recognized and evidenced by the spread of the post-carbon cities (PCC) concept, which is also shifting to a smaller scale of intervention through the creation of post-carbon districts (PCD) (Becchio et al. 2016). This constitutes something intermediate between a city and a building, representing the most appropriate scale to test the various transformation strategies of the urban system, making them more manageable, and to contain risks. However, it is important to highlight that district does not correspond to the simple sum of its buildings, but includes the whole of all parties that make up the urban system such as buildings, mobility, public lighting, open spaces, water and waste management.

In this context, the transformation measures must be programmed according to a long-term vision and all their impacts must be assessed, to ensure the achievement of the predefined objectives. The objectives are not limited to respecting the energy and environmental targets of the construction sector. The energy policies can lead to various positive social, environmental, and economic impacts that can bring added value to the choice of alternative strategies. To facilitate the transformation process, the benefits that can be generated by the requalification measures and the various impacts that they can cause for the whole community must be considered (Ürge-Vorsatz et al. 2014; Bisello and Vettorato 2018). Therefore, new support instruments and criteria are needed for considering these impacts, considered fundamental in a complex context such as the urban one, where several stakeholders with different interests are present (Wang et al. 2009).

In this study, we propose a model to assist planners, architects, and engineers in the field of energy and sustainable planning on a district scale in order to control the multi-dimensional problem in this domain. The model is based on an input/output approach. The inputs are made up of energy needs, renewable energy sources, mobility's fuel consumption, water expenditures, costs, and so on. The outputs consist of tangible economic benefits and intangible impacts. The balance of negative and positive impacts is assessed through the COmpoSIte Modeling Assessment (COSIMA) approach proposed by Barfod et al. (2011). In this approach, the Cost–Benefit Analysis (CBA) is extended by adding evaluations of the Multi-Criteria Decision Analysis (MCDA) to economic results through a value function calculated using a weighting procedure. According to MCDA theory, the criteria weighting makes it possible to consider stakeholders' opinion in the decision process, which otherwise would be omitted through a traditional evaluation procedure.

The following section illustrates the eco-district concept. The third section describes the methodological proposal. Results are presented in Sect. 4 and conclusions follow.

2 The Concept of an Eco-District

The examples of eco-districts and sustainable neighborhoods in Europe are quite vast. Concerning sustainable districts, the predominance of examples from Northern or Eastern Europe exists, while examples from Southern Europe are rarely mentioned in the literature. This vision could be supported by the fact that the sustainable neighborhood concept is defined as a Northern European model by the literature (Kyvelou et al. 2012).

The literature gives an extensive range of definitions of the so-called eco-district model. Bottero et al. (2019) and Marique and Reiter (2014) analyzed the characteristics of different real projects in order to develop some empirical insights into the relationships with sustainability in neighborhoods and its related models. In selected cases, energy aspects have certainly been identified as a priority in developing eco-districts. The transition to renewable energy sources (RESs), such as PV systems, heat pumps, or CHP, represents a crucial action to design a sustainable district (Becchio et al. 2018). The adoption of high-performance insulation systems and the development of the first attempts to develop passive houses and nearly zero-energy buildings are both more common (Barthelmes et al. 2016). Other urban sectors involved in the sustainable process of the districts are water and waste management. Various actions and technologies are applied in these sectors with the aims to collect, separate, and reuse them. An extensive range of initiatives occurs on the local scale, such as the reduction of water consumption in buildings and the improvement of municipal waste management.

Private and public transports represent another intervention area. Very similar actions were adopted in the various projects analyzed by the authors. The main goal of the measures proposed in this field of application is to discourage the use of private cars by implementing public mobility infrastructures. Besides, the use of private cars is discouraged by the rise of car-free parking in the peripheral areas and by a reduction in the number of parking lots in the center area of the cities. Carpooling and car-sharing initiatives are promoted if the cars are electric or hybrid.

Some projects envisage the energy efficiency of the public lighting system. The technologies implemented do not merely aim at replacing the lighting element with LED lamps. In pilot cases, smart poles provide for the installation of new measures in the field of Information and Communication Technologies (ICT) for the sharing of traffic data or data for monitoring urban air quality (GrowSmarter project 2020).

Some solutions were identified in the urban fabric design of the districts. The most frequently adopted actions covered the development of mixed-use buildings, enabling the combination of various services and facilities with the residential functions. High-density areas are also encouraged to reduce land use and to increase the number of green spaces.

The definition of eco-district covers all the sectors of the entire urban system. In addition to the buildings (B), which become an active part of the energy system, the sectors of water (W), waste management (WM), public and private mobility (M), public lighting (P) come into play (Fig. 1). From this perspective, a combined assessment model for supporting the decision-making process of alternative scenarios

Fig. 1 Eco-district dimensions

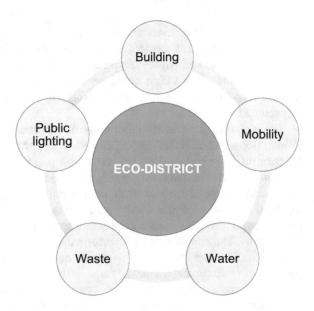

of sustainable transformation for a district is needed (Grujić et al. 2014). Starting from the most common approaches used in the field of investment evaluations, an approach that combines the potential of Cost–Benefit Analysis (CBA) and Multi-Criteria Decision Analysis (MCDA) is proposed to include financial and economic aspects, as well as intangible impacts generated by urban redevelopment projects.

3 Methodological Proposal

3.1 Evaluation Framework

The sustainable measures applied at the district and urban levels need innovative methodologies to consider the multi-dimensionality of the decision problem. The evaluation process has to go beyond simple reduction of consumption and investment costs indicators in order to address the full range of impacts involved.

As said before, the main purpose of the model proposed in this work is to support planners, architects, and engineers in the field of energy and sustainable planning at the district scale. The proposed methodological process for the assessment of districts transformation requires a series of steps that enable simultaneously considering the different urban elements that make it up (e.g., buildings, mobility, public lighting, water and waste management) (Fig. 2).

First of all, it is necessary to consider the state of the art (SOA) of each element of the urban system to study the starting point and to identify the retrofit actions to be applied.

With reference to energy consumption of buildings, it is unthinkable to analyze them individually. As shown by Ballarini et al. (2014), an archetype-based approach can help identify the reference buildings (RBs) when working on a large scale representing the heterogeneity of a city's building stock by dividing it into specific classes. Geographic Information System (GIS) could play a crucial role for classifying RBs in an existing real district (Mutani et al. 2016; Delmastro et al. 2016). Each class is based on features (e.g., date of construction and geometrical and thermophysical features) to which energy needs and consumptions, expressed in kWh/m^2y and estimated by the modeling of the representative RBs, are linked. In this way, the real buildings in the district are grouped into clusters according to the identified classes. In this way, the real buildings of the neighborhood are grouped into clusters according to the identified classes. First, an energy consumption is associated with each group of RBs, and then the overall consumption of the whole district will be determined.

Regarding waste management, municipal collection plans can provide information to both analyze the current state and define future scenarios.

The mobility sector can be reviewed starting from the local mobility plans that have the task of providing the public administration with the appropriate tools to address, in the logic of anticipation, the new needs of citizens and businesses.

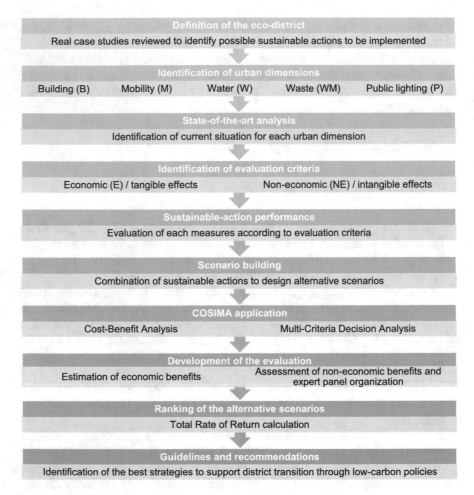

Fig. 2 Flowchart of the proposed method

Once the SOA is defined, it is necessary to plan the retrofit actions for each sector. New measures need to be designed for energy consumption, mobility, waste collection and water treatment. The measures subsequently need to be combined in alternative scenarios to identify guidelines and recommendations for the municipality.

In order to compare the alternative scenarios, the next step is the macroeconomic assessment in which economic and non-economic benefits are considered. Benefits must be identified and quantified for each scenario. To do this an assessment methodology that allows to compare the different scenarios through an aggregated index is necessary. The CompoSIte Modeling Assessment (COSIMA) analysis (Barfod et al. 2011; Barfod and Salling 2015) can be a support tool in this phase. It can be briefly explained as an analysis that combines the CBA (European Commission 2014) and the MCDA (Keeney and Raiffa 1993; Figueira et al. 2005). Indeed, COSIMA measures the performance of each scenario through an index that aggregates the results of both CBA and MCDA, namely the Total Rate of Return (TRR), represented as follows (1):

$$\mathrm{TRR}(A_k) = \frac{TV(A_k)}{C_k} = \frac{1}{C_k} \cdot \left(\sum_{i=1}^{I}(p_i \cdot X_{ik}) + \alpha \cdot \left(\sum_{j=1}^{J}(w_j \cdot Y_{jk}) \right) \right) \quad (1)$$

where A_k is alternative k, C_k is the total investment costs; p_i is the unit price for the CBA impact i; X_{ik} is the quantity of the CBA variable i; w_j is the weight for the criterion j of MCDA; Y_{jk} is the value score of alternative k under criterion j; and α is an indicator that expresses the trade-off between the CBA and MCDA.

Therefore, COSIMA considers co-benefits expressed in monetary terms (as in the case of CBA) and non-monetary benefits that are defined through both quantitative and qualitative Key Performance Indicators (KPIs).

One of the main potentials of the COSIMA method compared to traditional CBA is the possibility of considering the opinion of various stakeholders in the assessment. Indeed, the decision-makers, as subjects interested in the evaluation of the alternatives and in the choice, can define the degree of importance of the aspects that characterize the project by assigning the weights to the criteria. In order to define the weight w_j of each non-economic criteria j, the Simple Multi-Attribute Rating Technique Extended (SMARTER) procedure could be used (Barron and Barrett 1996). SMARTER derives weights from a simple classification of criteria and is very effective compared to direct weighting. The proposed methodological framework thus set up aims to provide guidelines for future actions in terms of energy and urban sustainability at the district scale, identifying the potential of each measure in the overall performance according to the different actors connected to the process.

4   Results

Since the COSIMA analysis includes non-monetary criteria in addition to monetizable benefits, it is useful to define a list of parameters to be considered in the evaluation starting from a review of the literature on the co-benefits generated by the regeneration process. This phase was useful for defining the evaluation variables to consider all the urban sectors (B, M, P, W, WM) involved in the district-scale transformation processes. The criteria selected from the review analysis are shown in Table 1. The criteria range from environmental aspects to energy, economic, and social ones (Bertolini et al. 2018; Gabrielli et al. 2019; Dell'Anna et al. 2021). Each of them is evaluated using a quantitative or qualitative indicator. The quantitatively

Table 1 Economic and non-economic impacts in redevelopment project at the district scale

Evaluation criteria	Unit	Criteria type		Urban sectors				
		E	NE	B	M	P	W	WM
Energy saving	€/kWh	×		×		×		
CO_2 emission avoided	€/CO_2ton	×		×	×	×		×
PM_{10} emission avoided	€/PM_{10}ton	×		×	×	×		×
Real estate market value increase	€/m²	×		×			×	
New green jobs	€/new green job	×		×			×	
Fuel costs avoided	€/kg or €/l	×			×			×
Increase in public transport passengers	Passenger/km		×		×			
Reduction of drinking water usage	l/per capita		×				×	
Covering renewable energy sources	%		×	×				
Visual impact	Qualitative scale		×	×	×		×	×
Reliability of technology	Qualitative scale		×	×	×		×	×
People acceptance	Qualitative scale		×	×	×		×	×

assessed impacts will be identified through a unit of measurement, while the qualitative ones will be identified through a level scale that will vary according to the criterion considered.

As shown in Fig. 3, the results of the CBA and MCDA analysis can be examined separately, to identify the performance of different alternative scenarios based on

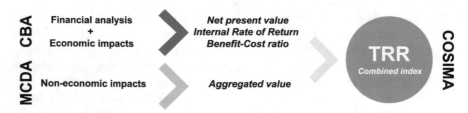

Fig. 3 Interpretation of the results

economic (E) and non-economic (NE) values. Furthermore, by combining the results of the two analyzes according to the COSIMA approach, the complete performance of the scenarios is calculated considering the entire range of impacts through the TRR index. By introducing the results of the MCDA into the TRR calculation, the different points of view of the actors participating in the decision-making process are taken into account. In this way, the COSIMA analysis allows to build a participatory, iterative and transparent decision-making framework by analyzing the different urban sectors affected by the transformation.

5 Conclusions and Future Development

According to the new European standards, the inclusion of co-benefits in the decision-making process has significant importance in the field of defining energy policies on an urban and district scale to better describe the performance of alternative projects and choose the one that maximizes the co-benefits. In this study, a multi-step methodology for the evaluation of alternative retrofit scenarios of a hypothetical district has been proposed. The proposed methodology foresees a preliminary analysis of the urban sectors that characterize an eco-district and subsequently identifies the useful evaluation criteria to highlight their potentiality in the sustainable transition. To evaluate the feasibility in terms of economic and non-economic benefits generated by the redevelopment project, the integration of the COSIMA method in the evaluation framework is suggested. The proposed multi-step approach consists of integrating the purely economic analysis of the costs and benefits with a quantitative and qualitative criteria that considers the non-monetizable impacts. This multi-step approach enables overcoming the difficulties related to the application of the manual-based CBA, that admits the consideration of the only benefits that can be monetized, excluding others that have the same importance for the environmental, economic, and social development of a neighborhood. The co-benefits reviewed in this framework appear to be suitable to represent the complexity of the problem under consideration, and the synthetic index of the Total Rate of Return seems to be useful for informing decision makers on the priorities of alternative retrofit scenarios.

Given the fragmentation of stakeholders within the urban context, this participatory and iterative decision framework may guide the choices in redevelopment and new sustainable measures considering their opinions (Sarnataro et al. 2020; Capolongo et al. 2019; Cerreta et al. 2019). The integrated framework seems to be suitable to respond to the needs of public administrations for a decision support system capable of addressing the issues that come into play at the urban level from a sustainable perspective (Assumma et al. 2019; Napoli et al. 2020). The implementation of the proposed methodology will make it possible to evaluate alternative retrofit scenarios of real world case studies concerning the redevelopment of districts considering different urban sectors (buildings, waste, water, mobility). Subsequently, the results of the evaluation will support the definition of future actions and policies for urban regeneration in a sustainable perspective.

References

Assumma V, Bottero M, Datola G, De Angelis E, Monaco R (2019) Dynamic models for exploring the resilience in territorial scenarios. Sustainability 12:3. https://doi.org/10.3390/su12010003

Ballarini I, Corgnati SP, Corrado V (2014) Use of reference buildings to assess the energy saving potentials of the residential building stock: the experience of TABULA project. Energy Policy 68:273–284. https://doi.org/10.1016/j.enpol.2014.01.027

Barfod MB, Salling KB (2015) A new composite decision support framework for strategic and sustainable transport appraisals. Transp Res Part a: Policy Pract 72:1–15. https://doi.org/10.1016/j.tra.2014.12.001

Barfod MB, Salling KB, Leleur S (2011) Composite decision support by combining cost-benefit and multi-criteria decision analysis. Decis Support Syst 51:167–175. https://doi.org/10.1016/j.dss.2010.12.005

Barron FH, Barrett BE (1996) The efficacy of SMARTER—simple multi-attribute rating technique extended to ranking. Acta Physiol (Oxf) 93:23–36. https://doi.org/10.1016/0001-6918(96)00010-8

Barthelmes VM, Becchio C, Bottero M, Corgnati SP (2016) Cost-optimal analysis for the definition of energy design strategies: the case of a Nearly-Zero Energy Building. Valori e Valutazioni 16:61–76. https://hdl.handle.net/11583/2654216%0A

Becchio C, Bottero MC, Corgnati SP, Dell'Anna F (2018) Decision making for sustainable urban energy planning: an integrated evaluation framework of alternative solutions for a NZED (Net Zero-Energy District) in Turin. Land Use Policy 78:803–817. https://doi.org/10.1016/j.landusepol.2018.06.048

Becchio C, Corgnati SP, Delmastro C, Fabi V, Lombardi P (2016) The role of nearly-zero energy buildings in the transition towards Post-Carbon Cities. Sustain Cities Soc 27:324–337. https://doi.org/10.1016/j.scs.2016.08.005

Bertolini M, D'Alpaos C, Moretto M (2018) Do Smart Grids boost investments in domestic PV plants? Evidence from the Italian electricity market. Energy 149:890–902. https://doi.org/10.1016/j.energy.2018.02.038

Bisello A, Vettorato D (2018) Multiple benefits of smart urban energy transition. In: Urban energy transition. Elsevier, pp 467–490. https://doi.org/10.1016/B978-0-08-102074-6.00037-1

Bottero M, Caprioli C, Cotella G, Santangelo M (2019) Sustainable cities: a reflection on potentialities and limits based on existing eco-districts in Europe. Sustainability 11:5794. https://doi.org/10.3390/su11205794

Capolongo S, Sdino L, Dell'Ovo M, Moioli R, Della Torre S (2019) How to assess urban regeneration proposals by considering conflicting values. Sustainability 1(14):3877. https://doi.org/10.3390/su11143877

Cerreta M, Poli G, Regalbuto S, Mazzarella C (2019) A multi-dimensional decision-making process for regenerative landscapes: a new Harbour for Naples (Italy). In: Misra S et al (ed) Computational science and its applications – ICCSA 2019. ICCSA 2019. Lecture Notes in Computer Science, vol 11622. Springer, Cham. https://doi.org/10.1007/978-3-030-24305-0_13

Dell'Anna F, Bottero M (2021) Green premium in buildings: Evidence from the real estate market of Singapore. Journal of Cleaner Production 286:125327. https://doi.org/10.1016/j.jclepro.2020.125327

Delmastro C, Mutani G, Corgnati SP (2016) A supporting method for selecting cost-optimal energy retrofit policies for residential buildings at the urban scale. Energy Policy 99:42–56. https://doi.org/10.1016/j.enpol.2016.09.051

European Commission (2014) Guide to Cost-benefit Analysis of Investment Projects: Economic appraisal tool for Cohesion Policy 2014–2020. In: Publications Office of the European Union. https://ec.europa.eu/regional_policy/sources/docgener/studies/pdf/cba_guide.pdf. Accessed 12 Dec 2019

Figueira J, Greco S, Ehrogott M (2005) Multiple criteria decision analysis: state of the art surveys. Springer, New York, New York

Gabrielli L, Giuffrida S, Trovato MR (2019) Real estate landscapes and the historic city: on how looking inside the market. In: Calabrò F, Della Spina L, Bevilacqua C (ed) New metropolitan perspectives. ISHT 2018. Smart innovation, systems and technologies, vol 101. Springer, Cham. https://doi.org/10.1007/978-3-319-92102-0_29

GrowSmarter project (2020) From dream to reality. Recommendations from a European Smart City project. https://grow-smarter.eu/fileadmin/editor-upload/Reports/recommendations_for_Policy_makers_and_practitioners.pdf. Accessed 12 Jan 2020

Grujić M, Ivezić D, Živković M (2014) Application of multi-criteria decision-making model for choice of the optimal solution for meeting heat demand in the centralized supply system in Belgrade. Energy 67:341–350. https://doi.org/10.1016/j.energy.2014.02.017

Keeney RL, Raiffa H (1993) Decisions with multiple objectives: preferences and value trade-offs. Cambridge University Press, Cambridge

Kyvelou S, Sinou M, Baer I, Papadopoulos T (2012) Developing a South-European eco-quarter design and assessment tool based on the concept of territorial capital. In: Curkovic S (ed) Sustainable development - authoritative and leading edge content for environmental management. IntechOpen

Marique AF, Reiter S (2014) A simplified framework to assess the feasibility of zero-energy at the neighbourhood/community scale. Energy Build 82:114–122. https://doi.org/10.1016/j.enbuild.2014.07.006

Mutani G, Delmastro C, Gargiulo M, Corgnati SP (2016) Characterization of building thermal energy consumption at the urban scale. Energy Proc 101:384–391. https://doi.org/10.1016/j.egypro.2016.11.049

Napoli G, Bottero M, Ciulla G, Dell'Anna F, Figueira JR, Greco S (2020) Supporting public decision process in buildings energy retrofitting operations: the application of a multiple criteria decision aiding model to a case study in Southern Italy. Sustain Cities Soci 102214. https://doi.org/10.1016/j.scs.2020.102214

Sarnataro M, Barbati M, Greco S (2020) A portfolio approach for the selection and the timing of urban planning projects. Socio-Econ Plan Sci 100908. https://doi.org/10.1016/j.seps.2020.100908

UNFCCC (2015) Paris Agreement, FCCC/CP/2015/L.9/Re.1. https://unfccc.int/resource/docs/2015/cop21/eng/l09r01.pdf. Accessed 12 Jan 2020

UNFCCC (2017) The Sustainable Development Goals Report. United Nations Publications. https://doi.org/10.18356/3405d09f-en. Accessed 12 Jan 2020

Ürge-Vorsaz D, Herrero ST, Dubash NK, Lecocq F (2014) Measuring the co-benefits of climate change mitigation. Annu Rev Environ Resour 39:549–582. https://doi.org/10.1146/annurev-environ-031312-125456

Wang JJ, Jing YY, Zhang CF, Zhao JH (2009) Review on multi-criteria decision analysis aid in sustainable energy decision-making. Renew Sustain Energy Rev 13:2263–2278. https://doi.org/10.1016/j.rser.2009.06.021

Open Innovation Strategies, Green Policies, and Action Plans for Sustainable Cities—Challenges, Opportunities, and Approaches

Mohsen Aboulnaga, Marco Sala, and Antonella Trombadore

Abstract Planning for smart and sustainable cities is vital since the world is currently witnessing a rapid increase in urban population. Cities globally are encountering colossal challenges in terms of high-energy use, transport, and traffic congestion resulting in high GHG emissions, in addition to huge consumption of resources such as water and materials, and high levels of air pollution, especially in megacities. Cities are facing the urban heat islands effect and other climate change impacts. Such vast increases of population put pressure on cities' infrastructures that are not resilient, as well as putting pressure on local governments and municipalities in planning and managing cities. Therefore, reducing energy use, mitigating CO_2 emissions, and improving air quality by utilizing open innovation, and green design strategies and policies (urban-framing innovation and technologies, and green facades), as adaptive tools for urban areas, are essential planning to achieve sustainability and liveability in megacities, yet mitigate and adapt to climate change (CC). This paper presents strategies and policies needed for developing sustainable cities. It also highlights the major challenges facing cities globally and the means to support local governments in tackling these challenges to bring about smart and sustainable cities. In addition, it presents policies and tools for greening cities, using opportunities in terms of city innovation and open innovation models. We review smart and green strategies that drive cities to solve such challenges through energy efficiency tools and mobilizing stakeholders. In this work, we present adaptive natural-based green solutions to ensure city resilience and we highlight strategies and policies related to sustainable energy, urban farming, and green facades. We also illustrate strategies,

M. Aboulnaga (✉)
Sustainable Built Environment, Faculty of Engineering, Cairo University, Giza 12613, Egypt
e-mail: mohsen_aboulnaga@yahoo.com; maboulnaga@eng.cu.edu.eg

M. Sala
Architectural Technology, School of Architecture – DIDA, University of Florence, Santa Teresa, Via della Mattonaia n.14, Florence, Italy
e-mail: marco.sala@unifi.it; marcosala48@gmail.com

A. Trombadore
Environmental Design, School of Architecture – DIDA, University of Florence, Santa Teresa, Via della Mattonaia n.14, Florence, Italy
e-mail: antonella.trombadore@unifi.it

policies, and action plans that drive green cities to lessen climate change risks. We present and discuss the actions and stakeholders' mobilization needed for liveable and smart urban areas for CC adaptation.

Keywords Smart and green cities · Climate change adaptation · Strategies and policies · Innovation urban farming

1 Introduction

Cities worldwide are encountering colossal challenges nowadays, thus innovation strategies and action plans supported by green policies are urgently needed to transform cities to be smart, green, and sustainable, yet also to capture opportunities for fostering such processes by utilizing sustainable energy action plans (SEAP) and innovative urban farms to make cities more liveable and generative. Sustainable planning for cities is vital because the world witnesses a rapid increase in urban population and migration from rural to urban areas searching for better incomes, decent jobs, and a better quality of life. Currently, 50% of the world's urban population resides in cities (UN-Habitat 2009; UN-Habitat 2017), and this will reach 70% by 2050 (United Nations 2018). Cities globally face huge challenges, including intensive energy and water consumption, transport congestion, and use of non-green materials, hence resulting in high GHG emissions. About 70% of global total energy is consumed in cities, which emit about 65% of the world's total GHG, mainly CO_2 (United Nations 2019; World Bank 2019; World Energy Recourses 2016; United Nations 2018); and cities are also responsible for 75% of the natural resources (UNEP 2019). Moreover, cities are experiencing additional challenges such as the urban heat island effect (UHIE) and other climate change impacts (Aboulnaga and Mostafa 2019; Aboulnaga et al. 2019). Such increases add pressure on cities' infrastructures, which in most cases are not sufficiently resilient to muddle through CC severe events, such as heat waves and floods. Furthermore, air pollution in megacities and urban areas is another challenge, and this is receiving global attention. According to the World Health Organization—WHO, at least 96% of the populations in megacities exposed to PM2.5—exceeding WHO air quality guidelines levels (Marlier et al. 2016 (2–15); Krzyzanowski et al. 2014 (1–85). About seven million deaths are reported annually worldwide (WHO 2017), thus improving air quality in cities by using urban-framing innovation (Sala et al. 2017, 2018, 2019), adaptive green designs in cities (Trombadore 2017, 2018), and urban public spaces' strategies (Battisti et al. 2018) are essential tools to achieve sustainability in megacities and mitigate CC.

This work addresses how smart and green strategies can drive cities to deal with such challenges by urban-farming innovation, energy efficiency tools, and mobilizing stakeholders (CES-MED 2018; Aboulnaga 2016). It presents cities' strategies, policies, and answers a major question how strategies, policies, and actions can drive green and smart cities to lessen challenges and adapt to CC. The work also tackles three important topics: (a) city challenges: strategies, actions, stakeholders' mobilization for liveable and smart urban areas, and CC adaptation (www.ces-med.eu); (b) innovative policies and tools for greening cities—urban-farming opportunities (www.centreoabita.unifi.it); and (c) adaptive design of green facades and natural-based solutions for resilient architecture and environment (www.unifi.it). Finally, it draws conclusions to support local governments, policy makers, developers, architects, researchers and the public in developing smart and sustainable cities of the future.

2 City Challenges

Climate change (CC) is one of the major challenges facing cities nowadays at both the local government and municipalities' level. Cities worldwide face economic, social, and environmental challenges. Impacts of CC also pose fundamental and unprecedented threats to all sectors, regions, and societies, as well as countries (Fig. 1). Climate change is not limited to a nation or a region, but rather stops at no borders. According to a statement of Angela Merkel, Chancellor of Germany:

> Climate change knows no borders. It will not stop before the Pacific islands and the whole of the international community here has to shoulder a responsibility to bring about a Sustainable Development, [Angela Merkel, Chancellor of Germany]. (Safi 2014)

Open innovation strategies (OIS), policies, and climate actions can drive sustainable and green cities to adapt to climate risks. This is a crucial question that needs to be addressed by the federal, central, and the local governments and municipalities to meet Paris Climate Agreement targets before it is too late.

2.1 Policies and Tools for Greening Cities—Urban Farms Opportunities

How and when could we practice and apply the concept of Green Architecture, to buildings and cities (urban strategies)? It is not merely an aesthetic approach,

(a) Storm flooding, Cairo, October 2019   (b) Floods in Ras Ghareb, October 2016   (c) Discoloured River Nile, October 2016   (d) Snow storm in Jeddah - January 2020

Fig. 1 Climate change severe impacts in Egypt and KSA. *Images' Source*. Fig. 1b Westwards 2016

even if many people and architects still seem to want just to paint over the existing architecture, 'upgrading' it with a powerful marketing strategy. The National Institute of Building Sciences (NIBS) suggests that the interrelationships of these design objectives *must be understood, evaluated, and appropriately applied*—accessibility; aesthetics; cost-effectiveness; functional or operational—"The functional and physical requirements of a project"; historic preservation; productivity (comfort and health of the occupants); security and safety; and sustainability" (NIBS 2020). Jackie Craven wrote that in *Primer on Green Architecture and Green Design: 'Green Architecture is More Than a Color'* (Craven 2019). Rather, it strives to minimize the use of resources and the energy balance embedded in materials and in construction processes, for both buildings and cities, where GA symbolizes the sustainability of modern cities (Wines 2008; Philip 2018).

To attain Sustainable Development Goals (SDGs), it is also imperative to adopt more innovative solutions to green our cities. Innovative urban farm (UF) policies and tools are essential to encourage local municipalities to transform existing buildings and urban areas to be generative. The role of rooftop urban farming (RTUF), in reducing temperature and mitigating urban heat islands (UHI) as well as lowering air pollution, is becoming more valuable and efficient as big data shows the immense benefits of UF in cities.

2.1.1 Urban Farms' Significant

Examples worldwide illustrate that urban farms (UFs) can be implemented in different types of buildings (households, public, educational buildings, and social gardens). In fact, UFs are part of the urban ecological systems, and they play an important part in the urban environmental management system (Lori et al. 2017). UFs are widely implemented in Asia, where 85% of fresh vegetables generated in 14 big cities in China (Wageningen University 2018). A research study carried out by the European Union shows that rooftop gardens in cities could possibly generate as much as three quarters of vegetables used up in these cities. Another study indicates that rooftop urban farms (RTUFs) can possibly provide more than three quarters of all consumed vegetables, e.g., if rooftop farms are exploited in Bologna, they can supply vegetables amounting to 12,500 tons, which corresponds to about 75% of the city's requirements (Urban Agriculture in Europe 2017).

2.1.2 Innovative Urban Farm Policies and Tools

Fostering urban farm policies lead to: (a) lowering the heat island effect during summer; (b) reducing energy consumption for air conditioning; (c) mitigating CO_2 and produce oxygen; (d) improving thermal and human comfort; (e) reducing solar overheating on roofs and façades; and (f) producing fresh, clean, and zero kilometers food. The classification of general policies that rule rooftop urban farms lists four types (Carter and Fowler 2008). The City Council of Ghent, Belgium, has approved

Fig. 2 Europe's and the world's largest rooftop urban farm in Paris, France. *Image Source* https://www.forbes.com/sites/alexledsom/2019/08/29/worlds-largest-urban-farm-to-openon-a-paris-rooftop/#7fd750d69305

in a Climate Plan in 2015 to make Ghent a climate–neutral city by 2050. Such a policy embraces actions to promote UFs by space for urban agriculture and provide guidelines for schools to commence UFs projects with social employment, generate healthy and affordable food supply (Ghent Climate Plan 2014–2019; 2015). The French government also enacted a policy, supported by a law in March 2015 that mandates the partial coverage of all new buildings, in commercial areas, by either UFs or green roofs. A direct upshot of such a policy is that Paris started in August 2019 constructing the largest urban rooftop farm ever in Europe and the world (Urban Agriculture in Europe 2017). The largest rooftop urban farm in Paris covers an area of 14,000 m^2, and it cultivates more than 30 different plant species and produces around 1,000 kg of fruits and vegetables daily (Fig. 2), yet it uses entirely organic methods (Urban Agriculture in Europe 2017).

2.2 Tools for Climate Mitigation

Nature-based solutions, such as adaptive design of green facades for resilient architecture and environment, are effective tools to deal with city challenges and make buildings' envelopes energy efficient. Introducing vegetation elements as productive plants into architecture is the newest tendency that integrates holistic design principles into sustainable architecture. For eco-design or eco-friendly architecture, there are benefits associated with vegetation (Akadiri et al. 2012). Vegetation in architecture means health contributions, air purification, and psychological and social well-being; above all, it reduces the stress of living in a fully artificial environment.

2.2.1 Green Cities and Green Facades

There are many reasons to justify the new trend of implementing nature-based solutions in cities:

- *reducing UHIE during summer*: The growing of peck temperature in towns is not only a consequence of climate change, but also primarily the result of urbanization. It is also due to the absence of green areas and parks, as well as introduction of vegetation, trees, gardens, and vertical farming in the urban tissues (Aboulnaga and Mostafa 2019), and
- *improving thermal comfort and human behavior:* Water that plants absorb into the soil from irrigation or rain ultimately returns as water vapor by direct evaporation/transpiration within plants' leaves improves microclimates in open and interior spaces and human comfort.

Vertical greening of architecture in regions with a Mediterranean climate provides cooling effects for the building surface, which is vital during summer periods in hot and temperate climates. A greening building's envelope reduces the peak temperature. Some studies demonstrate that greening building technology for roofs and façades can increase the dynamic thermal characteristics of the wall-surfaces temperature to maintain good thermal behavior (Pisello et al. 2015, 2016). Beautiful trees make liveable cities because a dynamic urban forest supports a healthy community, economy, and environment. Trees are also integral to the urban design of any city or town (Arnold 1980). Hence, there is an urgent need to develop and apply strategies for sustainable cities, including forests and street tree plantings.

This work presented a workshop conducted at NOI Center in Bolzano, Italy, 2019. This includes three main parts: (a) OIS for sustainable energy actions and mobilizing stakeholders (SEAP/SECAP); (b) Urban farming innovation policies; and (c) Climate mitigation for public spaces.

3 Objective

The objective of this work is to highlight the role of city innovation and open innovation strategies in driving smart and sustainable approach to enhance cities' liveability and to provide tools to local governments/municipalities for meeting city challenges by utilizing sustainable energy plans, urban farming innovation, and green facades to help make cities resilient and by adapting to CC.

4 Method

The methodology depends on inductive and deductive approaches to review and assess cities in driving them to be sustainable and smart. The study includes two

Fig. 3 Open Innovation Strategy's main points of the sustainable energy and climate action plans

fields that cover the city of Rotterdam, The Netherlands, and the city of Hurghada, Egypt. In this study, we review and utilize open innovation strategies (OIS) to develop sustainable energy and climate action plans (SECAP) for the selected city. The first field focuses on a review of adopted OISs to solve problems of the city of Rotterdam (flood management, CO_2 mitigation, and air pollution), whereas the second fold highlights the results of how OIS and action plans can drive cities to meet sustainability measures and strategically save energy. This work was part of the European Union funded project CES-MED, contract ENPI 2012/309-311/EuropeAid/132630/C/SER/MULTI, Egypt. In the GHG emissions calculation, there are many recognized methods/tools developed by international institutions; thus, mapping these methods/tools was vital (Aboulnaga, Amin, and Rebelle 2020). The European Commission Joint Research Center developed methods/procedures for conducting Baseline Emission Inventory, and we use it to calculate the GHG emissions in the city of Hurghada (Aboulnaga et al. 2020). We used data gathered from the governorate and other national and governmental entities in the scope and methodological principles of the GHG emissions. Figure 3 shows the OIS points adopted in SECAP of the city of Hurghada, Egypt.

5 Innovative City

Becoming an 'innovative city' is vital to drive cities to meet current challenges and foster opportunities in terms of resources. The 'Open innovation' concept is increasingly accepted as a new tool for governments to tackle global challenges, as well as city challenges. The type of OI varies based on the nature of the entity, whether public or private. Traditional OI used in private organizations distinguishes three types: (a) outside-in, where an organization or a company utilizes the resources outside the entity to benefit the process; (b) inside-out—resources within the organization are put outside its boundaries; and (c) coupled processed—a and b are combined to work with partners and stakeholders. In the public and governmental contexts, OI can be classified into three different types: (a) Citizen Ideation and Innovation—citizen's access online platforms to utilize the creativity and knowledge of the crowd; (b) Collaborative Administration—common administrative tasks are improved through involving external actors in a systematic way; and (c) Collaborative Sharing—larger groups of people and experts are involved in policy making and implementation. To ensure significant innovation outcomes, stakeholders (government, policy makers, entrepreneurs, academics, citizens, bankers, NGOs) should be engaged in the process. This section highlights examples of open innovation strategies, OIS (public type) in two cities: Rotterdam and Hurghada in relation to innovation, strategies, and policies. It also discusses green façades as a tool for green cities. The OIS models used in the City of Rotterdam and in City of Hurghada, the latter is part of CES-MED project (2011–2018) which are shown in Fig. 3.

6 Strategies, Policies, Tools, and Actions for Smart and Sustainable Cities

This section presents strategies, policies, tools, and actions for developing smart and sustainable cities.

6.1 Open Innovation Strategies, City of Rotterdam, the Netherlands

The city of Rotterdam adopted OISs to solve problems such as flood management, CO_2 mitigation, and reducing air pollution. The OISs led to the creation of seven initiatives: (a) Blijstroom—Cooperative for shared solar panel energy; (b) The s' Gravendijkwsstraat; (c) Happy Streets; (d) FLIP Traffic Light Interventions; (e) Benthemplein Water Square; (f) DakAkker Rooftop Urban Farm; and (g) StraaD—Water-Sensitive Rotterdam (a climate-conscious street plan). This section highlights Initiative no. 6—The DakAkker rooftop urban farm that aimed at improving sustainability and liveability in the city. The initiative resulted in having the largest rooftop urban farms built on an abandoned office building, where it provides a food supply and residents can eat local clean food and enjoy life in a clean air environment (van Genuchten, Calderón Conzález, and Mulder 2019). Figure 4 shows the planning, stakeholders, implementers, and outcomes of climate action.

Fig. 4 Open Innovation Strategy, DakAkker Rooftop Urban Farm Initiative, City of Rotterdam, The Netherlands. *Image Source* https://www.foodurbanism.org/dakakker-urban-rooftop-farm/

6.2 Sustainable Energy Action and Mobilizing Stakeholders—City of Hurghada

6.2.1 Open Innovation Strategy for SECAP

In developing the Sustainable Energy and Climate Action Plan (SECAP) for the City of Hurghada in the Red Sea Governorate, we adopted OIS (Fig. 9c—Appendix). In the three conducted workshops, 17 municipality representatives of local government officials, headed by Minister Ahmed Abdallah, Governor of the Red Sea Governorate (2014–2019) were involved. These include local council, environment, urban, technical, marine science, health and safety, and energy—the Ministry of Electricity as well as tourism management, media, public relations, and HEPCA waste-recycling company responsible for managing city's waste. For experts and academia, five persons of the CES-MED project participated (team leader, international energy consultant, SECAP senior expert, local municipality development expert, and researcher, plus a co-creation expert for media and awareness raising and stakeholders' mobilization. It is vital to state some of data about the city such as municipality's area of 111.30 km^2, governorate area of $203,685 \text{ km}^2$, and inhabited area of $27,300 \text{ km}^2$, as well as mass populated area of 1284 km^2 that were used in OIS and SECAP development. This also includes the overall population of the city and governorate (2015) of 279,684 persons and 472,203 persons respectively. In terms of electricity consumption and electricity/capita, these are 1234.747 GWh/year and 4415 MWh/year, respectively. The energy consumption and energy use/capita are 3338 GWh/year and 11.9 MWh/year, and the GHG emission/yr and per capita are 1342 ktCO_2eq and 4.8 tCO_2eq, respectively.

6.2.2 Climate Action

In the OIS session on climate action, the Governorate of the Red Sea, with the support of a team of experts, collaborated to assess the impacts of each climatic-hazard type of vulnerable and impacted sectors. These impacted sectors are: (a) population (public health); (b) infrastructure (transport, energy, water, and social);

(c) built environment (buildings' stock and materials); (d) economy (tourist and agriculture); and (e) biodiversity (coastal zones ecosystem: green zones and forests). A fourth OIS workshop was conducted and focused on related topics, including (a) introduction to climate change (general); (b) a national strategy on climate change adaptation; (c) regional strategy on climate change adaptation; (d) climate data; (e) climate projections; (f) risks; (g) vulnerability; and (h) adaptation actions.

7 Results and Discussions

Figures 2–5 present the results of open innovation strategies (OIS) during the development of Sustainable Energy and Climate Action Plans—SECAP (Appendix). Figure 10 in Appendix illustrates the annual energy consumption and GHG emissions of eight sectors (residential buildings, tertiary buildings, public lighting, transport, industry, agriculture, water and wastewater, and tourism) in the city of Hurghada. It is clear from Fig. 2a that the city's sectors using highest amounts of energy per year are transport 39%, tourism 36%, and residential buildings 13%, followed by tertiary buildings 9%. This is due to intense tourism activities including aviation in the two airports. Figure 6b in Appendix exhibits the same patterns for GHG emissions in these sectors. It is important to mention that in Fig. 7, 8 and 9 (Appendix), the governorate buildings' energy consumption rates are included under tertiary buildings. These detailed data can be found in the dedicated chapter and in the BEI Excel spreadsheets (www.ces-med.eu; 2018: 34–37). Emissions include combustion in accordance with IPCC Guidelines 2006 and upstream emissions for producing and delivering energy based on UNFCC Guidelines. The emission factor used for electricity is the recommended one by the GHG protocol factor for 2012 ($443.76 \, gCO_2eq/kWh$). In addition, Fig. 7 (Appendix) shows the business-as-usual (BaU) scenario 2030 for the city of Hurghada. It is clear that energy consumption and GHG emissions of all sectors will steadily increase due to developmental factors until 2020, and then decrease in one or two sectors and stabilize in most of the sectors until the year 2030. Results show that the highest emitting sectors in the city are mainly transport and tourism; hence, the planned actions are vital to mitigate emissions. Transport is one of the key concerns in the city that needs attention since it is the first energy consumer sector with 1,301,117 GWh/year (39%) and the second GHG emitting sector with 351,746 $ktCO_2eq/year$ (28%). Therefore, to reduce energy use and mitigate GHG emissions in the city, a set of seven priority actions resulted from the OIS workshops to drive the city to be smart, green, and sustainable (Fig. 10—Appendix). The following section highlights the high priority actions (transport and tourism).

7.1 Transport Planned Actions

Transport has two main developed actions:

- Action 1—Improvement of management and awareness among transport operators could lead to a 10% reduction in energy consumption beginning in 2020
  (without significant investment),
- Action 2—Provision of Sustainable, Clean and Green Urban Mobility Master
  Plan implementation to start in 2020 at the latest and would demonstrate a 50%
  progress in implementation in 2030. The expected impact of such a plan should
  make a 30% reduction by 2030, including

  – New engine technologies will allow securing a 10% reduction;
  – Active mobility development (cycling and walking) allowing percent reduc-
    tion;
  – Urban planning allowing traffic optimization that will reduce consumption by
    10%; and
  – Public transport system and RTB service to the airport resulting in another 5%
    reduction.

The results of the action plan in the transport-sector reduction in energy and
GHG emissions indicate that the expected cut in transport energy use and GHG
emissions by 2030 are 521,160 MWh/year and 140,891 tCO_2eq/year. This will lead
to a situation in 2030 where GHG emissions from the BAU scenario of 453,753
tCO_2eq/year would be cut by 31%. For each sub-action, the reduction in energy and
GHG emissions is:

I. Common charter for improving transport service	130,290 MWh/year	35,223 tCO_2eq/year
II. Sustainable urban mobility plan:		
– Promoting new engines	130,290 MWh/year	35,223 tCO_2eq/ year
– Active mobility development	65,145 MWh/year	17,611 tCO_2eq/year
– Traffic optimization by urban planning	130,290 MWh/year	35,223 tCO_2eq/year
– Public service and RTBl	65,145 MWh/year	17,611 tCO_2eq/ year

7.1.1 Tourism Planned Action

The tourism sector is also one of the prime sectors that urgently need to be addressed
because it possesses the highest rank in energy consumption with 43% of the total
use, converted into 1196 GWh/year. It is measured first in GHG emissions with 44%
of the total emissions, translated to 491 $kteCO_2$/year (33%).

Results on tourism sector to improve the 2015 situation on energy use indicate
action plans as follows:

- Awareness raising for hotels and resorts staff (10%) based on energy use of
  1,114,026 MWh/year;
- Hotel and resort refurbishment (overall energy consumption of hotels after 10%
  cut due to awareness raising efforts) according to energy use of 1,002,624
  MWh/year;

- Awareness raising and occupancy optimization in diving boats due to energy use 50,489 MWh/year;Awareness raising and occupancy optimization in diving boats due to energy use 50,489 MWh/year;
- Divers boats refurbishment (50% after all awareness efforts conducted on energy efficiency), which is based on the calculated energy consumption of 45,440 MWh/year.

Ideally, a city committed to reduce energy consumption, GHG emissions such as Hurghada would develop and implement a Sustainable and Green Tourism Plan (SGTP) based on smart eco-solutions:

- Structure the tourism sector to empower leisure and tourism authorities take action for improving tourism activities, i.e., close collaboration between the Municipal Council and Egyptian Tourism Authority and the evolution of rules, which would assist in reducing energy consumption;
- Develop communication and participatory strategies aiming at facilitating behavioral changes;
- Convert tourist boats to operate with natural gas—CNG instead of fuel, when appropriate and be powered by electric engines using PV cells;
- Integrate renewable energy in seaports, marinas, and diving centers along the shore of the city;
- Promote a charter for responsible tourism that engages the tourists to take care of the fragile environment they enjoy when visiting the City of Hurghada.

The results of the action plan of the tourism sector's reduction in energy and GHG emission show that the expected cut in tourism energy use and GHG emission by 2030 are 439,958 MWh/year and 218,077 tCO_2eq/year. This will lead to a situation in 2030 where GHG emission from the BAU scenario of 633,280 tCO_2eq/year is reduced by 34%. For each sub-action in tourism, the energy reduction and GHG mitigation are:

– Awareness raising for hotels and resorts staff– Awareness raisin	111,402 MWh/year111,402 MWh/year	46,084 tCO_2eq/year
– Hotels and resorts' refurbishment– Hotels and resorts' refurbishmentg for hotels and resorts staff	300,787 MWh/year	146,564 tCO_2eq/year
– Occupancy optimization in diving boats	5049 MWh/year	1340 tCO_2eq/year
– Divers boats refurbishment	22,720 MWh/year	6089 tCO_2eq/year

For climate action, it was vital to carry out a Climate Risk Assessment (CRA). Table 1 lists ten receptors according in terms of their impacts and probabilities of occurrence. For impact and probability, a weight is given to each receptor on a scale 5 to 1, where 5 is very high and 1 is very low. Figure 11 (Appendix) presents the results of CRA. It is clear from Fig. 7 that costal zones (R9) are the most vulnerable receptor

Table 1 City of Hurghada's climate risk assessment—receptors

Receptors	Impact	Probability
R1—Public health	1	2
R2—Transport	1	1
R3—Energy	1	1
R4—Water	1	2
R5—Social	1	1
R6—Buildings' stock and material	2	2
R7—Tourists	2	2
R8—Agriculture	0	0
R9—Costal zones/ecosystems	2	3
R10—Green zones/forests	0	0

from climate risks followed by buildings' stock (R6) and tourists (R7). In addition, public health (R1) and water (R4) come after in terms of climate risks (Table 1). Therefore, these receptors at the city level need actions and immediate measures to reduce climate risks.

8 Conclusions

Open Innovation Strategies (OIS) outlined above have proven effective tools in assisting local governments and municipalities in meeting challenges and driving sustainable, smart, and green cities. Green policies, action plans, and approaches play a vital role and path toward sustainable cities and in counterbalancing challenges for transforming them to opportunities utilizing green architecture and urban farm innovation and greenery in urban spaces. The OIS, including DakAkker Rooftop Urban Farm Initiative in the city Rotterdam, The Netherlands, and sustainable energy and climate action plan (SECAP), city of Hurghada, Egypt, were productive tools. It is imperative to state that priority actions, particularly in the transport and tourism sectors resulting from OIS workshops to drive the city to be green and sustainable were effective and useful to reduce energy use and mitigate GHG emission in the city of Hurghada. Green architecture, through many green features contributes to enhancing liveability, well-being in cities, and climate mitigation. Nevertheless, we need more strategies, policies, and city innovation to meet Paris Climate Agreement targets and attain SDGs.

Acknowledgements The authors express their appreciations to Dr. Adriano Bisello—EURAC Research and SSPCR Editorial Boards for compiling and editing this book and for the organization of the SSPCR 2019 conference held at NOI Park in Bolzano in South Tyrol, Italy. The first author would like to thank the European Union for the funded CES-MED Project, DG-NEAR, where SECAP was an excellent sustainable action plan for other cities to adopt in meeting challenges, mainly energy saving, CO_2 mitigation and climate action

Appendix

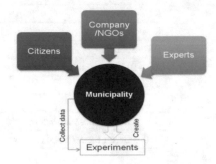

(a) Open innovation strategy on the city level 1, where a municipality creates the experiment/project

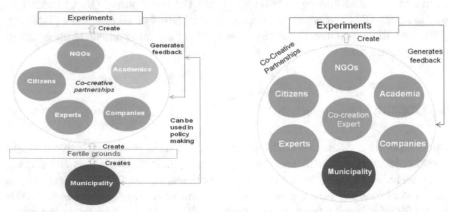

(b) Open innovation strategy on the city level 2, (Municipality performs as an Infrastructurer)

(c) Open innovation strategy on the city level 3, (Municipality is taking part in theco-creative process)

Fig. 5 Open innovation models used in the City of Rotterdam, The Netherlands, and in the City of Hurghada, Egypt

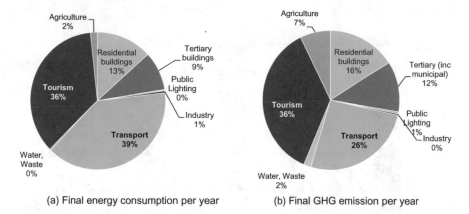

(a) Final energy consumption per year   (b) Final GHG emission per year

Fig. 6 Baseline emissions inventory and total energy use of city's eight sectors—City of Hurghada

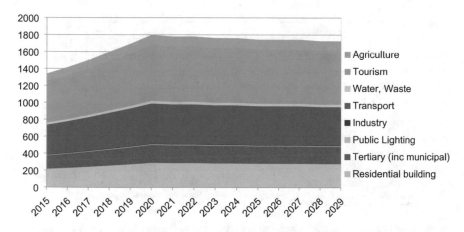

Fig. 7 The Business-as-usual scenario 2030—City of Hurghada

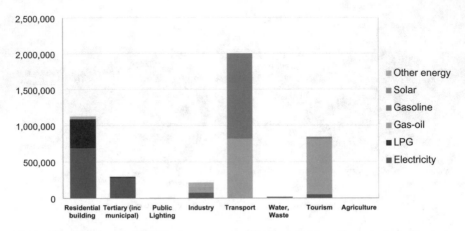

Fig. 8 Energy consumption per sector and type of resources in the City of Hurghada

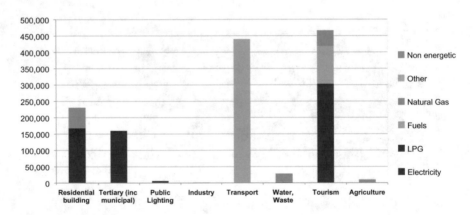

Fig. 9 Greenhouse gases (GHG) emissions per sector and per energy in the City of Hurghada

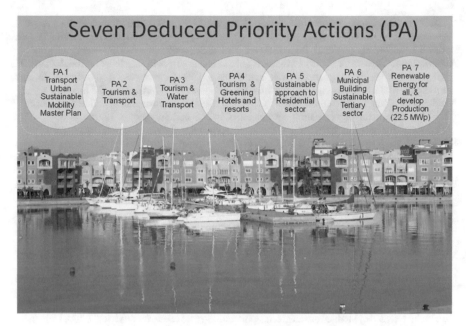

Fig. 10 A set of priority actions for City of Hurghada resulting from OIS

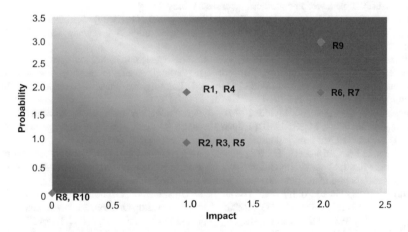

Fig. 11 Results of climate risk assessment based on climate data availability—City of Hurghada

References

Aboulnaga MM et al (2019) Introduction, 'urban climate change adaptation in developing countries – policies, projects and scenarios', pp 1–18. Springer, Berlin. Available at: https://www.springer.com/gp/book/9783030054045/

Aboulnaga M (2016) Recommended national sustainable energy savings' actions for Egypt, Cleaner Energy Saving Mediterranean Cities, CES-MED (2018), ENPI

2012/309 311, EU Funded Project, EuropeAid/132630/C/SER/Multi, Hulla & CO Human Dynamics – KG, October 10, 2016; https://ces-med.eu. Available at: 180917_CES-MED_National_Report_Egypt_FINAL2rev.pdf/

Aboulnaga M, Mostafa M (2019) Mitigation of heat island effect in megacities through districts' prioritization for urban green coverage applications: Cairo – as a case study, Renew. Energy Environ Sustain 4(5). Available at: https://www.rees-journal.org/. Accessed 22 July 2019

Akadiri PO, Chinyio EE, Olomolayie PO (2012) Design of sustainable building: a conceptual framework for implementing sustainability in the building sector. Buildings 2:126–152. 10-33902020126, ISSN 2075309. Available at: www.mdpi.com/journal/buildings

Battisti A, Laureti F, Flavia Zinzi M, Volpicelli G (2018) Climate mitigation and adaptation strategies for roofs and pavements: a case study at Sapienza University Campus. Sustainability [RIVISTA IN CLASSE "A"], pp 1–31, ISSN: 2071-1050. https://doi.org/10.3390/su10103788

Battisti A, Santucci D, Volpicelli G, Auer T (2018) Rigenerare la città storica, Activating Munich Outdoor Resilience Regenerating the historic city, Activating Munich Outdoor Resilience. Urbanistica Dossier, pp 176–179, ISSN: 1128-8019

Borneman E (2014) Urban Heat Island, [Online]. available at: https://www.geographyrealm.com/urban-heat-islands/. Accessed 28 Feb 2020

Carter T, Fowler L (2008) Establishing green roof infrastructure through environmental policy instruments. Environ Manage 42(1):151–164

Carven J (2019) A Primer on Green Architecture and Green Design. Thought Co. [Online]. Available at: https://www.thoughtco.com/what-is-green-architecture-and-green-design-177955. Accessed 27 Feb 2020

Ceccherini NL (2018) Mediterranean green architecture- *research and* innovation. DIDA University Press – Firenze, ISBN 978-88-3338-024-7

Cities - United Nation Sustainable Development Action 2015. [Online]. Available at: https://www.un.org/sustainabledevelopment/cities/. Accessed 1 May 2019

Design Objectives, Whole Building Design Guide—WBDG (2020) National Institute of Building Sciences (NIBS), USA, [Online]. Available at: https://www.wbdg.org/design-object ives/. Accessed 27 Feb 2020

FoodMetres-Urban-Organic-Waste-Management-in-The-Hague.pdf. Wageningen University, Academic Consultancy Training, 2014–05–01 [Online], available at: https://www.foodme tres.eu/wp-content/uploads/2014/06/FoodMetres-Urban-Organic-Waste-Management-in-The-Hague.pdf/. Accessed 13 Jan 2020

Genuchten EV, Conzález AC, Mulder I (2019) Open innovation strategies for sustainable urban living. Sustainability, 11-3310, MDPI, June 15, 2019, [Online]. Available at: https://doi.org/10.3390/su11123310, www.mdpi/journal/sustaiablity/. Accessed 28 Oct 2019

Ghent Climate Plan 2014–2019. The Sustainable Energy Action Plan (SEAP) puts Ghent on course for climate neutrality in 2050, [Online]. Available at: https://stad.gent/en/city-structure/ghent-cli mate-plans/ghent-climate-plan-2014-2019/. Assessed 13 Jan 2020

Jodidio P (2018) Green Architecture (Bibliotheca Universalis), Taschen; Multilingual edition, June 15, 2019, ISBN-13: 978-3836522205. Available at: https://www.Green-Architecture-Phi lip-Jodidio/dp/3836522209/ref=pd_sbs_14_t_0/147-1554264-2546960?_encoding=UTF8&pd_ rd_i=3836522209&pd_rd_r=b0d336cc-ef15-4e85-9819-aada168bfc3f&pd_rd_w=etnxz&pd_ rd_wg=Ofsy4&pf_rd_p=5cfcfe89-300f-47d2-b1ad-a4e27203a02a&pf_rd_r=PBBF829MGJGY A23CEGD6&psc=1&refRID=PBBF829MGJGYA23CEGD6/. Accessed 26 Feb 2020

Krzyzanowski M, Apte JS, Bonjour SP et al (2014) Air pollution in the meg-cities. Curr Environ Health Rpt, pp 1–85. Available at: https://doi.org/10.1007/s40572-014-0019-7. Accessed 21 July 2019

Ledsom A (2019) World's Largest Urban Farm to Open on a Paris Rooftop [Online]. Available at: https://www.forbes.com/sites/alexledsom/2019/08/29/worlds-largest-urban-farm-to-openon-a-paris-rooftop/#7fd750d69305/. Accessed 14 Jan 2020

Lori SH, Lori H, Toner E (2017) Urban agriculture: environmental, economic, and social perspectives. Horticultural Rev 44, 1st ed. Janick J (ed) Wiley-Blackwell, Wiley. Available at: https://www.researchgate.net/publication/308543504_Urban_Agriculture_Environmental_E conomic_and_Social_Perspectivs. Accessed: 14 Jan 2020

Marlier ME, Jina AS, Kinney PL et al (2016) Extreme air pollution in global megacities. Curr Clim Change Rep, pp 2–15. Springer, Berlin. Available at: https://doi.org/10.1007/s40641-016-0032-z/. Accessed 21 July 2019

McEldowney J (2017) European parliamentary research service urban agriculture in Europe Patterns, Challenges and Policies. https://www.europarl.europa.eu/RegData/etudes/IDAN/2017/614641/EPRS_IDA(2017)614641_EN.pdf/. Accessed 12 Jan 2020

Pisello AL, Pignatta G, Piselli C, Castaldo VL, Cotana F (2015) Effect of dynamic characteristics of building envelope on thermal performance in summer conditions: in field experiment. In: 5th CIRIAF National Congress Energy, Environment and Sustainable Development: Perugia, Italy, April 2015. Available at: https://www.researchgate.net/publication/275098300_Effect_of_dynamic_characteristics_of_building_envelope_on_thermal_performance_in_summer_condit ions_in_field_experiment

Pisello AL, Pignatta G, Piselli C, Castaldo VL, Cotana F (2016) Investigating the dynamic thermal behavior of building envelope in summer conditions by means of in-field continuous monitoring. Am J Eng Appl Sci 9(3):505–519. https://doi.org/10.3844/ajeassp.2016.505.519/

Resource Efficiency and Green Economy, United Nations Environment Programme UNEP (2019)

Sala M, Alcamo G, Ceccherini L (2017) Energy-saving solutions for five hospitals in Europe. In: Sayigh A (eds) Mediterranean green buildings & renewable energy – selected papers from the world renewable energy network's med green forum. Springer, Cham, pp 177–189, ISBN: 978-3-319-30745-9. https://doi.org/10.1007/978-3-319-30746-6_1

Sala M, Casazza C (2018) Ecovillages and friendly city - new alliance for a better green future, Friendly City 4 'From Research to Implementation for Better Sustainability', IOP Conf. Series: Earth Environ Sci 126(1):012193. IOP Publishing, pp 1–6. Available at: https://doi.org/10.1088/1755-1315/126/1/012193

Sala M, Trombadore A (2018) Design of seaside buildings in China. In: Sayigh A (eds) Seaside building design principles and practices. Springer, Cham, pp 217–229, ISBN: 78-3-319-67948-8. https://doi.org/10.1007/978-3-319-67949-5_10

Sala M, Nelli LC, Donato A (2019) A university master's course and training programme for energy managers and expert in environmental design in Italy. In: Sayigh A (eds) Sustainable building for cleaner environment. Innovative renewable energy. Springer, Cham, pp 347-359, ISBN: 978-3-319-94595-8. Available to: https://doi.org/10.1007/978-3-319-94595-8_29

Sustainable Cities, Challenges and Context 2008, [Online]. Available at: https://unhabitat.org/wp-content/uploads/2010/07/GRHS.2009.1.pdf/. Accessed 19 July 2019

Sustainable Energy and Climate Action Plan (SECAP) - City of Hurghada, Cleaner Energy Saving Mediterranean Cities (CES-MED), ENPI 2012/309–311, EU Funded Project, EuropeAid/132630/C/SER/ Multi, Hulla & CO Human Dynamics – KG, March 2, 2018; https://ces-med.eu. Available at: 02032018_Hurgaha SECAP_FINAL.pdf/, Accessed: July 20, 2019.

The World Bank Group, Urban Development, [Online]. Available at: https://www.worldbank.org/en/topic/urbandevelopment/overview

The World Bank, PART III 'Cities' Contribution to Climate Change', p 14, [Online]. Available at: https://siteresources.worldbank.org/INTUWM/Resources/340232-1205330656272/476 8406-1291309208465/PartIII.pdf/. Accessed 1 May 2019

Trombadore A (2017) Multidisciplinary energy efficiency think tank for supporting a multilevel governance model in energy policies and measures: MEETHINK Energy Project: Topic 6. In: Mediterranean Green Building & Renewable Energy – Selected Papers from the World Renewable Energy Network's MED Green Forum. Springer, Cham, pp177–189. ISBN: 978-3-319-30745-9. https://doi.org/10.1007/978-3-319-30746-6_12/

Trombadore A (2018) Green design for a smart Island: green infrastructure and architectural solutions for ecotourism in mediterranean areas. In: Sayigh A (eds) Seaside building design principles and practices. Springer Cham, pp 163–194. ISBN: 978-3-319-67948-8. https://doi.org/10.1007/978-3-319-67949-5_8

United Nation Habitat, UN-Habitat, World City Report 2016 – Urbanization and Development – Emerging Future [Online]. Available at: https://wcr.unhabitat.org/wp-content/uploads/2017/02/WCR-2016-Full-Report.pdf/. Accessed 16 May 2020

United Nations – Department of Economic and Social Affairs (DESA), 16th May 2018. [Online]. Available at: https://www.un.org/. Accessed 20 July 2019

United Nations Habitat, UN-Habitat – Energy, [Online]. Available at: https://unhabitat.org/urban-themes/energy/. Accessed 1 May 2019

Urban Agriculture in Europe - Patterns, challenges and policies: Briefing European Parliamentary Research Service- In depth Analysis, December 2017 — PE 614.641, Members' Research Service, ISBN: 978-92-846-2506-2. https://doi.org/10.2861/413185, [Online], World Energy Recourses 2016, World Energy Council, world-Energy-Resources-FullReport-2016.pdf/. Available at: https://www.worldenergy.org/wp-content/uploads/2016/10/World-Energy-Resources_FullReport_2016.pdf/https://www.worldenergy.org/wp-content/uploads/2016/10/World-Energy-Resources_FullReport_2016.pdf/. Accessed 1 May 2019

Velazquez A (2020) World's Largest Urban Farm to Open on a Paris Rooftop [Online], available at https://www.greenroofs.com/2019/08/28/worlds-largest-urban-farm-to-open-on-a-paris-rooftop/. Accessed January 14, 2020.

Wines J, Jodidio P (2019) Green architecture, Taschen - c2008, London (June 15, 2019). ISBN: 9783836503211. Available at: https://trove.nla.gov.au/version/33003918/. Accessed 26 Feb 2020

World Health Organisation—WHO (2017) Available at: https://www.who.int/airpollution/en/. Accessed 2 July 2019

Governing and Planning Local Climate-Change Adaptation in the Alps

Luca Cetara, Marco Pregnolato, and Pasquale La Malva

Abstract Several studies show that mountain territory has special vulnerabilities and is subject to specific impacts of climate change. This article investigates governance of climate change in mountain areas, focusing on the case of Italy where national and regional instruments address adaptation to climate change at various scales. The study intends to provide a methodological framework aimed to elicit adaptation actions applicable in mountain areas and consistent with specific impacts and adaptation objectives, supported by indicators and at the same time suitable for being perfected for use in other territories. The research stems from the "Budoia Charter," a voluntary declaration of commitment to implement adaptation measures to climate change in Alpine municipalities. This study illustrates the methods used in support of the application of the charter, particularly for a set of five pilot areas exposed to climate impacts, from the western to the eastern Alps in Italy. The units under investigation are described based on geographical characterization, available sources of knowledge, local climate context and impacts, vulnerabilities, preparedness on specific vulnerabilities, governance type, resources and key planning instruments, and public awareness of climate change. Scientific information is then coupled and weighed with local demands and perceptions through a mixed model (qualitative and quantitative). An analysis of coherence and comparison between the impacts, objectives and measures of adaptation to climate change was conducted through national, regional and sub-regional plans and other sources. Aiming at identifying a set of site-specific adaptation measures applicable to mountain areas, the alignment of all actions to overarching and/or legally binding plans has been carefully checked to designate adaptation measures coherent with the national and regional adaptation frameworks. The main goal of the study is to provide a standard procedure aimed at

L. Cetara (✉)
EURAC Research - Representing Office Rome, via Ludovisi 45, Roma, Italy
e-mail: luca.cetara@eurac.edu

M. Pregnolato
Fondazione Lombardia per l'Ambiente – FLA, Largo 10 Luglio 1976, 1, Seveso, Italy

P. La Malva
Department of Psychological Sciences, Health and Land (DiSPuTer), Unit of Earthquake and Environmental Hazards, G. D'Annunzio" University of Chieti-Pescara (UNICH), Via dei Vestini, 32, Chieti, Italy

© The Author(s), under exclusive license to Springer Nature Switzerland AG 2021
A. Bisello et al. (eds.), *Smart and Sustainable Planning for Cities and Regions*,
Green Energy and Technology, https://doi.org/10.1007/978-3-030-57332-4_5

identifying suitable adaptation actions for sub-regional geographical units in a mountain environment as a basis for framing a background to a decision support system (DSS) for sub-regional adaptation in mountain areas, aligned with the overarching planning and policy context.

Keywords Adaptation · Climate change · Spatial planning · Mountain regions

1 Introduction

Governance of climate-change adaptation has received growing attention in recent times. Much focus is often placed on the level at which adaptation takes place, depending on the affected administrative units, governance structures (e.g., countries, regions, municipalities) and planning/programming instruments (Bonzanigo et al. 2016).

However, territorial and geographical differences can be dramatic even between similar administrative units, depending on the characteristics that shape territorial units not corresponding to institutional and administrative borders as set by law (Hanssen et al. 2013; Hamilton and Lubell 2018). In this study, we focus on the mountain territory of the Italian Alps. When climate-change adaptation policies are discussed, mountain regions deserve to be handled as a special territorial case. In fact, studies show that mountains as a system have special vulnerabilities and are subject to specific impacts of climate change (Kohler and Maselli 2009). In the mountains, topography is marked and complex, consequently climate conditions can change considerably over short distances. This factor increases the difficulty to make climate-change projections with a higher level of confidence. In fact, reliable long-term and high-altitude records of mountain climate that enable verification of regional climate models were made only in very few areas, such as the European Alps.[1] Unlike lowlands and plains, where the general climate conditions tend to vary mainly along the horizontal dimension, in the mountains the vertical dimension plays a far more important role. At different altitudes, climate zones may be characterized by different impacts of climate change. Areas at the snow line or freezing line will be affected more heavily, and many mountain towns with ski resorts designed as winter tourist destinations are forced to adapt and promote summer tourism (Scott and McBoyle 2007). Another typical impact in the mountains concerns the melting of glaciers and permafrost. Permafrost thawing might trigger in turn slope instability, hence an increase of gravitational hazards: rock fall, debris flows, mud flows, soil

---

[1] In general, Europe has shown a greater warming trend since 1979 compared to the global mean, and the trends in mountainous regions are still higher (Böhm et al. 2001). Regional climate projections indicate warming of about 1.5 times the global average, with greater warming in summer. Precipitation is projected to decrease in summer and on an annual average, and to increase in winter. General warming is expected to lead to an upward shift of the glacier equilibrium line by between 60 to 140 m per °C temperature increase (Oerlemans 1992), along with a substantial glacier retreat during the twenty-first century. The duration of snow cover is expected to decrease by several weeks for each °C of warming at middle elevations in the Alps region (Kohler and Maselli 2009: 9).

erosion, etc. A specific risk is the build-up of glacial lakes and the threat of lake outbursts. Finally, in general, the mountains exert their influence on the regional and global climates. Changes in, for instance, atmospheric wind flow patterns may induce large and locally varying precipitation responses in mountain areas, which could be much stronger than average regional climate change (IPCC 2007).

Geographical factors make mountain communities around the world particularly exposed to both the direct effects of climate change such as floods, increased fire risks (Westerling et al. 2006), and indirect effects such as the loss of biodiversity and impacts on tourism, agriculture, availability of water resources, increased risk for infrastructures, also threatening community resilience, and ecosystem services (Ballarin et al. 2014).

For all these reasons, mountains deserve a distinctive approach when dealing with climate change and specific vulnerability assessments.

Moreover, mountain territories are characterized in many countries by the presence of unconventional territorial governance schemes, an object of long-standing research (McDowell et al. 2014). Notwithstanding their distinctive characteristics, mountains are usually subject to the same governance approach and the same set of spatial planning tools that apply to any other part of the territory in a country (Rudaz 2009). Also, the institutional capacity of mountain communities is often unfit to adequately address climate impacts and therefore the implementation and financing of local policies and measures depend on organizational, knowledge-based and financial support of higher government levels (Storbjörk 2007).

Yet, today, with the increasing recognition of the global importance of mountain areas (Debarbieux and Price 2008) and the international interest stimulated by the Alpine Convention, there appears to be a multiplication of initiatives for the "mountains" as an original entity (Debarbieux and Rudaz 2008). Mountain areas are thus acquiring a special status in policy-making at various levels, calling for an adequate form of organization (Snow et al. 2004) required for the coordination of the collective action of the mountain as an entity.

The analytical framework adopted in this research puts therefore the accent on the profiles and roles of the various actors involved (organizations, associations, institutions) as well as on the relations and methods of organization among actors (networks) that actually participate in initiating the action hereby described.

In this work, we focus on the application of the "Budoia Charter"[2] in some sub-regional sites across the Italian Alps. The "Budoia Charter" is a voluntary declaration ascribed to municipalities or groups of municipalities, aiming at adopting and implementing sub-regional adaptation measures. The charter was launched jointly by Italy

---

[2]Reference of the Budoia Charter are the United Nations Framework Convention on Climate Change (UNFCCC), the Paris Agreement signed as a conclusion of COP21, the Strategy for Adaptation to Climate Change of the European Union (2013), the Italian National Strategy (2015), and the technical process, in progress, of a National Plan for Italy. Additionally, a few other experiences of strategies, documents and regional planning instruments have been taken as reference, such as the Strategy and the Action Plan from region Lombardy.

and the Network of Municipalities "Alpine Alliance" in 2017, following the guide-lines on local adaptation to climate change in the Alps developed under the Italian presidency of the Alpine Convention (2014).

Based on the significant background briefly already presented concerning inno-vative governance approaches to climate-change adaptation by non-state actors, we aim to set up a workable framework of a standard procedure for identifying suitable adaptive actions for sub-regional geographical units in a mountain environment. As a result, we expect to identify a modular procedure suitable to be refined by use in other locations and supported by a set of quantitative and qualitative indicators aimed at framing a background serving as a decision support system (DSS) for actual decision-making concerning adaptation policies and measures at the appropriate level (Junier and Mostert 2014).

2 Materials

In line with the need to work on a multi-level governance structure, the sources addressed different levels and sectors.

The National Adaptation Strategy (NAS) and National Adaptation Plan (NAP) for Italy are the key reference documents for the national level. They are used not only as primary sources to identify adaptation measures in various sectors, but also to trace back the conditions of the pilot areas to the national adaptation framework. In fact, the NAP (finalized in 2017, submitted to public consultation, and yet to be formally approved) defines six climate macro-regions, characterizing the local climate at the present state, and then analyzes the projected anomalies in RCP4.5 and 8.5 scenarios. Eventually, it aggregates the areas foreseen to be facing the same effects from climate change (in terms of climate anomaly indicators) in homogenous climate zones.

The Alpine Convention Guidelines for climate-change adaptation at the local level (2014) provide a solid processed overview of the main impacts, objectives and measures for climate adaptation in prominent sectors of the alpine area, in line with the policy contents of the Alpine Convention and its protocols. The guidelines advise sub-regional and local-level institutions (mainly municipalities, groups of munici-palities and mountain communities) on climate adaptation, introducing criteria for the analysis of relevant policies; for impact, vulnerability and risk assessments; for planning, developing and ranking adaptation options and their implementation; and for monitoring in various sectors. They proved a useful knowledge base to infer from the NAP those objectives, policies and measures more suitable to be applied in the Alpine environment and to set up a descriptive and coherent framework for action in the Alps.

The project on mapping risks in Italian municipalities ("Mappa dei Rischi dei Comuni Italiani") by the Italian National Statistical Institute (ISTAT) set up a database (including ISPRA, INGV and CLC data) of geophysical and socioeconomic indica-tors for the whole territory in Italy at LAU 2 level, from which data for each pilot area have been collected. Indicators by ISTAT, available for all the national territory, allow for a coherent characterization of the pilot areas against the national framework and provide interesting information on their physical and social vulnerability.

At an intermediate level of governance (NUTS 2), we analyzed institutional and technical sources, available at the regional level, to characterize the pilot areas with detailed and recognized knowledge layers. For all areas, we retrieved documents partly or totally dedicated to climate change, including reports of the regional administrations or environmental agencies, and products of sub-regional and local actions (Friuli-Venezia Giulia 2018; Lombardia 2018; Piemonte 2019a, b; Valle d'Aosta 2016; Piemonte 2019a, b; Regione Lombardia 2014; Regione Lombardia 2016; Valle d'Aosta 2018).

The analysis of existing regional and local processes provided focused knowledge on the pilot areas and the "status quo" concerning impacts, mainstreaming approaches and adaptation measures planned or in force at the regional or local level.

Finally, we collected information from the local level through direct interaction with pilot areas, partly through meetings and interviews with experts based on project CLISP (Pütz et al. 2011), partly through an ad hoc questionnaire intended to obtain information as answers to the following five questions:

- (1) Are you observing/experiencing impacts from climate change in your territory?
- (2) What impacts are you observing/experiencing in your territory?
- (3) What impacts do you believe are going to get worse in the near future in your territory?
- (4) What sectors do you consider the most vulnerable to climate impacts?
- (5) What policy and planning instruments do you utilize in your work to manage the sectors you mentioned and the impacts over them?

3 Procedure

The analysis has focused on five pilot areas ranging from the western to eastern Alps in Italy including: Espace Mont Blanc in Valle d'Aosta; Alte Valli in Piedmont; the city and immediate surroundings of Morbegno (mandamento) and Bassa Valtellina in Lombardy; the municipality of Capizzone and Valle Imagna in Lombardy; and Alto Livenza or Pedemontana in Friuli-Venezia Giulia. All the areas under investigation but Valle d'Aosta (NUTS3) are at a lower than NUTS 3 level and include more municipalities.

The procedure defined for the development of the analysis includes six steps.

The first step aims to characterize the pilot areas based on some standard indicators covering geophysical, land use, physical risk, demographic and socioeconomic variables from the ISTAT database, as shown in Table 1, and georeferenced in local maps. The values of the selected indicators are compared with the respective figures for the whole province (NUTS 3), aiming at assessing the relative differences detectable for the pilot areas in comparison to the surrounding territorial units.

Table 1 Categories of data collected for each pilot area (ISTAT 2019)

Data categories	
Population	Population in areas with high flood risk
Population density	Population in areas with high geologic risk
Variation in population	Social and material vulnerability indices
Age index [% population >65yo over population <14yo]	Altitudinal zones
Structural dependency [% inactive people (<14yo and >65yo) over active population (between 14 and 65)]	Land-use patterns

The second step aims to position the pilot areas in the national framework for climate-change adaptation as defined by the NAS and particularly by the NAP.

From the NAP, we used the climatic zoning for the pilot areas by identifying their localization by climatic macro-region, each characterized by relative predicted weather anomalies (under RPC 8.5).

From ISTAT, we extracted the altitudinal zones and, through the CLC, the land-use patterns for pilot areas.

The third step refers to the identification of specific climate objectives and impacts applicable to the pilot areas. After an harmonization procedure of identified impacts and adaptation objectives from the selected literature at the national and international level (LGA and NAP), regional and sub-regional sources have been used for identifying local impacts for the areas under investigation aiming to set up a climate profile for each pilot region. Additionally, the results of questionnaires and workshops with local stakeholders in the pilot regions enabled identifying the local perception concerning existing and future impacts, as well as the likely involved sectors, according to local communities.

The fourth step aims to collect adaptation measures from the PNACC and LGA framed to respond to the identified impacts for each pilot region (as in step 3), with the purpose to define a set of consistent actions to be implemented on the local level to tackle the regional climate-change impacts.

The fifth step aims to survey a selection of spatial planning and other programming instruments thematically linked to the impacts and sectors, as provided by local and regional experts interviewed for each pilot region.

The sixth step aims to apply a qualitative methodology for assessing the fitness of the selected planning and programming instruments to address climate-change impacts and to set up appropriate responses and adaptation measures.

The procedure we set up is expected to achieve three principal goals: (a) providing experts and spatial planners with adequate information concerning climate-change impacts, adaptation objectives and measures being locally relevant; (b) listing spatial planning and programming instruments assisting the pilot regions to address sectors and areas likely to be impacted by climate change and assessing their suitability to address the foreseen impacts; (c) ensuring adaptation planning and a selection of adaptation objectives and measures for the local level that are consistent with overarching policies and plans as set at the international, national and regional levels.

4 Methods and Analysis

Given the lack of homogeneity of collected data, a mixed approach (qualitative and quantitative) was used in the application of the procedure and data analysis.

The first step includes calculating the value of quantitative indicators from ISTAT database (including from ISPRA and INGV) for each pilot area, and georeferencing them in local maps (see Table 1).

The values for the selected indicators are compared with the respective figures for the whole province (NUTS 3) aiming at assessing the relative differences detectable for the pilot areas, in comparison to the surrounding territorial units. We aimed at providing a detailed characterization of the pilot areas based on indicators qualitatively linked to specific mountain features and vulnerabilities.

The second step consists of assigning the climatic zones from the NAP to the pilot areas, localizing them under a climatic macro-region (MR), each characterized with relative predicted weather anomalies (under RPC 8.5). Each pilot area was georeferenced across the following four (out of six) national climatic macro-regions (see Table 2).

In consideration to the predicted climatic anomalies, we also extracted the homogeneous clusters from the NAP for each pilot area (see Table 3).

The third and fourth steps of the procedure adopted a qualitative methodology for identifying the adaptation measures suitable for mountain areas. After an harmonization procedure of the identified impacts (by sector) from the selected literature at the national and international level (LGA and NAP), regional and sub-regional sources were used for identifying local impacts for the areas under investigation aiming to set up a climate profile for each pilot region.[3] Subsequently, from the NAP, we identified the possible opportunities for pilot areas from the five climatic MRs. Then, we compared the adaptation objectives of NAP with the ones of the Alpine guidelines for each sector. The resulting objectives for pilot areas derive from a comparative analysis of the adaptation objectives of the NAP and those inferable from the LGA for the various sectors consistent with the sectors used by NAP. In NAP, the objectives are general in nature; in the LGA, they are more specific but are expressed with a varying level of depth.

Additionally, the results of questionnaires and workshops with local stakeholders in the pilot regions made possible to identify the local perception concerning existing and future impacts, as well as the likely involved sectors, according to local communities.

---

[3]It is to be noticed that currently a Regional Adaptation Strategy is only available for Lombardy. However, the process could be replicated for other regions in the future.

Table 2 Climatic macro-regions selected and respective description (NAP 2017)

Macro-regions from NAP and pilot regions	MR short description
MR1 (Capizzone-Valle Imagna; Alto Livenza): pre-Alps and Northern Apennines	Pre-Alps and northern Apennines: intermediate values as regards the cumulative values of winter and summer rainfall; high values, compared to other areas, for extreme rainfall phenomena (R20 and R95p); after MR2, it appears to be the area of northern Italy with the highest number of summer days (maximum temperature above 29.2 °C)
MR3 (Alte Valli): Central-southern Apennines and some limited areas of north-western Italy	Central-southern Apennines and some limited areas of north-western Italy: reduced summer rainfall; winter rainfall medium–high values compared to other MR; maximum number of consecutive days without rain (CDD) intermediate
MR4 (Espace Mont Blanc; Alte Valli; Capizzone-Valle Imagna; Morbegno-Bassa Valtellina)—Alpine area	Alpine area: minimum average temperature value (5.7 °C) and maximum number of frost days; less abundant winter rainfall (143 mm, medium–high value); summer rainfall: the most significant (286 mm) compared to all other MRs
MR5 (Espace Mont Blanc; Alte Valli; Alto Livenza)—northern Italy	Northern Italy: high precipitation values both in terms of average winter values (321 mm) and extremes (R20 and R95p); average high summer rainfall; dry maximum consecutive days (CDD): lower value than other MR and summer days: average low value

Table 3 Pilot areas with homogeneous clusters and respective climatic anomalies (NAP 2017)

Clusters from NAP and pilot regions	Climatic anomalies short description
Espace Mont Blanc (4A and 5A)	Reduction in summer rainfall and an increase of winter rainfall. General reduction of both frost days and snow cover (RCP8.5)
Alte Valli (4A)	Reduction in summer rainfall and an increase in winter rainfall. Furthermore, general reduction in both frost days and snow cover (RCP4.5)
Morbegno-Bassa Valtellina and Capizzone-Valle Imagna (4E and 4A)	Reduction of extreme events, frost days and snow cover (4E, RCP4.5) and reduction of summer and winter rainfall, reduction of frost days and snow cover (4A, RCP8.5)
Alto Livenza (1C, 1E and 5A)	Reduction in summer rainfall and an increase in winter rainfall. In general, there is a reduction in frost days, more relevant than the CPR4.5 (1C and 1E); based on the reference period with the most significant precipitation values, it is characterized by an increase in winter rainfall and a reduction in summer rainfall (5A)

Table 4 Criteria used to consistency analysis and respective description

Criterion	Description
Geographical	Excluding objectives and measures relating to radically different regions compared to Alpine/mountain areas
Legal administrative	Considering only the measures that can be implemented at a regional and preferably sub-regional level, including pressure by higher administrative levels on "subordinate" administrative levels
Impact coherence	Coherence of measures to the territorial impacts identified for the pilot areas by LGA and other regional and sub-regional sources for the pilot area
Local perception	Adjustment of the shortlist composition through the results from the questionnaires and a local audit scheme administered to stakeholders in the pilot areas

Each measure listed in the NAP corresponds to an adaptation goal of the NAP itself. As a consequence, once the relevant objectives are found, a "long list" of relevant measures is automatically established. This list is the overall reference from which specific adaptation measures for each pilot area have been identified ("short list"), by applying the listed criteria (see Table 4).

We extracted the corresponding measures from the NAP and for a minimum share from the LGA, since the latter includes very few measures.

The last step of the procedure was a survey and analysis of policy and planning documents that either stakeholders from the pilot areas indicated for the sectors and impacts by answering Question 5 of the questionnaire (see above) or that we identified from the analysis of the available regional and sub-regional literature. To make the analysis as objective as possible, we produced a simple but fairly complete matrix, as later described.

5 Results and Discussion

The analysis carried out in order to meet the three objectives declared at the end of the procedure section produced the following results.

Firstly, the elaboration and aggregation of local territorial data (ISTAT 2019) enabled characterization of the pilot area and set a common reference baseline for sub-regional analysis and comparison and selection of coherent adaptation actions. For example, Fig. 1 shows the aggregated figures of the indices applied to municipal data (LAU 2) in a pilot area compared to the respective data for a higher territorial unit (NUTS 3). These data, together with the climate zoning data from NAP, made possible demarking our five pilot areas within a national framework. Furthermore, they provided coherent territorial information for identifying impacts, and adaptation objectives and measures by paralleling international, national and regional sources

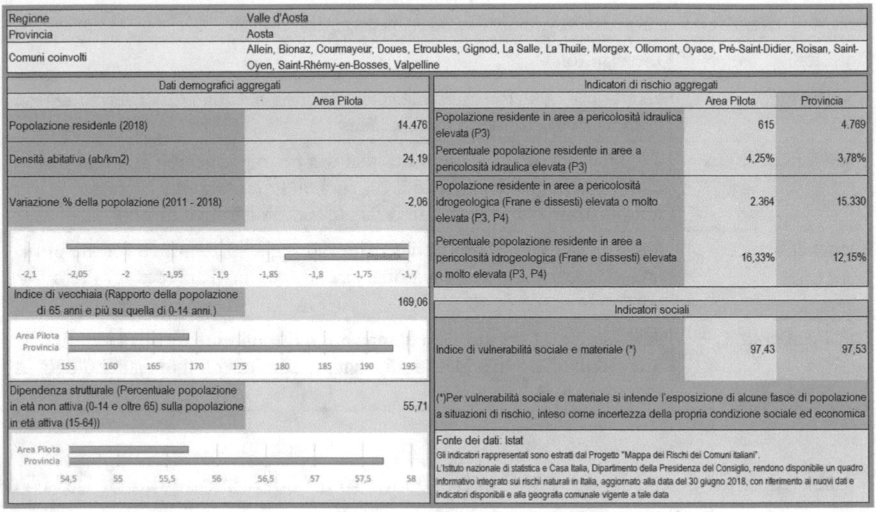

Regione	Valle d'Aosta
Provincia	Aosta
Comuni coinvolti	Allein, Bionaz, Courmayeur, Doues, Etroubles, Gignod, La Salle, La Thuile, Morgex, Ollomont, Oyace, Pré-Saint-Didier, Roisan, Saint-Oyen, Saint-Rhémy-en-Bosses, Valpelline

Dati demografici aggregati		Indicatori di rischio aggregati		
	Area Pilota		Area Pilota	Provincia
Popolazione residente (2018)	14.476	Popolazione residente in aree a pericolosità idraulica elevata (P3)	615	4.769
Densità abitativa (ab/km2)	24,19	Percentuale popolazione residente in aree a pericolosità idraulica elevata (P3)	4,25%	3,78%
Variazione % della popolazione (2011 - 2018)	-2,06	Popolazione residente in aree a pericolosità idrogeologica (Frane e dissesti) elevata o molto elevata (P3, P4)	2.364	15.330
[chart: -2,1 to -1,7]		Percentuale popolazione residente in aree a pericolosità idrogeologica (Frane e dissesti) elevata o molto elevata (P3, P4)	16,33%	12,15%
Indice di vecchiaia (Rapporto della popolazione di 65 anni e più su quella di 0-14 anni.)	169,06	Indicatori sociali		
Area Pilota / Provincia [chart: 155 to 195]		Indice di vulnerabilità sociale e materiale (*)	97,43	97,53
Dipendenza strutturale (Percentuale popolazione in età non attiva (0-14 e oltre 65) sulla popolazione in età attiva (15-64))	55,71	(*)Per vulnerabilità sociale e materiale si intende l'esposizione di alcune fasce di popolazione a situazioni di rischio, inteso come incertezza della propria condizione sociale ed economica		
Area Pilota / Provincia [chart: 54,5 to 58]		Fonte dei dati: Istat. Gli indicatori rappresentati sono estratti dal Progetto "Mappa dei Rischi dei Comuni Italiani". L'Istituto nazionale di statistica e Casa Italia, Dipartimento della Presidenza del Consiglio, rendono disponibile un quadro informativo integrato sui rischi naturali in Italia, aggiornato alla data del 30 giugno 2018, con riferimento ai nuovi dati e indicatori disponibili e alla geografia comunale vigente a tale data		

Fig. 1 Indicators from ISTAT (excluded: altitudinal zone and land-use pattern) for Espace Mont Blanc

and factoring in the local perceptions collected through meetings, interviews and ad hoc questionnaires in pilot areas.

Specific matrices were framed for systematically analyzing and comparing impacts, objectives, and measures from the NAP and LGA and (where relevant) Regional Adaptation Strategies and Plans. As a result, for each sector, distinctive impacts, objectives and measures for mountain areas were collected. From a set of 356 adaptation measures, by applying the site-specific criteria mentioned above, we extracted 123 measures. Most of them suffer from having been conceived at the national level. Thus, they need to be accurately adjusted to fit the needs of specific sites. However, the comparison between the NAP and LGA can already be a starting point when framing regional adaptation strategies and plans.

A matrix for analyzing policy instruments at the regional and sub-regional level has been set up and applied in the pilot areas. It includes five sections: sectors, impacts, objectives, measures, and horizontal governance and coherence. For each section, specific questions are to be answered to analyze each policy document (see Fig. 2). For the sections on impacts, objectives and measures, we also indicated whether references to climate change in general, mitigation and adaptation existed in the analyzed instruments.

The methods used and the results achieved would seem to be a sound basis for a decision support system (DSS). The data extrapolated from the process can support and guide policy-makers and planners in the selection of climate-change adaptation measures and their integration within sub-regional policies. The method presented aims to support an increase in resilience to climate change at the local level and improve the decision-making capacity of regional and local administrations in adapting to, and managing, climate-change impacts in the mountains.

SECTION/ Question		Type	Note				
1. SECTORS							
	a. Is the instrument inter-sectoral? Cross-sectoral?	Y/N	Complementary to 1.b				
	b. Is the instrument sectoral?	Y/N	Complementary to 1.a				
	c. Which sectors?	(list)					
2. IMPACTS						**CLIMATE IMPACTS**	
	a. Does the instrument include a focus analysis on environmental or socio-economic impacts or effects?	Y/N	+ comments?			Y/N	+ comments?
3. OBJECTIVES				For each of these sections:		**ADAPTATION OBJECTIVES**	
	a. Does the instrument include objectives/goals?	Y/N	+ comments?	• ...specific for CC?		Y/N	+ comments?
	b. Are those general or specific objectives/goals?	(specify)	+ comments?	○ ...specific for CCA? ○ ...specific for CCM?		(specify)	+ comments?
4. MEASURES						**MEASURES for ADAPTATION**	
	a. Does the instrument include Measures/Actions?	Y/N	+ comments?			Y/N	+ comments?
5. HORIZONTAL GOVERNANCE AND COHERENCE							
Does the instrument include a formalized evaluation of:							
	a. Internal Coherence	Y/N	+ comments?				
	b. External coherence	Y/N	**Especially with NAP or the SDGs**				
	c. Horizontal Governance for implementation	Y/N	Are **coordination structures** with other sectors/offices going to be set up?				

Fig. 2 Matrix for analyzing policy instruments

6 Conclusion

This work aimed at providing an instrument to critically assess the coherence and integration between local climate-change adaptation measures and regional/national plans, specifically applicable to mountain territories. We identified possible adaptation measures consistent with scientific knowledge and local perceptions in various sub-regional sites, and we provided local administrations with information on the fitness of regional to local planning instruments to address climate change. We focused on the sections of sub-regional policies, programs and plans more suitable for modification, by integrating consistent adaptation measures intended to increase local resilience to climate change. Furthermore, the methodology developed enables assessment of the coherence of the local approach to climate change in the pilot areas against impacts, objectives and measures set by national and alpine-wide documents (the NAP and LGA), as well as by regional and local planning instruments.

Finally, although this study highlighted opportunities for action in the pilot areas involved, it would be advisable to improve and refine this DSS to make it more flexible and better link its diverse outcomes, also aiming to make it applicable to other mountain areas, such as the Apennines.

References

Ballarin-Denti A, Cetara L, Idone MT, Bianchini A, Bisello A, Petitta M et al (2014) Guidelines for climate change adaption at the local level in the Alps. Italian presidency of the Alpine Convention

Böhm R, Auer I, Brunetti M, Maugeri M, Nanni T, Schöner W (2001) Regional temperature variability in the European Alps: 1760–1998 from homogenized instrumental time series. Int J Climatol: J R Meteorol Soc 21(14):1779–1801

Bonzanigo L, Giupponi C, Balbi S (2016) Sustainable tourism planning and climate change adaptation in the Alps: A case study of winter tourism in mountain communities in the Dolomites. J Sustain Tourism 24(4):637–652

Debarbieux B, Price MF (2008) Representing mountains: from local and national to global common good. Geopolitics 13(1):148–168

Debarbieux B, Rudaz G (2008) Linking mountain identities throughout the world: the experience of Swiss communities. Cultural Geographies 15(4):497–517

Friuli-Venezia Giulia ARPA (2018) exploratory study of climate change and some of their impacts in Friuli-Venezia Giulia

Hamilton M, Lubell M (2018) Collaborative governance of climate change adaptation across spatial and institutional scales. Policy Stud J 46(2):222–247

Hanssen GS, Mydske PK, Dahle E (2013) Multi-level coordination of climate change adaptation: by national hierarchical steering or by regional network governance? Local Environ 18(8):869–887

IPCC (2007) Climate Change 2007: THE PHYSICAL SCIENCE BASIS. Contribution of working group I to the fourth assessment report of the intergovernmental panel on climate change. In: Solomon S, Qin D, Manning M, Chen Z, Marquis M, Averyt KB, Tignor M, Miller HL (eds) Cambridge University Press, Cambridge

Junier S, Mostert E (2014) A decision support system for the implementation of the water framework directive in The Netherlands: process, validity and useful information. Environ Sci Policy 40:49–56

Kohler T, Maselli D (2009) Mountains and climate change. From understanding to action. Geographica Bernensia

Lombardia ARPA (2018) Rapporto sullo Stato dell'Ambiente 2018 (RSA in Lombardy)

McDowell G, Stephenson E, Ford J (2014) Adaptation to climate change in glaciated mountain regions. Clim Change 126(1–2):77–91

Ministero dell'Ambiente della Tutela del Territorio e del Mare (2007) "Piano Nazionale di Adattamento ai Cambiamenti Climatici PNACC", first draft for public consultation

Oerlemans J (1992) Climate sensitivity of glaciers in southern Norway: application of an energy-balance model to Nigardsbreen, Hellstugubreen and Alfotbreen. J Glaciol 38(129):223–232

Piemonte ARPA (2019a) "Relazione sullo Stato dell'Ambiente 2018" (RSA in Piedmont)

Piemonte, Cuore delle Alpi (2019b) Technical documentation of the project "Cuore Resiliente" (Interreg ALCOTRA PITER)

Pütz M, Kruse S, Butterling M (2011) CLISP climate change fitness checklist. ETC Alpine Space Project CLISP

Regione Lombardia (2014) "Strategia Regionale per l'Adattamento ai Cambiamenti Climatici" (SRACC)

Regione Lombardia (2016) "Documento di Azione Regionale per l'Adattamento ai Cambiamenti Climatici" (DARACC)

Scott D, McBoyle G (2007) Climate change adaptation in the ski industry. Mitig Adapt Strat Glob Change 12(8):1411

Snow DA, Soule SA, Kriesi H (2004). Mapping the terrain. The Blackwell companion to social movements, pp 3–16

Storbjörk S (2007) Governing climate adaptation in the local arena: challenges of risk management and planning in Sweden. Local Environ 12(5):457–469

Valle d'Aosta, AdaPT Mont-Blanc (2018) Changements climatiques dans le massif du Mont-Blanc et impacts sur les activités humaines (Interreg ALCOTRA)

Valle d'Aosta ARPA (2016) "XI Relazione sullo Stato dell'Ambiente" (RSA in Valle d'Aosta)

Valle d'Aosta ARPA (2018) "Il cambiamento climatico in Valle d'Aosta" (position paper)

Westerling AL, Hidalgo HG, Cayan DR, Swetnam TW (2006) Warming and earlier spring increase western US forest wildfire activity. Science 313(5789):940–943

Projections of Electricity Demand in European Cities Using Downscaled Population Scenarios

Gianni Guastella, Enrico Lippo, Stefano Pareglio, and Massimiliano Carlo Pietro Rizzati

Abstract This work projects future residential electricity demand derived from cities and municipalities' population and residential land-use projections. Starting from national-level energy intensity data, we derived statistically downscaled residential electricity consumption with the aim to disaggregate residential electricity at the local administrative unit level for all EU member states in the year 2050. The intensity in 2050 is obtained from population density that, in turn, depends on the evolution of population and residential land-use. Residential land-use is projected to 2050 according to a model linked to population trajectories at the LAU level via the share-of-growth method. Finally, country-level intensity multiplied by the projected value of LAU residential area returns the electricity demand for every LAU. The results suggest that the amount of electricity required by cities depends on their land-use patterns, but with an evident between-and-within-country heterogeneity. The national average temperature does not provide significant effects over the evolution of electricity demand, highlighting the need for more detailed climate-related variables. This evidence poses significant challenges for the planning of future cities because it points out how the current patterns of land-use will need to be properly categorized with respect to the development of future electricity requirements.

Keywords Electricity intensity · Residential land-use · Downscaling · Population scenarios · Local urban areas

1 Introduction

In Europe, the urban population is growing rapidly, and it is expected to reach 83.7% of the total population by 2050, which amounts to nearly 600 million people over the

G. Guastella · E. Lippo · S. Pareglio · M. C. P. Rizzati (✉)
Fondazione Eni Enrico Mattei, Corso Magenta, 63, 20123 Milano, Italy
e-mail: massimiliano.rizzati@feem.it

G. Guastella · S. Pareglio
Department of Mathematics and Physics (DMF), Università Cattolica del Sacro Cuore (UCSC), via Musei, 41, 25121 Brescia, Italy

© The Author(s), under exclusive license to Springer Nature Switzerland AG 2021
A. Bisello et al. (eds.), *Smart and Sustainable Planning for Cities and Regions*,
Green Energy and Technology, https://doi.org/10.1007/978-3-030-57332-4_6

projected total of 716 million (United Nations 2019). This increase in urbanization will shift the demand for energy, and electricity in particular, in cities. However, the real extent of this change will depend on several factors determining the electricity demand of the main recipients, such as the residential and the transport sectors. Many of these factors are the objects of dedicated planning policies, which means that the projected energy demand, conditioned to future economic and demographic scenarios, could provide a piece of highly relevant information for urban planners and policy-makers (Murakami et al. 2015). This is especially useful in the context of a better matching of demand and supply for renewable energy, which requires heterogeneous technologies and location based on physical characteristics of the areas (Ramachandra and Shruthi 2007; Aydin et al. 2010).

In this work, we aim to focus on, and to clarify the interconnections between the population and the residential land-use coevolution and their effect on future electricity demand. During the period 1990–2014, the residential land-use area has grown almost everywhere in Europe, losing the connection with the demographic trend (Guastella et al. 2019). In some cities with shrinking populations, the density has been observed to decline and the discontinuity to increase (although the two appear to be unrelated) (Pesaresi et al. 2019). A local population density decline seems therefore linked with a shrinking population, which points to the decoupling between built-up area and population, a condition that would call for specific policy measures tailored on the examined area.

Population and land-use are strongly related phenomena, but with great spatial variation, and the use of spatially disaggregated data is relevant to capture the causal effects and the consequences for the electricity demand. Unfortunately, electricity demand data are seldom published at levels such as the municipality one (Murakami et al. 2015), and pairing downscaling techniques to project variables enables obtaining data and information at finer levels. The reason to prefer downscaling techniques applied to aggregate data compared to the estimation or simulation of bottom-up information is that the latter often needs very specific statistics and technical information, whose degrees of complexity and availability greatly complicates its application to units of references of wider scope (Swan and Ugursal 2009).

We propose therefore an analysis to better characterize the influence of residential land-use and population growth scenarios on the future electricity demand of European cities. To do so, we extend to a multi-year setting a modelling framework to estimate residential electricity intensity at the municipality level (Seya et al. 2016) and apply it to European municipalities (here broadly defined as Local Administrative Unit, LAUs as per the 2018 unified European Commission definition). The framework seeks to understand how population growth drives urban spatial expansion and employs this information alongside downscaled population scenarios to predict future land-use patterns, projecting the electricity demand for the targeted year. This makes possible to explore the changes in the spatial distribution of residential electricity demand, with the objective of reaching a sufficient, secure and sustainable renewable-energy supply that can account for the heterogeneity of needs once a finer geographical unit is taken into account (Aydin et al. 2010; Barrington-Leigh and Ouliaris 2017; Ramachandra and Shruthi 2007). Moreover, we consider also a climate change-related variable to condition the projection results to predict the change of climate that might affect future electricity demand. There are potentially many additional variables that could concur to influence the scope of our work, such as for instance technological evolution and the resources pricing. While relevant, we will keep the focus on the interaction between land-use and population evolution, considering therefore the evolution of those variables as given.

The proposed empirical framework is rooted in the existing approaches in the literature. Murakami et al. (2015) downscale prefecture-level data to municipality-level data using population and built-up area weights and perform geographically weighted regressions. Seya et al. (2016) estimated municipality-level intensity starting from survey data on electricity expenditures and employed statistical methods to obtain the spatial interpolation of data, producing projections based on the changing spatial distribution of floor space. Yamagata et al. (2015) estimated grid-level electricity demand for the Tokyo metro area assuming a fixed country-level intensity (energy per floor area) and test the sensitivity of their result against the choice of the population scenario.

The contribution of this work to the existing literature is twofold. First, we carry out the empirical analysis at the European level, for which high-resolution studies producing evidence on the changing spatial distribution of electricity demand are still missing. Second, we attempt to consider how climate change may affect electricity consumption. Damm et al. (2017) looked at the effect of an increase of 2 °C increase on electricity demand for 26 European countries, finding that the increase in temperatures reduces consumption, due to a decrease in heating system usage that more than offsets the expected rise in cooling ones. Globally, De Cian and Wing (2017) estimate an effect of 7–17% change in energy consumption, depending on the degree of warming. Van Ruijven et al. (2019) study the effects of the changing climate on the amplification of energy demand. They develop a top-down approach for long-term projections of electricity, petrol and natural gas demands. Moreover, they model the change on hot and cold days as a climatic shock (a way to extend the degree-days information to tropical countries). They combine income elasticities for energy consumption alongside GDP per capita projections to 2050.[1] This model is then tested over an array of projections, with 21 earth models and two different emission scenarios. They found that scenarios of moderate warming might increase the world energy demand by 11–27%, while vigorous warming might cause an increase of 25–58%.

The remainder of the paper is organized as follows. The method section illustrates the methodological framework and the empirical approach; the sources for the data used in this analysis are described in the data section. In the result section, we present and summarize the empirical results and illustrate the evidence about the projected values. A final section proposed some concluding remarks alongside future research directions.

---

[1] 2050 is the target year of the European Union objective for reaching net-zero greenhouse gas emissions, as set also in the Paris Agreements.

2 Method

2.1 Modelling Framework

Let $I_{c,t} = \frac{E_{c,t}}{A_{c,t}}$ be the electricity intensity at country level c and period t, the ratio between residential electricity consumption $E_{c,t}$ in GW/h and country-level residential land-use $A_{c,t}$ in m^2. The following model relates the observed intensity to population density and climatic conditions at the country level:

$$I_{c,t} = \alpha_{0,c} + \alpha_1 \left(\tfrac{P}{A}\right)_{c,t} + \alpha_2 \left(\tfrac{P}{A}\right)^2_{c,t} + \alpha_2 (T)_{c,t} + \delta_t + \varepsilon_{c,t} \tag{1}$$

where the term $\left(\tfrac{P}{A}\right)_{c,t}$ is the ratio of the country population and residential area, or population density, $(T)_{c,t}$ is the annual average temperature in the country, $\alpha_{0,c}$ is a country-specific effect, δ_t is a time trend and $\varepsilon_{c,t}$ is the error term. We project these values to 2050 with the following approach. Following Murakami et al. (2015), we apply a model to explain residential land-use at the local-administrative-unit level, following this specification:

$$A_{i,t} = \beta_0 + \beta_1 P_{i,t} + \beta_2 P^2_{i,t} + u_{i,t} \tag{2}$$

Here we regress the value of current local land-use at LAU level for the ith unit against the local population $P_{i,t}$ value using a quadratic specification. This estimation provides us with the β coefficients value.

Subsequently, we develop downscaled projections for the variables of this specification for the target year 2050, by following the share of growth method also reported in Yamagata et al. (2015). First, we generate a weight for the ith LAU-level population share over the relative country, based on the difference between the actual populations in the years 2018 and 2000, the beginning and the end of our sample. We then multiply this weight by the delta of the country c population from the year 2018 to the Eurostat projections of the country population in year 2050. This enables obtaining the downscaled delta of population at the LAU level that, added to the figure for 2018, returns the projected value at 2050 at the LAU level:

$$\Delta \tilde{P}_{i,2050} = \left(\frac{P_{i,2018} - P_{i,2000}}{P_{c,2018} - P_{c,2000}} \right) \left(P_{c,2050} - P_{c,2018} \right) \tag{3}$$

$$\tilde{P}_{i,2050} = P_{i,2018} + \Delta \tilde{P}_{i,2050} \tag{4}$$

By employing the estimated coefficients for the available years $\hat{\beta}$ and the projections values for the delta of the explanatory variables added to the estimated value $\hat{A}_{i,t}$ we can then declare the model to estimate the residential area value at LAU level for 2050:

$$\tilde{A}_{i,2050} = \hat{A}_{i,t} + \hat{\beta}_1 \Delta \tilde{P}_{i,2050} + \hat{\beta}_2 \Delta \tilde{P}^2_{i,2050} \tag{5}$$

Notice that we assume that the population effect on land-use is bound at zero, meaning that a decreasing population in 2050 will not provide a negative value for future land-use.

Aggregating the LAU values of Eq. (4) and applying its same procedure to the estimation of the intensity value of Eq. (1), we can obtain the projected intensity for 2050. This is done by using our projected residential area together with downscaled population projections to define a projected population density for the year 2050:

$$\left(\frac{\tilde{E}}{\tilde{A}}\right)_{c,2050} = \hat{\alpha}_{0,c} + \hat{\alpha}_1 \left(\frac{\tilde{P}}{\tilde{A}}\right)_{c,2050} + \hat{\alpha}_2 \left(\frac{\tilde{P}}{\tilde{A}}\right)^2_{c,2050} + \hat{\alpha}_3 \left(\tilde{T}\right)_{c,2050} + \delta_c \tag{6}$$

As a final step, multiplying the intensity at country level for 2050 with the projected value of the local residential area permits us to obtain the downscaled value for electricity consumption in 2050:

$$\tilde{E}_{i,2050} = \left(\frac{\tilde{E}}{\tilde{A}}\right)_{c,2050} \tilde{A}_{i,2050} \tag{7}$$

2.2 Data

The data are divided into variables at the country level and variables at the LAU level to populate our models for the intensity and residential area. These constitute two-panel datasets, in which we collect the data about population, residential area, electricity intensity and their projected values.

Country variables are collected from Eurostat, including total country population figures, total country population projections to 2050 and total country residential electricity consumption.

The population at the LAU level is obtained from the correspondence table LAU–NUTS 2016 from Eurostat. Residential area at the LAU level is elaborated from the EEA Corine Land Cover GIS maps, by intersecting continuous and discontinuous urban fabric (code 111 and 112) with the LAU borders provided by GISCO Eurostat. The intersected values are then aggregated to have a single value for every LAU of the sample. The national residential area is then computed for each country by aggregating the LAU figures. The land-cover maps are available for the years 2000, 2006 and 2012 and 2018, and therefore, we adopt these years as a reference for our panel datasets, with 2050 as the target year for our projections.

Temperature data are taken from the NUTS aggregation of the ERA-NUTS C3S ERA5 reanalysis (De Felice and Kavvadias 2019). Per capita income data for the robustness checks are taken from Eurostat. The per capita income in purchasing power parities are taken from the JRC LUISA Trend Scenario 2016. Excluding countries due to missing data for the population at the LAU level leave 15 countries and 77,880 LAU units.

3   Results

Tables 1 and 2 summarize the estimation results of the two models for the electricity intensity at the country level and residential area at the LAU level. In both cases, time-invariant effects respectively at the country and LAU level are assumed.

Table 1 Effect of population on residential areas, LAU level

Variables	(1) Fixed effects
P	0.0542***
	(0.000811)
P2	$-1.12e-05$***
	(2.20e$-$07)
Constant	1.495***
	(0.00434)
Observations	228,566
Number of Id	59,394
Country FE	YES

Standard errors in parentheses, *** $p < 0.01$, ** $p < 0.05$, * $p < 0.1$

Table 2 Effect of population density on energy intensity, country level

Variables	Fixed effects
D	1.476***
	(0.520)
D2	-0.206***
	(0.0613)
T	-0.0779
	(0.149)
Constant	3.051*
	(1.557)
Country FE	YES
Year FE	YES

Standard errors in parentheses, *** $p < 0.01$, ** $p < 0.05$, * $p < 0.1$

Table 3 Summary statistics of the percentage change in population, residential area and residential electricity demand at the LAU level given the 2050 projections

Country	% Change population		% Change Res. Area		% Change electricity		Freq.
	Mean	Std. Dev	Mean	Std. Dev	Mean	Std. Dev	
AT	5.4357277	15.596375	2.1963425	22.946263	−0.85838486	5.2105119	1,906
BE	10.765884	6.4328867	1.6964483	4.5882101	−1.6875052	2.8395033	589
CZ	0.04607658	0.10207952	24.957174	1634.139	−4.575177	2.4484565	6,250
DE	0.54506491	3.1264657	124.69316	8540.1877	−3.3293885	6.3913515	10,960
EE	7.9527236	22.552821	0.38021248	1.8067244	−6.6709623	2.2971286	26
ES	−6.9245028	20.271249	12.275142	641.74881	−5.3640855	9.126629	8,111
FI	1.2070258	2.9476771	10.346033	13.944143	−5.0467428	0.59022266	311
FR	4.1427791	10.020528	349.82028	56,648.86	−0.12627585	58.31654	35,329
IT	2.0217813	26.801316	59.388605	2711.7693	1.2322305	20.559825	7,721
LT	−31.745817	18.025229	0.43945381	1.4905473	−37.322267	0.65421246	60
LU	63.750788	18.521631	6.7213093	6.7951843	2.2957024	3.5652929	102
LV	−25.144188	20.827449	0.01168061	0.09680639	−18.255164	0.65715506	119
FO	−8.5612383	17.793422	11.607797	252.23986	−33.944735	2.3994744	3,180
SE	8.092992	35.631373	1.6719962	5.265666	3.3298103	5.2533209	290
SK	15.440499	154.06503	559.86064	23,299.772	−4.6249408	111.50577	2,926
Total	1.942196	33.289498	207.03902	38,566.816	−2.9866709	45.985719	77,880

The residential area model in Table 1 identifies a positive effect of population levels on residential area use, with an overall better performance by the model specification that includes a quadratic population component with a negative coefficient sign. In the Appendix Table 4, we present additional robustness checks with other specifications and estimators, which overall confirm the sign of our coefficients. The Hausman test supports the choice of the fixed-effects estimator ($p = 0.0000$).

In Table 2, we present the results for the second model on electricity intensity, which shows a positive effect by population density on electricity intensity, with also a better fit of a quadratic model for population density with a negative sign, with coefficients within the 1% significance threshold. Average temperatures do not display significant coefficients. In the Appendix Table 5, we propose the same robustness checks also on the second model. Moreover, we test two additional specifications with GDP per capita and GDP per capita in purchasing-power-parities variables, which however have non-significant coefficients.

We couple the estimated coefficients with the projected population and residential area values to project and then downscale electricity demand levels for the year 2050 at the European LAU level.

This results in a dataset with the following projections for the year 2050 at the ith LAU level, and on which we focus to discuss the results: (a) population projections, obtained from country-level projections and downscaled via the share of growth method; (b) residential area projections, as obtained from our residential area model; and (c) residential electricity demand by crossing the information about the evolution of population and residential areas, namely the projection of population densities.

To provide some visual feedback to our analysis, we present maps of the results for the three main variables in the appendix.

4 Discussion

In Table 3, we summarize the results by presenting the means and the standard deviations of the percentage change in the population, residential area and residential electricity demand between 2018 and 2050, by country. The figures shown in this table should be taken carefully because some outlier values (such as the mean value in percentage change in electricity for Romania) are mainly driven by small LAUs, where the gap between 2018 and 2050 is so small that differences in decimals might cause wide variations in percentage terms. However, we present this table to display the seemingly counterintuitive relationships between the three variables signs.

These results are explained through two main channels: The first is dependent on the model used to project residential area and population, and the second is based on the values of the intermediate projections used to build the final ones.

In the framework of this model, the reasons for a negative electricity-demand growth despite an increased population density must be searched in the negative signs of the coefficients of the two models. Indeed, in the residential area model, the squared population term has negative coefficients that in some LAUs can dominate the positive effect of the first population term. In the second model used to estimate the coefficients for population density effect on electricity intensity, the effects of squared density and of the average temperature also have a negative sign, resulting therefore in decreased electricity demand despite the growing density.

As we have tested only one series for the projections of population, and given some limitations coming from our data sources (e.g. some countries have exited our sample due to missing figures for LAU population, and the CORINE land-cover data provide LAUs with zero as value for residential areas, despite this being not plausible), we do not consider these projections as ultimately realistic forecast values, which would require at least batteries of simulations to account for uncertainty and measurement issues. However, the result that the negative change in electricity demand could be driven by the squared characterization of the explanatory variables is quite interesting, if confirmed, could be used to classify LAU in different "phases" of population and land-use contribution. In a first phase, population and land-use could be increasing electricity as expected, while going forward, this effect will decrease at the margin, up to a point where the contribution of density to electricity intensity is minimized and dominated by other variables, such as the squared terms or the average temperature in our model. This effect could also be relevant for additional variables that might be included in future studies, such as additional climate variables (e.g. degree days, humidity) and socioeconomic variables (e.g. technology factors and the price of resources).

The aforementioned considerations lead us to another interesting feature of the resulting projections: There is a noticeable spatial heterogeneity both between and within the European countries in the sample (observable in the maps in the Appendix), which could mean that the different phases might coexist within even the same regional or even provincial context. Given these results, the main policy message is that the availability of downscaled projections can help the policy-maker to take more informed actions. For instance, accurate projections can inform the choice of investments in the electricity sector (Dilaver 2017). Costly and potentially distorting measures such as imposing carbon taxes or improving efficiency standards could be then applied more cautiously only where most needed. Moreover, the policy-making decision criteria taken on single factors or only aggregate variables could end up with local unintentional discrepancies and consequences.

5   Conclusion and Future Research

The results suggest that the amount of energy required by cities do depend on their land-use patterns but does so in various different ways. As we can observe for many LAUs, the increasing relationship between population, residential area and electricity demand that one could expect is not present, here following the use of a quadratic model to obtain the coefficients needed for the downscaling.

Validating the reasons for these differences through the analysis of additional LAUs and country-specific variables (such as ones related to technology and resource pricing) could result in a valuable exercise.

The downscaled projections of electricity demand display a remarkable level of heterogeneity, both between and within countries. Being able to spatially disaggregate the variables of this analysis to finer geographical units could help to provide policy makers with more precise information in the quest to contrast climate change. This is especially true when we also consider the supply of renewable electricity, constrained by the local availability of resources or plants and that will display similar patterns of disaggregation: Our results could then guide policy considerations by helping a better match of supply and demand.

The analysis should be refined by testing different population projections to the year 2050, as the share of growth method is but one of the available techniques. We are currently working on a version that matches the world grid population-downscaling projection by Gao (2017), based on the SSP scenario middle of the road, with LAU units. This procedure could moreover lead to acquiring enough population figures to again include the missing European countries in our sample. Alternative projections for the year 2050 should then be used as additional checks. Another direction is to test a better characterization of the climate part of the model, including seasonal figures and degree-days, and a battery test to simulate various temperature scenarios. An interesting direction could be the application of downscaled climate model variables (Poggio and Gimona 2015).

To conclude, the spatial characterization of the units will be employed in defining inverse distance-weighting matrices and to test spatial econometrics models.

6 Appendix

We depict statistics taken from the projected variables in Figs. 1, 2, 3, 4, 5, 6, 7, 8 and 9 with the aid of choropleth maps. Figure 3, 4, 5 and 6 depicts the deltas for the three projected 2050 variables with their 2018 counterparts, namely the delta of population (Fig. 3), the delta of residential land-use as obtained from our projection model first step (Fig. 4) and the delta of residential electricity demand (Fig. 5). Moreover, to better capture national effects, we represent in Fig. 6, 7 and 8 the deviations from national means of the three delta variables for each LAU, which should allow us to capture local outliers. The maps at the European level are still perhaps too wide, which means that finer geographical units should present a better view for the presentation of the results. In Figs. 9, 10 and 11, we provide as a sample a zoom of our maps at the NUTS 2 level for the Italian region of Lombardy (ITC4).

See Appendix Tables 4 and 5 for robustness checks on the LAU land-use model and on the Country electricity intensity model respectively.

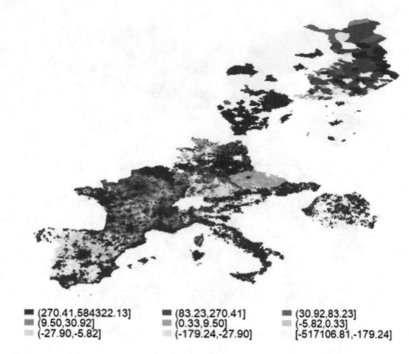

Legend:
- ■ (270.41,584322.13]
- ■ (9.50,30.92]
- ■ (-27.90,-5.82]
- ■ (83.23,270.41]
- ■ (0.33,9.50]
- ■ (-179.24,-27.90]
- ■ (30.92,83.23]
- ■ (-5.82,0.33]
- ■ [-517106.81,-179.24]

Fig. 3 Absolute change of projected variables for the years 2018–2050, population

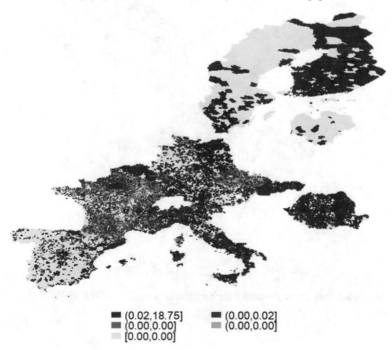

Legend:
- ■ (0.02,18.75]
- ■ (0.00,0.00]
- ■ [0.00,0.00]
- ■ (0.00,0.02]
- ■ (0.00,0.00]

Fig. 4 Absolute change of projected variables for the years 2018–2050, residential area

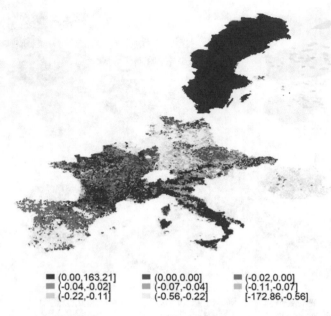

Fig. 5 Absolute change of projected variables for the years 2018–2050, electricity demand

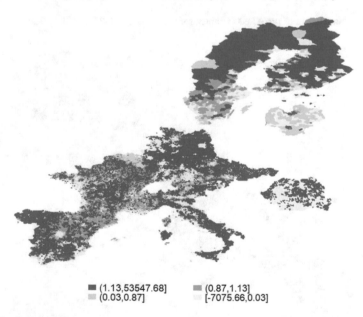

Fig. 6 Deviations from country averages for the delta of population (2018–2050)

Fig. 7 Deviations from country averages for the delta of residential area (2018–2050)

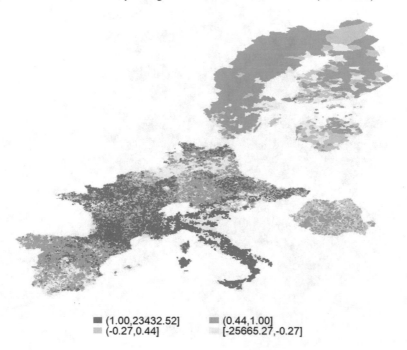

Fig. 8 Deviations from country averages for the delta of electricity demand (2018–2050)

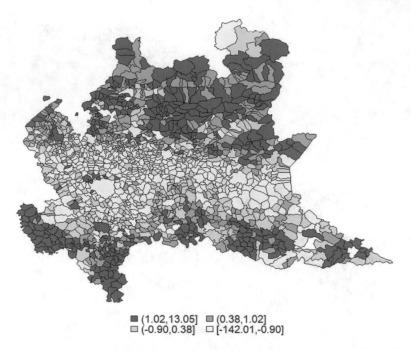

■ (1.02,13.05] ▨ (0.38,1.02]
▢ (-0.90,0.38] ▢ [-142.01,-0.90]

Fig. 9 Deviations from country averages for the delta of population, Lombardy (2018–2050)

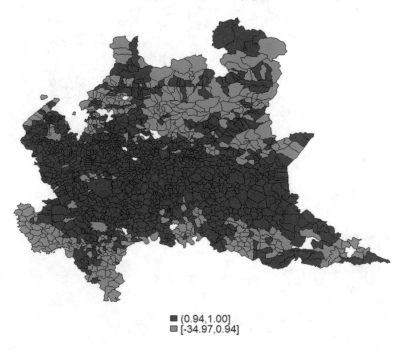

■ (0.94,1.00]
▨ [-34.97,0.94]

Fig. 10 Deviations from country averages for the delta of population, Lombardy (2018–2050)

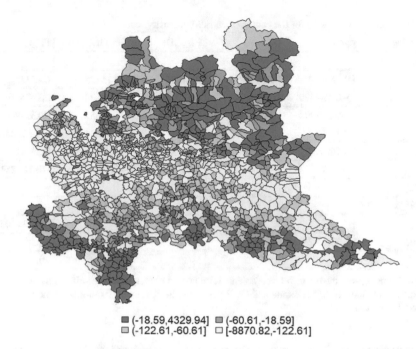

■ (-18.59,4329.94]   ■ (-60.61,-18.59]
□ (-122.61,-60.61]   □ [-8870.82,-122.61]

Fig. 11 Absolute change of projected variables for the years 2018–2050, population

Table 4 Robustness checks for the LAU land-use model

Variable	Pooled	RE	FE	FE + Cube
Year				
2006	−0.00739171			
2012	0.03111095*			
2018	0.02996443			
P	0.14631422***	0.13054914***	0.05423047***	0.06849765***
P2	−0.00002024***	−0.00002064***	−0.00001117***	−0.00003016***
P3				3.960e−09***
_cons	0.98445743***	1.059328***	1.4946633***	1.4320506***
N	228,566	228,566	228,566	228,566

The columns present the estimated coefficients for a pooled model, a random-effects model, the final fixed-effects model, and a fixed-effects model with a cubed term for population

Table 5 Robustness checks for the country electricity intensity model

Variable	pooled	RE	FE_GDP	FE_GDP_ppp	FE_Cubes
Y					
2	0.058984	0.140571	0.119291	0.121325	0.151158
3	−0.194690	0.133692	0.112538	0.101145	0.143579
4	−0.247994	0.003854	0.073072	0.052652	0.133249
D	0.157762	1.427579*	1.440447*	1.488641**	−0.322238
D2	0.168703	−0.176056**	−0.205162**	−0.208508**	0.237100
T	−0.379558**	−0.035899	−0.077425	−0.082682	−0.101851
GDP_pc_	0.000065***	0.000010	0.000001		
GDP_pc_PPP_				0.006179	
D3					−0.031305
_cons	4.332263*	2.308160	3.192379	2.968461	5.336486
N	59	59	59	60	60

The columns present the estimated coefficients for a pooled model, a random-effects model, fixed-effects models with GDP per capita and GDP per capita in PPs, and a fixed-effects model with a cubed density term

Acknowledgements We wish to thank the two anonymous Referees for their suggestions and comments.

References

Aydin NY, Kentel E, Duzgun S (2010) GIS-based environmental assessment of wind energy systems for spatial planning: A case study from western turkey. Renew Sustain Energy Rev 14(1):364–373

Barrington-Leigh C, Ouliaris M (2017) The renewable energy landscape in Canada: a spatial analysis. Renew Sustain Energy Rev 75:809–819

Damm A, Köberl J, Prettenthaler F, Rogler N, Töglhofer C (2017) Impacts of +2 °C global warming on electricity demand in Europe. Climate Services 7:12–30

De Cian E, Wing IS (2017) Global energy consumption in a warming climate. Environ Resource Econ, pp 1–46

De Felice M, Kavvadias K (2019) ERA-NUTS: time-series based on C3S ERA5 for European regions (Version 1980–2018). Zenodo

Dilaver Z, Hunt LC (2011) Modelling and forecasting Turkish residential electricity demand. Energy Policy 39(6):3117–3127

Gao J (2017) Downscaling global spatial population projections from 1/8-degree to 1-km grid cells. Technical Notes NCAR, National Center for Atmospheric Researcher, Boulder, CO., USA

Guastella G, Oueslati W, Pareglio S (2019) Patterns of urban spatial expansion in European cities. Sustainability 11(8):2247

Murakami D, Yamagata Y, Seya H (2015) Estimation of spatially detailed electricity demands using spatial statistical downscaling techniques. Energy Procedia 75:2751–2756

Pesaresi M, Florczyk A, Schiavina M, Melchiorri M, Maffenini L (2019) GHS-SMOD R2019A - GHS settlement layers, updated and refined REGIO model 2014 in application to GHS-BUILT R2018A and GHS-POP R2019A, multitemporal (1975-1990-2000-2015). European Commission, Joint Research Centre (JRC) [Dataset]

Poggio L, Gimona A (2015) Downscaling and correction of regional climate models outputs with a hybrid geostatistical approach. Spatial Stat 14:4–21

Ramachandra T, Shruthi B (2007) Spatial mapping of renewable energy potential. Renew Sustain Energy Rev 11(7):1460–1480

Seya H, Yamagata Y, Nakamichi K (2016) Creation of municipality level intensity data of electricity in japan. Appl Energy 162:1336–1344

Swan LG, Ugursal VI (2009) Modeling of end-use energy consumption in the residential sector: a review of modeling techniques. Renew Sustain Energy Rev 13(8):1819–1835

United Nations, Department of Economic and Social Affairs, Population Division (2019) World Population Prospects 2019, Online Edition. Rev. 1. https://population.un.org/wpp/Download/Standard/Population/. Accessed 1 Sept 2019

Van Ruijven BJ, De Cian E, Wing IS (2019) Amplification of future energy demand growth due to climate change. Nature Commun 10(1):2762

Yamagata Y, Murakami D, Seya H (2015) A comparison of grid-level residential electricity demand scenarios in japan for 2050. Appl Energy 158:255–262

Integrated Building Data for Smart Regions and Cities—An Italian Pilot

Ezilda Costanzo and Bruno Baldissara

Abstract Regional and local decision-makers still require relevant information and training in order to establish long-term strategies and to contribute to national and supranational energy and climate targets. As an example, a widespread participation of local authorities to comply with the Italian long-term building renovation strategy has not occurred so far. Thus, the overall target, annual 1% floor area of new or deeply renovated buildings to the nearly zero-energy building (nZEB) standard by 2020 (PanZEB 2015), proves to have been disregarded to date. Evidence-based, data-enabled assessment of the building stock and of its relationship with the energy system as a whole at a capillary level is crucial to this extent. In Italy, various building databases are already being used with the ultimate purpose of EPBD implementation and to track and record incentives for public and private building renovation. These datasets have an untapped potential for local energy planning that could be released from wider integration, also including energy consumption data and smart-metering data. Moreover, the regulatory landscape is changing toward an interaction of the building with the user, the energy grid and other buildings in a dynamic and functional way. Within this context, the paper will investigate how integrated data could unlock the value of a more evidence-based planning starting from the DIPENDE integrated dataset, a REQUEST2ACTION (IEE 2014–2017) pilot project combining data from energy performance certificates (EPCs) with bottom-up information on building renovation, and other data in order to support decision making at different territorial scales.

Keywords Big data · Long-term building renovation strategies · Smart energy planning

E. Costanzo (✉)
Department of Energy Technologies, Energy and Sustainable Economic Development, ENEA, Italian National Agency for New Technologies, Rome, Italy
e-mail: ezilda.costanzo@enea.it

B. Baldissara
ENEA, Studies, Analyses and Evaluations Unit, Rome, Italy
e-mail: bruno.baldissara@enea.it

1   Introduction

Buildings are responsible for 40% of the EU's total energy consumption and 36% of CO_2 emissions in the EU, severely contributing to global warming emissions in Europe.

Member states have set ambitious targets for their building stock by 2050, mainly based on energy-efficiency improvement, electrification, flexibility and enhanced renewables integration as a result of new storage solutions.

De-carbonization will require science-based information and scenario approaches that have not been frequently applied in practice so far. The role of end-use reduction (building envelope insulation, the use of more performance equipment, passive solutions, etc.) is often disregarded in these scenarios, especially at the higher local scales. The widespread engagement of local authorities in contributing to national and supranational energy and climate target (3293 municipalities have signed the Covenant of Mayors agreement in Italy!) requires improved capillarity of building and energy data.

The regulatory landscape is changing, producing an increased availability of energy consumption data, including information on the interaction of the building with the user and the energy grid and a more holistic approach. Just think of the 2018 Energy Performance in Building Directive (EPBD) amendment[1] supporting smart readiness of buildings and recharging infrastructures for electric vehicles.

Nevertheless, nowadays, energy data in buildings are still too fragmented and lack interoperability. Moreover, little evidence is given on the market trend toward deep renovation. Achieving nearly zero-energy buildings is one of the targets for a low-carbon economy by 2050, but the new construction rate is at present very low and the existing stock is obsolete and low performing. EPBD repositories, originally created to log energy performance certificates (EPC) and inspection data, are increasingly being used not only for control and compliance goals (2010 EPBD, Art. 18),[2] but also to complement other sources with the aim of enabling evidence-based policies and monitoring the energy efficiency of buildings.

Although EPC data in its raw format are insufficiently accurate, its usefulness can be reinforced by refining and combining EPC information with other data such as census, inspections, cadastre, incentives, gas registers, energy networks, bills, revenue agencies, and consumption.

Several EU projects (EPISCOPE, ZEBRA, ENTRANZE, ODYSSEE_MURE)[3] have collected key static data that fed a European Building Stock Observatory (BSO). To give an idea of the lack of standardized and comparable building data, only 13% of the 250 indicators of the BSO that include beyond building physics, energy poverty, embodied energy, indoor, and comfort data have been populated in the BSO database.

---

[1] Directive (EU) 2018/844 amending Directive 2010/31/EU on the Energy Performance of Buildings (EPBD) and Directive 2012/27/EU on energy efficiency (EED).

[2] Directive 2010/31/EU on the energy performance of buildings (recast) - 19 May 2010.

[3] EU EPISCOPE, ZEBRA, ENTRANZE, ODYSSEE_MURE projects.

Thorough the REQUEST2ACTION project, in 2017, agencies have set up services to provide accurate, trustworthy building performance data that improve insight. The Italian energy agency experience in this project is illustrated in this paper.

Nowadays, analysts are starting to process dynamic building data coming from sensors and meters. Integrating building static and dynamic data from multiple organizations and diverse environments is a severe challenge that entails a new "collaborative" approach that is being studied at ENEA within research on smart cities.

2 Method

The Lombardy region of Italy has the only totally public open database of energy performance certificates (EPC) in Europe. This data, including, for example, localization, building type, heated, and cooled volume and surface, energy services, thermal performance of the envelope, calculated energy consumption per energy sources per energy carriers, energy performance index (primary energy) and class, are combined with the regional cadastre of technical heating systems of buildings and the ground-source heat pumps cadastre into a wider information system that feeds and updates the regional energy balance and the regional emission-monitoring system. Nevertheless, the consistency of the certified building sample, the combination with socioeconomic territorial data and information of actual progress due to retrofit interventions is missing in the system.

Within the REQUEST2ACTION project,[4] ENEA elaborated and cross-referenced energy performance certificates (EPC) open data in the Lombardy Region with other datasets. This enabled further analyses, for example, establishing consistency and relationships between performance, ownership and occupation, building-stock age and typology, household density, geographic/climatic issues, technical building plants and elements retrofit, identifying erroneous records.

Information from energy certificates (EPCs) was brought up to date with unregistered performance levels variations by overlaying bottom-up envelope components and technical-equipment installation data coming from the national renovation incentives (ENEA 2012).

The main requirements of the pilot integrated database (named DIPENDE) were to provide targeted support to public (but also private) decisionmakers by easy-to-interpret categories and to enable replication and wider use on the national territory.

The various kinds of social, territorial, building, and energy data in DIPENDE are shown in Fig. 1.

Records are aggregated at the urban level. The database covers more than 1500 municipalities in Lombardy, including approximately 70 fields for each record. The DIPENDE tool facilitates:

---

[4]https://ec.europa.eu/energy/intelligent/projects/en/projects/request2action.

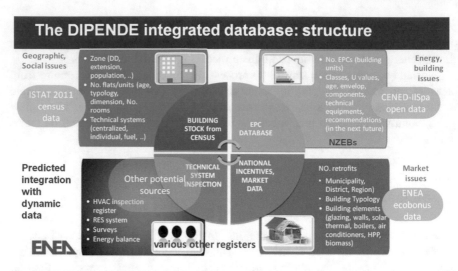

Fig. 1 DIPENDE—kinds of data in the integrated database

- Standardization of the analysis (through key queries in the Excel version);
- Selection of a list of common indicators that are applicable to different territorial contexts (Table 1);
- Easy-to-understand output formats.

DIPENDE is the result of a process of stakeholder consultation. The methodology has followed a product-service systems (PSS) approach. PSS represents an evolution of traditional generic and standardized services through co-creation with the potential users to attract and retain them as active contributors to service performance (Vasantha et al. 2012).

Table 1 DIPENDE—example of key indicators

Climate and building typology	Energy performance	Retrofit—market
Climatic zones—degree days of municipalities	Average U value of the building envelope	–
Housing density	No. of residential units with an EPC compared to the stock	Retrofits undertaken (distribution per type) related to the performance of the stock
Building typology related to number of dwellings in the building	No. of residential units with an EPC and distribution per energy class	
Age of the building units within the stock	EPC (*APE*) classes according to the age of the buildings and average U value of the envelopes	–

3 Results

DIPENDE integrated database is available on the web in the form of GIS maps that represent ten complex indicators that can be displayed by province. (Fig. 2).

By comparing these maps, users can identify areas where building renovation is a priority. This can facilitate energy performance policies such as voluntary certification, incentives, and energy audits.

The maps in Fig. 3 show the average transmittance of building envelopes (U values, giving an evaluation of the envelope thermal quality) compared to climate

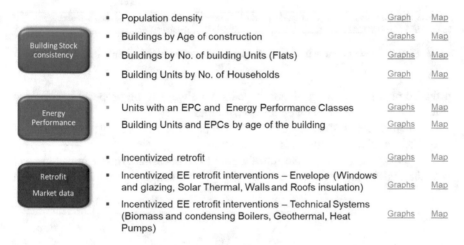

Fig. 2 DIPENDE—Standardized analysis indicators

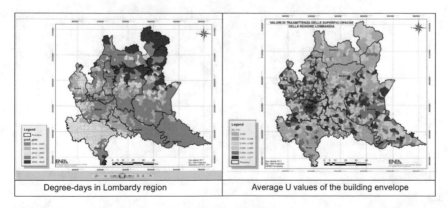

| Degree-days in Lombardy region | Average U values of the building envelope |

Fig. 3 Average transmittance of the building envelope (thermal characteristics) compared to climatic zones

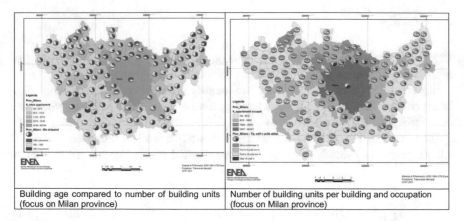

| Building age compared to number of building units (focus on Milan province) | Number of building units per building and occupation (focus on Milan province) |

Fig. 4 Distribution of building units per age, per number of units in buildings and occupation

in the Lombardy region: insulation appears not always to be properly located and actions are still to be taken on building envelopes.

The maps in Fig. 4 show the distribution of building units per age, per number of units (households) in buildings and occupation.

The maps in Fig. 5 illustrate retrofits that have been undertaken (envelope and solar thermal panels installed) compared to the percentage of building units that

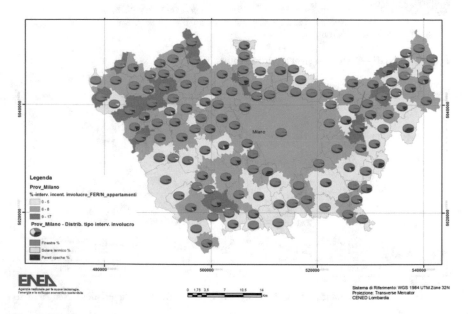

Fig. 5 Distribution of the type of retrofit (windows, solar thermal panels and opaque building components, walls, roofs) compared to the total percentage of units undergoing any kind of retrofit in Milan province

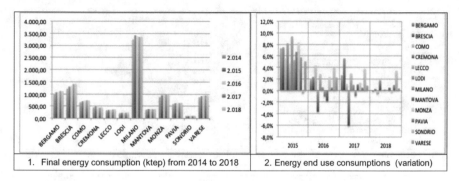

| 1. Final energy consumption (ktep) from 2014 to 2018 | 2. Energy end use consumptions (variation) |

Fig. 6 Final energy consumption (ktep) in buildings from 2014 to 2018 in Lombardy provinces

were renovated in 2012, thus guiding possible marketing actions by producers and installers. It is evident that window replacement and deeper renovation, in spite of very different degrees of cost-effectiveness, are almost independent of the building typology. This evidence should orientate local policies.

Other geo-referenced maps have been produced based on a standard elaboration of the dataset.

A webGIS (https://dipende-ba.casaccia.enea.it/pmapper/map_default.phtml) was generated from the same integrated dataset, enabling multi-criteria analysis. ENEA delivered the Lombardian ILSpa agency an Excel version of the dataset, with macros that allow to produce 53 default and pivot tables. Users can choose between six or seven types of graphs to display the results, but they also make free queries to explore other output configurations. From the stakeholder's consultation at the end of the project,[5] the maps and the analysis outputs were estimated as original and full of new pieces of information.

4 Discussion

The DIPENDE data service can be enriched with sectoral final energy consumption data. For many consumption data, it is now possible to have an infra-annual (monthly) detail, which allows us to grasp.

the weight of the various energy services, for example, the need for summer and winter air-conditioning and related potential CO_2 emissions. Figure 6 shows the evolution of final energy consumption in buildings, both residential and non-residential, for each of the Lombardy provinces during the last five years (ENEA elaborations on MiSE, Terna, GSE, GSE data).

Consumption data are to be analyzed (also in terms of year-on-year variations, see Fig. 6.2) in light of the trend of main drivers such as primarily demographic

---

[5] Workshop "Pianificazione energetica a livello locale: esperienze e condivisione" (Local Energy Planning: experience and knowledge sharing), 22th May 2017; ENEA Rome.

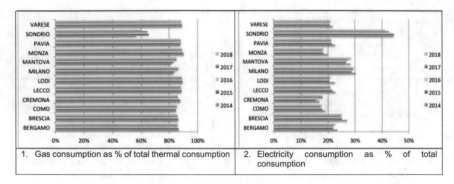

Fig. 7 Consumption, per gas and electricity carriers, as % of total building consumption

evolution, climatic variables and sectoral added value of services, which we also intend to integrate in the DIPENDE database. This will make it possible for us to analyze energy consumptions as an impact of the incentive measures for energy technologies in the sector.

Details on the final energy consumption of buildings by source and energy carriers (electricity, gas, LGP, etc.) will also enable assessing the spread (and the saving potential) of some energy technologies in the building stock (see Fig. 7).

For example, variations in the percentage of summer air-conditioning systems and the replacement of traditional heat boilers with electric heat pumps indicate the level and trend of electrification, as well as the integration of renewables. Temperature data with infra-monthly detail at the provincial level is also key to understand consumption and the elasticity of the summer and winter air-conditioning requirements to temperature variations, in the different areas of the region.

Present availability of data can enable us to replicate the application of DIPENDE for other Italian regions. Data on the building stock, climate, population and territory characteristics, incentivized EE interventions (and in some cases to the EPCs) are in fact accessible and published by public sources such as ISTAT, MiSE, ENEA, GSE, and TERNA.

The main goal of the further development of DIPENDE is to better interpret how energy demand in the sector is evolving at a more integrated and macro-scale territorial level, in quantity and mix, and to evaluate whether targeted energy and climate saving policies are producing, or not, the expected results.

This exercise might also let us detect exemplary cases of policy and regulation and learn what worked better in the different contexts.

5 Conclusions

In spite of the increased availability, building energy data are often fragmented and unmatched (in time, in formats); moreover different levels of competence among territories are often competing. Consistency and crossed-referenced analysis of this data are missing, as well as easy-to-interpret visualization allowing user friendly access and evidenced-based insight from public and private stakeholders.

DIPENDE enabled integration of EPC with bottom-up data (from the 2011 Census and governmental renovation incentives) establishing a relationship between estimated energy performance, recent systems and product installation, climatic and social data.

This exercise resulted in cleaning and processing data, easier-to-interpret categories (complex indicators), and evidence on how the building stock is performing and where to focus action.

The future predicted combination with real consumption data can permit the estimation of the impact of those renovation measures that have been undertaken so far and to provide insight into which procedures and technologies promote improvement of building performance.

Various stakeholders have estimated the DIPENDE tool as very useful to develop and monitor Sustainable Energy and Environment Plans (SEAPs) within the Covenant of Mayors and to update Regional Energy and Environment Plans. Possible replication beyond Lombardy could also facilitate the coordination of regional policies toward national targets in the sector.

Collected feedback from private actors (ESCos, associations of enterprises) proved the usefulness of the data service in facilitating marketing and business strategies.

The smart-city vision holds out the promise of integrating data from multiple organisations, diverse environments, and a wide variety of intelligent devices. Nevertheless, to process static and dynamic data from multiple and diverse sources gathered through various means, a new approach is needed. This "collaborative" building data processing for decision support has not been mainstreamed and systematized yet. The EU is presently asking for a common vision toward innovative big data approaches[6] that integrate compatibility with smart meters, sensors, IoT devices, and BEMS outputs.

References

Costanzo E (2019) Wider use of EPBD databases. Enabling monitoring and policy making. EU CA EPBD factsheet, https://epbd-ca.eu/archives/2765

Costanzo E, Weatherall D et al (2017) Can big data drive the market for residential energy efficiency? In: ECEEE 2017 Summerstudy, (8-295-17). Belambra Presqu'île de Giens, France

Costanzo E, Baldissara B, Rao M (2016) DIPENDE—a tool for energy planning of building districts based on energy performance certification. In: Input 2016 9th international conference on innovation in urban and regional planning conference proceedings. Torino, pp 245–250. Book ISBN 978-88-9052-964-1

DIPENDE - Portale 4E (2018) https://www.portale4e.it/centrale_dettaglio_pa.aspx?ID=1

---

[6]H2020 LC-SC3-B4E-7–2020 Call, "European building stock data 4.0" https://ec.europa.eu/info/funding-tenders/opportunities/portal/screen/opportunities/topic-details/lc-sc3-b4e-7-2020

EU Building Stock Observatory (2016) https://ec.europa.eu/energy/en/topics/energy-efficiency/ene
rgy-performance-of-buildings/eu-bso

Gouveia JP, Giannakidis G, Seixas J (2018) Smart city energy planning: integrating data and tools.
Researchgate 1:345–350. https://doi.org/10.1145/2872518.2888617

Lazarova-Molnar S, Mohamed N (2019) Collaborative data analytics for smart buildings: opportunities and models. Cluster Computing. 22:1. https://doi.org/10.1007/s10586-017-1362-x.Resear
chgate

REQUEST2ACTION EU Project (2014–2017) Removing barriers to low carbon retrofit by
improving access to data and insight of the benefits to key market actors. https://www.buildup.
eu/en/explore/links/request2action-project-6

RICS, BSO 2nd Stakeholder Workshop, 2019. https://aldren.eu/wp-content/uploads/2019/02/EU-
BSO-Workshop_2_-2019.pdf

Vasantha A et al (2012) A review of product-service systems design methodologies. J Eng Des
23:635–659. https://doi.org/10.1080/09544828.2011.639712

Thermal Performance Evaluation of Unshaded Courtyards in Egyptian Arid Regions

Hatem Mahmoud and Ayman Ragab

Abstract Effective architectural features such as courtyards can mitigate heat stress in hot arid regions. Appropriate configuration of the courtyard leads to significant improvement in the thermal performance of the building and directly influences the behavior of its users and the functionality of the space. The aim of the study was to evaluate the thermal performance of various courtyards by examining various sky view factors (SVF) and courtyard orientations so that suggestions could be offered for future guidelines for construction designs in Egyptian arid regions. The study was conducted on the new Aswan University campus built in the desert region of new Aswan city. Field measurements and simulations were used to evaluate the thermal conditions of these courtyards. Thermal comfort was measured by physiologically equivalent temperature (PET). The study determined that unshaded courtyards should be oriented in a north–south direction to mitigate the effects of solar radiation intensity. Whereas SVF can be lowered to less than 0.2 for courtyards in a north–south orientation, further SVF reduction for courtyards in other orientations might result in heat-trapping. Adding a greening area improves the thermal performance of courtyards.

Keywords Hot arid climate · Thermal comfort · ENVI-met · University campus · Egypt · Courtyard · PET · SVF

H. Mahmoud (✉)
Department of Architecture, Faculty of Engineering, Aswan University, Tingar, Egypt
e-mail: hatem.mahmoud@aswu.edu.eg

Department of Environmental Engineering, Japan University of Science and Technology, New Borg El-Arab City, Alexandrian, Egypt

A. Ragab
Department of Architecture, Faculty of Engineering, Aswan University, Tingar, Egypt
e-mail: egayman.ragab@aswu.edu.eg

1 Introduction

Because university campus buildings account for a large amount of energy consumption (Ma et al. 2015), campus energy sustainability has become an issue of global concern for university policy-makers and planners. Indeed, many universities today can be regarded as "small cities" (Alshuw-aikhat and Abubakar 2008). Cairo University, for example, has a student and staff population exceeding one hundred thousand (CU on Spot—CU Statistics—Bachelor Level Students 2009–2010, undated). The urban and architectural design and the various activities taking place on universities campuses have direct and indirect environmental impacts on energy consumption; these frequently lead to a reduction in thermal comfort in outdoor and indoor spaces. A courtyard, essentially an outdoor open space surrounded by walls or buildings, is common to the building designs, especially in hot arid regions (Sibley 2004). It is one of the world's oldest vernacular architectural elements that can be traced back 5000 years in the Middle East and China (Soflaei et al. 2017). Appropriate courtyard designs contribute meaningfully to improving the thermal performance of buildings, consequently affecting user behavior and usage of the spaces (Manioğlu and Oral 2015; Almhafdy et al. 2013b). Hence, the study of courtyard-design parameters that affect thermal performance is important, especially in regions with hot climates (Aldawoud 2008).

An abundance of literature states that the aspect ratio of a courtyard, its orientation, and shape are the most effective design parameters that affect its thermal performance (Soflaei et al. 2017). Nowadays, the shapes of courtyards, having assumed more dynamic forms compared to the traditional rectilinear shapes, make it more difficult to evaluate the effect of courtyard shape on thermal performance and energy efficiency. However, the sky view factor (SVF), which represents the percentage of the unobstructed sky at specific locations (Lin et al. 2010), has been found useful in identifying the effect of several courtyard shapes on the microclimate conditions (Forouzandeh and Richter 2019). As SVF represents the observable extent of the sky as a proportion of the total possible sky hemisphere (Watson and Johnson 1987; Taleghani et al. 2015), undesirable thermal conditions in courtyards can be circumvented by controlling this factor. Several techniques have been developed to calculate SVF, such as those involving the fisheye camera and numeric models (i.e., SkyHelios), which generate virtual fisheye pictures (Hämmerle et al. 2011). Moreover, several studies have assessed the thermal performance of courtyards using site observations and computer simulations to evaluate its influence on the microclimate (Almhafdy et al. 2013a, b; Muhaisen 2006) that impacts thermal comfort and energy efficiency (Manioğlu and Oral 2015; Aldawoud 2008; Safarzadeh and Bahadori 2005; Cantón et al. 2014). At the same time, other studies have investigated heat mitigation strategies (material, vegetation, shading, etc.) to improve the thermal performance of courtyards (Taleghani et al. 2014, 2015). Extensive literature has covered parametric studies on multiple morphologies, SVF and the proportions and orientations of courtyards (Martinelli and Matzarakis 2017; Soflaei et al. 2017; Sharmin et al. 2017). Nevertheless, there is scarce literature (Soflaei et al. 2017; Berkovic et al.

2012) focusing on the courtyard in hot arid regions, especially in Egyptian public buildings. The objective of this study was to carry out a numerical analysis of the SVF and orientation effects on the thermal efficiency of unshaded university courtyards, with the aim of arriving at suggestions for better designs of environmental spaces in Egyptian hot arid zones.

2 Research Methods

In this study, we examined the impact of SVF and the orientation of various types of unshaded courtyards on the occupants' thermal comfort by relying on field measurements and ENVI-met tools for microclimate assessment. ENVI-met is considered by many as the best tool for urban planners and architects to assess the impact of geometric parameters on microclimate in the courtyard. Based on the extracted data from the simulation processes for each type of courtyard, we assessed how microclimate conditions were affected by SVF and orientation. Subsequently, we would offer guidelines for courtyards design considerations in the early stages of design planning.

2.1 Site Description

The study was conducted in New Aswan city. Based on Köppen–Geiger climate classification, the city is located in a hot arid deserted region of Egypt (24.085296°N 32.904779°E) (Aswan Climate Aswan Temperatures Aswan Weather Averages, no date), where the average maximum temperature in June is 42 °C (World Maps of Köppen-Geiger Climate Classification, undated). Aswan University New Campus was chosen as a case study owing to the city and campus being in its early stages of construction. This would allow the study findings to be applied during stages of future development. The location of the campus is as shown in Figs. 1 and 2.

3 Analysis of the Courtyard Characteristics

The study selected eleven unshaded courtyard alternatives in six buildings of the new Aswan University campus as simulation spaces. The investigated points within the case study were divided into two groups, with on-site observation points in group "P", and the points that were extracted from the simulation model in group "C". The five on-site points were used only for calibrating the sensitivity and accuracy of the simulation model. A single point was chosen as a receptor in each courtyard, and the eleven receptors were numbered from $C1$ to $C11$. Table 1 presents the geometric variables of the studied courtyards, while Fig. 3 shows the locations of the observation

Fig. 1 Aswan University campus location in New Aswan City. (Image source: Google Earth, June 2017, edited by the authors)

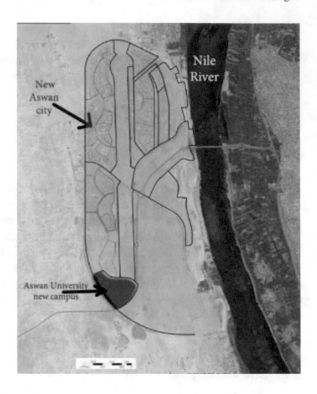

and simulation points. Some courtyards were neglected due their small areas because they are not accessible for the public for activities. The first point (*P*1) was in the outdoor area between faculties; it was considered a reference point. Two points (*P*2 and *P*3) were also located in outside spaces. Another two points (*P*4, and *P*5) were located inside the courtyards of two buildings. *P*2 was chosen to be located over grass, while *P*3 was sited over sand. Numerical simulation was applied to extract simulated points. The model was evaluated using statistical analysis between on-site measures "*P*" and simulated records "*C*", before evaluating the expected microclimate in the other courtyards. It should be noted that not all receptors were located in the same place as the measuring points. Nevertheless, *C*1 was in the same position as *P*5, and *C*9 was in the same position as *P*4 (Fig. 3).

4   Results and Discussion

4.1   Site-Measurements

Air temperature and relative humidity were monitored using Hobo *U*12 data loggers placed in hand-made solar radiation shields fixed at a height of 1.5 m at all observation

Fig. 2 Aerial view of Aswan University new campus in New Aswan City. (Image source: Google Earth, June 2017, edited by the authors)

points. The measurements were carried out for 48 h from 12 pm on July 12, 2018 to 11 am on July 14, 2018. Air temperature and relative humidity for $P1$ to $P5$ from 12 p.m. on July 12, 2018 to 11 a.m. on July 14, 2018 are shown in Fig. 4.

4.2 Model Validation

The simulation model was validated by comparing the site observations and simulated data outputs. To simulate 24 h in ENVI-met on July 13, 2018, ENVI-met was run to simulate weather conditions from midnight on July 13, 2018 to 7 a.m. on July 14, 2018, i.e., over a period of 32 h. The first eight hours in the simulation were omitted to ensure greater accuracy in the results because the ENVI-met model required about four simulation hours to stabilize. An ENVI-met forced type simulation was set by inserting the hourly measured "24 inputs" for air temperature and relative humidity to obtain more accurate results. Result calibration for the model output was undertaken from 8 a.m. on 13 July to 7 a.m. the following morning.

To validate the model, the accuracy of simulated values was evaluated by comparing them with observed readings. For this purpose, Willmott (1981) suggested

Table 1 Geometrical configuration of the campus courtyards

Building	Building plan	Court no	Simulation point	SVF	Orientation
Faculty of arts		1	C1	 0.154	N-S
		2	C2*	 0.474	N-S
Faculty of languages		1	Unused		
		2	C3	 0.193	–
		3	C4	 0.127	N-S
College of nursing		1	Unused		
		2	C5	 0.489	N-S
		3	Unused		

(continued)

Table 1 (continued)

Faculty of tourism and hotels		1	C6	 0.163	–
		2	C7	 0.108	N-S
		3	C8	 0.129	–
		4	C9	 0.200	E-W
Admin building		1	C10	 0.142	NE-SW
Student housing		1	C11	 0.236	NE-SW

*Note Concerning greening, only C2 contains grass, with an average height 15 cm

calculating the average difference between observed and simulated values with the root mean square error RMSE representing the general model performance (Middel et al. 2014) and the index of agreement (*d*). This indicates the degree to which the modeled values are error-free, with $d = 1.0$ indicating that the simulated value equals the observed value (Willmott 1981). Moreover, the suite of different measures used

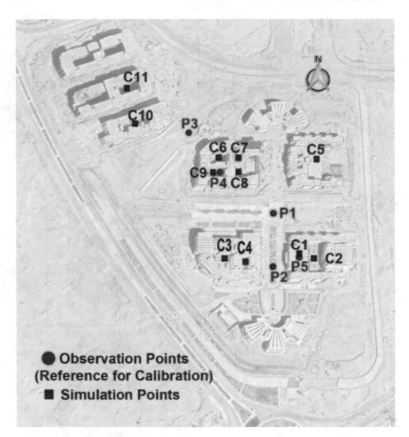

Fig. 3 Location of observation and simulation points

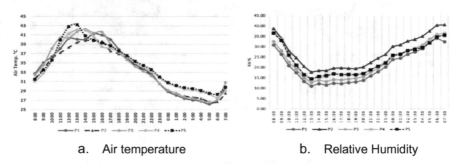

a.   Air temperature   b.   Relative Humidity

Fig. 4 Air temperature and relative humidity at the measuring points

Table 2 Validation of simulation model

	P1	P2	P3	P4	P5
MBE:	−0.05	−0.59	−0.91	−1.2	−0.83
MAD	0.82	0.65	1.26	1.49	1.33
MSE	1.06	0.91	2.21	2.72	2.17
RMSE	1.03	0.95	1.49	1.65	1.47
MAPE	2.62	2.1	3.99	4.62	3.99
d	0.98	0.98	0.98	0.96	0.97

MBE Mean bias error, *MAD* Mean absolute deviation, *MSE* Mean square error, *RMSE* Root mean square error, *MAPE* Mean absolute percentage error, d: Index of agreement

included the mean bias error (MBE) and mean absolute deviation (MAD). In the simulation process, five points were selected for comparison with the observation points (P1, P2, P3, P4 and P5). These simulation points were positioned in the same location as the observation points. The patterns of air temperature between the observations and the simulation were about the same; the observed peak Ta (ambient temperature) was 1–2 °C higher than the simulated value at all the points. Simulated data gave generally good agreement with field-observed data, despite small under or over estimations, with both the index of agreement d (from 0.96 to 0.98, as shown in Table 2), and the correlation ($r^2 > 0.96$) indicating that our simulation captured the measured diurnal temperature trends successfully. As shown in Table 2, MBE, MAD and MAPE were of acceptably low magnitudes. RMSE of the air temperature between simulation and observations ranging from 0.95 at P2 to 1.65 at P4. These values were an improvement on those reported in other studies (Chow and Brazel 2012; Middel et al. 2014). Hence, we considered the validation parameters adequate for the microscale simulation.

4.3   Factors Influencing Thermal Comfort

The study demonstrated the impact of different courtyard geometrical characteristics on thermal comfort. It was found that almost all results reflected the relationship between PET values and solar access to courtyard surfaces. In this regard, a PET map distribution in two-hour steps was obtained using Envi_met. Figure 5 presents the microclimatic conditions in terms of PET in the courtyards at the Aswan University campus. Eleven points were defined as thermal receptors in the simulation model. In general, it was observed that PET values generally followed the same behavior at all measurement points. It increased dramatically in all courtyard receptors during sun exposure hours in the courtyards, except for C2 which showed normal thermal behavior as shown in Table 3. Therefore, it could be said that all the studied courtyard spaces were thermally uncomfortable, with high PET values over 50 °C due to the high solar radiation intensity and long duration of direct solar radiation. In addition,

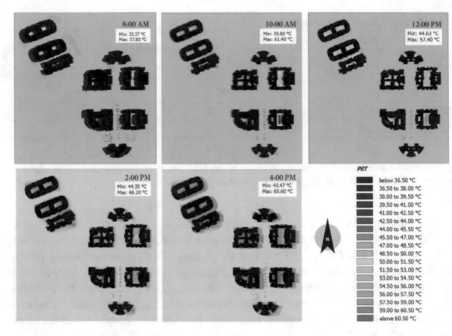

Fig. 5 Extracted PET slice at 1.5 m

Table 3 Diurnal values of PET (°C) for the studied courtyards.

Simulated points	C1	C2	C3	C4	C5	C6	C7	C8	C9	C10	C11
SVF	0.154	0.474	0.193	0.127	0.489	0.163	0.108	0.129	0.2	0.142	0.236
Courtyard orientations	N-S	N-S	-	N-S	N-S	-	N-S	-	E-W	NE-SW	NE-SW
Greening type	-	Grass	-	-	-	-	-	-	-	-	-
8:00 AM	35.26	33.55	34.85	35.91	35.69	35.48	35.25	35.06	35.91	35.68	52.83
9:00 AM	39.85	37.87	39.27	40.81	49.40	40.36	39.99	39.67	40.81	58.40	55.89
10:00 AM	43.93	41.02	42.47	56.32	49.57	56.45	43.81	54.77	56.32	57.20	56.05
11:00 AM	51.48	43.29	50.84	53.87	49.53	53.94	53.30	52.34	53.87	55.00	53.78
12:00 PM	49.95	44.75	49.12	51.86	49.46	51.93	51.17	50.43	51.86	52.80	51.59
1:00 PM	53.54	45.96	47.82	55.83	50.15	55.69	54.78	53.86	55.83	57.80	55.55
2:00 PM	48.49	46.10	47.36	59.45	52.72	59.46	48.22	47.99	59.45	49.40	58.99
3:00 PM	47.48	45.68	46.81	61.49	48.88	47.66	47.34	47.20	61.49	48.00	48.49
4:00 PM	45.60	44.11	45.11	45.84	46.50	45.61	45.46	45.30	45.84	45.52	46.13
5:00 PM	42.31	41.11	42.00	42.35	42.84	42.25	42.17	42.07	42.35	42.22	42.54

PET °C
40   45   50   55

it was found that large courtyards oriented in the north–south direction had the lowest PETs of between 36–40 °C until the middle of daylight hours. While small courtyards became hotter in the early hours, all courtyards returned to their lowest PET values of between 36–45 °C at 4 p.m. In order to illustrate the relationship between PET values and the factors that affected them, the results of this study are divided here into the following subsections: SVF analysis, the effectiveness of courtyard orientations and the utilization of greening.

4.3.1 SVF Analysis

As shown in Table 3, the best PET readings were observed at C2 (SVF: 0.474). In this regard, factors such as orientation and greening need to be taken into account. In this courtyard, the maximum PET values did not exceed 46.1 °C despite its relatively high SVF. While small courtyards such as those associated with $C1$, $C3$, $C4$, $C6$, $C7$, $C8$, $C9$, $C10$ and $C11$ had SVFs < 0.25, they were hotter than the large courtyards, $C2$ and $C5$, due to their propensity to heat-trapping. Accordingly, the temperatures in nine courtyards with SVF values between 0.108 and 0.236 were raised above 50 °C during the solar exposure hours. Hence, thermal comfort was not determined by SVF alone.

4.3.2 Effectiveness of Courtyard Orientations

Another observation regarding PET values was that even large courtyards, with SVF > 0.4, but oriented in a north–south direction, were more acceptable than the smallest courtyards with SVF < 0.25 and oriented in east–west or northeast–southwest directions. Thus, it was noticed that $C10$ and $C11$ were very hot from the early hours of the day until the final hours of daylight. This increase in PET showed how courtyard orientation affected solar exposure in the courtyard. It can hence be concluded that the worst courtyard orientation was the east–west direction or a direction close to this. Table 3 highlights the different values of PET during the diurnal hours in the studied courtyards.

4.3.3 Utilization of Greening

It is known that adding greenery can improve thermal comfort levels (Ghaffarianho-seini et al. 2015). So, a comparison between $C2$ and $C5$ was made to examine the effect of the greening. It was observed that $C2$ was thermally more comfortable than $C5$ when the courtyard had a similar orientation and an approximately similar SVF. This result could be attributed to the green area at $C2$. Increasing green areas in the courtyard increase the relative humidity of the vicinity.

5 Conclusion

Based on the case analysis of thermal performance of the courtyards, the findings can serve as a basis for guidelines in the design of future extensions for the university campus or other public buildings in Egyptian arid cities. Such guidelines should take cognizance of the following:

- The most significant impact on the thermal performance of the courtyard is solar radiation intensity. Hence, controlling its level is crucial to improving thermal comfort and reducing heat stress on building facades. Accordingly, the courtyard should be designed with dimensions and orientation that minimize the effect of solar radiation (n–s direction, the smaller dimension perpendicular to the sun's path). In the present study, C3 (SVF = 0.193) had a relatively good thermal performance, with minimum PET of 34.85 at 8 a.m. and a maximum PET exceeding 50 °C only for one hour at 11 a.m. Therefore, C3 had better thermal performance in comparison with C10, a courtyard with approximately the same area, but oriented in a northeast–southwest direction.
- SVF can be lowered to less than 0.2 in courtyards facing north. However, for other orientations, greater reduction in SVF might result in heat-trapping which significantly decreases the level of thermal comfort.
- Adding a greening area improves the thermal performance of courtyards and prevents a large increase in PET during the hours of solar exposure. Greening tends to keep the PET curve smooth through daylight. As an important thermal mitigation strategy for courtyards, its impact was clear in the improvement of PET in C2, with a reduction of about seven PET degrees as compared with C5.

Acknowledgements This work was a part of the research project "Adapting Sustainable Urban Planning to Local Climate Change in Egyptian New Towns (Case Study New Aswan)", ID: 23022, in the framework of the program German Egypt Research Fund "GERF4". The study was funded by the Science and Technology Development Fund (STDF), Egypt, and the Federal Ministry of Education and Research, Germany (BMBF).

References

Aldawoud A (2008) Thermal performance of courtyard buildings. Energy Build 40(5):906–910
Almhafdy A, Ibrahim N, Ahmad S, Yahya J (2013a) Courtyard design variants and microclimate performance. Procedia 101:170–180
Almhafdy A, Ibrahim N, Ahmad S, Yahya J (2013b) Analysis of the courtyard functions and its design variants in the Malaysian hospitals. Procedia 105:171–182
Alshuwaikhat H, Abubakar I (2008) An integrated approach to achieving campus sustainability: assessment of the current campus environmental management practices. J Cleaner Prod 16(16):1777–1785
Aswan Climate. https://en.climate-data.org. Accessed 12 May 2019

Berkovic S, Yezioro A, Bitan A (2012) Study of thermal comfort in courtyards in a hot arid climate. Sol Energy 86(5):1173–1186

Cairo University statistics Online. https://cu.edu.eg. Accessed 19 June 19 2019

Cantón M, Ganem C, Barea G, Llano J (2014) Courtyards as a passive strategy in semi dry areas. Assessment of summer energy and thermal conditions in a refurbished school building. Renewable Energy 69:437–446

Chow W, Brazel A (2012) Assessing Xeriscaping as a sustainable heat island mitigation approach for a desert city. 47:170–181

Forouzandeh A, Richter T (2019) Accurate prediction of heating energy demand of courtyard's surrounding envelopes using temperature correction factor. Energy Build 193:49–68

Ghaffarianhoseini A, Berardi U, Ghaffarianhoseini A (2015) Thermal performance characteristics of unshaded courtyards in hot and humid climates. Build Environ 87:154–168

Hämmerle M, Gál T, Unger J, Matzarakis A (2011) Comparison of models calculating the sky view factor used for urban climate investigations. Theoret Appl Climatol 105(3):521–527

Manioğlu G, Koçlar GO (2015) Effect of courtyard shape factor on heating and cooling energy loads in hot-dry climatic zone. Energy Procedia 78:2100–2105

Martinelli L, Matzarakis A (2017) Influence of height/width proportions on the thermal comfort of courtyard typology for Italian climate zones. Sustain Cities Soc 29:97–106

Ma Y, Lu M, Weng J (2015) Energy consumption status and characteristics analysis of university campus buildings. In: Proceedings of the 5th international conference on civil engineering and transportation

Middel A, Häb K, Brazel A, Martin C, Guhathakurta S (2014) Impact of urban form and design on mid-afternoon microclimate in Phoenix Local Climate Zones. Landscape Urban Plann 122:16–28

Muhaisen A (2006) Shading simulation of the courtyard form in different climatic regions. Build Environ 41(12):1731–1741

Ping Lin T, Matzarakis A, Lung Hwang R (2010) Shading effect on long-term outdoor thermal comfort. Build Environ 45(1):213–221

Safarzadeh H, Bahadori M (2005) Passive cooling effects of courtyards. Build Environ 40(1):89–104

Sharmin T, Steemers K, Matzarakis A (2017) Microclimatic modelling in assessing the impact of urban geometry on urban thermal environment. Sustain Cities Soc 34:293–308

Sibley M (2004). The courtyard houses of North African medinas: past, present and future. pp 89–107

Soflaei F, Shokouhian M, Zhu W (2017) Socio-environmental sustainability in traditional courtyard houses of Iran and China. J Renew Sustain Energy Rev 69:1147–1169

Soflaei F, Shokouhian M, Abraveshdar H, Alipour A (2017) The impact of courtyard design variants on shading performance in hot-arid climates of Iran. Energy Build 143:71–83

Taleghani Taleghani M, Sailor D, Tenpierik M, Dobbelsteen van den A (2014) Thermal assessment of heat mitigation strategies: the case of Portland State University, Oregon, USA Build Environ 73:138–150

Taleghani M, Sailor D, Tenpierik M, van den Dobbelsteen A (2014) Heat in courtyards: A validated and calibrated parametric study of heat mitigation strategies for urban courtyards in the Netherlands. Sol Energy 103:108–124

Taleghani M, Kleerekoper L, Tenpierik M, van den Dobbelsteen A (2015) Outdoor thermal comfort within five different urban forms in the Netherlands. Build Environ 83:65–78

Watson I, Johnson G (1987) Graphical estimation of sky view-factors in urban environments. J Climatol 7(2):193–197

Willmott C (1981) on the validation of models. Phys Geogr 2(2):184–194

World Maps of Köppen-Geiger Climate Classification. https://koeppen-geiger.vu-wien.ac.at/. Accessed 19 April 2020

Societal, Research and Innovation Challenges in Integrated Planning and Implementation of Smart and Energy-Efficient Urban Solutions: How Can Local Governments Be Better Supported?

Judith Borsboom-van Beurden and Simona Costa

Abstract This paper reports on a special session organized during the SPPCR 2019 conference by Urban Europe Research Alliance and the Action Cluster Integrated Planning, Policy and Regulations of the European Innovation Partnership on Smart Cities and Communities, in collaboration with several key networks, projects, and programs. The aim of this special session was to discuss how integrated planning and implementation can help to boost the transition to low-carbon cities and how good examples and best practices of such an integrated approach can foster wider replication of smart and energy-efficient solutions in cities across Europe. European Commission policy officers updated the audience on expected changes in smart sustainable city policies in the new programming period between 2021 and 2027. Ambitious and comprehensive on-going European smart city projects highlighted how they designed and deployed a holistic, integrated perspective during the phases of project preparation and execution, often in a living lab approach, and the challenges they had to overcome in doing so. Cities and researchers committed to replication of smart and energy-efficient solutions, shared their needs and dilemmas, in particular regarding the role of holistic, integrated approaches. However, it was also pointed out that specific constraints, make that replication has to explicitly designed for in the very early stages of any project and should be part of any holistic approach. Others presented different avenues for stepping up the efforts to create smart and energy-efficient cities in an integrated way: from providing low-threshold roadmaps for developing customized strategies and building collective transformative capacity through pooling of national research resources, to setting up learning energy communities and deploying Interreg funds for innovative decision-making on energy-efficient public buildings. Interactive sessions explored persistent knowledge gaps regarding integrated planning and implementation and how new projects

J. Borsboom-van Beurden
Locality, Rozenstraat 8, 3971 CN Driebergen-Rijsenburg, The Netherlands
e-mail: judith.borsboom@locality-eu.nl

S. Costa (✉)
Tour4EU, Rond Point Schuman, 14 - 1000 Bruxelles, Belgium
e-mail: s.costa@tour4eu.eu

can be built for positive energy districts. It is concluded that the value of a holistic or integrated approach is generally acknowledged and several good tools have been developed in EU-funded projects helping to concretize it, but that the concept needs to be better articulated, made accessible to time-pressured city administrations, geared toward specific urban situations and contexts, and translated into specific local processes and procedures to better support local governments.

Keywords Smart cities · Energy efficiency · Integrated planning and implementation · Local governments

1 Introduction

While the importance of a holistic, integrated approach to planning and implementation of smart and energy-efficient solutions in cities is widely acknowledged (e.g., European Commission 2013; Bahr (ed) 2014; IEA 2017; Mission Board on Climate-Neutral and Smart Cities 2020), in practice multiple societal, research, and innovation challenges make this quite complicated. How can time-pressured, often siloed local governments be better supported in this across Europe? How can the good practices demonstrated by several European projects be more easily taken up by other municipalities, and how should this information be disclosed? On December 10, 2019, an event dealing with this key issue was organized by the Urban Europe Research Alliance (UERA), an organization representing 55 universities and research institutes across Europe related to Joint Programme Initiative Urban Europe (JPI Urban Europe), at the Smart and Sustainable Planning of Cities and Regions 2019 Conference in Bolzano/Bozen, hosted by EURAC Research.

This event was jointly organized with a wide range of highly active key players as JPI Urban Europe, the SET-Plan Action 3.2 on Positive Energy Districts, the Action Cluster Integrated Planning, Policies and Regulation of the European Innovation Partnership on Smart Cities and Communities (EIP-SCC), Horizon 2020-funded smart city lighthouse projects Smarter Together, REPLICATE, +CityxChange and STAR-DUST, the European Energy Award (EEA), and Tuscan Organization of Universities and Research for Europe. The event brought together a wide range of key stakeholders around integrated planning and implementation of smart and energy-efficient solutions in cities, to discuss not only how efforts to decarbonize cities can be stepped up through wider replication, but also which agenda for research and innovation should be defined to build more collective intelligence and transformative capacity for local governments, under Horizon Europe, national policy agendas and JPI Urban Europe's Strategic Research and Innovation Agenda 2.0 (JPI Urban Europe 2019).

Many important questions were discussed at the event. What can we learn from the experiences of lighthouse projects and similar projects? What withholds urban ecosystems from replication and which role is played in this by lacking or deficient holistic, integrated approaches? How can we make better use of lessons already learned by others? And last but not least, which messages should be delivered to the newly established mission boards regarding needed research and innovation?

The lively and successful event was attended by around 140 participants. Key messages, not only for new policy agendas, but also for a new research and innovation agenda centering around holistic, integrated approaches, have been collected and will be transferred to national policy-makers, the Mission Board on Climate neutral and Smart Cities, JPI Urban Europe and SET-Plan Action 3.2 on Positive Energy Districts. Here, a summary of the event and its most important outcomes are presented.

2 Smart Cities in the Future Programming Period of the European Commission

The opening session of the event welcomed all participants and discussed European research and innovation policies for smart and energy-efficient cities, thus setting the scene for Horizon Europe. Georg Houben, Policy Officer, DG Energy Dir. *C*2, discussed the European Commission's (EC) plans on smart cities policies and funding schemes in the future programming period. Key tenets are the continuation of funding opportunities for smart city lighthouse projects in Horizon Europe, and in the Green Deal under preparation, which will publish calls for making cities more energy-efficient in a smart way in September 2020. In the Horizon Europe program, the intervention area 5 "Communities and Cities" will be part of the pillar 2 "Global Challenges and European Industrial Competitiveness" in the cluster 5 "Climate, Energy and Mobility." Two co-funded partnerships are proposed: "Driving Urban Transition to a Sustainable Future and Clean Energy Transition," whereas a Mission on Climate-Neutral and Smart Cities has been launched in July 2019, chaired by Hanna Gronkiewicz-Waltz.

Following, Jérôme Böhm, Program Manager Italy and Malta of DG Regio Dir. G 4, updated the audience on the architecture of the European Regional Development Fund (ERDF) and Cohesion Fund in the new programming period 2021–2027. Within a total budget of 373 Billion Euros, important changes are a higher budget for less developed and transition regions, a lower budget for more developed regions and cohesion, beside new regional eligibilities. Apart from a more connected and more social Europe, a smarter, greener, low-carbon Europe and a Europe closer to its citizens are key objectives of Cohesion Policy from 2021 to 2027. For a smarter Europe, interregional cooperation in value chains will promote innovative and smart economic transformation. For a greener Europe, 30% of the ERDF operations is expected to contribute to climate objectives through energy transition, the circular economy, climate adaptation, and risk management. For Europe closer to its citizens, integrated territorial investments and community-led local development will be important tools, while 6% of ERDF is earmarked for promoting sustainable urban development through local development partnerships. Here the European Urban Initiative will come into play, specifically to tackle the fragmentation of current support offered to cities by capacity-building, innovative actions, knowledge policy development, and communication with the overall aim of strengthening an integrated and participatory approach to urban development and provide a stronger link to EU policies, in particular Cohesion Policy. Lastly, the EC plans more cooperation and synergies between Interreg (cross-border or transnational cooperation), ESF + (Integrated development in deprived neighborhoods and social inclusion), EAFRD and EMFF (integrated strategies in rural and coastal areas), Horizon Europe, InvestEU (market-based support for integrated strategies) and technical assistance.

3   Innovative Approaches to Integrated Planning and Implementation of Smart City Solutions—Real-Life Examples from EU-Funded Projects

After the position of smart cities in the future programming period of the EC had been outlined, four EU-funded projects shared their real-life examples of innovative approaches to integrated planning and implementation of smart city solutions.

Adriano Bisello of EURAC, senior researcher involven in the SINFONIA project funded under 7th Framework Program, highlighted the importance of co-benefits in integrated planning and implementation of smart energy district projects in Bolzano. Like when beholding an iceberg, avoided CO_2 emissions and energy savings are often well-recognized outputs of projects, but many co-benefits of solutions remain largely invisible. A co-benefit is defined as any socioeconomic and environmental positive effect related to the execution of a project, exceeding the primary goal, regardless if intentional or not (Bisello et al 2017; Bisello and Vettorato 2018; Bisello 2020) In the SINFONIA project, Bisello (2020) categorized these co-benefits and developed a systemic method which reveals the contribution of co-benefits to overall quality of life for citizens in four specific investigations, possibly leading to a different assessment of deep retrofitting options. At first, individual participants scored relevance and likelihood of different smart city components and ranked their priorities, what provided input to a world café setup where knowledge and group thinking on these complex topics was elicited. Following, a hedonic price method helped to determine how energy efficiency would influence the value of residential properties. After that, a contingent evaluation method on smart points was applied to understand the willingness to pay for a new infrastructure. Subsequently, an analytic hierarchy process on the decision on deep energy retrofitting helped to better understand the priorities of households regarding various types of sub-criteria, such as design and spatial quality, economic benefits, and acoustic and thermal comfort (Bisello and Vettorato 2018; Bisello 2020). Therefore, making co-benefits visible is of crucial importance for integrated planning and implementation of smart and energy-efficient solutions, not only by providing a better balanced assessment of different options, but also by creating a better understanding of each other's priorities while developing a communal strategy, thus securing the "buy-in" of different stakeholders.

Fig. 1 Different approaches to urban living labs (*Source* Smarter Together)

After this, Etienne Vignali of Lyon Confluence, Project Manager of Horizon 2020-funded Smarter Together, explained how inclusiveness has played a central part during the implementation of smart solutions in Urban Living Labs in the districts of Simmering North-West (Vienna), Neuaubing-Westkreuz (Munich) and Confluence area in Lyon (Smarter Together 2020). Main outcomes of the project are 17 MW of newly installed renewable capacity, 152,000 m^2 of renovated floor space in refurbished housing, 89 electric vehicles and 64 charging stations, and three Urban Data Platforms, while 8500 citizens have been involved in this process. Inclusiveness through engagement of and co-creation with stakeholders is at the heart of the project. However, per Urban Living Lab, this took place in a different way. While in Vienna a mobile information bus which changed location every five to six weeks proved to be a much more effective means for engaging citizens and other stakeholders in the district than more traditional hearings, in Lyon and Munich a more stationary approach was taken at a centrally located co-creation place (see Fig. 1). While different tools were used for engagement of and co-creation with citizens, each Urban Living Lab also faced different realities and choices in smart solutions for energy retrofitting of buildings and electric mobility. Main focus in Vienna was on tenants in public and social housing and users of public facilities as schools, integrated with e-carsharing, e-bikes, e-forklifts, and mobility station schemes. Lyon targeted in particular social housing operators, groups of private owners and office spaces, e-cars, autonomous shuttles and electric charging stations. Munich worked mostly with owner-occupier associations in multiple-residency buildings, in combination with e-cars, e-bikes and district sharing boxed to improve sustainability of the last mile logistics (Smarter Together 2020). Smarter Together learns that integrated planning and implementation cannot be done using a blueprint, but that different approaches have to reflect local different realities and contexts.

Subsequently, Alessandra Barbieri, Project Manager at the City of Florence, shared the main features and holistic approach chosen in the Horizon 2020-funded REPLICATE project (REPLICATE 2020). Florence has a longstanding commitment to sustainability since signing the Aalborg Charter in 1998, manifesting itself in a Sustainable Energy Action Plan (2011), a long-term Smart City Plan until 2050 (2016), and a Sustainable Urban Development Plan (2016). Florence wants to regenerate the city through developing a polycentric structure for a more sustainable, compact, and socially affordable resilient city. Four specific domain strategies underpin this: (1) redevelopment of dilapidated urban areas combined with rewarding energy-efficient systems to prevent further urban expansion; (2) promotion of public transport, low emission and electric vehicles, walking and cycling; (3) more ICT platforms and apps, digitalization of public services and e-government, beside extension of the Wi-Fi network; and (4) sustainable construction, extension of the ecological network, smart lighting, sensor systems, green procurement, bio-schools, and the action plan on climate change mitigation and adaptation. The REPLICATE project made it possible to accelerate the transition to a smart city and the deployment of not only innovative technologies but also of novel organizational and economic solutions, in a lighthouse project with San Sebastian and Bristol. Smart solutions applied in Florence's lighthouse project are renewable energy production and thermal insulation, implementation of smart grids with sensors enabling remote control, smart lighting also offering surveillance for traffic control, a public tender licensing 70 e-taxis and four fast recharge stations for taxis, and participatory methods based on systems thinking for engaging citizens. Data are seen as a core enabler for acting, as sharing of data can help to simplify life for citizens, to improve communication not only between experts but also between citizens, and to avoid reinventing the wheel. Integrated planning means sharing skills, data, infrastructures and services and the Smart City Control Room makes this possible (Bellini et al 2018).

The role of the municipality in this can be compared to that of a conductor of an orchestra: a center of aggregation that takes care of collaboration and synergies between different bodies and utilities, while the individual parties play their part. This federated model is enabled by a Big Data platform, developed together with the University of Florence, containing data on a wide array of topics but also having a physical location (see Fig. 2). The municipality is the only one who can promote a joint vision and a holistic, integrated approach, where different operators and service providers can make decisions and manage public services much

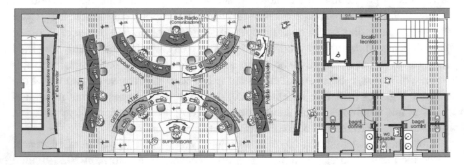

Fig. 2 Smart City Control Room: a physical place to better manage the city for a better life (*Source* REPLICATE)

faster sharing systems and real-time data. REPLICATE shows how bringing together urban data in a big data platform supports not only technical but also organizational collaboration.

Lastly, Annemie Wyckmans, Norwegian University of Science and Technology, Project Coordinator of Horizon 2020-funded + CityxChange, explained how integrated planning is being used to create positive energy districts in Lighthouse Cities Limerick and Trondheim (+CityxChange 2020). This will be achieved along two lines. At first, accelerating the energy transition through jointly developing a vision, engaging stakeholders, experimenting in the innovation playgrounds, modeling of potential impact, scrutinizing legal frameworks in regulatory zones, and fine-tuning investment plans and risk-sharing models with stakeholders from public and private sector, academia and civic society. At second, increasing the energy system integration of Positive Energy Buildings (PEBs), microgrids, electric mobility as a service, local trading and developing a market for flexibility. The barriers for replication of good smart city practices have been well-documented earlier, e.g., by Vandevyvere (2018), Borsboom-van Beurden et al (2019) and DoGA (2019), and + CityxChange wants to address them by setting up an extensive process of open innovation and co-creation with all 32 project partners and local associates. Its aims are: to build a common project culture across sectors and countries, to clarify expectations, to make mutual dependencies explicit, to integrate lessons learned into every organization, and to communicate, share and evaluate frequently (+CityxChange 2020). A concrete example contributing to better integrated planning and implementation are the glossary workshops, where a project-wide common understanding of specific terms and expected outputs provided a very useful basis for multi-disciplinary collaboration. In the same vein, the Bold City Vision provided guidelines for balancing long-term strategic planning aims for the cities with short-term incremental actions in the project. Other examples are the inventory of how regulatory frameworks should be adjusted to enable the desired innovations, and a framework for the innovation playgrounds organized in Limerick that combined both physical and digital instruments. Lastly, the CommunityxChange Framework for PEB Innovation Labs creates "a permeable culture for co-creation "in the city, while storytelling workshops explore ways toward citizen-led energy transitions (+CityxChange 2020).

4   How Can the Uptake of Smart Sustainable City Solutions Be Accelerated in the Future?

After lunch Marc Dijk, Research Fellow at the University of Maastricht, kicked-off with a keynote on how the project SmarterLabs, funded by JPI Urban Europe, worked out guidelines for better anticipating constraints on upscaling of inclusive Urban Living Lab results. Many Urban Living Labs suffer from two major pitfalls: There are unforeseen constraints on large-scale change in socio-technical urban systems, and social groups not matching the required "smart citizen" profile are excluded. To overcome these pitfalls, SmarterLabs aimed to define guidelines for this, based on literature review and retrospective analysis in the four engaged cities (Bellinzona, Brussels, Maastricht, Graz), and to subsequently test this approach through action research projects in these cities, complemented by a testing workshop in three other cities (Dijk et al. 2018). Ten typical constraints to upscaling of living lab results were found:

1. Citizens lack financial, intellectual, and time resources to participate in the Living Lab
2. Relevant stakeholders remain outside the Living Lab
3. Groups and impacts outside the Living Lab context are overlooked
4. Existing power structures are reproduced inside the Living Lab
5. The Living Lab's potential for learning is underexploited
6. The Living Lab is disconnected from broader societal debate
7. The Living Lab consensus is not reflected in policy and society
8. Stakeholders and institutions are highly fragmented
9. Urban assemblage is sticky and locked-in
10. The Living Lab meets low institutional receptiveness.

Per constraint, measures to anticipate this constraint, such as participatory visioning, have been collected, elaborated and validated in the cities, resulting in a set of valuable recommendations to ensure not only inclusiveness but also upscaling and replication after the project's end-of-life (see Dijk et al 2018). SmarterLabs shows that designing for inclusiveness, replication, and upscaling by addressing these typical constraints on beforehand, should be part and parcel of any smart city and energy efficiency project's architecture and its preparation and execution.

Marc Dijk's keynote was followed by an animated panel discussion with Horizon2020 SCC-01 fellow cities Lecce (Serena Pagliula, Project Manager Horizon 2020), Gdansk (Joanna Tobolewicz, Mayor's office) and Marc Dijk, moderated by Simona Costa, EIP-SCC/TOUR4EU. Key topic was how the uptake of successfully demonstrated smart city solutions for integrated planning and implementation could be accelerated in the future. The panel discussion zoomed in on challenges, needs, and how to accelerate the market uptake of smart city solutions through practice, research, and innovation. The panelists agreed that the collaboration between practitioners, city administration, solution providers, and civil society is key for successful smart city projects, but not yet common practice. To be successful in this collaboration, the panelists deemed it extremely important to make mutual interdependencies between different stakeholders explicit and clarify expectations. However, the panelists also indicated that working in an interdisciplinary way across different domains is not easy, and they saw much room for improvement in terms of making this collaboration more efficient and getting everybody on the same page. Further, panelists remarked that not the same problems are experienced everywhere regarding integrated planning and implementation of smart city projects, and that local situation and context play a very significant role. The research community could moderate such a collaboration process and get everybody on the same page during preparation of plans, for example by developing a glossary at EU level specifying the unambiguous meaning of terms (see, e.g., + CityxChange 2020). Other next steps research can contribute to, are monitoring if (common) barriers are really taken away, analyzing how to adjust national/region/local procedures toward implementation of smart city projects, and performing an in-depth analysis of regulatory frameworks.

5 Stepping Up the Efforts to Create Smart and Sustainable Cities in the Near Future by Practice, Research and Innovation

The final plenary session presented several ways for stepping up the efforts to create smart and energy-efficient cities in the near future by practice, research, and innovation. The session was moderated by Georg Houben, Policy Officer, DG Energy Dir. C2.

Judith Borsboom-van Beurden, UERA/EIP-SCC/NTNU, presented the Smart City Guidance Package (SCGP), an inspirational document and self-help guide for integrated planning and implementation of smart city and energy efficiency projects, produced by the Action Cluster Integrated Planning, Policy and Regulations of the EIP-SCC with the help of nearly 100 city administrations, businesses, research institutes and universities from the wider smart city community. It bundles experiences of cities and helps other cities in avoiding common pitfalls when preparing and carrying out their plans, with a main focus on a recommended process, not on technologies. The SCGP is an easily accessible introduction into integrated planning and implementation, primarily meant for mayors and politicians, staff supporting them, such as strategists and advisors, directors of unit, project managers, and other local authorities, e.g., utilities. However, it can also be used to facilitate communication with partners in the cities' local ecosystem, such as energy network and transport operators, real estate developers and facility managers, housing associations, citizens and local businesses. It was developed because it was observed that while there is a high level of complexity in these transformative actions, city administrations are often under pressure and lack the time to explore common repositories as the Smart City Information System, and applicable ISO Standards are often perceived as rather heavy in terms of implementation. The roadmap has been composed by integrating information from a series of EIP-SCC workshops, review of hundreds of smart city projects, 29 interviews conducted, feedback from cities and city networks, and five validation workshops in Santa Cruz de Tenerife, Sofia, Vaasa, Brno and Parma. Key features of a holistic approach facilitated by the SCGP are: (1) integration of a long-term perspective when deciding upon short-term actions: (2) integration of domains and disciplines; (3) integration of multiple technologies in one territory; (4) integration of different stakeholders and commitments; (5) integration of financial aspects and co-benefits. The result is a sequence of seven different stages for integrated planning and implementation of smart city projects (see Fig. 3).

Fig. 3 Different stages in SCGP roadmap (*Source* Borsboom-van Beurden et al. 2019)

Subsequently, each stage is elaborated using the same elements: the key question of that specific stage, a checklist of to do's, the main output than can be expected after ticking the boxes for all to do's, and suggestions for tools and standards than can help to facilitate each to do. More details on each to do indicate what needs to be done, give examples of good practices, and explain why this particular step is needed (Borsboom-van Beurden et al. 2019). Next steps foreseen are: (1) developing a summary for local politicians in EU languages; (2) developing an Open Access web-based version; (3) supporting the development of local smart and energy-efficient cities strategies; and lastly (4) recruitment of new testbeds and gradual refinement of the method.

Next, Christoph Gollner and Susanne Meyer, FFG and AIT, explained how currently collective transformative capacity is built in the SET-Plan Action 3.2 and Positive Energy Districts (PED) Program, led by JPI Urban Europe. The mission is to enhance capabilities of cities, industry and research to make Europe a global role model and market leader in technology integration for and large-scale implementation of PEDs, while taking into account inclusiveness. By 2025, 100 PEDs in Europe by 2025 should be synergistically connected to the European energy system, exporting related technologies. The SET-Plan Action 3.2 makes this mission concrete by innovation actions establishing PED Labs, developing PED Guides and Tools, fostering PED replication, and setting up monitoring and evaluation for existing PEDs and the new PED Labs. Focus will be on the problem-owners: public authorities which have to achieve energy and climate targets, where PEDs should be part of holistic urban strategies helping to avoid silo-thinking and fostering cross-domain collaboration (SET-Plan Action 3.2 2018). Main elements in the work plan for 2019–2020 are the publication of an overview of current PEDs, development of a common PED framework, guidelines for urban stakeholder engagement, the launch of two calls, cross-cutting cooperation with other initiatives as the EIP-SCC and development of a European partnership. The first PED call (Spring 2020) focuses in particular on enabling factors, e.g. legislation, different contexts for feasibility of plans, and the setup of PED Labs. AIT analyzed specific features of current 52 PED/Toward PED projects, such as size, new or old buildings, land use, mean project investment, financing, and energy technologies (JPI Urban Europe 2020). Most PEDs appear to be less than 10 hectare, have both new and old buildings on mixed land use, mostly costing around 100 million Euros financed by at least three different sources and integrate four to five different energy technologies (JPI Urban Europe 2020).

Following, Chiara Tavella and Mariadonata Bancher representing the eea, shared the experiences gained in more than 20 years in terms of standardized processes and learning communities. The EEA is implemented in more than 1500 cities and communities in Germany, France, Switzerland, Austria, Belgium, Italy, Croatia, Poland, Romania, Bulgaria Serbia, Greece, and Luxemburg, counting between 150 and more than 1 million inhabitants. EEA helps to implement local climate change policies through a proven process, several instruments and qualified advisors, covering six themes: mobility, spatial planning, internal organization, supply and disposal of energy and waste, municipal buildings and facilities. EEA is attractive for local authorities for several reasons. It uses a result-oriented yet iterative approach and well-structured process, including the establishment of an interdepartmental Energy Team, where local climate and energy policies are continuously improved, and energy targets are often achieved faster. Coaching by an external expert and the certification scheme lead to higher quality of plan preparation and execution. Further, cities and communities get access to many country-specific tools and checklists provided by the national EEA organizations. In addition, awarding proves to be an important incentive for improving climate and energy policies, while it simultaneously enhances visibility and promotes the city or community marketing. What is more, the frequent exchange of experiences and offered training foster peer-to-peer learning, with national EEA organizations learning about common barriers they might need to address. While EEA can be carried out independently, it is also a tool for implementing Sustainable Energy (and Climate) Action Plans (SECAPs), and, like the SCGP, aligned with ISO standards 50,001 and 37,107. To bring about more cross-fertilization with adjoining initiatives, the current CoME Easy project provides common tools, e.g., an Emission Path tool, SE(C) AP creator, buildings database, KPIs dashboard, guidelines for stakeholder engagement and best practices library (see Fig. 4).

Two examples of EEA are highlighted. For implementing the Climate Pact in Luxemburg, the national government provided financial support and technical assistance, and supported further development of tools, while municipalities implemented EEA including its energy accounting system. In South-Tirol, EEA helped to execute the ambitious regional climate strategy until 2050 where 18 municipalities (among others Bolzano) implemented measures as promotion of renewable energy, renovation of the public building stock and introduction of energy-efficient lighting. In South-Tirol, the entry awarding step proved very useful to mobilize municipalities and ensure active participation from the start, while the region contributed to the costs of the external expert. In terms of integrated planning, the municipalities mentioned several benefits of EEA. It helped to overcome siloes by appointing an energy team and allocating specific roles, what leads to better exchange between departments. The catalogue proved very helpful in answering what-if questions, supporting the choice for the solutions with the highest impact, and ensuring the buy-in of different stakeholders. The scheme also improves communication with and engagement of citizens. Lastly, the external auditing and competition element of awarding allow for comparison and raise the overall ambition level.

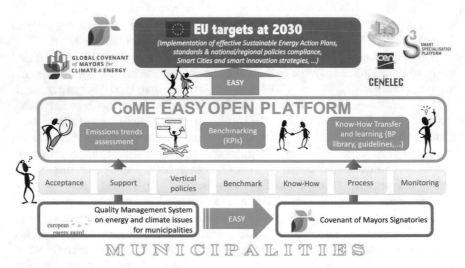

Fig. 4 Mutual reinforcement of different initiatives in CoME Easy project (*Source*: EEA)

Lastly, Carmelina Cosmi of CNR-IMAA, presented PrioritEE, an Interreg Project to support decision-making on energy-efficient buildings, which showcases how EU structural funds can be used to bring about innovation in several EU territories (PrioritEE 2020). Barriers to energy efficiency come from a complex socio-technical system in which institutional and technical aspects as well as actors are deeply linked. To address these barriers, PrioritEE aimed to strengthen the policy making and strategic planning competences in energy management of public buildings of local and regional public authorities in five Mediterranean countries. Cities and regions participating were Karlovac, Potenza, Teruel, CIMLT, and West-Macedonia. A twofold approach consisted of (1) developing decision-making tools as the Prior-itEE toolbox and common strategies for developing energy consumption manage-ment plans for municipal buildings; next to (2) stakeholder engagement through case study visits with peer-to-peer learning, local living labs engaging the wider commu-nity, and technical workshops for gap analysis and training. The PrioritEE toolbox contains a technology analytics database, How-to-Briefs and a repository of good practices. It helps authorities to determine the best value for money, and how the available budget can be optimally spent on which technologies. Carmelina Cosmi concludes that the availability of the decision-making support tools in each country's languages was crucial for their success and that having a web application did hugely increase the usefulness of the decision support tools. In Karlovac, the tools were inte-grated in Karlovac's energy recovery strategy; while in Potenza's region Basilicata, they contributed to local action plans on energy efficiency and SECAPs, promoting the transition to a low-carbon regional economy. Regarding stakeholder engagement, she concludes that a multi-disciplinary environment is essential for facing energy issues, and that non-conventional sharing of knowledge can help to generate change, e.g., by making citizens protagonists of the energy transition.

6 Research and Innovation Collaboration for Smart Sustainable Cities: Recommendations for Horizon Europe

The event ended with two interactive sessions, where the main research topics in integrated planning and implementation, including possibilities for funding and finance, were discussed. Session 1, moderated by Judith Borsboom-van Beurden (UERA/EIP-SCC/NTNU), Carmelina Cosmi (CNR-IMAA), and Paolo Nesi (University of Florence), focused specifically on recommendations for Horizon Europe. Researchers from Smarter Together SCC-01 project, Urban Europe Research Alliance, Tuscan Organization of Universities and Research for Europe, PrioritEE project and others discussed current knowledge gaps and barriers to widespread innovation.

The group observed that a proper common definition of an integrated or holistic approach toward smart and energy-efficient cities is still lacking. Usually, the concept emphasizes cross-domain aspects and gravitates toward energy efficiency of districts. Participants mentioned several other aspects which should be part of the concept, in particular (1) the role of ICT and big data as enabler, e.g., through modeling of spatial objects and functionality; (2) financial and governance aspects; (3) the accumulated performance of the entire territory (e.g., highly energy-efficient new buildings compensating for less efficient old buildings in a district); (4) the role of mobility and transport; (5) urban planning as main integrative instrument for transforming specific parts of the city toward specific end goals; (6) inclusiveness of citizens; and (7) the integration of both energy aspects and the human factor— between functional and social structures—and how they contribute to the overall urban quality and livability of a neighborhood.

Several suggestions were made for addressing current knowledge gaps and barriers: (1) setting up quick projects with a time horizon of three to five years reflecting different contexts, aiming to bring down CO_2 emissions and energy consumption; (2) develop better data-driven, evidence-based, citizen-centric proce-dures for urban planning, e.g., with the help of smart city dashboards for what-if questions; (3) fostering collaboration with national energy agencies; (4) making clear which technologies are applicable, which stakeholders should be engaged, which dilemmas can be expected; (5) investigating how to make smart city plans ready for procurement; and (6) researching the role of flexibility in energy systems and demand.

7 Replication in Europe: How to Build New Projects for Positive Energy Districts

The second interactive session discussed how to build new projects for PEDs through replication. It was moderated by Simona Costa and Susanne Meyer and focused on upcoming calls in Horizon Europe, JPI Urban Europe and ERDF, which will offer new possibilities for replication of successfully implemented smart city solutions across Europe. What will it take to build new projects and capacity for Smart Cities and Positive Energy Districts and learn from the experiences in other projects?

The group stated that a Positive Energy District should not only deal with energy aspects but should also address other aspects of the inhabitants' lives. Starting from a PED lab, first of all by choosing the community focus and the aspects in which the PED lab could increase the quality of life for that particular community. For example, it can be more focused both on inclusiveness and on social aspects whenever there is a particular community that requires it, or more related to air quality in a particular area with much traffic and pollution. So, first it is important to decide the community's priorities, then what to do in order to improve the quality of life. The group agreed that the PED lab should identify the research aspects of the priorities, because it requires a great deal of monitoring and background analysis. Furthermore, this could be an interesting starting point for interacting with university campuses and university areas.

8 Conclusions and Next Steps

It can be concluded that Europe provides many excellent examples of integrated planning and implementation of smart city and energy efficiency projects, and a range of useful concepts and tools have been developed supporting such a holistic approach. While all approaches and accompanying tools stress the paramount importance of cross-domain working, multidisciplinarity and profound stakeholder engagement for making districts or cities effectively smarter and more energy-efficient, a commonly accepted definition of integrated planning and implementation is still lacking and some key aspects of smart and energy-efficient cities are not yet incorporated, such as the eventual contribution to local quality of life, inclusiveness and potential replication, or urban data as an enabler. The development of such a comprehensive definition could facilitate the positioning and possibly even integration of different concepts and tools in future, enlarging their joint applicability for users as city administrations. To better support local governments, it would also help to make these concepts and tools better accessible for time-pressured city administrations, adapt them toward specific urban situations and contexts, and translate them into specific local processes and procedures.

Valuable suggestions have also been made for how research could address persistent knowledge and innovation gaps, to name but a few: making clear which technologies are when applicable, which stakeholders should be engaged then, and which dilemmas can be expected; or: investigating how to make smart city plans ready for procurement.

The event also made clear that besides the continuation of smart city and urban energy efficiency projects in Horizon Europe program and upcoming Green Deal, ERDF funding will be more and more attractive for achieving local innovation, the more because in the next programming period 2021–2027 synergies between ten European direct funding programs (including Horizon Europe, Digital Europe and Invest EU and ERDF) will be possible to promote the entire chain from low to high Technology Readiness Level. This will stimulate not only SME participation but also provide opportunities for researchers and practitioners. Meanwhile, the Mission on Climate-Neutral and Smart Cities in Horizon Europe is intending to ensure the commitment of 100 cities through city contracts, what would mobilize citizens, policies and actors well beyond research and innovation. As such, in principle partnerships may contribute to the achievement of this mission.

References

+CityxChange (2020) https://cityxchange.eu. Accessed 8 June 2020

Bahr V (ed) (2014) Energy solutions for smart cities and communities–lessons learnt from the 58 pilot cities of the CONCERTO Initiative. European Commission-DG Energy, Stuttgart

Bellini P, Cenni D, Marazinni M, Mitolo N, Nesi P, Paolucci M (2018) Smart city control room dashboards: exploiting big data infrastructure. https://doi.org/10.18293/DMSVIVA2018-020

Bisello A, Grilli G, Balest J, Stellin G, Ciolli M (2017) Co-benefits of smart and sustainable energy district projects: an overview of economic assessment methodologies. Green Energy Technol 11(1):127–164. https://doi.org/10.1007/978-3-319-44899-2_9

Bisello A, Vettorato D (2018) Multiple benefits of smart urban energy transition. In: Droege P (ed). Urban energy transition–renewable strategies for cities and regions, 2nd edn. Elsevier, pp 467–490

Bisello, A. (2020). Assessing multiple benefits of housing regeneration and smart city development: the european project SINFONIA. Sustainability, 12, 8038. https://doi.org/10.3390/su12198038

Borsboom-van Beurden J, Kallaos J, Gindroz B, Costa S, Riegler J (2019) *Smart city guidance package. a roadmap for integrated planning and implementation of smart city projects.* Norwegian University of Science and Technology/European Innovation Partnership on Smart Cities and Communities, Action Cluster Integrated Planning, Policy and Regulation, Brussels

Dijk M, De Kraker J, Hommels A (2018) Anticipating constraints on upscaling from urban innovation experiments. Sustainability 10(8), [2796]

Dinges M, Borsboom J, Gualdi M, Haindlmaier G, Heinonen S (2020) Foresight on demand: climate-neutral and smart cities. Services to support the mission board "climate-neutral and smart cities" under the framework contract 2018/RTD/A2/PP-07001–2018-LOT1. Austrian Institute of Technology, Vienna

DoGA (2019) Nasjonalt veikart for smarte og bærekraftige byer og lokalsamfunn-En guide for kommuner og fylkeskommuner-utarbeidet av Design og arkitektur Norge, Smartbyene og Nordic Edge i samarbeid med utvalgte aktører. DoGA, Oslo

EIP-SCC (2013) Strategic implementation plan, draft 8 October 2013. European Innovation Partnership on Smart Cities and Communities, Brussels

Gollner C, Hinterberger R, Bossi S, Theierling S, Noll M, Meyer S, Schwarz HG (2020) Europe towards positive energy districts-first update. A compilation of projects towards sustainable urbanization and the energy transition. JPI Urban Europe/Austrian Research Promotion Agency (FFG), Vienna

IEA-EBC (2017) Implementation of energy strategies in communities (Annex 63). Volume 2: development of strategic measures energy in buildings and communities programme. Salzburg Institute for Regional Planning and Housing, Salzburg

JPI Urban Europe (2019) Strategic research and innovation agenda 2.0. JPI Urban Europe, Vienna

JPI Urban Europe (2020) Europe towards positive energy districts. A compilation of projects towards sustainable urbanization and the energy transition. First update February 2020. JPI Urban Europe, Vienna

Mission Board on Climate-Neutral and Smart Cities (2020) 100 climate-neutral cities by 2030-by and for the citizens. Interim report of the mission board for climate-neutral and smart cities. European Commission, Brussels

PrioritEE (2020) https://prioritee.interreg-med.eu. Accessed 10 June 2020

REPLICATE (2020) https://replicate-project.eu/. Accessed 10 June 2020

SET-Plan Action 3.2 (2018) SET-Plan Action no. 3.2 Implementation Plan. Europe to become a global role model in integrated, innovative solutions for the planning, deployment, and replication of Positive Energy Districts. JPI Urban Europe, Vienna

Smarter Together (2020) https://www.smarter-together.eu/. Accessed 5 June 2020

Vandevyvere, H. (2018). Why may replication (not) be happening? Recommendations on EU R&I and Regulatory policies. D32.3A. EU Smart Cities Information System. Smart City Information System, Brussels

Urban (Big) Data: Challenges for Information Retrieval and Knowledge Discovery

Transposing Integrated Data-Driven Urban Planning from Theory to Practice: Guidelines for Smart and Sustainable Cities

Viktor Bukovszki, Ahmed Khoja, Natalie Essig, Åsa Nilsson, and András Reith

Abstract The smart city (SC) discourse is a dynamic academic field pushed by a drive for urban digitalization and the convergence of urban planning and management with advances in information-communication technologies (ICT). A standardization of the professional profile "smart city expert" would benefit both the academic discourse, by positioning the field and outlining a competence set for education and cities, by characterizing the disruption of the SC paradigm to their current practices. One approach to define such a profile is through a literature review; however, a bottom-up approach deriving standards from practice is equally important. This study presents a step-by-step methodology for smart urban planning, defining accompanying activities to a conventional planning process. These activities use ICT to accessorize planning with improving the efficiency of evidence generation, increasing the range of actions based on evidence, and the transferability of knowledge expanding participation scope. The planning process is validated in four case studies on planning challenges with complex, multi-sectoral implications. The practice-based approach demonstrated the impact of the chosen ICT-accessorized planning process, and the feedbacks outlined the necessity of a professional with a broad knowledge of urban planning that is supplemented by specific data science skills and an understanding of the ICT market.

Keywords Smart urban planning · Smart city · ICT · Practice-based approach · Decision-support system

V. Bukovszki (✉) · A. Reith
ABUD Kft, Váci Street 99, 1139 Budapest, Hungary
e-mail: bukovszki.viktor@abud.hu

A. Reith
e-mail: reith.andras@abud.hu

A. Khoja · N. Essig
ESSIGPLAN GmbH, Untere Sandstr. 4, 96049 Bamberg, Germany

Å. Nilsson
IVL Svenska Miljöinstitutet, Valhallavägen 81, 114 27 Stockholm, Sweden

1 Introduction

The smart city (SC) paradigm—an emergent, procedural approach to urban planning and management (UPM) that, exploits technological advancements—is a common and popular goal for European cities (European Parliament 2014), with the promise of transforming city management processes to better respond to the volatile problem space of the twenty-first century (United Nations 2009; US Department of Homeland Security 2015). As multiple fields claim competence in smart cities, which has been, historically, partially driven by market actors (Yigitcanlar et al. 2019), the smart city discourse, and even the definition of smart city is not yet standardized (Chourabi et al. 2012). The conventional approach to outline the research field of smart cities is through a systematic review of related research. Indeed, there are multiple reviews conceptualizing smart cities from both technological and social science perspectives (e.g. Ruhlandt 2018; Israilidis 2019).

However, there is an emergent outline for a currently ill-defined professional profile from the evident gap between an ICT solution supply and urban administrators on the demand side (Klopp and Pettretta 2016; Poole 2014; Robinson 2015; Barns 2018). This gap has prompted a market niche for smart city consultancy that seeks and identifies entry points for ICT tools in UPM, translates urban requirements into developer-friendly specifications and vice versa translates ICT affordances as opportunities for cities. Appropriately specifying the competencies required for a successful, full-stack consultancy would provide a bottom-up approach to standardize the discourse on smart cities, at the very least by delineating the core knowledge that is transferable to practice. This bottom-up approach does not replace literature reviews but is supplementary and potentially provides validation from a different perspective. Moreover, a professional standard for SC consulting also defines and streamlines the disruption that the SC paradigm implies for urban administrations.

The goal of this study is to contribute to specifying the activities and necessary competences for a professional standard of SC consultants and to position both the SC discourse in academia and the SC practice in UPM. This goal is defined as three research questions: (1) Where in the planning process are practical and impactful entry points for ICT?; (2) How do the entry points disrupt the planning process?; and (3) Which skills are necessary to deliver these entry points?

2 Methodology

A practice-based approach is taken, generating inductive answers by designing and deploying a new practice in case studies (Candy 2006). The subject of investigations is a new planning process fitting the SC paradigm, leveraging ICT advancements—henceforth called the smart urban planning process. This process is deployed in real-life case studies, emulating the provision of consulting services to cities. The research question is answered by: (1) specifying the range and deployment of ICT

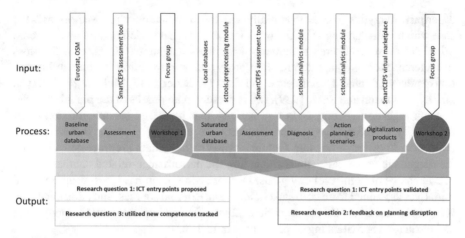

Fig. 1 Overview of the study process

tools; (2) documenting the skills and competences beyond that of a conventional urban planner; and (3) documenting necessary changes in institutional design, public servant routines and procedures, newly acquired resources and previously missing capacities (Fig. 1).

2.1 Smart Urban Planning Process

A planning cycle is defined as a linear progression through four generic phases. In preparation, the planning team is assembled, the scope of consulting and the high-level targets/symptoms of the city are defined, and the depth of stakeholder involvement is agreed upon. In the diagnosis phase, a comprehensive knowledge base is built and progressively scaled up to investigate areas of urban performance and return a set of actionable targets. The action planning phase consists of the articulation and selection of interventions. Finally, the implementation and follow-up phase monitors realized interventions against targets. While progression through the phases is linear, the key phases themselves (diagnosis and action planning) are iterative.

In this scope, accessory activities are recurrent supporting activities with different goals and depths in the various phases. Accessories make the planning cycle "smart" by incorporating in decision-support systems and multi-stakeholder collaboration to the process. Four are introduced: descriptive analytics, data ingestion, validation and analytics for decision support.

Descriptive analytics (DA) is the process of communicating a narrative with data (Hagerty 2016). It entails organization, classification, structuring of datasets, production of queries, visualizations and providing their interpretation to describe what currently "is", according to the underlying data. The importance and complexity of descriptive analytics grow exponentially with the introduction of more actors, as is the case with urban planning. The DA workflow can vary and is left open; however, core principles must be adhered to audience specificity and simplicity; information-loss management across data; interpretation, story and representation; accountability and transparency of context, confidence and risks (Kienzler 1997; Mulrow 2012; Bryan 2009; Brockmann and Rook 1989). DA is present in all phases of the planning cycle.

Data ingestion encompasses any activity that increases the magnitude and/or complexity of information channelled into the planning process, including data collection, knowledge transfer, feature engineering and data preprocessing. During the diagnostic phase, this entails expanding the performance-related database along causal chains of high-level city goals. During action planning, ideas, proposals and initiatives are mined from both stakeholders and external sources, and additional data is ingested that is needed for scenario analysis. During implementation, monitoring and evaluation activities ingest project-specific data.

Validation refers to the involvement of stakeholders, beyond sources of information, to validate both decisions and interpretations of evidence behind decisions. In the planning cycle, this includes high-level goals, areas of intervention, interventions to implement and assumptions made by analytics. It is not the scope of this study to review methodologies.

Decision-support analytics refer to diagnostic analytics, the investigation of causal chains that lead to the observed situation and predictive analytics and a forecast of impacts of certain events (Hagerty 2016). During diagnosis, it provides a guide to scale the urban database to specific needs. During action planning, it supports the selection of competing scenarios and interventions. The developed analytic workflow on preprocessed data is as follows: (1) fitting parameters for the underlying model; (2) compiling the model for a specific analytic task; and (3) interpretation of results. A standard model architecture is provided (see Sect. 2.3) as planners are expected to be appliers and not developers of analytic pipelines.

2.2 Case Study Design

The case study methodology is an execution of the most important steps in the action plan—as described in Sect. 2.1. In a regular use-case, the planning cycle is expected to take several months and rely on intensive back-and-forth interactions with the municipality. To decrease the burden on partner cities, a simplified "virtual" planning cycle of the platform prototype was defined without sacrificing core innovative features of both the planning methodology and the platform, decreasing the points of contact to cities to three (Table 1).

Four medium-sized European cities (pop. 50,000–250,000) were recruited (or selected) as case studies: Bamberg, Veszprém, Aarhus and Debrecen. The criteria for involvement were: meeting the population threshold; having at least two cases closer to the lower and two to the higher threshold; and having two cases where the focus group is the municipality and two where it is a consulting firm working with

Table 1 Case study timeline

Event	Content
Introduction workshop	Presentation of the software, functionalities, proposed consulting services (including the planning cycle), planning problem definition and feedback on consulting services
Data collecting period	Data collection of different aspects of the city for the analysis of performance indicators, preparation of various reports
Closing workshop	Presentation and feedback on the results

Table 2 Case-study cities and their challenges

Veszprém	The city identified a shortage of parking spaces in the inner city and is considering alternative solutions to address it: increasing parking capacity or increasing public transport coverage
Debrecen	The city is interested in how smart cities identify a problem, diagnoses what is behind it and how a solution is offered and supported by evidence. The planning problem, in this case, was specified during the analysis from the bottom-up
Bamberg	The city is interested in the dynamics of the urban network of its region, and its role within this network
Aarhus	The city requires evidence to support investments in green infrastructure

the municipality. The latter was the case for Aarhus and Debrecen. In all cities, the workshops were held in a focus-group format, with the planning cycle played out in the first half, followed by a guided discussion and feedback session focusing on the research questions. The municipal focus group was a mix of decision-makers and technical staff, while the consultant focus groups were purely technical from the field of urban planning. Group sizes varied between four and six.

In the period between the introduction and closing workshops, the consortium delivered various consulting services, which are adaptations of the general planning-process steps. The specific range of services was tailored to the needs of the partner city that was based on the specific planning problem defined during the introductory workshop (Table 2).

The services were then derived from a general portfolio based on the action planning methodology (see Table 3).

2.3 Deployed Decision-Support Tools

The prototype SmartCEPS[1] desktop application was used to collect and preprocess urban data, generate KPIs and conduct quick assessments. The SmartCEPS virtual marketplace was consulted for relevant digitalization solutions. Selected third-party

---

[1] smartceps.com.

Table 3 Range of consulting services tested in the case studies

Service (case-study adapted planning process)	Bamberg	Debrecen	Veszprém	Aarhus
Service: urban database saturation (preparation phase) Data collection for baseline dataset Missing data management	x	x	x	x
Service: comparative quick assessment (preparation phase) Quick assessment: interpretation of core key performance indicators (KPI) Quick assessment: inspecting urban network	x	x	x	–
Service: diagnostic report (diagnostic phase) Priorities extraction for diagnosis Diagnostic analysis: database scaling	–	x	x	x
Service: digitalization market guidance (implementation phase) Collecting relevant solutions, products Solution specification	–	x	x	–
Service: action plan (action-planning phase) Impact assessment of interventions Predictive analytics: multi-criteria decision analysis of intervention scenarios	–	x	x	x

databases were scraped using their dedicated APIs (Eurostat, Open Street Map), and the remaining data was filled in by hand from other third-party sources (TEIR, KSH, Atlas Bayern, Genesis, Open Data DK), or imputed. The imputation strategy was: (1) last-observation-carried-forward (LOCF) for incomplete time series with entries not older than three years; (2) k-nearest-neighbor imputation, where the "k" dimensions were the city KPIs and contextual metadata, for data that was at least 50% complete; and (3) substitution by regional-scale data in the remaining cases. Diagnostic and predictive analytics were conducted using the sctools.analytics private Python module developed specifically to analyze SmartCEPS causal graphs (see the documentation of deployed algorithms in appendix). For a description of the causal model of decision support, see Bukovszki et al. (2019). In the current analytic tasks, an unweighted graph, with a global attenuation factor of 0.5 was used. For descriptive analytics, the built-in representational tools of the SmartCEPS application were used, while the built-in survey tool for the same software was used for validation. This was restricted to validate the assumptions determining impacts of interventions on KPI. Finally, the SmartCEPS virtual marketplace of urban digitalization solutions was used to demonstrate market support.

3 Results

The outcome of the research is an account for the changed planning process due to implementing accessory activities and technologies as listed. As the answer to the research question based on this account is an interpretation of said process—both by the researchers and by focus groups—this section is restricted to describing the planning process, and reflection on the role and value of ICT: the disruption in current practices is presented in the discussion.

3.1 Database Saturation

An average 44% data saturation for core SmartCEPS KPI inputs was achieved through scraping, 36% for Hungarian cities, 39% for Aalborg and 46% for German cities. This saturation level includes LOCF values. Only around 20% of total data was readily available and up to date. Fifty percent of the remaining inputs were imputable, while the rest (14 inputs) were filled in by hand, or by regional data. For comparative analysis, the cities of Bayreuth, Landshut, Miskolc, Szeged, Pécs, Székesfehérvár and Kecskemét were added to the database. The cities were selected by the clients but have a similar population and legal status.

3.2 Assessment

Quick and comparative assessment of cities was shown on standardized dashboards. On a thematic level, for Debrecen, environment and economy are the strongest performing areas, while built environment, governance and inhabitants are the weakest. Aarhus has the highest overall smartness index, and is generally well rounded, with room for improvement in mobility. Bamberg would benefit from focusing on economic and governance indicators while protecting their strong infrastructural and environmental performance. Finally, Veszprém has been found to be a country leader in built environment with a strong showing on economic measures, with the highest improvement potential in mobility and infrastructure. Benchmarks were customized to represent top performers of the country, and the highest difference in KPI scores to benchmark identified recommended high-level goals. These KPIs are: traffic safety for Veszprém; building vacancy for Bamberg; waste production for Aarhus; and traffic safety for Debrecen.

Table 4 Recommended areas of intervention and/or recommended database scaling directions. Results of the diagnosis

KPI	Debrecen	Veszprém	Aarhus
Urban compactness index	x	x	–
Greenhouse gas emissions	–	–	x
New business density	–	x	x
Digitalization maturity of mobility	x	–	–
Share of the informal sector	–	x	–
Green-space density	x	–	x
Transport network performance	x	–	–
Availability of energy-efficiency financing	–	–	x
Land-use diversity	x	–	x
Annual GDP growth	–	x	–
Government-mediated investment	–	x	–

3.3 Diagnosis

Based on these goals, diagnostic analytics was conducted for Aarhus, Debrecen and Veszprém to provide recommendations for database expansion. The focus groups validated the goals provided by the system in a survey defining areas to protect and areas to target for development. The automatically generated goals were included in all cases. The diagnostic algorithm returned the top 5 KPI by influence on the targeted goals. The targeted goals, based on automatically, and focus group generated targets: (1) all KPI-themed "mobility"; (2) all KPI-tagged "public services", (3) all KPI-tagged "investment attractiveness"; (4) the single KPI for greenhouse gas (GHG) emissions; and (5) all KPI-tagged "sustainable city". Table 4 shows recommended KPI ($n = 5$) to add to the urban data model for the three cities and also as recommended areas of intervention.

3.4 Digitalization Market Guidance

In the cases of Veszprém and Debrecen, digitalization solutions were screened for their respective areas of intervention. Based on a target budget set by the focus group, a real-time passenger information display was shown to Veszprém, and a smartphone app for crowdsourcing tax fraud protection for Debrecen. Both solutions were automatically listed by price and relevance to the KPIs "digitalization maturity of mobility" for the former, and "share of the informal sector" for the latter.

3.5 Action Planning

The simplified service involved generating alternative intervention scenarios for a single planning issue identified earlier and using the predictive analytic engine to provide objective decision analysis among them. Three scenarios were defined to solve inner-city parking shortage in Veszprém, and three to increase the penetration of e-government in Debrecen (Table 5). The direct impacts of each e-service scenario were set to uniform + 10-point change on their respective KPIs, as the scenarios were specifically chosen to influence different KPIs. For parking, all scenarios influence modal share: capacity increase promotes car usage, bus densification increases the comfort of bus journeys and bus-network expansion decreases distances using public transportation. The survey function of the app was used to probe local population on the likelihood of changing modes of transport, given each scenario. Table 5 summarizes all scenarios with their direct impacts.

A vector of changes in KPI scores due to direct impact of interventions was fed into the predictor engine to calculate indirect impacts. The prediction on e-services showed that both subsidizing broadband Internet and a digital skills' program result in a higher net increase in e-government penetration among citizens than promoting e-services directly—provided that each intervention results in a 10-point direct impact on their target KPI. Also, broadband Internet access is a precondition for both digital literacy and e-government penetration, meaning that, up to a certain point of diminishing returns, it yields impacts on all three areas of performance. Both Internet access and digital literacy also increase the chance of citizens utilizing e-business applications.

The parking scenarios were based on Veszprém's score. Apart from the negative impact on modal share itself, increasing the parking capacity is also shown to cascade down to a decline in air quality, traffic safety and eventually in life expectancy. The public transportation development options have opposite effects. Based on the results of the survey, expanding the network would yield more transfers to public transport trips than densifying the existing one. The team could recommend to the city to invest in new bus lines around the city, based on the potential benefits alone—with the caveat that costs were not included in the analysis.

Table 5 Scenarios defined for action planning

Scenario	Description	Impact	KPI affected
P1	Increasing parking capacity	17% public transport users switch	Modal share
P2	Densification of buses	12% car users switch	Modal share
P3	Expanding bus network	19% car users switch	Modal share
E1	Digital skills' education	10 pt increase	Digital literacy
E2	Broadband Internet subsidy	10 pt increase	Connected society
E3	Awareness campaign	10 pt increase	E-services for citizens

4 Discussion

4.1 The Role of Accessories and ICT Tools

One part of the research question focuses on entry points and added value of computational tools exploiting the digitalization of cities in the smart city paradigm. Their roles are summarized in Table 6.

In general, the efficiency of evidence generation, the range of use-cases where evidence can be applied and the transferability of this knowledge are the main areas where smart urban planning tools proved to be impactful.

Regarding evidence range, the focus group highlighted the previously unavailable option to predict indirect impacts and assess interventions beyond disciplinary boundaries, such as the impacts of addressing parking shortage in different ways for Veszprém. The limiting factor here is the quality of the underlying models. It is a necessary first step to explore urban performance in the context of its interacting subsystems; however, it leads to an entire strand of research to model and parameterize these interactions. It is not evident how likely people utilize e-services when acquiring broadband connection, for example, and how important this factor is in Bamberg, compared to Debrecen.

For evidence-generating efficiency, it must be highlighted how many of the tasks were semi- or fully automated. The data preprocessing pipelines more than doubled database saturation with a couple of clicks in the case studies. The targets interpreted from automatic assessment results generally corresponded to the targets defined in the focus group: mobility was returned as an issue for Veszprém as expected, while environment performed better for Aarhus than anticipated. However, there is much room to improve when it comes to the efficiency of transferring knowledge

Table 6 ICT entry points and their role

Output	Corresponding accessory activities
Data generation	Data ingestion
Data integration and processing	Data ingestion
Data security	Data ingestion
Knowledge generation	Analytics in decision support
Transparency and knowledge accessibility	Descriptive analytics
Capacity and awareness building	Descriptive analytics, validation
Solidarity and improved justice	Validation
Improved collaboration among actors	Descriptive analytics, validation
Substitution, synergies and optimization of tasks	Analytics in decision support

throughout the planning process. The case studies used only public databases and consulted the municipality, while future decision-support systems will have to receive inputs from a plethora of human- and machine-generated sources, such as third-party databases, big data, inputs from citizens, reference studies for urban interventions, product descriptions, etc. Only if these inputs are defined in a shared information architecture will the planning pipeline be efficient and straightforward to use.

Finally, regarding information transferability, the role of descriptive analytics cannot be understated. In the case studies, numerical and graphical representation of scores were powerful tools to show during assessment and for scenario impacts during action planning. However, it is unclear how this translates to a wider range of actors (citizens) who will be expected to both read decision-support systems and write into them. Data visualization and uniform dashboards might not be socially scalable, and transdisciplinary efforts involving social sciences and design sciences will be necessary to build dictionaries translating between the reality and value models of actor types and the models used by decision-support systems.

4.2 Disruption of Innovation: Cities

As the planning process itself was designed to accessorize a conventional planning cycle, little, if any procedural disruption is expected solely from the transition to smart urban planning. The scaling of participation and the scaling of data were identified as the two key drivers of the complexity of planning, and both precipitate to challenges for decision-support system design rather than for institutional design, meaning the burden of change appears to be more on the professionals than municipalities. From the feedback, two key areas require significant change: availability of data as a resource and availability of adequate competences in consultants.

In terms of resources, the focus groups highlighted data scarcity as a limiting factor, which to some extent was demonstrated to be managed. However, while incomplete databases are manageable, they were less frequent than almost completely empty databases during the study. Especially in the case of Hungarian cities, data scarcity is an obstacle, but in all cases, statistical data will always represent a centrally defined demand and not the range of data specific to the needs of each city. This is best illustrated by the fact that Aarhus was recommended to expand their database to solve their planning challenge regarding green infrastructure, and an action plan was not possible with the baseline dataset. Given the variety of potential data sources, both the authors and the focus groups see the role of the public sector as a network-builder, recruiting market actors and citizen science, serving as conduits for many individual data sources and providers. This will increase the significance of data marketplaces and consultants who are able to ingest data in actionable analytic models.

Table 7 Competences defining the professional profile of an SC consultant

Phase	Competences covered during case studies
Preparation	Data visualization, storytelling with data, feature engineering in urbanism, data quality assurance, vertical and horizontal collaboration
Diagnosis	Model building and tuning, associative data structure management, network analysis
Action planning	Data ingestion, interoperability in participatory planning, model building and tuning, data visualization, storytelling with data
Implementation and follow-up	Database scaling, monitoring and evaluation, digitalization market knowledge

4.3 Disruption of Innovation: Professionals

This leads to the skill set needed in consultancies. All focus groups objected to the idea of a smart city expert in-house and expected market actors to provide consulting services. The setting of the case studies, where the data analytic and urban digitalization competences were vested in the researchers, was deemed optimal. This also makes sense from a developer perspective: complex model development and analysis of urban digitalization markets are better met by a community of independent consultants who have access to multiple city partners. This consultancy, given that the services demonstrated in the case studies require a new competence profile, thus raised new demands in education, combining urban-planning knowledge with a deep understanding of decision-support systems, such as data science or its equivalent (Table 7). This is in accordance with the literature (Bibri 2019); however, it helps to focus research on knowledge on the application of data analytic, machine learning and database management methods in urbanism.

5 Conclusion

The study presented a step-by-step methodology for urban planning adopting the principles of smart cities. The new planning process took state-of-the-art decision-support systems leveraging urban data analytics to support evidence-based, integrated, participatory urban planning. Four case studies where different SC consulting services were emulated are presented to demonstrate the scope, value and impact of ICT in planning. It identified the SC consultant as an expert in urban planning, with selected supplementary skills in data science and an up-to-date understanding of the ICT market. This partially corresponds to the scope of SC field defined in the literature and puts an emphasis on the subfield of information technologies focusing on databases and data analytics.

It must be noted that, as all practice-based research, this study does not provide the only answer to the research questions, but one possible answer. In a field that is so close to the market, case studies and responses from the main beneficiaries of the SC transformation are crucial and should shape academic discourse. Further case studies testing variations of the activities and professional profile of SC consulting are necessary, using different systems, adopting different technologies, tweaking the planning process and testing out bolder disruption in urban administration and in citizens with deeper public participation.

Appendix Documentation of Deployed Network Analytic Tools

sctools.analytics.causal_network	
Call signature	analytics.**causal_network**(*self, nodes_csv=None, edges_csv=None*)
Constructs causal network as NetworkX DiGraph from list of nodes and edges. The constructor assigns default weight of 0.5 on all edges.	
Parameters:	**nodes_csv** : *str, path object or file-like object* List of nodes in the data structure of SmartCEPS App. The structure must include an "Id", "label", and "theme" as first three columns, and Boolean attributes for subsequent columns. **edges_csv** : *str, path object or file-like object* List of edges in the data structure of SmartCEPS App. The structure must include "SourceKPIId" and "TargetKPIId" attributes.
Returns:	NetworkX DiGraph object

sctools.analytics.Julia	
Call signature	analytics.**Julia**(*self, graph=None, subgraph=None, cutoff=5*)
Computes the influence of each node on a subgraph of interest. Used to sort nodes by influence on a selected subset of urban performance.	
Parameters:	**graph** : *NetworkX DiGraph object*
	Takes a directed graph, either constructed via causal_network, or by hand using NetworkX library. Must contain "weight" attribute on edges. **subgraph** : *NetworkX subgraph object* Takes a subgraph for the digraph specified under graph parameter. **cutoff** : *int* Limits the computation to a weighted topological distance from the nodes in the subgraph.
Returns:	"influence" : *float* NetworkX DiGraph node attribute quantifying the aggregate node influence on the subgraph.

sctools.analytics.justified_network	
Call signature	analytics.**justified_network**(*self, graph=None, source=None, cutoff=3*)
Construct justified graph from selected node. Justified graph is a subgraph in the causal network which includes nodes reachable from the source node.	

Parameters:	**graph** : *NetworkX DiGraph object*
	    Takes a directed graph, either constructed via causal_network, or by hand
	    using NetworkX library. Must contain "weight" attribute on edges.
	source : *NetworkX node object*
	    Takes a node from the digraph specified under graph parameter.
	cutoff : *int*
	    Limits the computation to a weighted topological distance from the source
	    node.

Returns:	NetworkX subgraph object
	"distance" : *float*
	    NetworkX DiGraph node attribute quantifying the topological distance of the
	    node from the justified graph source node.

sctools.analytics.score_network	
Call signature	analytics.**score_network**(*self, graph=None, city=None, score_csv=None*)
Imports KPI scores to the nodes in the causal network.	

Parameters:	**graph** : *NetworkX DiGraph object*
	    Takes a directed graph, either constructed via causal_network, or by hand
	    using NetworkX library.
	city : *str*
	    Name of the city.
	score_csv : *str, path object or file-like object*
	    List of nodes in the data structure of SmartCEPS App. Only "Id" and "score"
	    attributes.

Returns:	NetworkX DiGraph object "score" : *float* NetworkX DiGraph node attribute quantifying their scores.

sctools.analytics.basic_predictor	
Call signature	analytics.**basic_predictor**(*self, graph=None, score=None, score_diff=None*)
Computes the impact of changes in node scores on all other nodes based on a ripple effect defined by the graph edges. Every "n" difference on a node score ripples to subsequent nodes as "n" times "weight". Predicted scores are limited to a 100-point scale.	

Parameters:	**graph** : *NetworkX DiGraph object*
	    Takes a directed graph, either constructed via causal_network, or by hand
	    using NetworkX library. Must contain "weight" attribute on edges.
	score : *NetworkX node attribute, array-like object of floats*
	    Node scores are floating points on a 100-point scale from 0. Array length must
	    be equal to the quantity of nodes.

	score_diff : *array-like object of floats*
	    Node score differences are floating points from -100 to 100. Array length must
	    be equal to the quantity of nodes.

Returns:	NetworkX DiGraph object "score" : *float* NetworkX DiGraph node attribute quantifying their predicted scores.

References

Albino V, Berardi U, Dangelico RM (2015) Smart cities: definitions, dimensions, performance, and
    initiatives. J Urban Technol 22:3–21

Barns S (2018) Smart cities and urban data platforms: Designing interfaces for smart governance. City Cult Soc 12:5–12

Bibri SE (2019) Big data science and analytics for smart sustainable In: Urbanism: unprecedented paradigmatic shifts and practical advancements. Springer

Brockmann RJ, Rook F (1989) Technical communication and ethics. Society for Technical Communication, Arlington, VA

Bryan J. (2009). Down the slippery slope: ethics and the technical writer as marketer.

Bukovszki V, Apró D, Khoja A, Essig N, Reith A (2019). From assessment to implementation: design considerations for scalable decision-support solutions in sustainable urban development. IOP Conf Ser Earth Environ Sci 290:012112

Candy L (2006). Practice based research: a guide. Creativity and cognition studios report, 1

Chourabi H, Nam T, Walker S, Gil-Garcia JR, Mellouli S, Nahon K et al (2012) Understanding smart cities: An integrative framework. In: Proceedings of the annual Hawaii international conference on system sciences, 2289–2297

European Parliament, Directorate-General for Internal Policies P D A (2014) Mapping smart cities in the EU

Hagerty J (2016) Planning guide for data and analytics

Israilidis J, Odusanya K, Mazhar MU (2019) Exploring knowledge management perspectives in smart city research: a review and future research agenda. Int J Info Manag

Kienzler DS (1997) Visual ethics. J Bus Commun 34:171–187

Klopp JM, Petretta D (2016) Can we actually agree on indicators to measure urban development? Cityscope

Mulrow EJ (2002) The visual display of quantitative information. Technometrics 44:400–400

Robinson R (2015) 6 inconvenient truths about Smart Cities Urban Technol

Ruhlandt RWS (2018) The governance of smart cities: a systematic literature review. Cities 81:1–23

U.S. Department of Homeland Security—National Protection and Programs Directorate (2015) The future of smart cities: cyber-physical infrastructure Risk

United Nations Human Settlements Programme (2009) Planning sustainable cities: global report on human settlements 2009 ed N D Mutizwa-Mangiza. Earthscan/UN-HABITAT, London

City Indicators Visualization and Information System (CIVIS)

Álvaro Samperio-Valdivieso, Paula Hernampérez-Manso,
Francisco Javier Miguel-Herrero, Estefanía Vallejo-Ortega,
and Gema Hernández-Moral

Abstract Sustainable urban transformation is characterized by multiple factors (e.g. technical, socioeconomic, environmental and ethical perspectives) that should address the evaluation of alternative urban planning scenarios or opportunities. This defines a complex decision-making process that includes various stakeholders where several aspects need to be considered simultaneously. In spite of the knowledge and experiences during the recent years, there is a need for methods that lead decision-making processes. In response, a City Evaluation Framework is proposed in the TEC4ENERPLAN project aiming at supporting cities during the preliminary stages of the strategic planning process through the definition of an evaluation framework to assist city managers in transforming cities towards sustainability. This methodology, based on indicators allows assessing specific characteristics of the city, diagnoses challenges or discovers patterns through reliable metrics but also compares various aspects of the cities thanks to the standardization of the indicators and indexes defined. Additionally, there is a need for tools that support the city planners when handling and interpreting data, solving automatically the complex processes of data processing and storage, selection of indicators, and valorization of the city-related values. Inside the TEC4ENERPLAN project, a full web-based application called City Indicators Visualization and Information System (CIVIS) is being developed to fulfil the afore-mentioned goals. Combining programming in Shiny under R language and the usage of storage with PostgreSQL databases, with the integration of advanced statistical techniques, the multidisciplinary data from the cities can be evaluated within an evaluation framework defined by indicators extracted from related ISO standards.

Á. Samperio-Valdivieso (✉) · P. Hernampérez-Manso · F. J. Miguel-Herrero · E. Vallejo-Ortega · G. Hernández-Moral
Fundación CARTIF, Parque Tecnológico de Boecillo, 205, 47151 Boecillo, Valladolid, Spain
e-mail: alvval@cartif.es

P. Hernampérez-Manso
e-mail: pauher@cartif.es

F. J. Miguel-Herrero
e-mail: framig@cartif.es

E. Vallejo-Ortega
e-mail: estval@cartif.es

© The Author(s), under exclusive license to Springer Nature Switzerland AG 2021
A. Bisello et al. (eds.), *Smart and Sustainable Planning for Cities and Regions*,
Green Energy and Technology, https://doi.org/10.1007/978-3-030-57332-4_11

Currently, the tool is in its final stage of development, prior to final validation with data from the cities. The result will be a web-based application tool that will inform users about the city's progress towards sustainable development through the visualization of the city indicators aggregated per city pillar; it will even enable their comparison with the ones belonging to other cities. The application works this way as a valuable advisor during the decision-making process by highlighting in accordance with the resulting indexes which city assets need to be enhanced, which ones comply with the objectives set by the city, and the degree of development of those assets compared with similar municipalities. In a long-term view, the graphics and data would help to trace the evolution of the various different city strategies or politics, exposing clearly if those actions have had a positive or negative impact on the sustainable development. Considering this, the application should perform as a reliable measure for the city policies through time.

Keywords Sustainability · Indicators · Evaluation framework · Web application · Statistics · Data management

1   Introduction

The TEC4ENERPLAN project is an ambitious project aiming at supporting all relevant scales in energy planning through the development of advanced methods and algorithms. These technological advancements will enable effective management of data, automate the generation of simulation models and treat real data as well as map information in a user friendly way, in order to support decision-making at different scales.

The analysis and exploitation of multi-scale and multi-sectoral publicly available data will be paramount, as well as the use of relevant standards to ensure interoperability and replicability (IFC, CityGML, INSPIRE Directive, ISO standards on indicators, ISO standards on EPB). As a result, it is expected to be designed and developed so that a wide range of modules can be connected among each other and can be scalable. They will enable planners to address challenges at building, district, city and regional levels.

These modules will not only target planners at the different scales tackled, but all stakeholders dealing with energy aspects. To this end, special efforts are devoted in the project to the identification of users' needs and the collaboration and feedback of potential end users (collaboration with companies) to ensure well-oriented and usable techniques.

This paper will concentrate on one of the results of the TEC4ENERPLAN project, focusing on a specific set of stakeholders (city planners), that deploys their activities at the city level. The selection of actions to deploy in the city, the generation of plans and the assessment of the impact these measures have had over time are some of the crucial challenges to be overcome by these stakeholders. To support the decision-making process, the definition of indicators to measure these aspects based on robust evaluation frameworks is of the utmost importance.

There are many examples of evaluation frameworks in the literature and previous experiences (REMOURBAN website 2020; MAtchUP website 2020) that can help city planners when identifying the needs and opportunities of cities to achieve their objectives or the desired city model in specifics city fields (e.g. energy, mobility or smartness). However, when the awareness of climate-change issues is the order of the day, it is necessary that the focus of the evaluation framework would be in promoting the sustainable urban development. In response, a City Evaluation Framework and a user-friendly tool (City Indicators Visualization and Information System—CIVIS) are proposed to handle the huge amount of data needed to assess when facing all the city aspects to be considered when facing the sustainable development target.

2 Sustainable Evaluation Framework

Sustainable development is a common and contemporary goal of many urban development policies in various countries and cities. Thus, an evaluation framework based in sustainability is a key point to establish a way to rate the sustainable features of the urban development of cities and can help cities to identify strengths and weaknesses and consequently to set priorities for the selection of actions to be implemented to transform a city into a sustainable city.

Specifically, the city-level evaluation framework defined in TEC4NERPLAN aims to: (i) identify the city needs and challenges that help city managers in the decision-making process when prioritizing a city strategy; (ii) measure different aspects of cities to be aware how close a city is to becoming sustainable and smart; (iii) monitor the progress of the city to show to what extent sustainability goals have been reached; (iv) provide information in a comprehensive way to facilitate the communication of information to stakeholders and city planners; and (v) set a reference methodology for benchmarking and comparison of various aspects of cities.

Based on the outcome of the literature review, the structure of the Evaluation Framework is performed to steer and measure the performance of a city under four main levels:

- **Pillars**: Five pillars or strategic areas of the city were identified for the evaluation of sustainable development, with the aim of making cities and human settlements, through their application fields' improvement, became more inclusive, safe, resilient and sustainable.
- **Application fields**: Each pillar was subdivided into application fields that allow the aggregation of indicators for an individual evaluation.

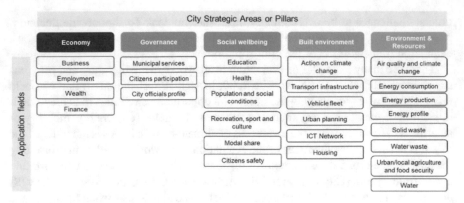

Fig. 1 Evaluation framework' pillars and application fields

- **Indicators**: 262 indicators have been selected after the analysis of previous standards, projects and initiatives devoted to this aim. Eighteen of them aim to characterize the city profile, 110 were identified as primary (core indicators) and 152 as secondary.
- **Variables**: In order to ease the data collection and analysis, each of the selected indicators was subdivided into the variables that enable their calculation.

Regarding the indicators included in the evaluation framework introduced, previous initiatives were reviewed, such as SCIS-EU Smart Cities Information System (EU Smart Cities Information System website, 2020), CITYKeys (European Project CITYKeys website 2015) or the Sustainable Development Goals (SDG) indicators (UN website 2020); however, it is well-known that applying ISO International Standards guarantees the safety, reliability and quality of the product normalized, thanks to the participation of experts worldwide in its drafting. *ISO 37120 Indicators for City Services & Quality of Life* (ISO website 2018) and *ISO 37122 Indicators for Smart Cities* (ISO website, 2019.) were chosen, therefore, as the main source of indicators when selecting the indicators for assess each application field for each pillar of TEC4ENERPLAN evaluation framework (Fig. 1).

3   Web-Based Tool Concept and Workflow

City Indicators Visualization and Information System (CIVIS) is a complete web-based tool designed as a consultancy tool for Sustainable Urban planning, with local authorities as target users. Its objective is to facilitate the decision-making process during the selection of the main fields where the city should focus when defining strategies for Sustainable Urban development and ease the periodical evaluation of targets consecution.

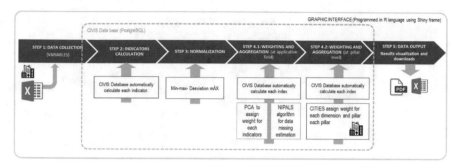

Fig. 2 CIVIS workflow. Stages and outcomes

Nowadays, the best way to reach the largest quantity of users and eliminate the necessity of making different versions for different devices is the deployment of a web application. It does not need a previous installation, and it can be accessed wherever a connection to Internet is available. Moreover, the updates to the tool are applied automatically so it eliminates the duty of downloading patches or reinstalling the software. For these reasons, it was decided to provide a web-based tool developed in R using the Shiny package that allows visualization and interaction of R elements in a website environment. Shiny is compatible with Google Chrome, Mozilla Firefox, Safari (available through Apple Software Update) and Internet Explorer, but the main visual tests have been done in Chrome and Firefox.

The general CIVIS workflow concept can be summarized in the following scheme (Fig. 2):

The process starts by collecting and uploading in the database the data corresponding to variables **(Step 1: Data Input)**. This can be performed either through an Excel template (.xlsx format file) filled by cities that can be uploaded in the website, or through direct data insertion since the application contains some fields that can be completed online. After each data imputation, indicators are calculated deploying the corresponding formulas **(Step 2: Indicators calculation)**, and then, the indicators are normalized, to make them comparable and to be able to calculate composite indexes **(Step 3: Normalization)**.

Next, the webpage loads data from normalized indicators and performs the last step in the process of construction of composite indexes, i.e. the aggregation stage, assigning different weights to every normalized indicator for each pillar **(Step 4: Weighting and Aggregation)**.

Two different levels of groupings were defined. Within each application field, particular weights are given to indicators included in that category which are aggregated, resulting in a partial index for every application field **(Step 4.1: W & A at the Application Field Level)**. The same process is carried out for each application field, obtaining intermediate indexes for every pillar as well, by aggregating indexes for each application field included in each pillar **(Step 4.2: W & A at Application at Pillar Level)**. Finally, these intermediate indexes for each pillar are again weighted

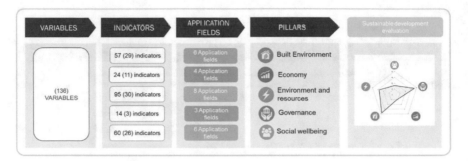

Fig. 3 Methodology for building indexes

to obtain the final evaluation of the sustainable development. This process is shown in the following Fig. 3.

Once the aggregation step is complete, the webpage saves composite indexes in the database, finishing the data management process. All of this process is done at once right after each data imputation to avoid calculation delays when the users navigate through the application.

CIVIS offers many different visualization possibilities **(Step 5: Data Output)**. Within the City Analysis tab, the tool displays the full evaluation of a particular city at the three levels of aggregation, letting the user choose whether to use weights provided by the city or not at the last aggregation step, and which list of indicators to use (between indicators from ISO standards or an extended list with indicators added from other kinds of sources analyzed). The user can also download the evaluation report (.pdf format) or even the indicators calculated from the variables provided (.xlsx format file).

The following Fig. 4 shows an example of the visualization of the evaluation of the indicators within "Housing" application field, which is part of "Built Environment" pillar.

Moreover, within the City Comparison tab, the tool displays the full evaluation of various cities, which can be chosen from a list filtered by GDP, population or climate, to compare results. At this tab, the user can choose which list of indicators to use and can also download the evaluation report or data. The weights applied to the KPIs are the same for all cities considering all aggregation steps (using PCA as it is explained in the next subsection), to render the results comparable. Figure 5 provides an example of the visualization of the comparison between two cities at the pillar level and at the application field level inside the "Built Environment" pillar (Fig. 5).

The three main aspects of the of the methodology for building indexes are described next in detail, summarizing not only the statistical methods applied during the data management, but also the data base needed to support the whole analysis performance.

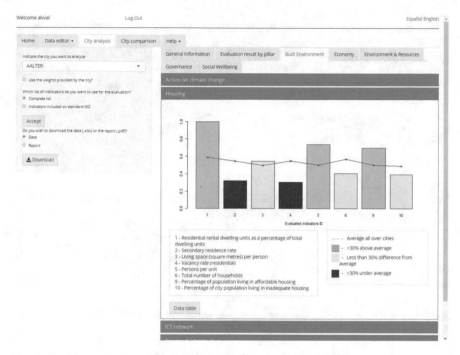

Fig. 4 Graphic output showing indicators inside one application field. (The figures were created from testing data in order to avoid legal issues)

4 Data Management

The evaluation of the sustainability of cities through the previously described framework involves the application of statistical and multivariate analysis techniques to a large amount of data, which allows CIVIS tool to obtain a very accurate picture of the most important aspects to be treated in cities on the road towards sustainable development.

In the following sections, the mathematical techniques used for this purpose are detailed, following the reference of the European Commission's guide for the calculation of composite indexes (Langedijk 2017).

4.1 Normalization and Outliers' Treatment

Once the indicators have been calculated from the corresponding formulas, it is crucial to detect and treat its outliers, that is, values of an indicator that are far from the values of that indicator for most cities; this is because they may be the consequence of measurements or unit errors of the data provided at variable level by the cities and are values that can have a serious impact in statistical analyses.

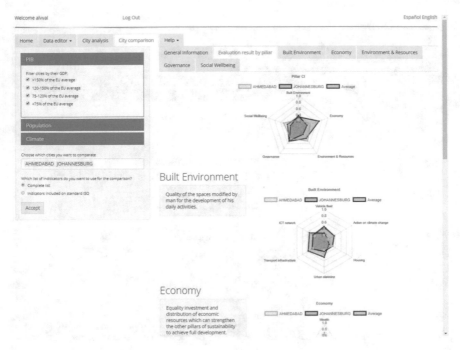

Fig. 5 Comparison graphics between city indicators. *(Ahmedabad and Johannesburg in orange and blue colours, respectively)* (The figures use testing data for privacy legal reasons)

The outliers' detection uses the modified z-scores method suggested in Iglewicz and Hoaglin (1993) which calculates superior (resp., inferior) threshold values for each indicator equal to its median across all cities plus (resp., minus) 3/0.6745 times its median absolute deviation (MAD) and considers as an outlier any value greater (resp., lesser) than this threshold. The outliers' treatment consists of setting values above the superior (resp., below inferior) threshold to it. This method has the advantage over the z-scores method (Langedijk 2017) based on the mean and standard deviation (SD) that the median and MAD are robust, not like the mean and SD, so the modified method is less affected by outliers than the non-modified one, outperforming it, especially in case there would be few cities.

After that, it is necessary to normalize the data of each indicator, in order to make different indicators comparable and to be able to calculate composite indexes or indexes afterwards. The normalization consists on the following linear map onto the [0, 1] interval using the threshold values calculated previously to detect outliers:

$$\text{normalized indicator} = \frac{\text{indicator} - \text{inferior threshold}}{\text{superior threshold} - \text{inferior threshold}}$$

4.2 Aggregation and Data Estimation

One of the goals of the defined evaluation framework is to allow the results of several cities to be comparable to each other, so it is necessary that the data imputation would be completed (by means of having all the indicators calculated, and if not, provide an estimation value for those not provided by the city), to be able to accomplish the three levels of aggregation for at least the priority indicators of the defined evaluation framework. This step is therefore crucial to be able to obtain composite indexes at both the "application field" and "pillar" levels, as well as to make them comparable among cities.

At this point, the European Commission guide (Langedijk 2017) proposes, firstly, to face the problem of missing data through applying an estimation processes to have a common group of indicators, and, secondly, to allocate weights to the data set once it is complete (estimated and originals) to calculate indexes as a weighted average. However, the main novelty of the CIVIS methodology is the simultaneous treatment of both issues by applying principal component analysis (PCA).

PCA (Jolliffe 2002) is a widely used procedure for estimating the dimension of datasets and reducing them. In each application field formed by p indicators, each city is considered as a point in the real p-dimensional space ($\mathbb{R}^p$). Therefore, there is a set of points in p dimensions, and the goal is to find the best possible summary of the points in one dimension, minimizing the loss of information in the process of passing from the values of p indicators to the value of a single field of application composite indicator.

PCA calculates this summary from the eigenvector[1] of the data-correlation matrix corresponding to its highest eigenvalue[2], and consequently, it uses its components as weights to calculate composite indexes or indexes as a weighted average.

The correlation among indicators in the same application field it is required to be positive, as recommended by Langedijk (2017). This condition guarantees that the eigenvector corresponding to the highest eigenvalue can be chosen uniquely with all non-negative components and sum 1, by Perron–Frobenius Theorem (Cheng et al. 2012.). This fact justifies using its components as weights.

In case there is no missing data, CIVIS computes PCA utilizing fast standard R's singular value decomposition (SVD) algorithm (R Documentation website 2020). In the opposite case, CIVIS uses NIPALS algorithm (Universidad Nacional de Colombia website 2016), which can handle missing data. NIPALS method consists of applying the Power Method to the correlation matrix, but without building it (thus avoiding increasing greatly the amount of missing data). The algorithm computes the eigenvalues one by one skipping missing values and leads to the matrix decomposition onto its projections on the eigenvectors, which can be used to estimate missing data.

---

[1] An eigenvector is a vector whose result of multiplying it by the matrix provides another vector that is proportional to it.

[2] Constant of proportionality associated with the eigenvector.

That way, one manage to determine the linear structure of the data despite missing data, and an estimation of it obtained, based on the linear structure itself, solving the problem of obtaining weights and the estimation of data simultaneously, instead of estimating data with a procedure that is not consistent with the linear structure of the data (which can change it significantly), and then determining it to obtain the aggregation weights.

Prior to the PCA-based aggregation, CIVIS submits data of each application field to a filtering process, removing data from some indicators and cities, leading to a situation in which there is a minimum of 60% of data available for each city and a minimum of 20% of data available for each indicator. It also submits data to a Kaiser–Meyer–Olkin (KMO) test (Langedijk 2017) that determines if the correlations between indicators are high and if it is appropriate to obtain an indicator composed of them by PCA. Those filtering percentages have been established as a consequence of an accuracy study of data estimation.

Once the aggregation step from indicator to application-field level has been complete, the user will have an evaluation result from 0 to 1 per application field, being able to easily identify which application field should be improved. Just after the first aggregation has ended, CIVIS lets each user choose between two options for the last aggregation step (up to pillar level). The first option is to use the same methodology employed in the previous step (PCA), and the second one is to use weights provided by each city, based on city's experts' knowledge. In both cases, a previous filtering and correlation study within each pillar is carried out, with the same criteria used as the previous aggregation step.

5   Database, Design and Structure

The database has been formulated in PostgreSQL as long as it was the technology that fit best for our case. To organize the information as faithfully as the real hierarchy of data, the database was organized into three main blocks (Fig. 6):

The first block (**Block 1: Data hierarchy**) contains the main tables of the system: the table for the definitions of the indicators paired with the table with the different indicators of the cities, and then equivalent pairs of tables for the application fields and pillars. The second block (**Block 2: Users and Security**) includes all the users, the types of users, logs to the system and elements belonging to the cities, such as images and ancillary data. It must be noted that the dark green arrows represent the processes to calculate indicators from the variables. Finally, the third block is a place for all the remaining elements (**Block 3: Inputs and Auxiliary Elements**).

All these blocks are interconnected, and the result is a complete connection between all the main tables with a certain degree of modularity and a structure

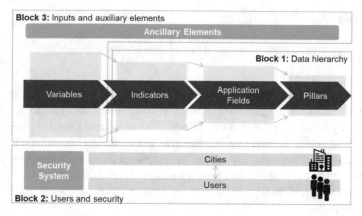

Fig. 6 Basic schema of relations for the database

that can be easily followed in case that something had to be changed, or new tables were needed.

5.1 Security and User Management

Nowadays, there is much concern about data protection issues, and even more so if we are talking about sensitive data, as is the case of some of the ones collected and treated in TEC4ENERPLAN methods and tools. In that case, special attention was given to cope with security issues regarding the application and the data stored in the database. Different user profiles, with authorization levels for certain functions and data values, have been designed: Administrators, Cities and Users.

The management of the users is done through a local application done in Java. It creates new users, edits existing ones and deletes users if necessary. It is a way to automatize the interaction with the database, avoiding the necessity of directly interacting with the database using PostgreSQL. This application is only accessed and handled by the managers of the tool (CARTIF). Although it is not directly related to the security, the configuration of the users would not be complete without the city administration. As long as the users of the type "cities" are associated with one city, these have to be previously pre-configured in a same fashion as the users themselves: There is a Java application handled by the administrators, to create, edit and delete cities in the system.

All these procedures serve not only to separate functionalities amongst the different kinds of users, but also to make sure that the data is not accessed by the wrong users. And, as long as these applications are only accessed internally, there is no risk of external intrusion.

6 Conclusion

6.1 Summary

The CIVIS tool aims at providing an easy way to visualize City Indicators, as has been presented. In addition, its main features have been explained: indicators are calculated automatically from variables that can be easily inserted, and they are presented and grouped in a graphic format that lets the user get a visual and realistic idea about the current city performance. The system also shows indicator comparisons between cities, facilitating the decision making for municipal stakeholders and policymakers, as well as steering and measuring its improvement over time by comparing the results of different years.

Inside the tool, several statistic methods have been implemented improving the quality of the results provided, to identify outliers, detect correlations between indicators and approach values that otherwise would not have been available.

6.2 Future Works

TEC4ENERPLAN is a two-year project, which will end by September 2020, so there are still some expected improvements of the CIVIS tool to be developed in future work:

- Firstly, a fine-tuning of the evaluation framework, based on a detailed correlation study among cities' indicators results
- Implement the functionally for evaluating cities progress over time which will detail the evolution of each city and the impact of certain actions taken to improve sustainability; and
- Last but not least, establish more accurate methods to fixed thresholds.

Once CIVIS had been validated with the real data from the client cities, it will serve as a tool for assessing the sustainability of cities in future projects. The singularities and requirements detected on these projects can lead to a broadening and improvement of the possibilities that CIVIS offers as well.

Acknowledgements The TEC4ENERPLAN project is currently financed by the ICE, Instituto de Competitividad Empresarial from Junta de Castilla y León with reference CCTT1/17/VA/0001, co-funded with European Union ERDF funds (European Regional Development Fund).

References

Cheng Y, Carson T, Elgindi MBM (2012) A note on the proof of the Perron-Frobenius Theorem

Langedijk S (2017) COIN 2017—15th JRC annual training on composite indicators & scoreboards 06-08/11/2017, Ispra (IT)

European Project CITYkeys Website (2015) http://www.citykeys-project.eu/citykeys/cities_and_regions/Performance-measurement-framework. Accessed May 2020

Iglewicz B, Hoaglin DC (1993) How to detect and handle outliers, vol 16 of ASQC basic references in quality control: statistical techniques. American Society for Quality Control, pp 10–13

ISO 37120:2018 Sustainable cities and communities—indicators for city services and quality of life. https://www.iso.org/standard/68498.html

ISO 37122:2019 Sustainable cities and communities—indicators for smart cities https://www.iso.org/standard/69050.html

Jolliffe IT (2002) Principal component analysis, 2 ed. Springer

MAtchUP project D1.1. (2020) Indicators tools and methods for advanced city modelling and diagnosis

R Documentation website (2020) https://www.rdocumentation.org/packages/base/versions/3.6.2/topics/svd. Accessed May 2020

REMOURBAN project. D1.20. 2020. Urban regeneration model refined

UN Sustainable Development Goals (2020) https://unstats.un.org/sdgs/indicators/indicators-list/. Accessed May 2020

Universidad Nacional de Colombia Website (2016) https://ciencias.medellin.unal.edu.co/eventos/seminario-institucional/images/presentaciones/pres_victor1.pdf. Accessed May 2020

Methodology and Operating Tool for Urban Renovation: The Case Study of the Italian City of Meran

Alice Schweigkofler, Katrien Romagnoli, Dieter Steiner, Michael Riedl, and Dominik T. Matt

Abstract Nowadays, hundreds of cities around the globe are undergoing smart urbanization: The concept of smart cities is rapidly gaining worldwide attention. The agenda for smart urban renovation is expected to solve a multiplicity of challenges, with the key aim of significantly increasing the urban quality of life. In this wide scenario, the majority of methods for becoming 'smart' imply the use of a strategic approach. Considering this, the following question arises: how should local authorities, small and medium enterprises, as well as utility providers, approach smart urbanization? This research work aims at contributing to this issue, through the development and application of a methodology for the smart city implementation. The assumption is made that holistic enabling tools designed to support smart urban development are more than ever crucial. Their adoption is necessary to establish new smart city services and to manage processes and data information flows, as well as to allow better communication and collaboration among stakeholders. Starting from these observations, the main objective of this work is to develop an operating tool that supports cities in defining a smart city roadmap for strategic decision-making that is in line with their defined city vision and that enables the implementation of smart services. The concept was developed during the research project 'OPENIoT4SmartCities', funded by the Operational Program for the European Regional Development Fund ERDF 2014-2020 (CUP: B11B17000720008) in Südtirol/Alto Adige. The tool was tested on the city of Meran, in the Italian province of South Tyrol. The smart services to be studied were chosen as a result of workshops held together with the city's key-stakeholders, during which brainstorming was done on the basis of the most important issues linked to urban development, i.e., energy, mobility, security and public lighting, environment and waste, and e-governance as transversal topic. In a first screening, all smart services were evaluated based on: (i) their importance according to the cities' stakeholders; (ii) their contribution to the

A. Schweigkofler (✉) · K. Romagnoli · D. Steiner · M. Riedl
Fraunhofer Italia Scarl, Via a. Volta 13A, 39100 Bolzano, BZ, Italy
e-mail: alice.schweigkofler@fraunhofer.it

D. T. Matt
Free University of Bozen-Bolzano, Bozen-Bolzano, Italy

Piazza Università, 39100 Bolzano, BZ, Italy

city vision; (iii) the service's smartness; and (iv) their estimated potential impact. All the services that contribute to the city vision and are considered to have a positive impact on the city's inhabitants are subjected to a pre-feasibility study in which the technical, economical and legal feasibility are evaluated in order to generate a strategic roadmap. The results are presented in the form of a classification of the urban services, together with an explanation of the smart city roadmap for the city of Meran.

Keywords Roadmap · Co-creation & open data · City dashboard · Urban liveability · Urban internet of everything

1 Introduction

The smart city concept has been the subject of intense debate for several years and is now giving rise to many initiatives (Husár et al. 2015). However, to this day, there is still no agreement as to how the concept of 'smart city' (Angelidou 2015) should be defined. Instead, debating goes on and causes the proliferation of sometimes conflicting definitions, none of which are able to fully grasp the concept's scope. Multiple attempts to analyze smart cities have been made, but tools are needed to understand their complexity and reflect the stakeholders' role in developing initiatives and their capacity to face urban challenges (Fernandez-Anez et al. 2018). Most challenges related to smart cities consist of the search for better quality services in various sectors (Chen and Han 2018; Daely et al. 2017), and this wide approach contributes to the idea that a city could be qualified as 'smart' only by introducing single, extemporaneous and non-coordinated actions. In addition, it appears that technology and Information and Communications Technology (ICT) solutions are primary facilitators if cities begin their process of 'smartization' as a path toward integration of technology in every aspect of the urban environment (Silva et al. 2018, Bifulco et al. 2015). Moreover, this industry is faced with a difficult translation of the ready technology solutions into complex and sustainable urban management plans. Therefore, the ability of smart city initiatives to provide an integrated and systematic answer to urban challenges is being constantly questioned (Fernandez-Anez et al. 2018).

Reflecting this situation, research activities in this field are mostly geared toward the study of models for smart city definition and estimation (Fraunhofer Gesellschaft 2019). Indeed, the impact assessment of various measures will have to be carried out in the short, medium and long term. In this context, there is an urgency to develop decision-making tools based on methods of optimization and prediction of effects to help define the priorities for urban design and planning actions. In fact, tool-based decision-making is one of the main results coming out of the analysis of various initiatives aimed at increasing the quality of public decisions (Karvonen et al. 2019; The British Standards Institution 2018). In fact, tool-based decision-making is one of the main results of the analyzed initiatives for increasing the quality of public decisions (Pereira et al. 2017).

According to this, we assume that such a model could help in policy-making processes as the starting point of discussion between stakeholders using more partic-ipative approaches (Padilla et al. 2016), as well as a decision tool for citizens to adopt measures and the best evaluated options (Heiskanen and Acharya 2017; Lazaroiu and Roscia 2012). The aim of this study is to develop a holistic, enabling and operating tool that facilitates the generation of a roadmap for the transition toward a smart city. We define here 'roadmap' as a strategic plan that contains the key steps and mile-stones needed to reach a defined goal or a desired outcome. In addition, a roadmap is a useful tool for the communication of purposes and for the support of strategic deci-sions needed to reach the desired goal. The goal of this operating tool is to generate a timeline that includes the major steps in this transition, while considering the complex technological interaction between infrastructure, sensors, data and services.

2 Methodology

A methodology was developed to implement an operating tool aimed at facilitating the creation of a roadmap for a smart city. The method is described in this section and consists of the following four steps:

1. Collecting ideas for urban services in collaboration with the city's key-stakeholders;
2. Quantifying the suitability of each idea, enabling discarding ideas that are not in line with the objective of the roadmap;
3. Conducting a pre-feasibility study for every selected idea, considering technical, financial and social acceptance, as well as legal aspects;
4. Developing the roadmap and structuring it within a time-based framework.

The activities were conducted based on the specific service needs of Meran, a city in South Tyrol, Italy, which had already been identified in previous analysis about the city context (Fraunhofer Italia 2018). With less than 50,000 inhabitants, Meran is the second biggest city in the mountainous region of the Province of Bolzano-Bozen. More precisely, this paper will present the results of the four-step methodology application, which involves 45 stakeholders belonging to public administration, local service providers, cooperatives, companies and the civil population of the city of Meran (Schweigkofler et al. 2019).

2.1 Collecting Ideas

The ideas for urban services were collected during seven workshops organized with the city's main stakeholders. Each workshop was attended by a minimum of eight and a maximum of 13 participants and was focused on thematic areas deemed critical for the city's development: 1. Security and public lighting; 2. Mobility: fluxes; 3. Environment and waste; 4. Mobility: means of transport; 5. Inclusion and accessibility; 6. Energy; and 7. E-governance. Each workshop was driven by a well-defined agenda based on the Design Sprint methodology (Knapp et al. 2016), combined with the 6-3-5 Design Thinking method (Rohrbach 1969) for the sketch-phase. The aim was to generate a large number of ideas for smart services for the city of Meran. Moreover, the ideas were evaluated during workshops in which the participants were asked to express their opinions regarding each idea during a voting phase.

2.2   Evaluation Criteria

The second methodological step enabled evaluating every idea based on the four criteria in order to generate a shortlist. The first evaluation criterion determines whether or not an idea would contribute to the city Vision Meran 2030 ('Yes', 'Partially', or 'No'). The second criterion evaluates the smartness of an idea in terms of Internet of things (IoT) needs ('Yes, needs platform, sensors, digital infrastructure'; 'Yes, needs platform and connection'; 'No IoT-needed'). The third criterion was used to identify how many user groups would directly interact with the designed smart service (see Table 1). Any user category 'X' could be assigned to any idea 'Y', as long as the answer to the question 'Would X use or directly be impacted by 'Y' was affirmative. This exercise was done using the persona method (de Boer 2010). The aim was to identify the ideas with a high impact on a city's stakeholders. The fourth and last criterion considers the votes given during the workshop by the city's stakeholders.

The ideas received a score between 0 and 20, according to the following rules:

- Vision: 'Yes' (5 points), 'Partially' (2.5 points), 'No' (0 points).
- Smart: 'Yes, needs platform, sensors, digital infrastructure' (5points); 'Yes, needs platform and connection (2.5 points)'; 'No IoT-needed' (0 points).
- Users: number of user groups divided by the total number of user groups, scaled to 5.
- Votes: number of votes divided by the total number of votes achieved during the workshop, scaled to 5.

The maximal score obtained was 14.93, and the minimum score was 2.86. Only the ideas with a score of at least 10 points were selected to be subjected to a pre-feasibility study.

Table 1 Definition of user categories

User category	Description
Commuters	From outside Meran
Elderly	Citizens no longer working and older than 65 years
Young adults	Citizens between 18 and 30 years old
Tourists excl. day-trippers	People living outside South Tyrol
Disabled	People with physical or mental disabilities
New-residents	Citizens less than five years in South Tyrol
Families	Parent(s) with child(ren) or teenagers
Low-income/unemployed	Citizens without a job or an annual income below €15 k
Working middle class	Citizens between 30 and 65 years old and an annual income above €15 k
Children	Citizens between 0 and 12 years old
Youth	Citizens between 13 and 18 years old
Day-trippers	South Tyroleans visiting Meran
Public administration	Including public services
Companies and merchants	People conducting or owning economic activities

2.3 Pre-feasibility Study

The pre-feasibility study was conducted to give an indication on the viability and complexity to develop each proposed smart service for the city Meran. This evaluation was done by considering seven criteria described below and summarized in Table 2. Four technical levels of implementation of a smart city were considered: 1. Infrastructure, 2. Sensors, 3. Service delivery platform and 4. Application and service (Ernst & Young 2016). At the same time, the Social Acceptance (SA) degree was evaluated (i.e., 'Would the idea be perceived as negative by certain user groups?'), the Legal Constraints (LC) were analyzed (i.e., 'Are there any (inter)national or local legal constraints that limit the idea's feasibility?'), and the Financial Aspect (FA)

Table 2 Pre-feasibility study evaluation options for the seven criteria and their impact (in years) on the roadmap timeline

Evaluation options	PT (yrs)	IT (yrs)	Description
Criterion 1: Technical: Level 1. Infrastructure			
Existing	0	0	Infrastructure is available
Extension	0	1	Extension of the existing infrastructure is needed
New	0	3	New infrastructure is needed
Criterion 2: Technical: Level 2. Sensors and actuators			
Existing	0	0	Needed sensors are available
Extension	0	0.5	Extension or re-allocation of existing sensors is needed
New	0	1	New sensors are needed
Criterion 3: Technical: Level 3. Service delivery platform			
Open	0	0	The platform is open to other users
Flexible	0	0	The platform contains partially sensitive data but partially consists of data that could open for other users
Closed	0	0	The platform contains sensitive data that needs to be treated accordingly
Criterion 4: Technical: Level 4. Application and service			
Integration	0	0	The service and/or application can be easily integrated in existing ones
New app	0	0.5	A new stand-alone application needs to be developed
New service	0	1	A new stand-alone service needs to be developed
Criterion 5: Social Acceptance (SA)			
✓	0	0	General positive acceptance to implement the idea or indifferent
!	0.75	0	Communication is needed to generate the needed level-playing field as the idea can raise limited negative feelings
✗	1.5	0	Communication is needed to generate the needed level-playing field as the idea can raise strong negative feelings
Criterion 6: Legal Constraints (LC)			
●	0	0	There are no legal constraints and the legal framework currently exists
●	2.5	0	Legal constraints can be amended at local level or legal gap can be resolved at local level
●	5	0	Legal constraints can be amended at supra-local level or legal gap can only be resolved at supra-local level
Criterion 7: Financial Aspect (FA)			
€	0	0	The development of an app or service is the only financial burden

(continued)

Table 2 (continued)

Evaluation options	PT (yrs)	IT (yrs)	Description
€€	3	0	Infrastructure and sensor investments below €10 M or a need for human resources during operation
€€€	5	0	Infrastructure and sensor investments exceeding €10 M

was taken into account (i.e., 'How feasible is the idea's financial aspect'). The social acceptance was determined by considering four key user groups:

- Young adults (18–30 year) and teenagers (13–18 year) (SA-1);
- Tourists (including day-trippers) (SA-2);
- Families (SA-3);
- Vulnerable groups comprising elderly, low-income or unemployed and disabled people (SA-3).

For every idea, all seven criteria were evaluated, and each time, one of three evaluation options was selected. According to Table 2, each evaluation option was tied to the duration of the preparatory phase to start the development of a service, i.e., preparation time (PT), and to an estimation of the timeframe required to develop and launch that idea, i.e., implementation time (IT). The total duration of the PT equals the sum of the PT for every criterion. The total duration of the IT equals the sum of the IT for every criterion +1 year, to have at least an IT of one year.

2.4 Operating Tool for Strategic Roadmapping

The operating tool developed is a spreadsheet and includes the main sections 'input', 'use case', and 'roadmap'. In the 'input' section, all the service ideas and parameters were organized in a database and served as input to automatically generate the 'roadmap' and 'use case' sections, which are used to visualize the results in a way that supports the decision-making process.

All the data collected in the previous methodology steps and described in Sects. 2.1, 2.2 and 2.3 contains the relevant input data for the generation of a roadmap. Step 2.2 influenced the ranking of the ideas, and step 2.3 influenced the roadmap timeline, i.e., the total time required to launch a service resulting from the PT and IT assigned to each criterion. Since PT and IT are determined by the SA, FA, and LC, they are the important parameters with which, eventually, a decision maker can determine the timeline of the roadmap according to every city's context. These key parameters can be updated in the 'input section'. A detailed technical report about the adopted process is available (Fraunhofer Italia Roadmap Toward a Smart City 2018) upon request by contacting the authors, together with the summary of all the collected service ideas.

3 Results

A total of 131 ideas were generated in the first step (Sect. 2.1), after which many similar ideas were merged together, thereby bringing the final output down to 100 distinct ideas. As a result of the second evaluation step (Sect. 2.2), from a total of 100 ideas, only five did not fit in with the Meran city vision, whereas 40 fit in only partially, whereas the majority totally fit in. Regarding to the 'smartness' criterion, 31 ideas were classified as 'No IoT-needed', which means they are not considered as 'smart'. Ideas that did not fit in with the vision or did not need any IoT were excluded from the roadmap. Consequently, the list was shortened from 100 to 66 items, and a unique ID was assigned to each one of them, starting with 1 for the highest total score and finishing with 46 for the lowest score.

As a result of the pre-feasibility study (Sect. 2.3) and the development of the operating tool (Sect. 2.4), a roadmap for the city of Meran was realized (see Fig. 1) based on the technological requirements of the various services.

Four different visualizations of the roadmap can be generated, according to:

1. Chronology: the projects that can start immediately appear first;
2. Thematic areas: grouped by subject (e.g., energy and environment, mobility, e-governance, etc.) (see Fig. 1);
3. The outcome of the ratification process: from the highest score (i.e., number 1) to the lowest score (i.e., number 46);
4. Infrastructure: presents the roadmap of the technology (i.e., 5G, optical fiber, etc.) needed to deliver the smart city services.

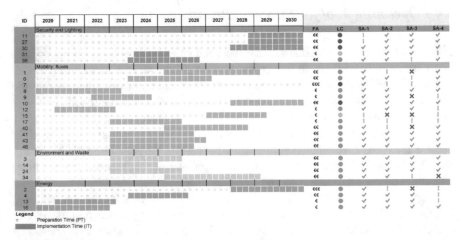

Fig. 1 Visualization of the section 'roadmap' in thematic areas. The first row shows the timeline between 2020 and 2030, and each following row indicates the temporal development of a use case grouped by thematic areas with different colors. Columns from left to right represent: ranking ID assigned to each idea; preparation time PT (dots); implementation time IP (colored cells); on the right, the respective outcomes of the feasibility study on costs (FA), legal constraints (LC), and social acceptance (SA) users groups are summarized (see Table 2)

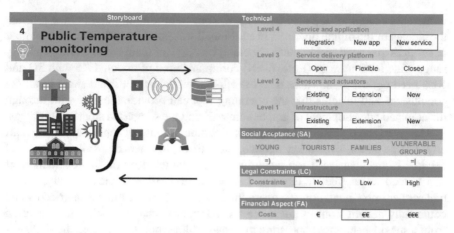

Fig. 2 Example of use case output with left the storyboard and right the output from the pre-feasibility study according to Table 2

Although the roadmap provides an overview of all smart services, it does not provide enough information to understand the precise scope of single smart city service. In addition, each service needs to be socially accepted, to be supported by a solid legal framework and be economically justifiable to the city dwellers. Therefore, the operating tool provides a dedicated section in which every smart service is summarized in a sheet (see Fig. 2) containing the following information:

- General information: Title, rank, short description of the idea, impacted thematic areas and user groups.
- Story boards (left side of Fig. 2), explaining the intended application or service in real life;
- Pre-feasibility output (according to Table 2) on all four levels including comments.

4 Discussion and Conclusion

The operating tool is based on an open and flexible model that can be adopted in various different contexts and in which input is processed in an automatic and parametric way. This tool enables city planners to alter parameters, to add or delete items, and to modify contents to automatically adjust the roadmap. Moreover, graphs that give valuable insights into the transition strategy toward smarter cities can be extracted. Which services are urgently needed? Which infrastructure must absolutely be developed? For example, for the city of Meran, it was possible to establish that more than 60% of the services should aim at improving mobility. Governance and digitalization were also found to be strategic thematic areas. Although most of the smart services are directed at young people, workers, and families, the more vulnerable categories are also considered, something that highlights how relevant the closing of the digital divide is. In addition, it was possible to estimate the number of services that need specific infrastructure, i.e., 28% of them could work with long-range wide-area network (LoRaWAN) communication protocol, 17% with 5G and 13% need optical fiber. These direct results represent only a small share of all the conclusions and insights that the operating tool can provide the user with through the automatic development of graphs summarizing the suggested smart city strategy.

However, the operating tool requires a certain amount of input data (i.e., smart city service ideas, evaluation options, etc.) to be inserted manually. This could become labor-intensive if too many ideas were gathered by the stakeholders or if this method was applied to larger and more complex cities. In the event of an eventual further implementation, the operating tool could become more user-friendly and could fully benefit from its adaptable and parametric characteristics, if transferred from a spreadsheet-based operating tool to an online tool. In addition, the standardized method developed in this project could be further tested and validated in other contexts and cities to improve its robustness.

It can be concluded that this holistic operating tool can effectively help public administrations, decision makers, and city planners in defining a strategy for urban planning.

Acknowledgements This paper has been supported by funding from the European Regional Development Fund ERDF 2014-2020 Südtirol/Alto Adige, in the project OPENIOT 1069 (CUP: B11B17000720008). The authors would like to thank the City of Meran, in particular the city Councilor for innovation Diego Zanella, as well as, the involved company Systems SRL for their valuable contribution to the current paper.

References

Angelidou M (2015) Smart cities: a conjuncture of four forces. Cities 47:95–106. https://doi.org/10.1016/j.cities.2015.05.004

Bifulco F, Tregua M, Amitrano CC, D'Auria A (2015) ICT and sustainability in smart cities management. Int J Public Sector Manag 29(2):132–147

Chen Y, Han D (2018) Water quality monitoring in smart city: a pilot project. Automat Constr 89:307–316

Daely PT, Reda HT, Satrya GB, Kim JW, Shin SY (2017) Design of smart LED streetlight system for smart city with web-based management system. IEEE Sens J 17:6100–6110

de Boer T (2010) Global user research methods. In: Handbook of global user research, Chapter 6. Elsevier

Ernst & Young (2016) Rapporto smart city index. http://www.comune.bologna.it/iperbole/piancont/archivionov/tabelle_grafici/EYsmartindex/2016-EY-smart-city-index.pdf. Accessed 10 Oct 2018

Fernandez-Anez V, Fernández-Güellb MJ, Giffingerc R (2018) Smart city implementation and discourses: an integrated conceptual model. The case of Vienna. Cities 78:4–16

Fraunhofer Italia (2018) MeranSmart—sn inclusive and digital city. https://www.fraunhofer.it/ Accessed 30 Jan 2019

Fraunhofer Gesellschaft (2019) Morgenstadt—City of the future Initiative. https://www.morgenstadt.de/en.html. Accessed Jan 25 2019

Heiskanen O, Acharya K (2017) Envisioning the future of public lighting with citizens for upcoming technologies. Design J 20:S1782–S1793

Karvonen A, Cugurullo F, Caprotti F (2019) Inside smart cities: place. Politics and Urban Innovation, Routledge, London

Knapp J, Zeratsky J, Kowitz B (2016) Sprint: how to solve big problems and test new ideas. In: Just five days, Simon and Schuster, 288 p. ISBN: 1501121774, 9781501121777

Lazaroiu GC, Roscia M (2012) Definition methodology for the smart cities model. Energy 47:326–332

Padilla M, Hawxwell T, Wendt W (2016) City Lab Lisbon—development of a smart roadmap for the city of the future. In: REAL CORP 2016 proceedings

Pereira GV, Cunhab MA, Lampoltshammera TJ, Paryceka P, Testac MG (2017) Increasing collaboration and participation in smart city governance: a cross-case analysis of smart city initiatives. Info Technol Develop 23:526–553

Rohrbach B (1969) Creative by rules—Method 635, a new technique for solving problems. German Sales Mag "Absatzwirtschaft" 12:73–75

Schweigkofler A, Follini C, Steiner D, Romagnoli K, Riedl M, Matt DT (2019) Processing of use cases for the development of an open platform to support the smart urban development. In: Conference proceedings of the SBE19 series conference. Helsinki, Finland

Silva BN, Khan M, Han K (2018) Towards sustainable smart cities: a review of trends, architectures, components, and open challenges. Smart Cities Sustain Cities Soc 38:697–713

The British Standards Institution, Smart city framework—Guide to establishing strategies for smart cities and communities. URL https://www.bsigroup.com/en-GB/smart-cities/Smart-Cities-Standards-and-Publication/PAS-181-smart-cities-framework/. Accessed 09 Oct 2018

Investigate Walkability: An Assessment Model to Support Urban Development Processes

Francesca Abastante, Marika Gaballo, and Luigi La Riccia

Abstract This chapter is about defining and testing a multi-methodological framework able to measure the "walkability" in the urban practice perspective, based on assessment indicators and Geographic Information Systems (GIS). Nowadays, cities are facing a complex challenge concerning sustainability, which is fueling the search for new development solutions. Among others, one of the most important problems is how to make cities sustainable and resilient, as stressed by the Sustainable Development Goal 11 (SDG11) highlighted by the United Nations through the 2030 Agenda. The topic of "walkability" appears in this framework: Walking has ecological, social, economic and political benefits. Moreover, designing walkable networks is important to create a functional and multi-modal city with transport choices and makes urban settlements sustainable and inclusive from the perspective that a sustainable city is also a walkable city. However, despite the positive impact of walkability on public space, it is still difficult to fully include it in governmental strategies because of its novelty in the scientific debate. The ongoing research proposed here aims at: (i) describing the problem, related to what trends and strategies have been implemented to face it; (ii) investigating walkability, understanding its definition in the scientific panorama, and how it is evaluated; (iii) understanding the current evaluation methods to assess the walkability of spaces; (iv) proposing a new multi-methodological framework based on existing methods that are able to measure the walkability degree from the perspective of better planning of cities. The multi-methodological framework has been tested through a case study: the Politecnico di Torino Campus (Torino, Italy).

Keywords Walkability · Sustainable urban mobility · Walkability evaluation index · Evaluation model

F. Abastante (✉) · M. Gaballo · L. La Riccia
Politecnico di Torino (DIST Department), viale Pier Andrea Mattioli 39, 1012 Torino, Italy
e-mail: francesca.abastante@polito.it

© The Author(s), under exclusive license to Springer Nature Switzerland AG 2021
A. Bisello et al. (eds.), *Smart and Sustainable Planning for Cities and Regions*,
Green Energy and Technology, https://doi.org/10.1007/978-3-030-57332-4_13

1 Introduction

Nowadays, cities are characterized by a large number of walking paths, whose design is often underestimated in urban transformation processes and budgets. However, designing walkable networks is not only important to create a functional and multi-modal city with transport choices, but also to make urban settlements sustainable and inclusive. We can call this particular intention "walkability" (Rogers et al. 2013), understood as the easiest, cheapest and socially equable form of "soft-mobility" (La Rocca 2010). This does not mean that other transport modalities are not also recognized, but that they must be integrated in a sustainable way, improving the liveability of the city (Blečić 2015). This appears in line with the objectives of the 2030 Agenda for Sustainable Development with particular reference to the Sustainable Development Goals 11 (SDG11: Make cities and other human settlements inclusive, safe, resilient and sustainable) SDG13 (Take urgent action to combat climate change and its impacts).

In this panorama, the ongoing research, here presented, aims at defining and testing a multi-methodological framework based on assessment indicators and Geographic Information Systems (GIS) (Yin 2017; Lombardi et al. 2017; Chiantera et al. 2018) that are able to measure the "walkability" in the urban practice perspective.

The multi-methodological framework is structured in some interactive and iterative phases. First, we provided a literature review considering a span-time of ten years (2010–2019) in order to understand which indices and indicators should be used to measure "walkability."

Second, we applied a qualitative method based on questionnaires in order to verify the sensibility on the indices and indicators identified.

Third, the results obtained by the questionnaires have been aggregated and analyzed according to statistical models (Eliou and Galanis 2011; Shatu and Yigitcanlar 2018). This step turned out to be fundamental since it enables us to identify the final weights to be assigned to each index and indicator considering both quantitative and qualitative aspects.

In parallel, we explored the possibility of using visualization tools as assessment tools by mapping the indices and indicators identified using the QGIS software.

The multi-methodological framework here proposed has been tested in the walkability assessment of the Politecnico di Torino Campus (Turin, Italy).

The chapter is organized as follows: Sect. 2 provides an overview of the "walkability" issue and the main assessment methods to evaluate it; Sect. 3 reports the multi-methodological framework proposed through the application to a case study, while Sect. 4 concludes the chapter with a discussion of the future development of the research.

2 Walkability and Assessment

The increasing level of unsustainability that gradually affected every city at the global level requires an alternate international path related to urban mobility and accessibility. Among the urgent calls defined by the United Nations SDGs, "walkability" has a fundamental role. In fact, under the SDGs, this concept is explored and analyzed according to various perspectives, such as economic, political, social, ecological, and health (Rogers et al. 2013).

Accordingly, the "walkability" topic is increasingly becoming central in the research field of urban mobility and sustainability (Jensen 2013; Rogers et al. 2013; Urry 2016), and it is understood to be one of the factors that make cities "inclusive, safe, resilient and sustainable" (United Nation General Assembly 2017). It is demonstrated that improving "walkability" could lead to significant results in terms of money and time saved, the reduction of noise and air pollution, democratization of mobility, social cohesion and reduction of obesity, and prevention of cardiovascular diseases (Kaczynski 2012). This means that planning "walkable" cities could produce smart cities and communities as stressed by the SDG11.

According to the literature (Cambra 2012; Moayedi et al. 2013; Blečić et al. 2015), "walkability" is first of all a tool to measure the degree of pedestrian uses of a certain area (Abastante and Gaballo 2020; Abastante et al. 2020b). Despite this shared vision, a proper widespread definition of this concept is still missing: Some researchers define walkability as "the safety, security, economy, and convenience of traveling by foot" (Krambeck 2006), while others highlight a qualitative perspective linking "walkability" with the "quality of a place" (Ewing and Handy 2009). Those differences are probably due to several factors: (i) The ambiguity of the action of "walking" makes it tricky to catalogue "walkability," since people walk in an urban context for many reasons (Solnit 2005); and (ii) "Walkability" impacts different spheres of reality such as planning, transport, economy, and society. Therefore, we can affirm that the variables of "walkability" are many, making this concept subjective and making its definition dependent on those who deal with it (Lo 2009).

Moreover, despite that "walkability" is a consolidated field of analysis in the international context (D'Alessandro et al. 2016; Keat et al. 2016), in Italy this topic is little explored and considered by academics and Public Administrations (PAs).

In addition, while the socio-demographic impacts of "walkability" have been widely probed in the scientific literature (Saelens et al. 2003), studies about its physical and environmental variables linked to the built environment are scarce. This could be due to the fact that the act of walking in an urban context is wrongly often taken for granted. On the contrary, it actually requires proper design and be included in the planning of sustainable cities as a crucial mode of transport from the perspective that good "walkability" planning controls the way people move and determines the way they will move in the future (Jacobs 1961). Furthermore, "walkability" is a way of looking beyond the presence, distribution and accessibility of urban facilities: The spatial quality and the ability to accommodate and promote pedestrian mobility

Table 1 Main methods to assess walkability

Methods		Purpose
Quantitative	Statistical models	Provide an objective state of art
	Weighing indexes and indicators	Needed to obtain a global index structured on the basis of the indices considered, divided into indicators
Qualitative	Surveys	Outline users' perception of nonphysical and objectively measurable characteristics
	Empirical investigation	Provide scientific robustness for analysis

within the urban environment influence the way in which people perceive and use the entire city (Leslie et al. 2005).

In this panorama, a question emerges: How can we assess walkability in a city planning perspective?

According to the literature (D'Alessandro et al. 2016; Keat et al. 2016), properly understanding and measuring the complexity of walkability is extremely challenging. This is due to the jointly interrelated presence of two "souls": tangible/objectives elements (i.e., pavement height) and intangible/subjective aspects (i.e., the comfortable sensation felt when walking through a space). While the first are easily measurable and quantifiable, the second are not.

Many methods have been proposed in the scientific literature to assess "walkability" (Table 1).

From Table 1, it emerges that: (i) the most widespread quantitative methods are statistical models and weighting of indices and indicators, in the perspective of analysing the current state of a territory; and (ii) the most used qualitative methods are surveys and empirical investigations to identify nonphysical aspects and verify the robustness of previous analysis.

It is important to notice that the two "souls" of walkability, tangible/objectives elements and intangible/subjective aspects, are separately addressed by current methods proposed in the literature.

In our opinion, this could risk leading the city projects in diametrically opposite directions, complicating the decision-making processes and producing ineffective results.

The research presented in this paper aims at contributing to the current scientific debate by proposing a multi-methodological framework based on the main current assessment methods found in the literature and integrated within a spatial evaluation.

2.1 The Multi-methodological Framework

The multi-methodological framework here proposed is structured into three main phases (Fig. 1):

Fig. 1 Multi-methodological framework structure

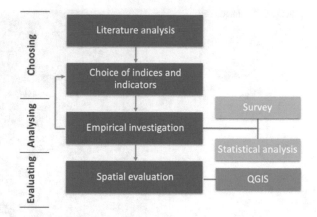

(i)   Choosing, related to the choice of indices, indicators and their range of weight through an in-depth analysis of the literature;
(ii) Analyzing, which includes qualitative and quantitative methods in order to test the results of the previous phase;
(iii) Evaluating, related to a spatial evaluation of phases 1 and 2.

To develop the multi-methodological framework, we approached the logic of so-called case study research (Stake 1995, p. 7) in which the case is understood as a "complex, functioning thing" useful to come to general understanding about the research question (Stake 1995, p. 2).

2.2   The Case Study: The Politecnico di Torino Campus

The case study that we "instrumentally" used to start developing the multi-methodological framework is based on the Politecnico di Torino Campus. In Fig. 2, the area under examination is depicted including the Politecnico di Torino Campus but also the main train station of the city and the major streets around the campus. This is due to the need to properly consider the accessibility to the campus.

The reasons for choosing this particular case study are many: (i) It is similar to a urban district in terms of territorial scale; (ii) it is the node of interesting actors' networks; (iii) it is usually the place in which ideas of urban transformations are produced together with the PA; (iv) a large amount of information is available, facilitating the development and test of the multi-methodological framework; and (v) it provides the chance to study, implement, and evaluate different aspects in the perspective of raising awareness among students, professors and staff about the crucial issues of our times.

Fig. 2 Case study area

2.3   Phase 1: Choosing

The first phase (Choosing) is fully based on literature analysis with the purpose of defining indices and indicators (Cambra 2012; D'Alessandro et al. 2016) that are able to reflect the complexity of the reality about the planning of walkability, considering both its tangible/objective elements and intangible/subjective aspects.

We therefore conducted a scientific literature review basing on Scopus and Google Scholar databases using four keywords: walkability, walkability measure, walkability indicators, and walkability indices. In turn, we decided to limit the analysis to the papers that were facing the "walkability" topic in an assessment perspective, coming up with 25 results. Those have been analyzed in depth to understand: (i) indices and indicators used to assess "walkability"; and (ii) the weights that usually are assigned to each index in terms of range (Table 2).

From Table 2, we can notice that the main indices found in literature are four (Safety/Security, Quality of paths, Comfort and Intermodality) which in turn are divided into 27 indicators. Moreover, the index safety/security seems to be the most important one with a range of weights among 25% and 50% followed by the quality of paths (21–40%), comfort (10–20%), and intermodality (10–20%). It is important to

Table 2 Main indices and indicators according to the literature review

Indices	Weights	Indicators	Frequency
Safety/Security	25–50%	Presence of busy roads	Cerin et al. (2011), Ford (2013), Lee and Talen (2014), D'Alessandro et al. (2016), Keat et al. (2016), Wibowo and Nurhalima (2018) Chiantera et al. (2018)
		Crossing equipped with traffic lights	
		No signaled pedestrian crossings	
		Separation of routes	
Quality of Paths	21–40%	Width of routes	Reid and Handy (2009), Cerin et al. (2011), Galanis and Eliou (2011), Cambra (2012), Ford (2013), Moayedib et al. (2013), Lee and Talen (2014), D'Alessandro et al. (2016), Keat et al. (2016), Wibowo and Nurhalima (2018), Shatu and Yigitcanlar (2018) Chiantera et al. (2018)
		Condition of the pavement	
		Non-sliding paths (with obstacles)	
		Well-connected paths	
		Slope	
Comfort	10–30%	Presence of trees	Reid and Handy (2009), Cerin et al. (2011), Cambra (2012), Ford (2013), Moayedib et al. (2013), Domokos et al. (2014), Lee and Talen (2014), D'Alessandro et al. (2016), Keat et al. (2016), Wibowo and Nurhalima (2018), Yin (2017), Chiantera et al. (2018)
		Adequate lighting	
		Presence of benches	
		Presence of baskets	
		Noise pollution	
		Covered routes	
		Presence of water points	
		Presence of tall buildings	
		Buildings with monotonous colors	
		Possibility to see the continuity of the routes	
		Refreshment points	
		Study points	
		Crowded spaces	
Intermodality	10–20%	Parking for private bikes	Cambra (2012), Ford (2013)
		Easy accessibility by public transport	
		Parking for private cars	
		Bike-sharing stations	
		Car-sharing stations	

underline that the intermodality indicator is addressed by only two included papers, diminishing the sensibility of the range here classified.

2.4 Phase 2: Analyzing

The second phase (analysing) is both qualitative and quantitative and pursues a double aim: testing the sensibility of the indices and indicators (Table 2) through the use of surveys and finding their weights through statistical analysis.

The interview sample selected for the surveys' analysis was composed of students, teachers and technicians of the Politecnico di Torino, for a total of 100 interviewees. It is fundamental to underline that, in this phase of the research, a small sample was sufficient to conduct experimental validation of the framework. The 100 interviewees were asked to answer to 36 questions using the five-point Likert scale of evaluation in which 1 means "strong disagreement" and 5 means "totally agree" (Likert 1932).

The questions were of the type:

- Considering the indices proposed (safety/security, quality of the paths, inter-modality and comfort) how much performing is each index in the Politecnico di Torino Campus? Provide a percentage according to your experience
- Considering the index "safety/security," the Politecnico di Torino Campus is characterized by the presence of busy roads. Provide an answer according to the Liker scale.

The results obtained by the surveys have been analyzed by calculating the modal value, the arithmetic average, the weighted average and the standard deviation to achieve a detailed statistical overview of the answers obtained.

A first result of this phase was related to the level of agreement in the responses: The answers collected were widely varied in relation to some indices compared to others, reporting a low degree of agreement for some indices.

Accordingly, Fig. 3 reports the calculation of the standard deviation of the answers.

Figure 3 highlights that the highest dispersion of the responses is related to the Intermodality index, showing the high subjectivity of this index. In fact, the transport modalities to reach the Politecnico di Torino Campus are many (by foot, train, subway, bus, bike, car, or shared transport), and this causes various perceptions of the intermodality efficiency depending on the transport modes that the interviewee usually uses.

Fig. 3 Standard deviation of the Indices

Fig. 4 Weighted averages of indices

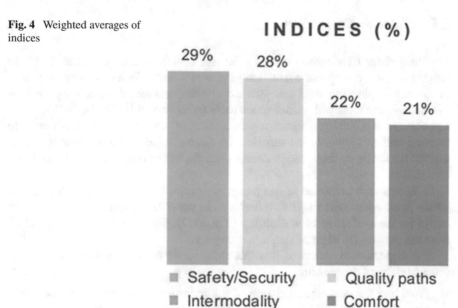

According to the results obtained by the surveys, Fig. 4 graphs the weighted averages calculated at the level of the Indices.

In terms of indices (see Fig. 4), it is possible to notice that the Safety/Security index has the highest weight followed by that of quality of the paths while the index with the lowest weight is the Comfort index, highlighting that Comfort is the least critical aspect in the Politecnico di Torino campus. With reference to this, Fig. 5 shows the weighted averages calculated at the level of the Comfort's indicators.

Figure 5 illustrates that the most problematic indicator is the presence of "Crowd spaces in Campus." This means that the spaces inside the Politecnico di Torino Campus need a better planning in this sense.

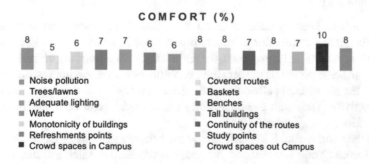

Fig. 5 Weighted averages of the comfort indicators

2.5   Phase 3: Evaluating

The third phase (Evaluating) involves the use of GIS (Feizizadeh et al. 2014) to provide a visual and spatial evaluation of phases 1 and 2. In recent years, research in the urban planning field has stressed the importance of combining qualitative/quantitative tools with visualization tools (Abastante et al. 2020a, 2017). Those demonstrate that having a visual representation of what is evaluated is useful to obtain a holistic picture of the situation, on the one hand, and to provide technical support for future strategic and/or design choices, on the other (Vennix 1996; Lami et al. 2014).

In the research presented in this paper, we decided to use the QGIS (qgis.com) which is an open-source software that makes possible calculating the characters useful for the evaluation of walkability (Yin 2017), thus improving understanding from the perspective of proposing urban projects.

It is important to underline that this third phase of the multi-methodological framework has a double aim:

1. Providing a spatial representation of the indicators assumed in the case study area, starting from selected tangible/objective territorial aspects;
2. Weighting the indicators with the percentages identified in the previous phases 1 and 2, to emphasize the importance to the intangible/subjective aspects.

Considering that this is ongoing research, in this paper we will briefly illustrate the logics undertaken to the first objective, while the integration of intangible aspects still in progress. All data have been spatially represented following three different means of representation (Okabe et al. 2009):

- Maps with "linear" distribution of data (e.g., cycling pathways—objective feature: width)
- Maps with "areal" distribution of data (e.g., public lighting—objective feature: wattage of the lamps)
- Kernel maps with statistical distribution of point data (e.g., traffic lights—not weighted data).

All data have been spatialized, given the value of a phenomenon, representing their diffusion and attenuation with a radius defined in relation to the phenomenon represented. Accordingly, we used the QGIS/GRASS algorithm named "*r.cost.*" even though some problems of software instability occurred. An interesting optional output of the algorithm is allocation, i.e., the identification of the area of influence of each activity: The result can be usefully interpreted if referred to destinations of the same kind.

Since the multi-methodological framework is developed starting from the Politecnico di Torino Campus case study, the space is modeled through a raster with 1 × 1 m cells. This dimension is smaller than the usual dimensions, but it perfectly fits the needs of the case study in terms of details.

An impedance (Kartshmit et al. 2020) is assigned to each cell, which is a cost of traveling on foot, more or less pleasantly and safely. The cells that cannot be traversed because they are included in areas destined for vehicular traffic or because they are nonpublic are excluded. The outputs/maps provided by this kind of analysis are often identified as Cost Rasters (Cittadino et al. 2019). The cost raster is understood as a concept which usually results by the application of a multi-criteria analysis (MCDA) (Abastante et al. 2017): It is the weighted sum of several raster whose cells values represent various aspects of cost (Fig. 6).

Figure 6 comprises the cost raster maps of the four indices considered: comfort, intermodality, quality of the path and safety/security. Those are the results of the weighted sum of the cost raster of the indicators (Table 2). Since indices and indicators could be measured in different scales of measure, the values have been normalized. In the maps, the green areas mean high values of indices (high practicability, high security, high pleasure) while the red ones mean low values (obstacles).

It is possible to notice that the index for comfort seems to be the most problematic one while the index for quality of the paths reports a very high evaluation.

3 Discussion and Conclusions

The "walkability" concept, as described in this research, could contribute to the construction of a more sustainable city by designing solutions that improve the possibilities of using public roads as public spaces, making a city more liveable.

Moreover, the role of the assessment appears fundamental in the context of walkability: Such evaluation methodologies are increasingly used as guide in order to support transformation processes addressing the actual challenges of the urban context (Bush and Doyon 2019).

Accordingly, the multi-methodological assessment framework developed in this research could be an aid for stakeholders who want to reason in terms of liveability as an element of growth and sustainability of the urban context or, more generally, who deal with the sustainable mobility issue.

The results that we reported in this chapter, despite their preliminary nature, constitute a strong basis for discussions useful to funding future steps of the research. First, we can affirm that phase 1 (Choosing) the multi-methodological framework proved to be fundamental to identifying the main indices and indicators in a tangible/objective perspective. However, the analysis of the literature also showed that those are the most varied, site-specific and closely related to the area examined.

Phase 2 (Analyzing) constitutes the core of the multi-methodological framework by being able to enrich overall analysis focusing on the intangible/subjective aspects. In our opinion, this second phase constitutes a fundamental contribution to the current international debate about walkability assessment because it puts people at the core of the urban planning by considering their needs, feelings and perception about the walkability of a place.

A strong limitation of this second phase has been the interviewed sample. Due to time constriction, the interviewed sample so far considered is very limited.

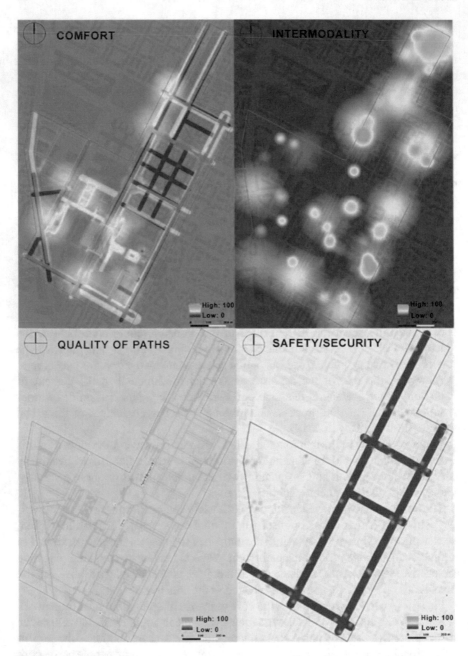

Fig. 6 Cost raster (indices)

In turn, we would be able to provide a robust phase 3 (Evaluating) by basing the analysis on a larger amount of data. In fact, despite the European principles of open by default (opendatacharter.net), we faced some difficulties in accessing data, particularly in reference to the absence of meta-documentation of some indicators.

In general, despite the limits showed by the multi-methodological framework proposed, it proved to have a lot of potential in terms of measuring the walkability characteristics, considering both tangible and intangible elements.

4   Future Developments

The future development of the research will involve not simply improving the high-lighted weaknesses but also taking steps forward. In the future development of the research, we aim at improve automation of some procedures to structure real-time decision processes (Lami and Tavella 2019) by integrating the calculation of cost distance maps (Chen et al. 2019). This will enable considering simultaneously the weights of indices and indicators objectively defined with the weights stated in the questionnaires.

Moreover, to improve the walkability of the site, we aim at defining alternative planning scenarios for the Politecnico di Torino Campus (Italy).

Finally, it would be interesting to provide a generalization of the results obtained to make possible translating them into strategic policy guidelines for public admin-istrations. This will enable consideration of walkability in developmental processes and their implementation in territorial governance tools in line with the sustainability objectives imposed by European document standards.

References

Abastante F, Lami IM, Lombardi P (2017) An integrated participative spatial decision support system for smart energy urban scenarios: a financial and economic approach. Buildings 7(4):103

Abastante F, Gaballo M (2020) How to assess walkability as a measure of pedestrian use: first step of a multi-methodological approach. In: International symposium of new metropolitan perspectives. Springer, Cham , pp 254–263

Abastante F, Pensa S, Masala E (2020a) The process of sharing information in a sustainable devel-opment perspective: a web visual tool. In: Values and functions for future cities. Springer, Cham, pp 339–350

Abastante F, Lami IM, La Riccia L, Gaballo M (2020b). Supporting resilient urban planning through walkability assessment. Sustainability 12(19):8131

Andersen DF, Richardson GP (1997) Scripts for group model building. Syst Dyn Interview 13(2):107–129

Blečić I, Cecchini A, Congiu T, Fancello G, Trunfio GA (2015) Evaluating walkability: a capability-wise planning and design support system. Int J Geogr Inf Sci 29(8):1350–1374

Bush J, Doyon A (2019) Building urban resilience with nature-based solutions: how can urban planning contribute? Cities 95:102483

Cambra P (2012) Pedestrian accessibility and attractiveness indicators for walkability assessment. Department of Civil Engineering and Architecture, Instituto Superior Técnico, Universidade Técnica de Lisboa, Lisbon

Cerin E, Chan KW, Macfarlane DJ, Lee KY, Lai PC (2011) Objective assessment of walking environments in ultra-dense cities: development and reliability of the Environment in Asia Scan Tool—Hong Kong version (EAST-HK). Health Place 17(4):937–945

Chen Y, She J, Li X (2019) Calculate accurate 3D cost distance efficiently. Abstracts ICA 1

Chiantera G, Cittadino A, Del Carlo G, Fiermonte F, Garnero G, Guerreschi P, Vico F (2018) Walkability della città: analisi raster per supportarne la progettazione e il suo incremento. In: Conferenza Nazionale ASITA 2018, pp 1–8

Cittadino A, Eynard R, Fiermonte F, Garnero G, Guerreschi P, Melis G, La Riccia L (2019) Improving the WALKABILITY for next-generation cities and territories, through the reuse of available data ad raster analysis. In: XXIV international conference LWC 2019, living and walking in cities, pp 1–38

D'Alessandro D, Appolloni L, Capasso L (2016) How walkable is the city? application of the walking suitability index of the territory (T-WSI) to the city of Rieti (Lazio Region, Central Italy). Epidemiol Prev 40(3–4):237–242

Domokos S, Tier A, Wiitala C, Villasis E (2014) Walkability on University Avenue

European Union. https://www.sustainabledevelopment.un.org. Accessed on 10 Nov 2019

Ewing R, Handy S (2009) Measuring the unmeasurable: urban design qualities related to walkability. J Urban Design 14(1):65–84

Feizizadeh B, Roodposhti MS, Jankowski P, Blaschke T (2014) A GIS-based extended fuzzy multi-criteria evaluation for landslide susceptibility mapping. Comput Geosci 73:208–221

Ford AM (2013) Walkability of campus communities surrounding Wright State University.

Galanis A, Eliou N (2011) Evaluation of the pedestrian infrastructure using walkability indicators. WSEAS Trans Environ Devel 7(12):385–394

Google Scholar. https://www.scholar.google.com. Accessed on 10 Jan 2020

Jacobs J (1961) The death and life of Great American cities. Random House, New York

Jensen OB (2013) Staging mobilities. Routledge, London

Kaczynski AT, Robertson-Wilson J, Decloe M (2012) Interaction of perceived neighborhood walkability and self-efficacy on physical activity. J Phys Activity Health 9(2):208–217

Kartschmit N, Sutcliffe R, Sheldon MP, Moebus S, Greiser KH, Hartwig S, Maier W (2020) Walkability and its association with prevalent and incident diabetes among adults in different regions of Germany: results of pooled data from five German cohorts. BMC Endocrine Disorders 20(1):1–9

Keat LK, Yaacob NM, Hashim NR (2016) Campus walkability in Malaysian public universities: a case-study of Universiti Malaya. Plann Malaysia 14(5)

Krambeck HV (2006) The global walkability index. Doctoral dissertation, Massachusetts Institute of Technology

Lami IM, Tavella E (2019) On the usefulness of soft OR models in decision making: a comparison of problem structuring methods supported and self-organized workshops. Eur J Oper Res 275(3):1020–1036

Lami IM, Abastante F, Bottero M, Masala E, Pensa S (2014) Integrating multicriteria evaluation and data visualization as a problem structuring approach to support territorial transformation projects. EURO J Decision Process 2(3–4):281–312

La Rocca RA (2010) Soft mobility and urban transformation. Tema J Land Use Mobil Environ 2

Lee S, Talen E (2014) Measuring walkability: a note on auditing methods. J Urban Design 19(3):368–388

Leslie E, Saelens B, Frank L, Owen N, Bauman A, Coffee N, Hugo G (2005) Residents' perceptions of walkability attributes in objectively different neighbourhoods: a pilot study. Health Place 11(3):227–236

Likert R (1932) A technique for the measurement of attitudes. Arch Psychol

Lo RH (2009) Walkability: What is it? J Urban 2(2):145–166

Lombardi P, Abastante F, Torabi Moghadam S, Toniolo J (2017) Multicriteria spatial decision support systems for future urban energy retrofitting scenarios. Sustainability 9(7):1252

Moayedi F, Zakaria R, Bigah Y, Mustafar M, Puan OC, Zin IS, Klufallah MM (2013) Conceptualising the indicators of walkability for sustainable transportation. Jurnal Teknologi 65(3)

Okabe A, Satoh T, Sugihara K (2009) A kernel density estimation method for networks, its computational method and a GIS-based tool. Int J Geogr Inf Sci 23(1):7–32

Open Data Charter. https://www.opendatacharter.net. Accessed on 10 Jan 2020

QGIS software. https://www.qgis.org. Accessed on 10 Jan 2020

Rogers SH, Gardner KH, Carlson CH (2013) Social capital and walkability as social aspects of sustainability. Sustainability 5(8):3473–3483

Saelens BE, Sallis JF, Frank LD (2003) Environmental correlates of walking and cycling: findings from the transportation, urban design, and planning literatures. Ann Behav Med 25(2):80–91

Scopus online. https://www.scopus.com. Accessed on 10 Jan 2020

Shatu F, Yigitcanlar T (2018) Development and validity of a virtual street walkability audit tool for pedestrian route choice analysis—SWATCH. J Transp Geogr 70:148–160

Solnit R (2005) Storia del camminare, vol 33. Pearson Italia Spa

Stake RE (1995) The art of case study research. Sage

United Nation General Assembly (2017) Global indicator framework for the sustainable development goals and targets of the 2030 agenda for sustainable development. United Nation, New York. Available at https://unstats.un.org/sdgs/indicators/indicators-list/. Accessed on 5 May 2020

Urry J (2016) Mobilities: new perspectives on transport and society. Routledge, London

Vennix JA, Akkermans HA, Rouwette EA (1996) Group model-building to facilitate organizational change: an exploratory study. Syst Dyn Rev J Syst Dyn Soc 12(1):39–58

Wibowo SS, Nurhalima DR (2018) Pedestrian facilities evaluation using pedestrian level of service (PLOS) for university area: case of Bandung Institute of Technology. In: MATEC web of conferences, vol 181. EDP Sciences, p 02005

Yin L (2017) Street level urban design qualities for walkability: combining 2D and 3D GIS measures. Comput Environ Urban Syst 64:288–296

Assessing the Level of Accessibility of Railway Public Transport for Women Passengers Using Location-Based Data: The Case of H2020 DIAMOND Project

Andrea Gorrini, Rawad Choubassi, Anahita Rezaallah, Dante Presicce, Ludovico Boratto, David Laniado, and Pablo Aragón

Abstract The chapter presents a data-driven approach based on the use of Geographic Information Systems and data analytics for assessing the level of accessibility for the women passengers of the railway network service managed by FGC— the Railway Agency in Catalunya (Spain). A series of geolocated open and proprietary datasets related to the land and sociodemographic and mobility characteristics of the Province of Barcelona and to the FGC's railway network has been analyzed and merged with disaggregated social-media data collected from Twitter. This was aimed at maximizing the diversity of station samples that will be observed, in order to ensure that the observed cases are representative of the different situations and locations of any single station. The selected stations are currently under investigation through on-site observations about universal design indicators and survey questionnaires focused on women passengers' needs and expectations. Within the objectives of the H2020 project DIAMOND, the final aim of the proposed research is to support the definition of guidelines and policies for the inclusion of women's needs in the design of future urban transport services.

Keywords Urban mobility · Location-based data · GIS · Data analytics · Inclusive transport system

1 Introduction

As highlighted by the European Charter for Women Rights in the City (1994) and by the 2030 Agenda for Sustainable Development adopted by all United Nations Member States in 2016 (i.e., SDG 5-Gender Inequalities; SDG 11.2-Sustainable

A. Gorrini (✉) · R. Choubassi · A. Rezaallah · D. Presicce
Systematica Srl, Via Lovanio 8, 20121 Milan, Italy
e-mail: a.gorrini@systematica.net

L. Boratto · D. Laniado · P. Aragón
Eurecat - Centre Tecnològic de Catalunya, Carrer de Bilbao, 72, 08005 Barcelona, Spain

P. Aragón
Universitat Pompeu Fabra, Roc Boronat 138, 08018 Barcelona, Spain

A. Bisello et al. (eds.), *Smart and Sustainable Planning for Cities and Regions*, Green Energy and Technology, https://doi.org/10.1007/978-3-030-57332-4_14

Transport for All), public transport should be designed to be gender-inclusive. Women experience and use transport systems differently than men (Duchene 2011). They prefer to use public transport, despite their more complex mobility patterns, since they are more concerned with economical aspects, accessibility, reliability, security issues related to aggression and harassment (International Transport Forum 2018).

As described in previous works already presented by the authors (García-Jiménez et al. 2020), the three main factors influencing the use of urban railway infrastructures by women are:

- Accessibility of the service (Turner 2012): operational details and quality of service related to different travel purposes of women (e.g., connection with other modes of transport, adequate travel and wayfinding information provision, flexible tariff policies, and discount fares);
- Design of the infrastructure (Peters 2013): architectural characteristics of the infrastructures (e.g., universal accessibility of the environment, furniture and facilities for the comfort of women and the other persons they accompany, cleanliness and maintenance of the infrastructures);
- Safety and security (Loukaitou-Sideris 2014): individual perception of security related to harassment, pickpocketing, emergency situations and overcrowding (e.g., security personnel, lighting, ventilation, CCTV).

In this regard, the H2020 DIAMOND research project[1] aims at transforming data from various sources into actionable knowledge for ensuring the inclusion of women's needs in the transport sector. The research follows a gender-sensitive approach that brings together urban and mobility experts, transport authorities, computer and data scientists, mobility economists, and social scientists.

According to the objectives of the DIAMOND project, the paper proposes a data-driven approach for investigating the level of accessibility of urban railway infrastructures considering the women's needs. This is based on the use of Geographic Information Systems (GIS) for the analysis of several geolocated open datasets related to the Province of Barcelona (Spain) and focused on: (i) land characteristics (urban fabric of land use; points of interest); (ii) sociodemographic characteristics of the inhabitants (population density; gender, age and nationality of the population); (iii) mobility characteristics (transport services; road infrastructures; parking facilities). This was aimed at identifying and characterizing a short list of suitable railway stations managed by Ferrocarrils de la Generalitat de Catalunya (FGC)—the Railway Agency in Catalunya,[2] as characterized by high and low levels of accessibility for women.

---

[1] Website: https://diamond-project.eu/.

[2] Within the objectives of the H2020 DIAMOND project, the research was focused also on the stations managed by Zarzad Transportu Miejskiego (ZTM) within the territory of the Metropolitan Area of Warsaw (Poland). Word-limit restrictions preclude the discussion of both scenarios of investigation.

Results of GIS-based analysis were merged with open and FGC's proprietary data related to the travel demand of the railway network and additional sources of user-generated social-media data from Twitter.

The shortlisted metro and urban railway stations are under investigation through on-site observations and questionnaires, to further characterize them according to universal design indicators and women's needs and expectations. The disaggregated data will be used to understand and trace the mobility patterns of women, to ask their opinion and to identify the factors important for them, in order to plan and design a fair, gender equitable and integrated urban public transport network.

The proposed methodology for the analysis of geolocated open data, proprietary data, and social media data is described in Sect. 2. The results achieved through the performed GIS analysis and data analytics are described in Sect. 3. The chapter concludes with final remarks and future works.

2 Methodology

The proposed methodological approach is based on analyzing and merging a series of large structured datasets (see Table 1), with reference to: (i) geolocated open data related to the land and sociodemographic and mobility characteristics of the Province of Barcelona; (ii) open and proprietary data related to the railway network managed by FGC; and (iii) user-generated social media data from Twitter.

The proposed approach was aimed at assessing the level of accessibility for the women commuting by the metro and urban railway infrastructures managed by FGC within the Province of Barcelona (82 stations in total), to identify and characterize a short list of ten suitable stations (five stations characterized by high levels of accessibility and five stations characterized by low levels of accessibility). A series of geolocated datasets were retrieved, sorted, and filtered from open-data repositories, national geoportals and census databases (see Table 1).

The indicators were analyzed to design a multi-layer map of the Province of Barcelona and to estimate the spatial distribution of each dataset considering the localization of the FGC's stations (see Fig. 1). From a general point of view, the analysis was based on various attributes and characteristics of the urban area surrounding each station. To do so, raw data related to the urban scale were extracted about surrounding areas of each station within a catchment area of 400 m (circular buffer areas of 0.5 km^2), commonly known as comfortable walkable distance (Buhrmann 2019).

Data were post-processed through: (i) density-based calculation on catchment areas; (ii) normalization of values (z values in a range between 0 and 1); and (iii) weighted formulas of normalized values to calculate the Territorial Data Index (TDI), Sociodemographic Data Index (SDDI), and Mobility Data Index (MDI). Quintile

Table 1 List of retrieved open data, proprietary data and social-media data that were analyzed and merged for the identification and characterization of a short list of relevant metro and urban railway infrastructures managed by FGC

Data typology	Indicator	Data source	Year
Preliminary data	Province of Barcelona	Cartographic Institute Catalunya	2019
	Census sections	National Statistics Institute	2019
	FGC stations	Barcelona's city hall data service	2019
Territorial data	Urban fabric of land use	Copernicus land monitoring	2012
	Points of interest	OpenStreetMap	2019
Sociodemographic data	Total population	National Statistics Institute	2011
	Gender of population	National Statistics Institute	2011
	Age of population	National Statistics Institute	2011
	Nationality of population	National Statistics Institute	2011
Mobility data	Public transports	OpenStreetMap	2019
	Road infrastructure	OpenStreetMap	2019
	Parking facilities	OpenStreetMap	2019
Railway network data	FGC Lines	FGC website	2019
	Tariff zones	FGC website	2019
	Under/above-ground	OpenStreetMap	2019
	Travel demand data	FGC's proprietary data	2017–2018
Social media data	Geospatial data	Twitter	2019–2020
	Time-based meta data	Twitter	2019–2020
	Users' profile meta data	Twitter	2019–2020

frequency distribution of results made possible the identification of the stations characterized by high levels of accessibility for the women (belonging to the highest quintile, $\geq$80th percentile) and the stations characterized by low levels of accessibility for the women (belonging to the lowest quintile, $\leq$20th percentile).

A group of 29 stations belonging to at least two highest or lowest quintiles among the TDI, SDDi, and MDI was identified. Then, the list was further shortened through the analysis of several open and FGC's proprietary data related to each station (see Table 1): (i) metro and urban railway lines; (ii) transport service tariff zone; (iii) underground or above-ground infrastructure; (iv) travel demand data (number of passengers per year). This helped to identify a short list of ten heterogeneous and

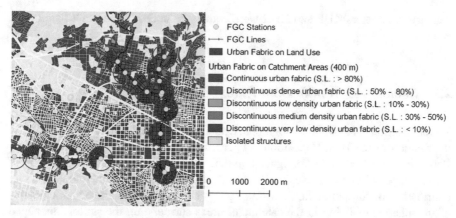

○ FGC Stations

⊢ FGC Lines

■ Urban Fabric on Land Use

Urban Fabric on Catchment Areas (400 m)

■ Continuous urban fabric (S.L. : > 80%)

■ Discontinuous dense urban fabric (S.L. : 50% - 80%)

□ Discontinuous low density urban fabric (S.L. : 10% - 30%)

■ Discontinuous medium density urban fabric (S.L. : 30% - 50%)

■ Discontinuous very low density urban fabric (S.L. : < 10%)

□ Isolated structures

0   1000   2000 m

Fig. 1 Spatial distribution of urban fabric on the territory of the Province of Barcelona and on catchment areas surrounding the stations managed by FGC

non-adjacent stations and characterized by positively and negatively relevant characteristics related to their level of accessibility. The selected ten stations were further characterized through the analysis of disaggregated social media data collected from Twitter, focusing on: (i) geospatial data; (ii) time-based metadata; and (iii) users' profile metadata.

2.1 Territorial Data Index

The calculation of the Territorial Data Index (TDI) was based on the density distribution of the urban fabric (UF_ca) and Points of Interest (PoI_ca) on the catchment areas surrounding the FGC's stations. Land-use dataset include the localization of continuous urban fabric, discontinuous dense urban fabric, and isolated structures. Data analysis was aimed at estimating the level of urbanization of the catchment areas since the level of accessibility of railway infrastructures for women greatly differs based on the urban or rural characteristics of the surroundings. Points-of-interest datasets include a series of heterogenous services and facilities (e.g., commercial activities, bars, supermarkets, playgrounds, sport facilities, nightclubs, university facilities, public services, tourist attractions, etc.). Data analysis was aimed at assessing the level of attractiveness of the catchment areas surrounding each station, considering the needs of different passengers' profile (e.g., commuters, students, tourists).

Equation 1 Territorial data index (TDI)

$$TDI = (KUF \cdot UF\_ca) + (KPoI \cdot PoI\_ca) \qquad (1)$$

TDI was calculated through the weighted summation of normalized density distribution values of urban fabric and points of interest on catchment areas (see Eq. 1). The constant parameters KUF and KPoI were equally balanced ($\sum$constant parameters = 1).

2.2 Sociodemographic Data Index

The calculation of the Sociodemographic Data Index (SDDI) was based on the density distribution of the Total Population (TP_ca), Female Population (FeP_ca), Elderly Population[3] (EP_ca), and Foreigner Population (FoP_ca) on the census section of the Province of Barcelona (Spain). Data analysis was aimed at estimating the density distribution of the population and the age, gender and nationality characteristics of the inhabitants living in the catchment areas surrounding the stations managed by FGC, as potential users of the railway transport service. The calculation of the SDDI relies on the density distribution of the population on the urban fabrics of the catchment areas surrounding the FGC's stations, to balance the population density between urban and rural areas.

Equation 2 Sociodemographic Data Index

$$SDDI = (KTP \cdot TP\_ca) + (KFeP \cdot FeP\_ca)$$
$$+ (KEP \cdot EP\_ca) + (KFoP \cdot FoP\_ca) \qquad (2)$$

SDDI was calculated through the weighted summation of normalized density distribution values of total population, female population, elderly population, and foreigner population on urban fabric of catchment areas (see Eq. 2). The constant parameters KTP (corresponding to 0.3), KFeP (corresponding to 0.3), KEP (corresponding to 0.2), and KFoP (corresponding to 0.2) were weighted to accentuate the impact of the density distribution of the total and female populations on SDDI ($\sum$constant parameters = 1).

2.3 Mobility Data Index

The calculation of the Mobility Data Index (MDI) was based on the density distribution of Public Transports (PT_ca), Road Infrastructure (RI_ca), and Parking Facilities (PF_ca) on the catchment surrounding the stations managed by FGC. Public transport datasets include the spatial distribution of the total number of metro and commuter railway stations, bus stops, tram stops, and taxi stations. Data analysis was aimed at estimating the level of connectivity of the railway infrastructures with other transport services, especially for the first- and last-mile connections. Data analysis of

---

[3]Elderly population dataset includes the spatial distribution of the inhabitants being over 64 years old.

road infrastructures and parking facilities datasets was aimed at analyzing the level of accessibility of the railway infrastructure by car, enabling multimodal mobility schemes.

Equation 3 Mobility Data Index

$$MDI = (KPT \cdot PT\_ca) + (KRI \cdot FRI\_ca) + (KPF \cdot PF\_ca) \qquad (3)$$

MDI was calculated through the weighted summation of normalized density distribution values of public transports, road infrastructure and parking facilities on catchment areas (see Eq. 3). The constant parameters KPT (corresponding to 0.4), KRI (corresponding to 0.3), and KPF (corresponding to 0.3) were weighted to accentuate the impact of public transport on MDI ($\sum$constant parameters = 1).

2.4 Railway Network Data

The above described methodology helped to define a list of 29 relevant stations belonging to at least the two highest or two lowest quintiles among the Territorial, Sociodemographic, and Mobility Data Indexes. The list was shortened through the analysis of a series of open datasets about the railway network (see Table 1), with reference to: (i) metro and urban railway lines (two branches, 15 lines in total); (ii) transport service tariff zones (six zones in total); (iii) underground or above-ground infrastructural characteristics of the stations; and (iv) characteristics of the settlements comprised in the catchment areas (city, town, village and countryside). This identified a short list of ten heterogeneous and non-adjacent stations, characterized by high and low levels of accessibility for the women.

Then, the selected stations were further characterized through the analysis of FGC's proprietary Travel Demand Data (TDD) (see Table 1) and focused on the average number of passengers per station during the years 2017 and 2018 (87 million of passengers in total in 2018).

2.5 Social-Media Data

The KALIUM tool (Napalkova et al. 2018) was used to retrieve data from the Twitter streaming API (see Table 1), focusing on: (i) geospatial data; (ii) time-based metadata; and (iii) users' profile metadata. All tweets geolocated in the Province of Barcelona and posted between December 3rd, 2019, and January 14th, 2020, were retrieved (about 60,000 tweets). Geospatial data analysis enabled to: (i) exclude the tweets with repetitive geographic information due to Twitter business accounts (about 52,000 tweets); (ii) extract a subsample of tweets localized within the catchment areas surrounding all the FGC's stations (881 tweets); and (iii) focus on the ten selected stations (104 tweets).

Then, the dataset was disaggregated considering the gender of the users estimated through the state-of-the-art tool M3inference (Wang et al. 2019). This tool relies on a multimodal deep-learning model trained on multilingual data to infer gender of users based on the username, short biotext, and profile picture. The tool returns the estimations of the probabilities of a user to be a woman or a man; to achieve reliable results, a threshold of 0.9 was employed (i.e., only users with estimated probability of being man or woman higher than 90%). This made possible investigating gender difference in the activity of Twitter users in the catchment areas surrounding the FGC's stations.

3   Results

The proposed methodology helped to identify those suitable FGC's stations belonging to the highest and lowest quintile of Territorial Data Index (see Table 2 and Fig. 2), Sociodemographic Data Index (see Table 2 and Fig. 3), and Mobility Data Index (see Table 2 and Fig. 4).

Results helped to define a list of 29 suitable stations belonging to at least two highest or lowest quintiles among TDI, SDDI, and MDI. The list was shortened

Table 2 Quintile frequency distribution of TDI, SDDI, and MDI values and related indicators

	Indicators/index	Highest quintile	Lowest quintile
UF	Km^2 urban fabric on catchment area	0.643 ($\pm$0.052 sd)	0.068 ($\pm$0.049 sd)
PoI	No. PoI on catchment area	144.938 ($\pm$132.270 sd)	9.938 ($\pm$10.016 sd)
TDI	Territorial data index	0.928 ($\pm$0.044 sd)	0.106 ($\pm$0.034 sd)
TP	Inhabitants per km^2 on UF	12,052.820 ($\pm$3038.486 sd)	32.453 ($\pm$28.922 sd)
FeP	Female inhabitants per km^2 on UF	6465.732 ($\pm$1678.948 sd)	16.255 ($\pm$14.594 sd)
EP	Elderly inhabitants per km^2 on UF	2442.324 ($\pm$786.926 sd)	4.722 ($\pm$4.806 sd)
FoP	Foreigner inhabitants per km^2 on UF	1677.630 ($\pm$543.122 sd)	3.254 ($\pm$5.204 sd)
SDDI	Sociodemographic data index	0.925 ($\pm$0.039 sd)	0.165 ($\pm$0.01 sd)
PT	No. public transports on catchment area	14.067 ($\pm$7.035 sd)	0.600 ($\pm$0.737 sd)
RI	Km road infrastructure on catchment area	14.982 ($\pm$5.936 sd)	6.315 ($\pm$1.732 sd)
PF	No. parking facilities on catchment area	5.867 ($\pm$3.944 sd)	0.067 ($\pm$0.258 sd)
MDI	Mobility Data Index	0.905 ($\pm$0.061 sd)	0.115 ($\pm$0.019 sd)

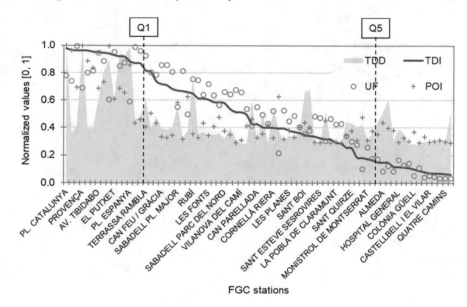

Fig. 2 Results of territorial data index (TDI) and travel demand data (TDD) related to the FGC's stations. Q1 and Q5 refer to the 80th and 20th percentile of quintile frequency distribution of TDI

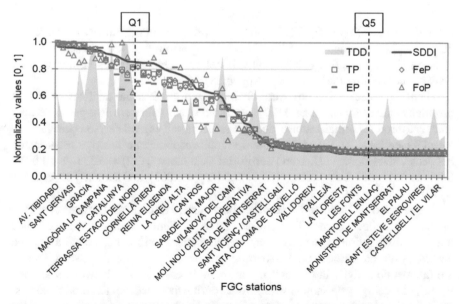

Fig. 3 Results of sociodemographic data index (SDDI) and travel demand data (TDD). Q1 and Q5 refer to the 80th and 20th percentile of quintile frequency distribution of SDDI

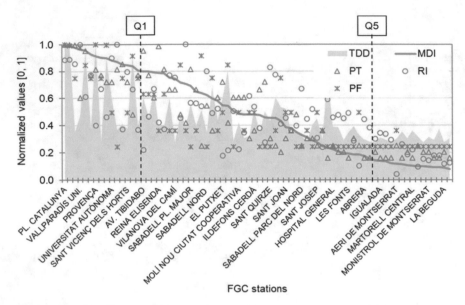

Fig. 4 Results of mobility data index and travel demand data (TDD). Q1 and Q5 refer to the 80th and 20th percentile of quintile frequency distribution of MDI

through a series of open and FGC's proprietary data about the railway network (see Table 1): This identified a short list of ten heterogeneous and non-adjacent stations, characterized by positively and negatively relevant characteristics related to their level of accessibility (see Table 3 in Appendix).

Then, the selected stations were characterized through the analysis of FGC's proprietary travel demand data (TDD) (see Table 1), and data were post-processed and analyzed through normalization of average values (z values) and Pearson correlations. Results (see Figs. 2, 3, and 4) show a significant moderate positive relationship between Travel Demand Data and Territorial Data Index ($r(72) = 0.531, p < 0.001$), Sociodemographic Data Index ($r(72) = 0.540, p < 0.001$), and Mobility Data Index ($r(72) = 0.568, p < 0.001$).

The selected ten stations were further characterized through the analysis of social-media data collected from Twitter (see Table 1). Figure 5 shows the combined temporal distribution of tweets during daytime hours (on the horizontal axis) and weekdays (on the vertical axis). Tweeting activity is concentrated between 7 and 2 am (almost no activity in the night), with more intense activity between 9 am and 9 pm. The weekly cycle shows lower activity during the weekend. Results are consistent with the circadian rhythms of human activity. Gender was inferred through the M3inference tool with probability higher than the 0.9 threshold for 76% of the users; among these, only 25.7% were women, and 74.3% were men.

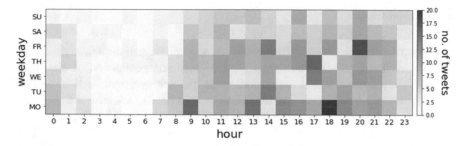

Fig. 5 Temporal distribution of the tweets geolocated within the catchment areas surrounding all the FGC's stations

4 Discussion

Urban informatics is becoming one of the most promising approaches for urban mobility and transport planning (Batty 2013). In this regard, the paper was based on the application of GIS and data analytics for maximizing the diversity of station samples that will be observed, in order to ensure that the observed cases are representative of the different situations/locations of any single station.

GIS-based analysis aimed to assess the level of accessibility for the women passengers of the railway stations managed by FGC in the Province of Barcelona through the investigation of: (i) the level of urbanization and attractiveness of the areas surrounding each station in terms of urban/rural contexts, available services, and facilities (TDI); (iii) the sociodemographic characteristics of the population living in the areas surrounding each stations (SDDI); and (iv) the level of connectivity of the railway infrastructures with other public transport services (MDI).

Results were merged with additional open datasets related to the railway network (e.g., metro and urban railway lines; transport service tariff zones; underground or above-ground infrastructural characteristics of the stations; and characteristics of the urban settlement comprised in the catchment areas). This enabled to identify and characterize a short list of ten heterogeneous and non-adjacent railway stations. Then, the analysis of proprietary data allowed to correlate the overall level of accessibility for women of the railway infrastructures with the number of passengers per year per station.

Data analytics technologies have been applied further to characterize the FGC's stations through disaggregated social media data collected from Twitter. Although the presented social-media data collection campaign was potentially biased since it overlapped with Christmas holidays, it represents a valuable example of the potential of this methodological approach. Indeed, the research work was aimed at investigating the possibility to analyze digitally widespread data sources as a valuable support of the activity of decision-makers by unveiling hidden patterns and specific target-users' needs.

5 Conclusion

The objective of the paper was: (i) to identify an appropriate sample of railway stations to be further investigated through ongoing social-media data collection, on-site observations focused on universal design indicators and survey questionnaires focused on women passengers' needs and expectations; and (ii) to seek relevance between location-based characteristics and overall accessibility level of the railway infrastructures for the women (after collecting data on site). Within the objectives of the H2020 DIAMOND project, the collected disaggregated data will be used to support the definition of guidelines and policies for the inclusion of women passengers' needs in the design of future transport services.

Acknowledgements This work was supported by the EU H2020 program within the DIAMOND project (grant No. 824326). The analyzed data were treated according to the General Data Protection Regulation (EU, 2016/679) and within the authorization of the DIAMOND Ethics Board. The authors thank Ferrocarrils de la Generalitat de Catalunya, partner of the DIAMOND project, for sharing proprietary data about the railway network. The authors thank Nicola Ratti (Systematica Srl) for his fruitful contribution.

Appendix

See Table 3.

Table 3 Results of the performed analysis aimed at selecting and characterizing a short list of suitable stations managed by FGC within the Province of Barcelona (Spain)

FGC stations	Metro/Railway lines	Tariff zones	Station Characteristics	Settlement Characteristics	TDI	SDDI	MDI	TDD
SARRIÀ	L6/L12/S1/S2/S5/S6/S7	1	Under ground	City	0,920	0,944	0,955	0,947
PÀDUA	L7	1	Under ground	City	0,962	0,961	0,846	0,414
PLAZA ESPANYA	L8/S3/S4/S8/S9/R5/R50/R6/R60	1	Under ground	City	0,883	0,856	0,991	0,987
SANT CUGAT	S1/S2/S5/S6/S7	2	Above ground	Town	0,917	0,586	0,834	0,920
CAN ROS	S3/S4/S8/S9/R5/R6	2	Above ground	Town	0,965	0,647	0,977	0,354
SANT JOAN	S2/S6	2	Above ground	Village	0,065	0,163	0,392	0,522
GUALADA	R6/R60	6	Above ground	Town	0,514	0,163	0,145	0,312
EL PALAU	S4/S8/R5/R6	2	Above ground	Village	0,071	0,164	0,107	0,318
MARTORELL ENLLAC	S4/S8/R5/R6	3	Above ground	Town	0,082	0,167	0,085	0,305
CASTELLBELL I EL VILAR	R5	5	Above ground	Village	0,073	0,163	0,149	0,276

References

Batty M (2013) The new science of cities. MIT Press

Buhrmann S, Wefering F, Rupprecht S (2019) Guidelines for developing and implementing a sustainable urban mobility plan, 2nd edn. Rupprecht Consult-Forschung und Beratung GmbH

Duchene C (2011) Gender and transport. OECD iLibrary

European Commission (1994) European charter for women in the city. Moving towards a gender-conscious city. https://www.hlrn.org/img/documents/1994-EuropeanCharterforWomeni ntheCity.pdf. Accessed on 5 Dec 2019

García-Jiménez E, Poveda-Reyes S, Molero GD, Santarremigia FE, Gorrini A, Hail Y, Ababio-Donkor A Leva MC, Mauriello F (2020) Methodology for gender analysis in transport: factors with influence in women's inclusion as professionals and users of transport infrastructures. In: Sustainability, 12(9), 3656. International transport forum (2018). Women's safety and security: a public transport priority. OECD Publishing

Loukaitou-Sideris A (2014) Fear and safety in transit environments from the women's perspective. Security J 27(2):242–256

Napalkova L, Arag´on P, Robles JCC (2018) Big data-driven platform for cross-media monitoring. In: Proceedings of IEEE 5th international conference on data science and advanced analytics (DSAA), pp 392–399

Peters D (2013) Gender and sustainable urban mobility, Nairobi. https://www.unhabitat.org/grhs/2013. Accessed on 21 Nov 2019

Turner J (2012) Urban mass transit, gender planning protocols and social sustainability—the case of Jakarta. Res Transp Econ 34(1):48–53

United Nation (2016) Transforming our world: the 2030 agenda for sustainable development. United Nations Secretariat, New York

Wang Z, Hale S, Adelani DI, Grabowicz P, Hartman T, Jurgens D (2019) Demographic inference and representative population estimates from multilingual social media data. In: Proceedings of world wide web conference, pp 2056–2067

New Value Propositions in Times of Urban Innovation Ecosystems and Sharing Economies

Assessing Integrated Circular Actions as Nexus Solutions Across Different Urban Challenges: Evidence Toward a City-Sensitive Circular Economy

Maria Beatrice Andreucci and Edoardo Croci

Abstract Cities across the world are actively exploring the circular economy concept, a key urban planning and design approach for the green transition, simultaneously enabling greater energy and material efficiency and lower pollution, as well as job creation, social inclusion, human health, and well-being. The city can be viewed as a complex socio-ecological system, in which infrastructures and urban forms have co-evolved along with sociocultural practices and the lifestyles of urbanites. Circular design and systemic thinking have not yet been incorporated into the planning and design of the urban built environment, and this limit has progressively created vulnerabilities and risks. Among the various urban resources, available land is often scarce, as it is natural landscape. Consequently, it is particularly important that vacant public space is re-functionalized and brownfield sites are restored. Equally, green infrastructure—urban forests, green roofs, green walls, permeable pavements, and constructed wetlands—provides critical ecosystem services (supporting, provisioning, regulating, and cultural services) at different scales: building, district, city, and region. Green elements and systems in the urban built environment regulate climate, air, and water quality; enable nutrient and water cycling and soil formation; provide space for growing food and for recreation. Using a mixed methods approach, including a literature review and case study analysis, the research identifies several opportunities and challenges to integrated circular actions, "nexus solutions" across various urban challenges, i.e., sociocultural, economic and financial, regulatory, political, institutional, ecological, environmental, and technological. The study then focuses on critical dilemmas faced when implementing nexus solutions. Providing an overview of selected international initiatives, the contribution, leveraging on an extensive interdisciplinary research, aims at showcasing how districts and cities are advancing the circular economy concept in practice. Evidence provided by projects and case studies—such as: Freshkills Park, a landfill reclamation project on Staten

M. B. Andreucci (✉)
Department of Planning Design, Technology of Architecture, Sapienza University of Rome, Via Flaminia 72, 00196 Rome, Italy
e-mail: mbeatrice.andreucci@uniroma1.it

E. Croci
Bocconi University, GREEN Research Center, Via Roentgen 1, 20136 Milan, Italy
e-mail: edoardo.croci@unibocconi.it

© The Author(s), under exclusive license to Springer Nature Switzerland AG 2021
A. Bisello et al. (eds.), *Smart and Sustainable Planning for Cities and Regions*,
Green Energy and Technology, https://doi.org/10.1007/978-3-030-57332-4_15

215

Island, in New York City; Royal Seaport, a major urban regeneration project in Stockholm; and Buiksloterham, a neighborhood and an urban living lab in Amsterdam North—are provided, aiming at testing and validating circularity at different scales. The outcomes of the conducted study identify in particular the impacts, both positive (benefits) and negative (trade-offs) of incorporating circularity into the urban planning and design processes, as well as how these can be assessed in order to stir robust systemic change in the long term.

Keywords Circular economy · Ecosystem services · Nexus · Well-being of citizens · Adaptive monitoring

1 Introduction

Urban areas currently host 55% of the world's population, projected to increase to 68% by 2050 (United Nations 2019) and are recognized as central nodes of the global economy. Urban areas spatially aggregate people, infrastructures, services, and financial resources, which generate great pressure on natural resources and evoke relevant social, environmental and economic challenges. Even though they occupy less than 2% of the Earth's surface, cities consume 60–80% of natural resources globally; they produce 50% of global waste and 75% of greenhouse gas emissions (Camaren and Swilling 2012), while the global urban footprint will triple over the years to 2030 (Seto et al. 2012).

Circularity is recognized as the paradigm that can enable more efficient production and consumption patterns, overcoming the unsustainability of the traditional linear economy. By "closing the loop" and adopting a regenerative perspective, the circular economy entails a more efficient use of resources, valuing and reusing waste in new production processes (Ellen Mac Arthur Foundation 2018).

Cities are ideal contexts to test circularity principles. The proximity of people and materials in the urban environment favors the possibility to reduce structural waste with a circular approach. Cities are also nodal points of innovation. The emergence of new technologies in sensoring, high connectivity, IoT, and big data is fostering new management practices and operations of urban systems and networks. Cities also hold the scale required to provide innovative business models and interaction patterns between firms, consumers, and local authorities.

City governments have relevant powers over spatial planning, mobility planning, building standards, and energy and water and solid-waste management, which can be leveraged to promote a more sustainable urban development thanks to circularity. Furthermore, city governments can integrate circular principles into their procurement criteria and practices, setting the example for other urban actors (URBACT 2016).

The need to foster the transition of cities toward more sustainable models is also advocated in the main policy frameworks, including the 2030 Agenda for Sustainable Development (2015), in particular Goal 11 mandating making cities and human settlements inclusive, safe, resilient and sustainable, the UN Habitat New Urban Agenda (2016) and the new European Urban Agenda (2016); the latter identifies the circular economy as one of the twelve critical urban challenges (Croci and Lucchitta 2018).

According to the Ellen Mac Arthur Foundation (2015), a possible definition of "circular city" implies the following main features that urban systems should pursue to increase efficiency, sustainability and minimize waste:

- built environment designed in a modular and flexible manner;
- energy systems that are resilient, renewable, distributed and enable effective energy use;
- urban mobility system that is accessible, affordable, clean and effective;
- urban bio-economy where nutrients will be appropriately returned to the soil;
- production systems that encourage "local value loops" and minimize waste.

Given the complexity of cities, all these systems are interconnected. Urban circularity should therefore be linked to the concept of urban metabolism, which looks as cities as living super-organisms in which there are continuous flows of inputs and outputs (UNEP 2017).

In spite of several sectoral applications of circularity in cities, an integrated assessment of sustainability of circular integrated experiences in cities is still in its initial stage (Marin and De Meulder 2018).

Measuring the circularity level in an urban context is urgently needed. Several monitoring approaches have been proposed to measure a city's performances (C40 and Climate-KIC 2018), leading to the development of several urban sustainability indicators and indexes (Science for Environment Policy 2018), which also consider the implementation of the Sustainable Development Goals at the urban scale (Sachs et al. 2019; SDSN and Telos 2019).

Nevertheless, common and standardized city-level indicators are still lacking (OECD 2019), even if some cities have started to develop their own circularity monitoring frameworks.

In the design of urban development policies, some key circularity principles have emerged: circular supply privileging recycled and bio-based materials; resource recovery along value chains; buildings and infrastructures' life extension and refurbishment; sharing models of services and spaces; and product as a services innovation favoring dematerialization.

The integration of policies aimed at resource efficiency, circularity, and resilience presents relevant innovation opportunities to generate high added values beneficial to multiple stakeholders, as demonstrated by the projects, currently in progress, hereby presented and discussed: Freshkills Park in New York, Royal Seaport urban regeneration project in Stockholm, and Buiksloterham, a neighborhood and an urban living lab in Amsterdam. Providing an overview of selected international initiatives, the contribution, leveraging extensive interdisciplinary research, aims at showcasing how districts and cities are advancing the circular economy concepts in practice.

2 Method

Using a mixed methods approach, including literature review and case studies analysis, the research identifies several opportunities and challenges to integrated circular actions and *nexus solutions* across various urban challenges, i.e., sociocultural, economic and financial, regulatory, political, institutional, ecological, environmental, and technological. The study then focuses on critical dilemmas confronting implementing nexus solutions, showcasing how districts and cities are advancing the circular economy concepts in practice.

Only recently (2018) did the European Commission, in the circular economy action plan, commit to develop a simple and effective monitoring framework, even if not specifically focused on cities. Beyond strengthening the logical value chains among the different circular concepts, the goal of the research is, therefore, to provide input and feedback to practitioners and decision-makers by highlighting how districts and cities can be made healthier, safer, smarter, and climate-resilient through evidence-based design and adaptive monitoring. This adaptive approach acknowledges that action is necessary or appropriate with imperfect knowledge and that initial actions can be refined as more information becomes available (Andreucci 2017, 2019).

Literature and research identify several opportunities and challenges to integrated circular actions across different urban aspects, i.e., sociocultural, economic and financial, regulatory, political, institutional, ecological, environmental, and technological. Understanding and making explicit the synergies and trade-offs across a variety of outcomes and actors has proved to be critical, aiming at structuring a decision-making process in which policy makers can consider multiple objectives simultaneously.

3 Results

3.1 Design Out Waste and Pollution

It was not until the second half of the twentieth century that artists and designers, with a growing awareness of environmental issues, focused attention on the multiple relationships between waste management, public awareness, functionality, and aesthetics. Artists and designers began to challenge traditional limits and started considering landfills as settings for artwork, sport and recreation facilities, eye-catching monuments in the landscape.

The emergence of contemporary projects dealing with waste landscapes generates innovative approaches that address key urban challenges and dilemmas, transforming existing waste and landfills into productive, safe, inviting and publicly accessible green infrastructure (Andreucci 2019). Post-industrial sites are capable of bringing people closer to landfills and facilities by integrating educational, sport and recreational activities within the everyday urban environment. This approach represents ways to develop workable strategies aimed at transforming the liability of waste management into socially attractive assets in which everyone participates.

Freshkills Park is a landfill reclamation project located in western Staten Island in New York City, along the banks of the Fresh Kills estuary. Prior to development, the area primarily consisted of tidal creeks and coastal marsh. Although the Fresh Kills landfill was originally activated in 1948 as a temporary solution to the city's increasing waste management challenges, it remained open for 53 years and received as much as 29,000 tons of trash per day during its peak years of operation, becoming the largest landfill in the world and "the only other manmade object, besides the Great Wall of China, which could be seen from space" (Trash Timeline n.d.).

Although officially closed on March 22, 2001, Fresh Kills was briefly reopened after the 9/11 World Trade Center attack in order to accept 1.2 million tons of debris from the fallen towers during a ten-month recovery effort. The 9/11 materials were screened and sifted for human remains, and the rest of the debris was placed in a 48-acre area on top of the West Mound, where a US flag in memory of the victims was installed (Freshkills Park n.d.) (Fig. 1).

The area is one of the few remaining vast open spaces in the City of New York, and consequently, the Department of City Planning and the architectural firm James Corner Field Operations developed a plan to convert this unique site into a contemporary urban park and thus alleviate the city's need for open space, while restoring the natural environment. Freshkills Park once completed, in 2036, will be almost

Fig. 1 Fresh kills, Staten Island, New York. Image credito: Maria Beatrice Andreucci

three times the size of Central Park and the largest park development project in the city for the past hundred years (Harnik et al. 2006).

3.2 Regenerate Ecosystems for People and the Environment

Royal Seaport is a major urban regeneration project in Stockholm, Sweden. This area is a brownfield redevelopment site, surrounded by valuable green and water ecosystems. The northern area of Hjorthagen is currently under construction, with several residential buildings completed. Industrial areas (e.g., a gas works plant and district heating plant) on site will soon be closed down. Two cruise ship terminals are still functioning successfully, attracting tourists and visitors.

The Strategic Urban Development Plan for the Stockholm Royal Seaport ultimately aims at providing health, safety, comfort, and better quality of life in the district. Such aims can only be achieved through a balanced portion of environmental, cultural-social, economic and governance sustainability within an ongoing business-oriented spatial development (Huang et al. 2017).

In order to tackle the above challenges and dilemmas, the existing plan for Royal Seaport has set environmental targets of reaching CO_2 emissions not more than 1.5 tons per capita by 2020 and to become fossil-fuel free by 2030. These targets are being supported by the applications of cutting-edge environmental technologies, including smart grids, smart communications, eco-cycle waste management, biogas and electric cars, and sustainable buildings in the form of active houses (Fig. 2).

It must be noted that Royal Seaport took a holistic people-oriented approach through the development of various activities that encourage more active engagement by the residents, not only as users or consumers, but as design activists during all stages of planning, design, and monitoring of the implementation toward highly ambitious circular economy targets and environmental standards.

Fig. 2 Royal Seaport, aerial view. Image credits: www.stockholmroyalseaport.com

Through the vision "The Way We Green", Stockholm is proposing that the Royal Seaport should not only become a world-class circular city district but most importantly should be transformed into a livable waterfront district, for its residents to live and work in good health, comfort, safety, and with high quality of life. Thus, the Stockholm Royal Seaport is designed with the focus not only on economic and environmental sustainability, but also putting emphasis on equity and inclusiveness.

3.3 Keep Construction Materials, Components, and Systems in Use

Buiksloterham is a neighborhood and an urban living lab in Amsterdam North. It is in a unique position to serve as both a living test bed and catalyst for Amsterdam's broader transition to becoming a circular, smart and bio-based city.

Within Amsterdam, Buiksloterham is a peculiar case: Though it has been treated as a functionally marginal district because of its industrial history, it is located close to the historical center of Amsterdam, across the IJ River (Gladek et al. 2014).

Unlike most other centrally located neighborhoods, Buiksloterham is a comparatively blank canvas with many empty spaces and almost no heritage buildings. This status creates space and offers flexibility for new developments (Gladek et al. 2014). Buiksloterham has been conceived as a device for the broader transition of Amsterdam. Its polluted lands and bare plots are gradually becoming the center of the implementation of innovative clean technologies and a hub for the closure of urban material cycles. People and activities needed to "close the loop" are critical resources and act as a driver for the local business, while establishing local social ecological systems. IT-based programs and devices will connect smart residents with one another and boost the effectiveness of resource flows. Urban biodiversity and climate-adaptation measures are conceived as an integrated strategy to bring long-term benefits to the area. Buiksloterham consequently can represent a blueprint and a living dynamic lab, i.e., a virtuous example for other marginal areas worldwide that can effectively be transformed into powerful engines for inclusiveness, diffused quality of life, and business regeneration in cities (Fig. 3).

4 Discussion

A circular economy promotes the adoption of state-of-the-art integrated systems of waste management leveraging on appropriate technological development (Schumacher 1973) and nature-based solutions, as well as the improvement of practices for the protection of nutrients and soil and the conservation of materials and energy resources, as opposed to relying on fossil fuels and to encourage indiscriminate consumption. A circular economy identifies and transforms the negative impacts of

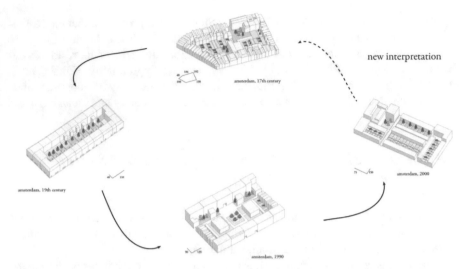

Fig. 3 Buiksloterham for circular Amsterdam by Studioninedots and Delva landscape architects

industrial activities that threaten human health and our natural capital, acknowledging that the majority of costs that we are directly or indirectly paying today are due to GHGs (36%), water consumption (26%), and improper land use (25%) (Trucost, 2013). Under this logic, Freshkills Park in New York City, once completed, will transform the liabilities of a degraded and polluted land into an attractive opportunity for inclusiveness and recreation. Addressing impacts from air, land, and water pollution will also result in significant environmental and social costs savings.

Circular economic model seeks to decouple real-estate development from soil consumption and natural capital depletion. A circular real-estate sector addresses resource-related concerns and promotes balanced solutions progressing business models and bottom-up initiatives able to generate sustainable growth, create jobs, and reduce environmental impacts such as carbon emissions. As the call for business models based on systems thinking grows louder, "design for people" practices leveraging the alignment of technological factors and social needs can empower the urban transition to a circular economy. The redevelopment of Royal Seaport in Stockholm—preserving and enhancing its natural capital by controlling the brownfield requalification and balancing business with quality of life—represents an inspiring example of how natural systems restoration combined with active people engagement can produce synergies at various different scales.

A circular economy favors activities that preserve existing values. With the right design approach, new systems and services to reuse materials and save energy, labor, and other resources used to generate them can be created. Virtuous production cycles characterize and differentiate the circular economy model from the "disposal and recycling" one, where large amounts of embodied energy and natural resources are inevitably lost. When aiming at "closing the loop", artifacts are designed using good quality materials and optimized for disassembly and components' reuse, making it easier and faster to transform or refurbish them. This implies designing for reuse, remanufacturing, and recycling to keep construction materials, components, and systems circulating in the economy, while advancing innovative businesses and processes also through the adoption of new business models. The economic co-benefits include increases in overall consumption and spending encouraged by lower prices. In "Buiksloterham for circular Amsterdam", making effective use of innovative clean technologies and closing local material flows will certainly act as a powerful driver for jointly promoting the local industry, the overall city image, and ultimately the quality of life of its inhabitants.

5 Conclusion

Cities across the world are just starting to implement the circular economy approach, as it has been only recently recognized as a key urban planning and design paradigm for a green new deal, simultaneously enabling greater production efficiency and lower emissions, as well as job creation, social inclusion, human health, and well-being, at all scales.

The implementation of an integrated circular vision in cities can bring significant economic, social and environmental benefits. It can foster the emergence of wealthy, robust, and inclusive cities. Such wealthy cities are where economic efficiency increases through improved mobility, designed out waste, and lower costs and where new developments and business opportunities support innovation, skills development, and job creation. Robust cities have improved quality of life and urban metabolism, reduced emissions and soil pollution, and enhanced social interactions. Inclusive cities prioritize caring for social needs while keeping materials and systems in use, thus reducing pressures on scarce resources and minimizing conflicts, engaging with both local communities and other relevant stakeholders, and taking advantage of digital technology (Ellen Mac Arthur Foundation 2019).

Multiple benefits can be achieved by changing how cities are planned, restored, financed, and managed. An integrated and balanced vision is needed to help address key urban challenges impacting social integration, housing, mobility, and sustainable economic growth. The circular urban economy advances the 2030 Sustainable Development Goals, as well as the UN Habitat New Urban Agenda and the new European Urban Agenda, supporting the reduction of greenhouse gas emissions and the adaptation to and mitigation of the effects of climate change.

Decision-makers and politicians are critically positioned vis-à-vis the success of a true transition to a circular economy—visionary administrators are the ones who can empower, manage, and engage key stakeholders from across the public and private sectors, using a wide range of policies and measures, encouraging innovative business models and inclusive approaches. Ultimately, the call is for a collaborative, holistic transition integrating resilience, efficiency and circular economy principles to benefit livable, inclusive, and competitive urban environments, as well as the people who animate them.

Acknowledgments Conceptualization, MBA; Methodology and case study development, MBA; Writing—original draft, MBA; Writing—review and editing, MBA, EC; Funding, MBA. All authors have read and agreed to the published version of the manuscript.

References

Andreucci MB (2017) Progettare Green Infrastructure. Tecnologie, valori e strumenti per la resilienza urbana. Wolters Kluwer, Milano, Italy

Andreucci MB (2019) Progettare l'involucro urbano. Casi studio di progettazione tecnologica ambientale. Wolters Kluwer, Milano, Italy

C40 and Climate KIC (2018) Municipality-led circular economy case studies. Available at https://www.climate-kic.org/wp-content/uploads/2019/01/circular-cities.pdf

Camaren P, Swilling M (2012) Sustainable resource efficient cities: making it happens. UNEP, Nairobi, Kenya

Croci E, Lucchitta B (2018) Introduction to special session nature-based solutions (NBSs) for urban resilience. Econ Policy Energy Environ 2:2–4

Ellen MacArthur Foundation (2015) Delivering the circular economy—a toolkit for policy-makers. Available at https://www.ellenmacarthurfoundation.org/assets/downloads/publications/EllenMacArthurFoundation__PolicymakerToolkit.pdf

Ellen MacArthur Foundation and Arup (2018) The circular economy opportunity for urban and industrial innovation in China. Available at https://www.ellenmacarthurfoundation.org/publications/chinareport

Ellen Macarthur Foundation and Arup (2019) Circular economy in cities. Project guide. Available at https://www.ellenmacarthurfoundation.org/assets/downloads/CE-in-Cities-Project-Guide_Mar19.pdf

European Commission (2015) Communication from the Commission to the European Parliament, the Council, the European Economic and Social Committee and the Committee of the Regions. COM (2015) 614 final Closing the loop—an EU action plan for the circular economy. Brussels 2.12.2105. Available at https://eur-lex.europa.eu/legal-content/EN/TXT/?uri=CELEX:52015DC0614

Freshkills Park: Site History (n.d.) Retrieved on March, 2019 from the New York City Department of Parks and Recreation web site. Available at https://www.nycgovparks.org/park-features/freshkills-park/about-the-site#tabTop

Huang H, Cai J, Gao X, Winarti A, Johani J (2017) Stockholm royal seaport. Thriving sustainable community. Available at https://www.academia.edu/19900873/Stockholm_Royal_Seaport_Redevelopment_Masterplan

Gladek E (Metabolic), van Odijk S (Metabolic), Theuws P (DELVA Landscape Architects), Herder S (Studioninedots) (2014) Circular Buiksloterham—transitioning Amsterdam to a circular city. Available at https://buiksloterham.nl/engine/download/blob/gebiedsplatform/69870/2015/28/CircularBuiksloterham_ENG_Executive_Summary_05_03_2015.pdf?app=gebiedsplatformand class=9096andid=64andfield=69870

Trash Timeline (n.d.) Retrieved on March 30, 2019 from the BFI Waste Services of Salinas web site https://www.bfi-salinas.com/kids_trash_timeline-printer.cfm

Harnik P, Taylor M, Welle B (2006) From dumps to destinations: the conversion of landfills to parks. Places, 18(1):83–88. Available at https://places.designobserver.com/press/pdf/From_Dumps_to__417.pdf

Marin J, De Meulder B (2018) Interpreting circularity. Circular city representations concealing transition drivers. Sustainability 10(5):1310. https://doi.org/10.3390/su10051310

OECD (2019) Regional statistics and indicators. Available at https://www.oecd.org/governance/regional-policy/regionalstatisticsandindicators.htm

Royal Seaport (2018) Sustainability report (2018). Available at https://www.stockholmroyalseap ort.com

Sachs J, Schmidt-Traub G, Kroll C, Lafortune G, Fuller G (2019) Sustainable development report 2019. Bertelsmann Stiftung and Sustainable Development Solutions Network (SDSN), New York

Science for Environment Policy (2018) Indicators for sustainable cities. In-depth Report 12. Produced for the European Commission DG Environment by the Science Communication Unit, UWE, Bristol. Available at https://ec.europa.eu/science-environment-policy

SDSN and Telos (2019) The 2019 SDG index and dashboards report for European cities (prototype version). Available at https://s3.amazonaws.com/sustainabledevelopment.report/ 2019/2019_sdg_index_euro_cities.pdf

Seto K, Güneralp B, Hutyra L (2012) Global forecasts of urban expansion to 2030 and direct impacts on biodiversity and carbon pools. Proc Natl Acad Sci USA 109:16083–16088

Schumacher EF (1973) Small is beautiful. A study of economics as if people matter. Blond and Briggs, London, UK

Sustainable Development Solutions Network (2019) Available at https://www.unsdsn.org/cities

Trucost PLC (2013) Natural capital at risk: the top 100 externalities of business. Available at www. trucost.com

UNEP (2017) UN environment annual report. Available at https://www.unenvironment.org/annual report/2017/index.php

United Nations, Department of Economic and Social Affairs, Population Division (2019) World urbanization prospects: the 2018 revision (ST/ESA/SER.A/420). New York, United Nations

URBACT (2016) Driving change for better cities. Available at https://urbact.eu/

Urban Agenda for the EU, Circular Economy Action Plan (2018)

Build or Reuse? Built Environment Regeneration Strategies and Real Estate Market in Seven Metropolitan Cities in Italy

Alessia Mangialardo and Ezio Micelli

Abstract To redevelop the existing city without consuming additional land, there are many regeneration strategies. The choice between the various strategies depends essentially on the expectations of developers. This study considers two in particular: the demolition and reconstruction of obsolete buildings and the reuse of existing assets. The research examines the feasibility conditions of the two strategies, highlighting the aspects that favor demolition and reconstruction over reuse with a model that holds together spatial and economic variables. The model is tested in seven metropolitan cities. The results show that demolition and reconstruction is an option that can only be pursued under favorable settlement and market conditions, forcing smaller cities to focus on recovery strategies for existing assets.

Keywords Demolition and reconstruction · Urban regeneration strategies · Building upcycle · Urban reuse · Building retrofit

1 Introduction

The urban regeneration approaches promoted by local administrators and private investors are different and depend on the economic feasibility of the owners and developers. Among these, the scrapping of the existing city through the replacement of obsolete building stock with new buildings is frequently used. Other strategies, more conservative, prefigure the reuse of the existing assets to enhance the material, energetic and social value still present in the buildings (Addis et al. 2004; Gaspar and Santos 2015; Thomsen and Flier 2008).

From an economic point of view, the choice between the two options depends on the preferences of the owners and developers, as well as on some hypotheses that academics and scholars have made the subject of a low-level investigation. The aim of the study is to deepen the feasibility conditions for the reuse of the existing city by highlighting the conditions that favor demolition and reconstruction processes

A. Mangialardo (✉) · E. Micelli
Department of Culture and Arts, IUAV University of Venice, Venice, Italy
e-mail: amangialardo@iuav.it

© The Author(s), under exclusive license to Springer Nature Switzerland AG 2021
A. Bisello et al. (eds.), *Smart and Sustainable Planning for Cities and Regions*,
Green Energy and Technology, https://doi.org/10.1007/978-3-030-57332-4_16

compared to interventions based on heritage recovery. To this end, an economic model developed by the authors in recent research has been taken up again (Mangialardo and Micelli 2019). This model considers property preferences toward the two options, and it has been tested in seven metropolitan cities as established by Law no. 56/2014.

The chapter is divided into three parts. Section 1 presents the model, developed in previous research by the authors, which makes it possible to identify the variables underlying the choice among the available options. Section 2 presents an application of the model in the seven metropolitan cities. Finally, Sect. 3 proposes an interpretation of the results of the survey.

2   The Economic Model to Measure the Benefits Between Demolition and Rebuilding and Reuse

The economic feasibility between demolition and reconstruction operations or reuse of the existing assets depends on the transformation conditions set by the urban planning instrument and the values of the real estate market.

Recent research conducted by the authors (Mangialardo and Micelli 2019) has shown that the property assesses the highest and best use of the assets and considers it advantageous to sell the property to a developer interested in demolition and reconstruction if:

$$V_p > V_e \tag{1}$$

where

Vp is the value of the building potential set by the urban plan;
Ve is the value of existing buildings.

Following a few steps (Mangialardo and Micelli 2019), this condition (1) can be transformed to (2) as follows:

$$ip/it > b/a \tag{2}$$

where

ip represents the building index. It is an economic parameter, expressed in a percentage or decimal, that establishes the building capacity of an area;
it represents the current density index of the area. It defines the ratio—expressed in a percentage or decimal—between the area and the built area;
a represents the area incidence coefficient;
b represents the ratio between the value of buildings in their current state and the value of new buildings.

The economic convenience of demolition and reconstruction occurs when the ratio ip/it, the area densification potential determined by urban planning choices, exceeds the land pressure index, defined by the ratio between the residual value of buildings and the area incidence of properties.

Intuitively, as the building potential set by the urban planning tools increases, as well as the positional quality of the property, the possibilities of transformation by demolition and reconstruction increase.

The b/a ratio—usually higher than 1—determines the need for an increase in density to verify the feasibility conditions for demolition and reconstruction: in other words, the scrapping of the city almost systematically foresees an increase in the density of the redevelopment areas.

This makes it possible to consider the role of the two other variables—b and it—of the model: As the existing density increases, as well as the residual value of the heritage increases, it will be more difficult to detect the feasibility conditions for demolition and reconstruction.

The research examines the real estate values of the regional capitals of the seven metropolitan cities established by law no. 56 of April 7, 2014: Milan, Genoa, Bologna, Florence, Rome, Naples and Reggio di Calabria.

The seven metropolitan cities can be considered representative of many economic and social situations and characterized by specific dynamics of the real estate market. If Milan and Rome represent the two metropolitan poles par excellence of the country and Naples has the profile of the lower-ranking metropolises, Genoa is representative of large cities—with almost 600,000 inhabitants each—while Bologna and Florence, although different in socioeconomic terms, are representative of medium-sized cities in terms of demographics. Reggio di Calabria, finally, is among the least populous metropolitan cities among those considered: Their populations do not reach 200,000 inhabitants.

3 The Development of the Model in the Twelve Metropolitan Cities

The estimation of the values of b was carried out thanks to specific analytical models able to return the values of the properties based on their different construction qualities and their positional characteristics.

The classification from a technological point of view distinguishes between new, used and inhabitable properties and those to be renovated, using a classification widely shared by supply-and-demand considerations. The survey then distinguishes three areas for each city—the center, semi-center, and suburb—again according to an established market classification (Table 1).

The sample used quantities, for each city, of more than 100 offer prices for properties for residential use. The offer prices were obtained from the main Italian real-estate

Table 1 Property values and a by location and building quality

Genoa	State of conservation/location	Center	Semi-center	Suburb
	Inhabitable	1.531	1.169	799
	Used	1.823	1.442	1.024
	New	3.645	2.277	1.763
	a	0.36	0.25	0.18
Florence	Inhabitable	3.904	3.443	2.585
	Used	4.398	3.634	3.152
	New	5.052	4.310	3.618
	a	0.31	0.26	0.21
Naples	Inhabitable	3.269	2.467	1.413
	Used	4.001	2.916	1.664
	New	5.994	3.402	2.061
	a	0.36	0.23	0.16
Milan	Inhabitable	5.445	3.421	1.765
	Used	6.662	4.052	2.564
	New	8.860	5.051	3.119
	a	0.44	0.33	0.25
Bologna	Inhabitable	2.455	1.657	1.301
	Used	3.112	2.703	2.055
	New	4.094	3.362	2.663
	a	0.36	0.29	0.21
Rome	Inhabitable	5.438	3.649	1.898
	Used	6.540	4.547	2.610
	New	9.184	5.904	3.464
	a	0.43	0.29	0.21
Reggio di Calabria	Inhabitable	981	759	585
	Used	1.208	951	818
	New	1.957	1.426	1.306
	a	0.23	0.17	0.12

Web sites and were discounted using a coefficient estimated on the basis of the most authoritative market sources (Nomisma 2018).

The values of the area incidence coefficient (a) were evaluated through the authoritative source of Il Sole-24 Ore and its magazine Consulente Immobiliare (2016).

The area incidence coefficients vary considerably as the positional quality of the area changes: They reach their maximum value in central areas and decrease in peripheral areas.

Based on the values of b^1 and a, it is therefore possible to identify the density multipliers (b/a) necessary to make the demolition and reconstruction process convenient. The ratio is minimum where the properties, located in central areas, have reached the maximum level of obsolescence. Based on the elaborations carried out, it varies from a minimum of 1.17 for Genoa and 1.39 for Rome and Milan and reaches 2.40 for the city from the less vibrant market, Reggio di Calabria and the suburb of Genoa, due to the different land pressure that inevitably distinguishes the different cities according to their size and attractiveness.

At the opposite extreme, the ratio will be maximum if the land pressure is minimal, as is the case in the suburbs, and when the assets maintain a significant residual value. If we consider how much volume is needed to carry out demolition and reconstruction work in the suburbs and when the heritage is used but perfectly fungible, the multiplicative coefficients of the existing density are important: In Milan, Bologna, Florence, and Rome's suburbs, the density must be multiplied by four times, but the value becomes even more important where the land component has less value as in the suburb of Genoa—where the multiplier is about 5.70—and in Reggio di Calabria, where it is 8.55 (see Table 2).

4 Is Reuse of the Existing Stock the Only Possibility for Urban Regeneration?

The economic conditions for the reuse of the city based on the scrapping of existing buildings are limited to certain parts of the territory. The elaborations carried out make it possible to detect how demolition and reconstruction operations are economically viable for the property only where urban planning procedures make possible significant increases in volume compared to existing settlement conditions.

The judgment on the cost-effectiveness of demolition and reconstruction operations deserves, however, to be graded according to the importance of the locations considered. In large metropolitan centers, of which Milan and Rome are the emblematic sample cities, the demolition and reconstruction of buildings are certainly more likely than in other locations. The different pressures of demand, maximum in the large metropolitan centers and minimum in the smaller localities of the country, determine the settlement pressure with a significant influence on the values of the model: In the central and semi-central areas of Milan and Rome, the expectations of land and residual low-density contexts favor urban transformation processes based on the replacement of buildings and urban areas.

Even in medium-sized cities such as Bologna and Florence, albeit in a less evident way, evidence shows that in central areas speculative pressure partially justifies demolition and reconstruction operations.

---

[1]The values of b were estimated by comparing, for each area, the real estate values in the current state of conservation (building to be renovated, used and new construction) to the values of new construction.

Table 2 b/a values for the different levels of obsolescence and the positional qualities of the three cities analyzed

City	State of conservation/location	Center	Semi-center	Suburb
Genova	Inhabitable	1.17	2.08	2.58
	Used	1.40	2.,56	3.31
	New	2.80	4.05	5.70
Florence	Inhabitable	2.48	3.07	3.44
	Used	2.79	3.24	4.19
	New	3.21	3.85	4.81
Naples	Inhabitable	153	3.10	4.22
	Used	1.87	3.66	4.97
	New	2.80	4.27	6.15
Milan	Inhabitable	1.39	2.08	2.23
	Used	1.70	2.23	3.24
	New	2.26	3.08	3.94
Bologna	Inhabitable	1.68	1.72	2.28
	Used	2.13	2.75	3.60
	New	2.80	3.42	4.66
Rome	Inhabitable	1.38	2.11	2.56
	Used	1.66	2.63	3.51
	New	2.33	3.42	4.66
Reggio di Calabria	Inhabitable	2.20	3.43	3.83
	Used	2.71	3.95	5.35
	New	4.40	5.92	8.55

In medium-sized and small cities, a inadequate demand is at the basis of land values that require particularly significant increases in density. In the cases under examination—Genoa and Reggio di Calabria, representative of average centers—such values can reach important thresholds whose practicability is questioned by a plurality of often concomitant factors.

First of all, the density of places is not always low. On the contrary, in the places of expansion in the 1950s and 1960s, consolidated densities are often important and make it difficult to replace buildings simply because the multiplication of volumes does not correspond to a multiplication of demand.

Finally, it is useful to underline how the areas of redevelopment and regeneration are characterized by a fragmented ownership structure, the result of decades of housing policies that have systematically promoted access to home ownership. The model has not deliberately taken into account the transaction costs of urban replacement processes, but they contribute to make such a development model even more fragile, especially in the case of interventions in areas characterized by medium and high settlement densities, which also correspond to dozens of properties that are not always aligned in terms of priorities and objectives (Antoniucci and Marella 2017; Brown 2018; Mangialardo et al. 2019).

The redevelopment processes of a large part of the Italian suburbs, especially the areas outside the large metropolitan centers, go through the regeneration of the existing assets without easy illusions with respect to radical replacements of public or private parts of cities (Farmer 2016).

The challenge concerns the ability to design and implement the upcycle of what remains of a heritage that is certainly obsolete, but still marked by a value that makes any hypothesis of demolition and reconstruction of buildings unlikely. Nevertheless, the exploitation of the energy and matter contained in the existing heritage makes it possible to pursue objectives of private and collective interests. The protection and enhancement of heritage are accompanied by development that is more consistent with the principles of the circular economy (Cheshire 2017; D'Alpaos and Bragolusi 2018; Mangialardo and Micelli 2018).

It is not by chance, therefore, that in several European countries the long-standing problem of low productivity in the construction sector is being tackled with solutions aimed at higher efficiency and therefore lower costs. The actions aimed at the reuse of the existing city through the use of new technologies are different: Retrofit interventions are aimed at making buildings less energy-intensive and can be carried out in conjunction with operations aimed at the reuse of existing building materials in a circular perspective. The common objective of all the innovative reuse experiments is to guarantee the possibility to intervene on the existing building materials in order to guarantee a wider audience of possible investors in this market with very significant potential (Mangialardo and Micelli 2018).

However, retrofit solutions based on industrialized building regeneration processes are not the only solutions to the problem of reusing the urban and building heritage of our cities. If the costs of off-site industrialized production prove to be decreasing due to significant economies of scale, urban areas, and particularly old building stock could be subject to a building replacement that seems economically unsustainable today.

5 Conclusions

The strategies for the redevelopment of the public and private heritage of cities are different and their viability is a function of economic feasibility, due to the convenience of owners and developers. The contribution therefore compared the economic viability of two alternatives widely considered in the debate: the demolition and reconstruction of buildings and parts of cities and the redevelopment of existing heritage.

The model developed and applied to seven metropolitan cities makes it possible to evaluate the convenience of the property to be demolished and rebuilt according to the specific urban planning rules that the plan provides for.

The results showed an important variation in the densification coefficient. If, in the city center of Milan and Rome, a modest increase in density already creates favorable conditions for demolition and reconstruction, in the suburbs of Reggio di Calabria, especially if the properties still have a significant residual value, the multiplication coefficient is considerable. Florence and Bologna are two exceptions.

In light of the results obtained, a large part of the suburbs, particularly small- and medium-sized towns, are condemned to regenerate their public and private heritage without being able to rely on radical transformations of buildings and neighborhoods. In fact, the multiplication of the density of existing buildings seems incompatible both with the characteristics of real estate markets, whose demand still seems modest, and for an inadequate fixed capital stock compared to a possible additional settlement load.

References

Addis W, Schouten J (2004) Design for deconstruction: principles of design to facilitate reuse and recycling. CIRIA, London

Antoniucci V, Marella G (2017) The influence of building typology on the economic feasibility of urban developments. Int J Appl Eng Res 12(15):4946–4954

Brown D (2018) Business models for residential retrofit in the UK: a critical assessment of five key archetypes. Energ Effi 11(53):1–21

Chesire D (2017) Building revolutions—applying the circular economy to the built environment. Riba Publishing, London

D'Alpaos C, Bragolusi P (2018) Buildings energy retrofit valuation approaches: state of the art and future perspectives. Valori E Valutazioni. 20:79–94

Farmer M (2016) Modernise or die. Construction Leadership Council, London

Gaspar PL, Santos AL (2015) Embodied energy on refurbishment vs. demolition: a southern Europe case study. Energy Build 87:386–394

Mangialardo A, Micelli E (2018) Rethinking the construction industry under the circular economy: principles and case studies. In: Bisello VA, Laconte D, Costa P (a cura di) Smart Sustain Plan Cities Regions. Cham (CH), Green Energy and Technology, Springer International Publishing AG 333–344

Mangialardo A, Micelli E (2019) Condannati al riuso. Mercato immobiliare e forme della riqualificazione urbana. Aestimum 74:129–146

Mangialardo A, Micelli E, Saccani F (2019) Does Sustainability affect real estate market values? Empirical evidence from the office buildings market in Milan (Italy). Sustainability 11(1):12

Nomisma (2018) Osservatorio sul mercato immobiliare. Nomisma, Bologna

Thomsen A, van der Flier K (2008) Replacement or reuse? The choice between demolition and life cycle extension from a sustainable viewpoint. In: Norris M, Slike D (a cura di) Shrinking cities, sprawling suburbs, changing countrysides. ENHR conference 2008. Centre for Housing Research, Dublin

Valier A (2019) The architectures of failure. Strategies and techniques for reusing unfinished projects. Territorio 87:113–121

Addressing the Problem of Private Abandoned Buildings in Italy. A Neo-Institutional Approach to Multiple Causes and Potential Solutions

Anita De Franco

Abstract The problem of having abandoned buildings in urban contexts is animating policy debates at various institutional levels (e.g., local, regional, and national). In Italy, as elsewhere, the often concerned tones concerning issues of "vacancy" or "decay" of buildings tend to overlap with "abandonment" problem. However, public discussions are not always able to identify and separate truly problematic states of built assets from totally legitimate states of affairs. In order to design viable policy strategies to tackle abandonment problem at various scales, it is important to understand: (i) Why private owners opt for the abandonment of their assets instead of using or selling them, and (ii) What can be done to bring properties back onto the regular market. The aim of this chapter is to explore the various causes and solutions to the abandonment problem in reference to Italian urban contexts and institutional settings. The study adopts a neo-institutional approach to address the problem of abandoned private buildings in Italy. The methods used for the investigation are mainly qualitative and based on (a) an extensive literature review, (b) official data found in official documents and reports from national and international agencies, (c) eleven interviews with relevant actors, (d) an analysis of local normative records and national regulations, and (e) on-site investigations in various Italian urban contexts. This paper intends to enlarge the analytical focus on urban abandonment processes, questioning how this empirical phenomenon is influenced by institutional and public policy settings at various levels of governance.

Keywords Abandoned buildings · Private properties · Neo-institutional approach · Urban regeneration · Regulation

1 Introduction

The problem of abandoned buildings in urban contexts is animating policy debates at various institutional levels (e.g., local, regional, and national). In Italy, as elsewhere,

A. De Franco (✉)
Department of Architecture and Urban Studies (DASTU), Politecnico di Milano, Via Edoardo Bonardi, 3 20133, 20133 Milan, Italy
e-mail: anita.defranco@polimi.it

the often concerned tones concerning issues of "vacancy" or "decay" of buildings tend to overlap with the "abandonment" problem. However, public discussions are not always able to identify and separate truly problematic from a totally legitimate state of affairs (see Moroni et al. 2020). This problem is further complicated by the lack of official figures on the magnitude of the abandonment problem in local contexts. Even in Italy, various quantifications of the number of abandoned buildings in the country often result from over- or underestimations.[1] This has critical effects in policy making because public discussions expand into too many ambiguous directions (as also noted by Gentili and Hoekestra 2018).

According to some authors, conceptual and practical perplexities on the abandonment problem require a more precise definition of the issues at stake (Morckel 2014; Haase et al. 2014). As suggested by Galster (2019), too little attention has been given to possible relationships between institutional processes negatively affecting abandonment problems in inhabited contexts. These institutional processes may be assessed by better understanding the types of properties under study, and how these can be transformed in various circumstances. Various authors identify the abandonment of buildings as a withdrawal of essential responsibilities of the property owners (e.g., use and maintenance) of their assets (Kraut 1999, p. 1140; Mallach 2006, p. 1). From this perspective, to understand the problem of abandonment, it seems necessary to address two basic aspects: Why private owners opt for the abandonment of their assets, and what can be done to bring properties back into regular market circuits, for instance, by making the use or transfer of the property profitable for the owner.

Considering the Italian case, the problem of abandonment relates to various problems affecting private property markets in the country. Unfavorable fiscal regulation eventually shortens the prospects of regeneration. For instance, the mere ownership and use of a building may be particularly costly for owners. In Italy, public revenues from property taxes increased from 0.6% in 2010 to 1.5% in 2015 (Agenzia delle Entrate 2015), and they almost tripled in 2015 (Confedilizia 2015). Moreover, changes in the national tax regulations made the use of any property that is not a primary residence more costly for many owners. Thus, since the introduction of I.M.U.[2] (i.e., a property ownership tax on structures), derelict properties increased 97% nationwide between 2011 and 2018 (Confedilizia 2019). Derelict properties are

---

[1] In Italy, for instance, census data from Istat (2014) suggests that 22.7% of dwellings are empty or occupied as other than primary residences. This estimation is general and over inclusive, making it impossible to distinguish truly problematic situations (e.g., "structural vacancy") from situations of under usage (e.g., second homes or holiday homes). Other data show that 5.2% of private real-estate stock is run-down, in ruin, or under construction (Istat 2014), although not even in this case is it possible to extract the precise number or share of abandoned buildings because this figure does not distinguish between deteriorated, derelict, and unfinished buildings. Other sources help to break down these estimations; for example, Agenzia delle Entrate (2017) indicates that there are 474,000 derelict properties in Italy, and CESCAT (i.e., Centro Studi Ambiente e Territorio) asserts that there are around two million abandoned properties in Italy. See https://www.assoedilizia.com/contenuto_exp.asp?t=comunicatiandid=375 (accessed February 2019).

[2] The I.M.U. was introduced by Legislative Decree no. 201 of 2011, converted into Law no. 214 of 2011.

de facto abandoned, since buildings left in ruins are not readily usable and are incapable of holding or producing income for owners. Also, other public rules in Italy (e.g., other municipal taxation) may inadvertently encourage owners to default from active use and maintenance of their assets.[3] Considering these introductory aspects, this chapter will discuss the causes for the abandonment of buildings in Italy and possible strategies for their regeneration. The chapter is structured in the following way: The second section presents the research methods and further clarifications, for the study of the abandonment problem though a neo-institutional approach. The following section presents a review of multiple causes for the abandonment of private buildings and the results of interviews with relevant actors. This is followed by a presentation of a potential solution to the problem of abandonment through a "regeneration-by-regulation" model, while discussing its implications for governance and institutional design. The chapter ends with conclusions and suggestions for future research.

2   Method

This study adopts a neo-institutional approach to understand how empirical phenomena (e.g., the abandonment of buildings) emerge and evolve over time within an institutional and policy framework. The hypothesis is that the abandonment of private buildings in Italy is influenced by institutional and public policy settings at various levels of governance. This view assumes that local, regional, and national governments are responsible for setting the "rules of the game" (van Karnenbeek and Janssen-Jansen 2018; Moroni 2010; Savini et al. 2015) within which agents make decisions on the future of their property assets. The "regeneration-by-regulation" model aims to enhance the opportunities to transform abandoned buildings in Italy though various suggestions for institutional design. The model emerges from a coordination of regulatory devices at various institutional levels—macro, meso, and micro—that in this chapter conventionally refer to governance authorities.[4]

Before entering into the discussion, a preliminary clarification is necessary to understand the phenomena under enquiry. In this study, "abandoned buildings" refers to abandoned private properties that are unhabitable or unserviceable without consistent works of refurbishment and upgrades. In this case, abandoned buildings are properties in a severe state of physical neglect and a subset of empty properties. At the same time, these are also different from run-down but physically sound buildings.[5]

---

[3] In the case where properties are unhabitable (i.e., "*edifici inagibili*"; DL 201/2011 Art. 13 co. 3, b) or in ruins (i.e., "*unità collabenti*"; D.M. 28/1998, D.Lgs. 504/1992 Art. 5, co. 6), tax discounts may go from 50% up to 100% on the premise that such properties are incapable of holding or producing any rent for the owners. See also: Court of Cassation verdict (n. 17,815 of July 19, 2017).

[4] See also Alexander (2005), Moroni (2010), and Alfasi (2018).

[5] These distinctions in the literature are not always present. Thus, various authors include abandoned buildings within larger categories of substandard properties (e.g., vacant, underused, or dilapidated buildings; see for instance Accordino and Johnson 2000; Pellegrini and Micelli 2019). Some authors

The methods used for the investigation are mainly qualitative, based on (a) an extensive literature review, (b) official data found in official documents and reports from national[6] and international[7] agencies, (c) eleven interviews with relevant actors, (d) an analysis of local normative records and national regulations, and (e) on-site investigations in various Italian urban contexts.

Relevant actors interviewed were selected on the basis of their representativeness of key socioeconomic institutions in Italy; they were one policy maker, two professionals (i.e., one real-estate developer and one architect), two representatives from national associations (i.e., owners, building contractors), three representatives of local administrations, and three academic experts.[8]

3   Multiple Causes of the Abandonment of Private Buildings

In this section, multiple causes for the abandonment of private buildings are systematically delineated and explored based on debates in the academic literature (see Table 1). In the review, four main causes emerged (i.e., economic, individual, procedural, and typological), which were discussed during the interviews with actors. Each interviewee gave a general opinion of the reasons for the abandonment of private buildings in Italy and then discussed the multiple causes more in depth. The term "causes" is used here, in a non-deterministic way, as an umbrella term for any kind of factor, trigger, driver influencing abandonment.

3.1   *Economic Causes*

Economic causes are the most relevant ones when discussing why owners abandon their properties. All interviewees highlight these economic causes as the most relevant ones considering the effects of the latest global economic crisis in various sectors in Italy. First, there are causes that strictly relate to the condition of the property (e.g.,

---

note, however, that there is a substantial difference in the types of institutional problems depending on how the public sees dysfunctional states of private properties in local contexts (see Henderson 2015; Hirokawa and Gonzalez 2010).

[6]Istat (2014), Banca d'Italia (2018), Agenzia delle Entrate (2015, 2017, 2019a, and 2019b), Ministry of Economics and Finances (MEF 2019), and private agencies for the construction sector (ANCE 2019), national association of property owners (Confedilizia 2015 and 2019) and Legambiente and CRESME (2013).

[7]Eurostat (2018), Buildings Performance Institute Europe (BPIE 2011).

[8]The interviews were semi-structured discussions lasting 30–45 min, focused on the abandonment of private buildings in Italian urban contexts. The discussions were conducted around three main questions: (a) Why are there so many abandoned private buildings in Italian urban contexts? (b) What makes their rehabilitation difficult? and (c) What kinds of solutions are necessary to tackle the problem of abandonment at the various levels of governance?

Table 1 Multiple causes for the abandonment of private buildings in Italian urban contexts

Causes	Hypothesis	References	Answers from actors
Economic	Financial issues (e.g., negative equity, disinvestments)	BPIE (2011), p. 12, Cho et al. (2013), Farris (2016), Foldvary and Minola (2017)	Crisis in the construction sector
			Crisis in the real-estate sector
			Crisis in property markets
Individual	Negative strategy by owners (e.g., disinterested, unwilling to use or sell the property)	Hirokawa and Gonzalez (2010), Henderson (2015)	Complicated ownership situations
	Positive strategy by owners (e.g., speculation on future revenues by holding the property unused)	Németh and Langhost (2014), Kraut (1999)	Unfavorable taxation system
Procedural	Bureaucratic issues (e.g., for property transformation and land development)	Hwang and Tan (2012), Meyer and Lyons (2007), Rosato et al. (2010), Tan et al. (2011), Williams and Dair (2007)	Administrative fines
			Heavy bureaucracy (e.g., difficulty in obtaining building and planning permits)
			Instability of the laws (i.e., recurrent changes of the laws)
Typological	Regulatory issues (e.g., technical regulations on building and construction activities, land-use prescriptions)	Brand (1995), Chan and Yung (2004), Olivadese et al. (2017), Ornelas et al. (2016), Winston (2010), Pellegrini and Micelli (2019), Németh and Langhorst (2014)	Distinction of activities (e.g., on new or existing assets)
			Functional classifications (e.g., uses, activities in urban areas)
			Design rules (e.g., density, shape, coherence)

property devaluation, negative equity, and functional obsolescence of the building). These usually emerge when costs of liens or the management of the property are higher than the value of the property. Second, there are more general problems of the stagnation of local markets and a severe crisis for the construction sector (ANCE 2019). On the one hand, there is the problem of foreclosures, which has particularly hit the private sector (e.g., dwellings, businesses, and construction firms). On the other hand, there is a more general problem of stagnating investments.[9] These issues point

---

[9] As financial institutions recover their credits, the immobility of local markets make the original value of the property drop, and negotiations between banks and property investment funds often end up in liquidation procedures (e.g., NPLs, see ANCE 2019; Banca d'Italia 2018, p. 6; MEF 2019, p. 17).

to a material deficit and lowered values for real-estate market assets in the country. Since the beginning of the global financial crisis, the values of urban dwellings in Italy have been decreasing compared to other trends in European urban contexts (Agenzia delle Entrate 2019a, p. 3; Eurostat 2018, p. 44).

3.2 Individual Causes

As argued in many studies, the owners' decisions are fundamental to understand the abandonment phenomenon. However, the reasons why owners abandon their buildings instead of conducting upkeep or selling them are not easily addressed in a neutral or unequivocal manner. The interviewees highlighted that there are two main attitudes that are inherently different but equally relevant in matters of property abandonment. A first possibility is that owners play a "negative game," in which they have ceased their interest in the property asset. Thus, owners may be unwilling to act because they lack entrepreneurial attitude or personal dealings (e.g., presence of co-owners or inheritances); such problems can exist or emerge anywhere. A second possibility is that owners play a "positive game," in which they intentionally postpone investments, in the expectation of better market conditions (e.g., speculation). In this case, owners cut expenses on the property, especially when regulations set high costs for the use of the property and low costs for holding it.

3.3 Procedural Causes

Procedural causes refer to the various bureaucratic barriers that emerge for the transformation of abandoned properties. The procedures change depending on the type of work required on the buildings. If an abandoned building is severely damaged in its physical state, the property may have several "code violations" (i.e., violations of safety or habitability standards). These violations are usually set by municipal authorities, often as fines on the manager (e.g., owner or developer) that inevitably raise the costs of the investments and periods for approval of local projects. Other problems may emerge within private contractual works (e.g., stock mobilities, property transfers, and acquisitions) but also because of administrative practices (e.g., acquiring building permits, compliance with technical standards, and land-use prescriptions). A final issue concerns general regulatory uncertainty concerning technical and construction activities set in various laws. Some research has shown that, in the last ten years, the regulations on building and construction activities in Italy changed almost once every 23 days.[10]

---

[10]Studies are based on changes in the *Testo Unico dell'edilizia* (Dpr 380/2001) without considering regional laws and national rules on bidding disciplines. (See https://amp.ilsole24ore.com/pagina/ AErKgimC, accessed November 8, 2019). A more up-to-date example in Italy is the so-called

These changes have been regarded positively by market operators because they accommodate renewable energy standards based on EU directives, although their effectiveness can be improved.[11]

3.4 Typological Causes

Typological causes refer to various standard costs attached to place-specific features of the building. Depending on the type of transformation envisioned (e.g., refurbishment, reuse, infill, and demolitions), typological issues may concern the building or its wider contexts. Building-related typological issues arise from the rigid distinction of activities for existing or new buildings in Italian regulations. Thus, construction operations in Italy tend to be very strict on existing assets and more flexible for new construction. Context-related typological issues usually depend on zoning regulations and functional classifications (e.g., residential, commercial, tertiary, and manufacturing). However, the problem is not the existence of zoning regulations, but the types of standard costs associated with the specific functions that are enforced by various authorities (e.g., national, regional, and local authorities), which usually add further specifications to standards for interventions (as noted by Pedro et al. 2010). For the interviewees, regulations are perceived as often out-of-date and too specific, making it particularly costly to switch from one building function to another. In their view, prescriptions on building and land-use standards make it difficult to adapt to emerging technologies, habits, and comfort demands in local markets. Further problems arise when public rules add other specifications on design and layout of the properties (e.g., heritage, building style, or architectural character). As is widely discussed in the literature, the conservation tradition of Italian planning instruments can be troubling for the rehabilitation of abandoned buildings (Ornelas et al. 2016; Pellegrini and Micelli 2019; Rosato et al. 2010). To make the property appealing and accessible in the market, an abandoned building may require radical changes in design, function, and layout. For instance, abandoned industrial buildings have specific, consistent standard costs for clearance and rehabilitation of the property (i.e., brownfield projects). Additional costs arise from other reconfigurations at wider scales to rehabilitate a site's potential within the urban market (e.g., changes on monofunctional land uses or low accessibility levels). For buildings in heritage areas, in

---

Bonus ristrutturazioni (restructuring bonus). In this case, building work for the renovation and refurbishment of individual units, or common areas of residential buildings, receive a tax break of 50%. Introduced by Legislative Decree no. 83 of 2012 (art. 11), this "bonus" has been subsequently extended. Another interesting example in this regard is a tax break (of 90%, and in place only for 2020) for restructuring facades (Law no. 160 of 2019). Initially conceived for all elements of facades, it was finally further circumscribed to certain features of them.

[11] As one interviewee observed, operators would prefer that fiscal incentives, certificates, and standard requirements remain stable in the long term (e.g., without amendments every fiscal year). In his opinion, for fiscal incentives to be truly advantageous, they should be accredited in a relatively short amount of time (e.g., five years instead of ten).

Italy the options for their rehabilitation may be inhibited or completely dismissed by assigned authorities (i.e., superintendencies).

4 Debating a "Regeneration-by-Regulation" Model for Abandoned Private Buildings in Italy

One of the main goals of this study was to understand the relationship between abandonment processes and institutional settings in Italian contexts. The review of the multiple causes and discussions with the actors suggest that direct public interventions to deal with the abandonment problem may be unsustainable and create dilemmas from a policy perspective (see Henderson 2015; Hirokawa and Gonzalez 2010). The Italian case suggests a relative interdependency of private and public issues on the abandonment problem (e.g., economic and individual conditions of the property or procedural and typological rules set by authorities). Considering these aspects, a regulatory approach seems a more viable and sustainable solution than an interventionist one. This argument is presented in the "regeneration-by-regulation" model (Table 2), suggesting various changes to stimulate the transformation of abandoned buildings in various ways. The model should be interpreted as an organic revision of some regulatory tools for urban development at different scales. For this framework, we reference three principal governance levels in Italy: national authorities ("macro"), regional authorities ("meso"), and local authorities ("micro").

The "regeneration-by-regulation" model suggests that various regulatory changes may be sited at various institutional levels. The selection of these institutional levels is based on the sphere of regulatory domain of fiscal rules, building and land activities, and planning institutions. Considering the relationship between taxation and property abandonment, macro-level institutions would decrease taxes to stimulate investments for the property (e.g., use, upgrades, and transfers of properties).[12] Meso-level institutions would deal with inter-organizational matters by easing complex and ordinary operations to favor resource exchanges across markets.[13] For instance, regional authorities may subsidize certain types of operations, instead of focusing only on specific types of building states or functions. More generally, all authorities should undertake the radical work of simplifying the procedural and typological aspects concerning technical building and construction rules. This would be mirrored in more accessible procedures for building permits and planning approvals that in Italy usually require a long time for approvals. Finally, micro-level institutions would directly deal with intra-organizational processes in local contexts. More active approaches

---

[12]This type of solution might also require more radical taxation reform, for instance, by taxing mainly (or exclusively) the value of the land, while diminishing (or totally erasing) taxation on property improvements. In this sense, even the least attractive areas would benefit from the concrete advantage of more accessible values and relatively low property taxes.

[13]As suggested in various studies; see Pedro et al. (2010), Hwang and Tan (2012), Chan and Yung (2004), Tan, Shen, and Yao (2011), and Winston (2010).

Table 2 A "Regeneration-by-regulation" model to solve the problem of abandoned private buildings in urban contexts, in reference to the Italian institutional frameworks

Institutional level	Regulatory solution	References
Macro-level institutions (e.g., central state)	Reduce taxation for rented properties (i.e., the "split incentive" problem)	BPIE (2011), p. 12, Cho et al. (2013), Farris (2016), Foldvary and Minola (2017)
	Decrease private taxes (i.e., the "land value taxation" mechanisms)	
Meso-levels institutions (e.g., supralocal authorities as regions, provinces, and counties)	Provide subsidies for complicated land operations (e.g., brownfield rehabilitation and demolitions, etc.)	Mallach (2006), Meyer and Lyons (2007), Williams and Dair (2007), Winston (2010), Olivadese et al. (2017), Pedro et al. (2010)
	Shorten the difference on types of operations (e.g., similar procedures for new and existing buildings, etc.)	
Micro-level institutions (e.g., city council, administrative offices)	Use "code enforcement" strategies (e.g., ask owners to remediate on the negative impacts of the property in the surroundings: risks of collapses, reduction in the value of neighboring buildings and disinvestment in the neighborhood)	Accordino and Johnson (2000), Schilling (2009), Alfasi (2018), Imrie and Street (2009), Tan et al. (2018)
	Use "urban coding" (i.e., set local building codes as principal regulatory source for land development)	

could incentivize property improvements or transfers, granting fair and sustainable compensation for all agents involved (Accordino and Johnson 2000; Henderson 2015; Schilling 2009). In this case, local building codes may play a pivotal role in regulating property development while challenging rational and comprehensive approaches in urban governance (Alfasi 2018; Imrie and Street 2009; Ross and Portugali 2018).[14] While being more stringent regarding certain aspects of the use of private properties (i.e., those creating risks and harms), it seems necessary to reduce certain nonessential constraints and standards that do not concern the fundamental issues of safety and security or well-defined externalities (e.g., disinvestment in neighborhoods). In conclusion, this "regeneration-by-regulation" model presumes that individual causes

---

[14]For instance, on renewable energy regulations and spatial strategies (BPIE 2011, p. 63; Legambiente and Cresme 2013; Moroni et al. 2016) and building management and sustainable development (Chang and Yung 2004, pp. 420–442; Hwand and Tan 2012). On the regulatory side, other radical solutions would look beyond a strict statutory approach on normativity (Lorini and Moroni 2020; Moroni and Lorini 2017).

are relevant factors, but governance institutions cannot deal directly with them. On the one hand, local, regional, and national policy makers may orient and stimulate beneficial decisions on property management mainly through normative incentives. On the other hand, private stakeholders may self-determine their strategies based on concrete uses and interests on property management, always in compliance with essential rules.[15] In this case, scale, scope, and outcomes of regeneration should be reconsidered in light of the ever-changing demands, needs, and customs of the society (Buitelaar and Sorel 2011; Olivadese et al. 2017; Tan et al. 2011).

5 Conclusions

This chapter has presented how the problem of abandoned buildings in urban contexts is framed as a general problem for the Italian case. The neo-institutional approach used in this study helped some crucial caveat that concurs to the emergence of abandonment phenomena (e.g., property markets, public regulations, and governance). The Italian case underlines how rising property taxes and bureaucratic obstacles for enterprises influence abandonment processes at various scales.[16] The multiple causes may be useful to open discussion of how certain negative spatial outcomes (e.g., the abandonment of private properties) may be influenced by public regulations. The main interpretation of this chapter is that a regulatory approach (rather than an interventionist one) may be a viable option to tackle the problem of abandonment at various levels of governance. The "regeneration-by-regulation" model expresses the hypothesis that public and private agents should challenge many statutory aspects that inhibit or narrow the opportunity for investment on private properties. Further contributions may also focus on abandoned public assets and interviews with citizens and nonprofit organizations. The analytical frameworks on multiple causes (Table 1) and the solution model (Table 2) can be used to compare the Italian case with other international contexts, while enriching the analytical categories to explore other types of solutions.

Acknowledgements The author would like to thank all the interviewees who participated in the research. All the written content should be considered as the author's interpretations of the answers provided by these individuals.

---

[15]In the case of direct subsidies for the owners, it may be necessary that the laws stipulating the funding procedures would specify specific codes of conduct. For instance, if owners do not effectively use the subsidies for property upgrades, then they would lose the financial benefit.

[16]Economic causes consider a wide range of implications concerning property management issues, which are common also in other southern European contexts (see Bogataj et al. 2016). Individual causes are generalizable in two main attitudes in owners' decisions to abandon their assets (i.e., negative and positive strategies). Procedural and typological causes may be more specific with regard to Italian issues (e.g., heavy bureaucracy, etc.).

References

Accordino J, Johnson GT (2000) Addressing the vacant and abandoned property problem. J Urban Affairs 22(3):301–315

Agenzia delle Entrate (2015) La tassazione immobiliare: Un confronto internazionale. Available at https://www.finanze.gov.it. Accessed on Nov 2019

Agenzia delle Entrate (2017) Statistiche catastali 2016. https://wwwt.agenziaentrate.gov.it. Accessed on December 2018

Agenzia delle Entrate (2019a) Rapporto Immobiliare 2019. Il settore residenziale. Pubblicazioni OMI. Published on May 2019. Data on 2018. Accessed 3 Nov 2019

Agenzia delle Entrate (2019b) Statistiche catastali 2018. Catasto edilizio urbano. Published on July 2019. Data on 2018. Available at www.agenziaentrate.gov.it. Accessed on 27 Jan 2020

Alexander ER (2005) Institutional transformation and planning: from institutionalization theory to institutional design. Plan Pract Res 4(3):209–223

Alfasi N (2018) The coding turn in urban planning: could it remedy the essential drawbacks of planning? Plan Theory 17(3):375–395

ANCE (2019) Osservatorio congiunturale sull'industria delle costruzioni. Data on 2018. Available at https://www.ance.it. Accessed on 15 Nov 2019

Banca d'Italia (2018) Indagine sui bilanci delle famiglie italiane. Statistiche Banca d'Italia. Data on 2016. Available at https://www.bancaditalia.it. Accessed on 21 Jan 2020

Bogataj D, McDonnell DR, Bogataj M (2016) Management, financing and taxation of housing stock in the shrinking cities of aging societies. Int J Prod Econ 181:2–13

BPIE (2011) Europe's Buildings under the microscope. a country-by-country review of the energy performance of buildings. Published in October 2011. Buildings Performance Institute Europe. Available at https://bpie.eu. Accessed on 18 Sept 2019

Brand S (1995) How buildings learn: What happens after they're built. Penguin, New York

Buitelaar E, Galle M, Sorel N (2011) Plan-led planning systems in development-led practices: an empirical analysis into the (lack of) institutionalisation of planning law. Environ Plan A 43(4):928–941

Chan EHW, Yung EHK (2004) Is the development control legal framework conducive to a sustainable dense urban development in Hong Kong? Habitat Int 28(3):409–426

Cho SH, Kim SG, Lambert DM, Roberts RK (2013) Impact of a two-rate property tax on residential densities. Am J Agr Econ 95(3):685–704

Confedilizia (2015) Dossier tassazione immobili. Available at https://www.confedilizia.it. Accessed on Feb 2016

Confedilizia (2019) Confedilizia notizie—anno 29, ottobre 2019 n. 9

Eurostat (2018) Living conditions in Europe. Statistical Books. Publications Office of the European Union, Luxembourg

Farris N (2016) What to do when main street is legal again: regional land value taxation as a new urbanist tool. Univ Pennsylvania Law Rev 164(3):755–777

Foldvary FE, Minola LA (2017) The taxation of land value as the means towards optimal urban development and the extirpation of excessive economic inequality. Land Use Policy 69:331–337

Galster G (2019) Why shrinking cities are not mirror images of growing cities: a research agenda of six testable propositions. Urban Affairs Rev 55(1):355–372

Gentili M, Hoekstra J (2018) Houses without people and people without houses: a cultural and institutional exploration of an Italian paradox. Housing Stud 3037:1–23. https://doi.org/10.1080/02673037.2018.1447093

Haase A, Rink D, Grossmann K, Bernt M, Mykhnenko V (2014) Conceptualizing urban shrinkage. Environ Plan A 46(7):1519–1534

Henderson SR (2015) State intervention in vacant residential properties: an evaluation of empty dwelling management orders in England. Environ Plan C Gov Policy 33(1):61–82

Hirokawa KH, Gonzalez I (2010) Regulating vacant property. Urban Lawyer 42(3):627–637

Hwang BG, Tan JS (2012) Green building project management: obstacles and solutions for sustainable development. Sustain Devel 20(5):335–349

Imrie R, Street E (2009) Regulating design: the practices of architecture, governance and control. Urban Stud 46(12):2507–2518

ISTAT (2014) Edifici e abitazioni. Available at https://www.istat.it. Accessed on Dec 2018

Kraut DT (1999) Hanging out the no vacancy sign: eliminating the blight of vacant buildings from urban areas. New York Univ Law Rev 74(4):1139–1177

Legambiente and CRESME (2013) ON-RE Osservatorio Nazionale Regolamenti Edilizi per il risparmio energetico. Available at https://www.legambiente.it. Accessed on Sept 2019

Lorini G, Moroni S (2020) Ruling without rules: not only nudges. Regulation beyond Normativity, Global Jurist, p 1

Mallach A (2006) Bringing buildings back: from abandoned properties to community assets: a guidebook for policymakers and practitioners. Rutgers University Press, New Brunswick, New Jersey

MEF (2019) Gli immobili in Italia. Ricchezza, reddito e fiscalità immobiliare. Data on 2016. Available at https://www.agenziaentrate.gov.it. Accessed on 21 Jan 2020

Meyer PB, Lyons TS (2007) Lessons from private sector brownfield redevelopers. J Am Plan Assoc 66(1):46–57

Morckel V (2014) Predicting abandoned housing: does the operational definition of abandonment matter? Commun Devel 45(2):121–133

Moroni S (2010) An evolutionary theory of institutions and a dynamic approach to reform. Plan Theory 9(4):275–297

Moroni S, Lorini G (2017) Graphic rules in planning: a critical exploration of normative drawings starting from zoning maps and form-based codes. Plan Theory 16(3):318–338

Moroni S, Antoniucci V, Bisello A (2016) Energy sprawl, land taking and distributed generation : towards a multi-layered density. Energy Policy 98:266–273

Moroni S, De Franco A, Bellè BM (2020) Vacant buildings. Distinguishing heterogeneous cases: public items versus private items; empty properties versus abandoned properties. In: Abandoned buildings in contemporary cities: smart conditions for actions. Springer, Cham, pp 9–18

Németh J, Langhorst J (2014) Rethinking urban transformation: temporary uses for vacant land. Cities 40:143–150

Olivadese R, Remøy H, Berizzi C, Hobma F (2017) Reuse into housing: Italian and Dutch regulatory effects. Prop Manage 35(2):165–180

Ornelas C, Guedes JM, Breda-Vázquez I (2016) Cultural built heritage and intervention criteria: a systematic analysis of building codes and legislation of Southern European countries. J Cult Heritage 20:725–732

Pedro JB, Meijer F, Visscher H (2010) Technical building regulations in EU countries: a comparison of their organization and formulation. In: CIB world congress 2010, building a better world: proceedings, 10–13 May 2010. Salford UK, University of Salford

Pellegrini P, Micelli E (2019) Paradoxes of the Italian historic centres between underutilisation and planning policies for sustainability. Sustainability (Switzerland) 11(9)

Radzimski A (2018) Involving small landlords as a regeneration strategy under shrinkage: evidence from two East German cases. Eur Plan Stud 26(3):526–545. https://doi.org/10.1080/09654313. 2017.1391178

Rosato P, Alberini A, Zanatta V, Breil M (2010) Redeveloping derelict and underused historic city areas: evidence from a survey of real estate developers. J Environ Plan Manage 53(2):257–281

Ross GM, Portugali J (2018) Urban regulatory focus: a new concept linking city size to human behaviour. R Soc Open Sci 5(5):171478

Savini F, Majoor S, Salet W (2015) Dilemmas of planning: Intervention, regulation, and investment. Plan Theory 14(3):296–315

Schilling J (2009) Code enforcement and community stabilization: the forgotten first responders to vacant and foreclosed homes. Albany Gov Law Rev 2:101–162

Tan Y, Shen L, Yao H (2011) Sustainable construction practice and contractors' competitiveness: a preliminary study. Habitat Int 35(2):225–230

Tan Y, Shuai C, Wang T (2018) Critical success factors (CSFs) for the adaptive reuse of industrial buildings in Hong Kong. Int J Environ Res Publ Health 15(7):1546

van Karnenbeek L, Janssen-Jansen L (2018) Playing by the rules? Analysing incremental urban developments. Land Use Policy 72:402–409

Williams K, Dair C (2007) What is stopping sustainable building in England? barriers experienced by stakeholders in delivering sustainable developments. Sustain Devel 15(3):135–147

Winston N (2010) Regeneration for sustainable communities? Barriers to implementing sustainable housing in urban areas. Sustain Devel 18(6):319–330

Unlocking the Social Impact of Built Heritage Projects: Evaluation as Catalyst of Value?

Cristina Coscia and Irene Rubino

Abstract To be sustainable, projects concerning built heritage resources need to take into account multiple dimensions, including the social one. More particularly, the implementation of initiatives combining either restoration or adaptive reuse with the achievement of social goals may be in some cases greatly recommendable: In fact, these types of interventions could be able not only to preserve and transmit the intrinsic and cultural components of built heritage but also to extend the relevance of the resources to larger segments of society and generate a multifaceted social impact overall. However, to effectively achieve social objectives, the adoption of evaluative thinking seems recommendable. Given this framework, this chapter strives to integrate the regeneration project of a system of historical farmhouses located in Volpiano, Italy, with actions aiming to favor the social inclusion of NEETS (i.e., youths not in education, employment, or training). Considering that the redevelopment of the system of the historical farmhouses was previously studied under the lens of corporate social responsibility, the integration of the social impact perspective represents an evolution in the discourse. By a methodological and processual perspective, the paper then proposes to follow the steps of logic models, while combining qualitative and quantitative evaluation approaches able to firstly describe and then quantify the multiple values engendered through the interventions. Finally, the contribution highlights that the application of evaluative thinking and evaluation procedures to built heritage projects with social objectives may facilitate both the definition and achievement of shared goals and thus function as a real catalyst of value.

Keywords Social impact · Well-being of citizens · Evaluation · Logic model · Built heritage

C. Coscia (✉) · I. Rubino
Department of Architecture and Design, Politecnico Di Torino, 39 Viale Pier Andrea Mattioli, 10125 Torino, Italy
e-mail: cristina.coscia@polito.it

1   Introduction

The multidimensional sustainability framework is currently acquiring increasing importance, and it is now informing both public and disciplinary debates, as well as decision-making processes at the global level. In line with this tendency, the theme of the sustainability of multi-scale interventions (e.g., at the building, urban and territorial scale) has undoubtedly started to be addressed, and now the economic, environmental, cultural, and social dimensions definitely need to be taken into account (Korkmaz and Balaban 2019; Lucchi et al. 2019; Coscia et al. 2018; Kohon 2018; Fregonara et al. 2016; Curto et al. 2014). This also applies to projects focusing on the *mise en valeur* of built heritage resources (Bottero et al. 2020), and in this context, in particular, an emerging topic is the achievement of sustainable and positive social impact through designed interventions. These may be represented, for instance, by retrofit actions on the historical built heritage (Lucchi et al. 2019)—which are definitely acquiring a strategic role for the revivification of these resources (Roberti et al. 2015)—but also by the integration of social functions in regeneration processes involving cultural heritage (Bottero et al. 2020; Coscia and Russo 2018). In fact, even if restoration and regeneration initiatives concerning built heritage resources are usually firstly performed for the intrinsic values attributed to them, the scarcity of financial means urges that the engendered value is maximized (Coscia and Rubino 2021; Coscia and Curto 2017). This is not only coherent with financial and economic considerations—including the emergence of new investment paradigms that aim at achieving both economic returns and social impacts (Alijani and Karotys 2019)—but it seems essential to: (a) make cultural heritage more relevant and meaningful for people; (b) conceive cultural heritage in the framework of the circular economy (Kee 2019; Fusco Girard and Gravagnuolo 2018; Foster 2020; https://www.clicproje ct.eu/); and (c) find strategies able to foster the maintenance and/or the existence itself of the buildings (Coscia and Chiaravalloti 2018). However, the design and implementation of projects able to effectively generate a positive social impact are far from being well-established procedures: firstly, social impact seems to still have multiple definitions in the field, making the discourse difficult; secondly, the achievement of positive impacts needs to be demonstrated and assessed, but evaluation practices are not always implemented because either considered as not important or deemed as an avoidable cost; thirdly, the singularity of the cases and the novelty of the paradigm make the identification of appropriate methods and metrics challenging.

In this framework, the goal of this paper is to shed some light on the relationships occurring between built heritage projects and social impacts, highlighting that the adoption of a social impact-oriented approach, together with the implementation of appropriate evaluation procedures, may function as catalyst of value. More particularly, in Sect. 2 we first explore the relationships occurring between built heritage and the social dimension, including the concept of social impact. In Sect. 3, we summarize how social impact has been evaluated so far. In Sect. 4, we then propose to apply a social impact-oriented approach to a project to be implemented in a peri-urban context, namely a system of historical farmhouses existing in Volpiano (Turin, Italy)

(Coscia and Russo 2018; Testù and Machiorletti 2016). The selection of a rural/peri-urban context as a case study was motivated by the consideration that extra-urban environments may be particularly interesting for the experimentation of projects that aim at achieving both social and economic goals. In fact, whereas urban areas are usually associated with high profits and competing economic interests (which might favor the pursuit of economic goals rather than social ones), initiatives with a social impact-oriented perspective might represent a particularly valuable opportunity for the redevelopment of less valued rural and peri-urban areas. Additionally, the redevelopment of the system of historical farmhouses located in Volpiano was previously studied under the lens of a corporate social responsibility approach (Coscia and Russo 2018), and the integration of the social impact perspective represents an evolution of the discourse. Section 5 offers final remarks and conclusions.

2 Built Heritage and the Social Dimension

The interrelationships between built heritage resources and the social dimension are multiple, and they may be regarded as: (1) the definition of built heritage itself; (2) the effects that built heritage resources (and valorization practices) may have on specific communities and society at large; and (3) the intentional inclusion of social goals into projects aiming to the *mise en valeur* of the resources.

With reference to the first point, it can be stated that built heritage resources are actually defined by the meanings that communities attribute to buildings and other elements of the built environment (Cerreta et al. 2014), as underlined for instance by the Faro Convention (Council of Europe 2005). Additionally, built heritage can be defined as such also in light of the social significance (e.g., capacity of a place/building to bind together members of the society, interpretation of a place/building as the reflection of the rules and beliefs shared by a given community, etc.) attributed by communities to specific places or buildings (ICOMOS Australia 1999).

With regard to the second point, it must be emphasized that the existence of built heritage is not neutral: in fact, it is known that its conservation state—as well as its physical and intellectual accessibility—engenders positive/negative effects (Amin 2018), together with socioeconomic consequences and various externalities (Throsby 2012; Al-hagla 2010; Rosato et al. 2008; Manganelli 2007). Moreover, the literature has also emphasized the role of built heritage in enriching the quality of life of people (Yung and Chan 2015), e.g., contributing to the fulfillment of the aesthetic, cultural, and leisure needs of a given community, fostering the development of social capital (Murzyn-Kupisz 2013) and social inclusion (Pendlebury et al. 2004), but also stimulating place attachment and sense of place (Jones 2017). Additionally, on a specular perspective, it has also been acknowledged that the recognition of social aspects is also fundamental in engineering assessments, retrofit interventions, or technical analyses (e.g., human comfort or energy efficiency) aiming to revitalize historical towns and buildings (Lucchi et al. 2019).

The need to maximize the value stemming from public and private expenditures, together with the goal of enabling social inclusion and finding sustainable solutions able to safeguard the preservation of less-known built heritage resources, has then favored the development of projects combining conservation, restoration and reuse of historical buildings with the achievement of social goals (third point). In this context, the adaptive reuse of buildings (Aigwi et al. 2019; Plevoets and Sowińska-Heim 2018) and the engagement of local communities are strategies that have been implemented so far to extend the relevance of built heritage for local targets and enable both the preservation of heritage and the socioeconomic sustainability of the interventions.

Overall, the development of awareness about built heritage as a possible agent of change has led to some reflections about the "social impact" engendered by projects concerning built heritage resources. However, it must be noted that multiple definitions of social impact exist, and they may vary according to the discipline and field of application, such as environmental studies (Burdge and Vanclay 1996), program evaluation (Kellogg Foundation 2017), and third sector (Zamagni et al. 2015). In line with the terminology mainly used in the program evaluation field, in this paper we will use the term social impact to indicate the medium/long-term effects of given interventions. In fact, even though the use of social impact in the cultural heritage literature may assume different nuances, it is possible to state that the term generally makes reference to the changes engendered by a project on individuals, communities, and even society at large. The types of effects explored by scholars are various, and they frequently include: residents and/or visitors' perceptions about projects aiming to enhance the local built heritage (or even perceptions linked to its decline, as described in Amin (2018)); the well-being and quality of life of residents (Korassani et al. 2019; Murzyn-Kupisz 2013); the degree and quality of community life; sense of place and attachment (Amin 2018); local community involvement and local capacity building (Korassani et al. 2019). Then, an alternative (or complementary) approach is the application of indicators and metrics, which are frequently expressed under the form of counts—e.g., number of participants in a given cultural activity, number of volunteers engaged, number of new jobs, etc.—, amounts—such as amount of euros collected form visits to a regenerated site—or percentages, including the percentage of increase in the number of visitors (Nocca 2017). However, in addition to single qualitative and/or quantitative approaches, the evaluation disciplines have then developed more specific—and sometimes hybrid—methods. In fact, with the introduction of social aspects and impacts, traditional approaches and "pure" quantitative assessment methods have entered into crisis.

3   Achieving and Evaluating Social Impacts of Built Heritage Projects: Some Approaches

Overall, the evaluation frameworks that are usually applied to the evaluation of social impacts are: (1) multicriteria (or dashboard) models, which are especially used when multiple dimensions and criteria need to be weighted and considered; (2) synthetic models, which tend to express into monetary terms the value created; (3) processual models, which are particularly recommended when new value chains and relationships among stakeholders are created (Camoletto et al. 2017).

Among the methods that take into account a variety of criteria, an interesting path is the one experimented by Korassani and colleagues (2019). Researchers evaluated the social impact of restoration works performed on a historic fortress in the context of a life-cycle management model: first, social themes (e.g., health and safety, wages, experiences, well-being, cultural development, access to tangible resources, employment, community involvement) and stakeholders (e.g., workers, local communities, consumers, society, and actors involved in the value chain) were identified; second, appropriate indicators (i.e., semiquantitative in their case) were selected, and third, a scoring system (a range from -2 to $+2$, where -2 indicates a not acceptable performance, 0 a performance aligned with international standards and $+2$ an ideal performance) was adopted; authors calculated "social topic scores," "stakeholder scores," and then the "total score."

From the point of view of synthetic and financial analysis models, it is useful to recall the experimentation of Social Return on Investment (SROI): this approach is frequently adopted, since it does not only enable expressing the value created into monetary terms, it is also suitable to be incorporated into multicriteria analyses (Camoletto et al. 2017).

As highlighted by the practices of organizations focusing on the conservation and valorization of the historical built heritage—such as The Churches Conservation Trust (https://www.visitchurches.org.uk/) and the Architectural Heritage Fund (https://ahfund.org.uk/)—the adoption of processual models from the very beginning of a project may be very fruitful, too. For instance, the Logic Model and Theory of Change frameworks prescribe to follow a *"plan backward, implement forward"* way of operating, recommending first to outline the desired impacts and then identify outcomes, outputs, and inputs (Kellogg Foundation 2017; Camoletto et al. 2017; Coryn et al. 2011). In addition, the Theory of Change also describes how and why an intervention or project fosters planned and unplanned changes in a given context, with reference to specific outcomes, targets and stakeholders (Morra-Imas and Rist 2009, p. 152; Funnell and Rogers 2011). In both the frameworks, a clear definition of impacts, outcomes, etc., from the very beginning of the project is essential not only to guide operational steps but also to inform evaluation (e.g., methods to be followed, metrics to be monitored, etc.), which is seen as an integral part of the whole process.

Coherently with this background, in the next paragraph we will describe how the application of a processual and evaluative approach to the regeneration of a system of historical farmhouses located in the nearby of Turin (Italy) could not only facilitate the collaborative definition of social objectives but also favor their achievement and the generation of additional value.

4 Enhancing Value Through an Evaluative and Social Impact-Oriented Framework: The System of Historical Farmhouses in Volpiano (Turin, Italy)

In 2015, the Municipality of Volpiano, i.e., a town of about 15,000 inhabitants located 16 km north-east of Turin (Italy), encouraged a collaborative agenda with some local stakeholders (such as the Politecnico di Torino university and bank foundations) to promote the cultural values of a system of historical farmhouses, while maintaining agricultural production and enabling economic sustainability (Coscia and Russo 2018; Testù and Machiorletti 2016). The analytical and decisional processes were conducted adopting an original and innovative perspective, i.e., integrating principles of corporate social responsibility (CSR). Management guidelines for the conversion of this system of farmhouses into a sustainable and multifunctional production system, in which the feasibility check was tested by a "hybrid" set of qualitative and quantitative evaluation methods, were then provided accordingly. In this paper, we report an advancement of the initial CSR approach, incorporating a social impact-oriented key.

To propose a feasible project consistent with a CSR perspective, preliminary considerations were performed and various evaluation methods were applied. More specifically, the evaluation approaches and methodological phases adopted to identify the most beneficial scenario were the following: (1) SWOT analysis; (2) stakeholder analysis and mapping; (3) community impact analysis (CIA); and (4) costs-revenues analysis carried out under a CSR perspective. SWOT analysis and techniques of stakeholder analysis and mapping were carried out paying particular attention to a set of contextual dimensions (e.g., accessibility, demographics, socioeconomic trends and conditions, agricultural and industrial activities, etc.): the interrelationship between the two instruments was considered essential not only for its ability to critically analyze the context but also to highlight the initial social pact between the subjects involved. For its suitability to inform the decisional process, SWOT analysis is in fact frequently integrated with other evaluation tools (as recently performed, for instance, by Bottero et al. (2020), who integrated SWOT analysis into a structured analytic hierarchy process). The stakeholder analysis took into account not only decision makers and institutions but also private entrepreneurs, the local population and temporary users of the areas under consideration; community impact analysis was performed, preliminarily identifying the social groups potentially affected by the interventions; finally, costs and revenues analyses aimed to verify the economic sustainability and profitability of the hypothesized interventions. Overall, evaluations led to a proposal focusing on the renewal of five farmhouses characterized by accessibility, the presence of agricultural production activities and the expected low costs of planned restoration works (Coscia and Russo 2018). More precisely, the selected project advanced to the integration of the cultural and economic dimensions with a social one, i.e., proposing the implementation of a multifunctional agriculture model that combined traditional and new cultivations (i.e., crops and hazelnuts respectively) with social farming activities (i.e., ortho-therapy, ortho-didactics, barefoot paths, pet therapy). On the one hand, investors took advantage of "green" incentives, but on the other they renounced a part of the profits and of the risk premium to favor the fulfillment of social functions that could benefit the community at large, coherently with a CSR perspective (Coscia and Russo 2018). In other words, objectives strictly linked to the enhancement of local agricultural production were combined with the intention of improving the quality of life in rural areas, increasing occupational opportunities and improving the attractive power of the areas through diversification.

If, on the one hand, this scenario definitely took into account the social dimension (with reference to both the methods followed, the identified objectives and the recommended functions); on the other one, it is possible to suggest that the shift toward a social impact-oriented approach could not only further strengthen the social outcomes of the project but also influence the planning processes and the evaluation strategies adopted. In fact, the adoption of processual methods and of a logic model finalized to the explicit achievement of social goals would entail the more robust and cross-cutting use of qualitative–quantitative assessment tools aimed at: (1) a deeper recognition of the needs to be fulfilled, especially with reference to specific targets: in fact, the recognition of the needs is a fundamental step to inform the definition of goals and strategies; (2) the identification of the desired impacts, followed by a more granular definition of outcomes, outputs and inputs; (3) the collaboration among stakeholders since the very beginning of the project, as to sharing responsibility in the decision-making process and in the definition of objectives and indicators; and (4) a clear definition since the initial phases of the evaluation strategies, methods, and metrics to be followed. In fact, the adoption of an impact-oriented approach would transform a simple collaborative agenda into a real partnership of various stakeholders (i.e., municipality, university, owners of farmhouses, local entrepreneurs, associations operating in the third sector, etc.) able not only to generate innovative solutions to previously identified problems but also create value (e.g., institutional, relational, reputational, etc.) rightly through collaboration. In this sense, the qualitative tools of the SWOT and above all of the stakeholder and network analysis (Coscia and Zanetta 2018) interrelated with the CIA can both strengthen the detailed analysis of the responsibilities and impacts by subject and inform economic–managerial analysis in a "social" sense. In more empirical terms, this would translate into the description of the processes triggered among stakeholders and into the monitoring of the outcomes of the activities, e.g., according to the indicators collaboratively defined in the decisional phases of the project. Overall, the inclusion of a specific social objective in the realm of the proposed interventions could offer the possibility to enhance the overall value of the project itself, especially for identified social targets. In fact, engaging specific segments into the activities organized in the historical farmhouses (and inspired to social agriculture principles) would enable making local built heritage more relevant for larger portions of the local community, possibly activating virtuous cycles of support.

For instance, the implementation of a specific program aiming to favor the social inclusion of youths not in education, employment, or training (also known as NEETS) could be integrated into the multifunctional model already proposed for the system of historical farmhouses of Volpiano. The NEETS phenomenon interests countries across all of Europe, and a recent report has illustrated that, among European countries, Italy has the highest percentage of NEETS (Eurostat 2019). According to the same report, NEETS are distributed in cities as well as suburbs, towns and rural areas, and the cost of their inactivity for the Italian state has been estimated in 36 billion euros in 2016 (Fagnani 2017). For several reasons, the elaboration of strategies and programs able to facilitate the socioeconomic inclusion of this particular segment of young people is thus critically needed and should inform national and local agendas. Recent examples show that social agricultural programs have been implemented in Italy to mitigate NEETS' personal discomfort, enhance their well-being and facilitate their integration into the job market (Centro Nazionale di Documentazione e analisi per l'infanzia e l'adolescenza 2018; Finzi and Romero Aranda 2016). Additionally, the promotion of local cultural heritage has been identified as a promising initiative to engage NEETS, suggesting that the experimentation of social agriculture programs in the context of the *mise en valeur* of historical farmhouses could be particularly fruitful. In this view, the adoption of a collaborative and impact-oriented approach would make possible both to help public actions and enrich the CSR approach, which usually largely relies on the attitudes of single entrepreneurs. In this case, the responsibility and achievement of social goals would be shared among multiple stakeholders, instead. In the case of Volpiano, the social theme of the NEETS could first be highlighted in the SWOT and subsequently related to the analysis of the impacts and of the stakeholders, both at the scale of the farm system and at the enlarged one of the peri-urban area. Figure 1 graphically shows the suggested process, making reference to the specific intervention concerning NEETS and the historical farmhouse system of Volpiano.

If qualitative and processual analyses could be performed to map the value created by the collaborative approach, indicators such as the percentage of participants who found a job after six months from the completion of the program, the number of volunteers adhering to local social agriculture initiatives, the increase of awareness about the historical value of the farmhouses, etc., could be employed to evaluate the social outcomes and impacts engendered by the project. Additionally, an estimation in monetary terms of the value created by the program could be performed, while considering the costs avoided for public finance thanks to the potential overcoming of the NEET status by some of the participants in the program. Finally, it must be underlined that such estimates will influence the subsequent quantitative phase: in fact, they will provide new factors to be introduced in the items of the financial–management analysis and in the identification of the threshold values of the profitability indicators of the management Discounted Cash Flow Analysis.

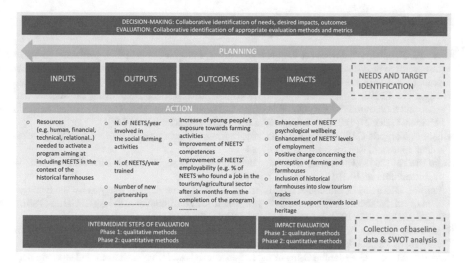

Fig. 1 Enhancing the social value of built heritage projects: applying a social impact-oriented approach in the strategical phase (*Source* authors' own elaboration)

5 Conclusions

The inclusion of evaluation procedures from the very beginning of a built heritage project, throughout its development and after its completion, necessitates no doubt a cost. However, the adoption of evaluative thinking could be overall considered more a type of investment rather than a simple cost since it is useful to: (1) clearly define the goals, impacts, outcomes, outputs and inputs of a project; (2) foster collaboration among stakeholders from the very beginning, encouraging shared responsibility and a cooperative definition of the desired objectives and actions; (3) incorporate monitoring activities throughout the project, so as to timely check whether intermediate and final objectives have been achieved or not and then perform prompt corrections if needed; (4) describe and estimate the change generated by the project, also in light of the accountability framework; (5) possibly express in monetary terms the value engendered by the performed actions; and (6) not only build human and institutional capital but also provide useful data for future local planning and development. As shown in the case study of the historical farmhouses of Volpiano (Italy), the adoption of a social impact-oriented approach that explicitly includes the achievement of social goals may not only strengthen the CSR approach but also overall extend the relevance of built heritage projects, thus functioning as a catalyst of value. Additionally, it should be added that the accomplishment of social goals may not only represent a benefit in itself but also enable well-being conditions favoring the activation of positive behaviors and support toward heritage. Given this framework, the next steps of future research could be represented by the investigation of these new, more extended value chains, so as to better understand and quantify the added value of built heritage projects encompassing social goals.

References

Aigwi IE, Egbelakin T, Ingham J, Phipps R, Rotimi J, Filippova O (2019) A performance-based framework to prioritise underutilised historical buildings for adaptive reuse interventions in New Zealand. Sustain Cities Soc 48:101547. https://doi.org/10.1016/j.scs.2019.101547

Al-hagla KS (2010) Sustainable urban development in historical areas using the tourist trail approach: a case study of the cultural heritage and urban development (CHUD) project in Saida Lebanon. Cities 27(4):234–248. https://doi.org/10.1016/j.cities.2010.02.001

Alijani S, Karyotis C (2019) Coping with impact investing antagonistic objectives: a multistakeholder approach. Res Int Business Fin 47:10–17

Amin HMTM (2018) The impact of heritage decline on urban social life. J Environ Psychol 55:34–47. https://doi.org/10.1016/j.jenvp.2017.12.002

Bottero M, D'Alpaos C, Marello A (2020) An application of the A'WOT analysis for the management of cultural heritage assets: the case of the historical farmhouses in the Aglié Castle (Turin). Sustainability 12(3):1071. https://doi.org/10.3390/su12031071

Burdge RJ, Vanclay F (1996) Social impact assessment: a contribution to the state of the art assessment. Impact Assess 14(1):59–86. https://doi.org/10.1080/07349165.1996.9725886

Camoletto M, Ferri G, Pedercini C, Ingaramo L, Sabatino S (2017) Social Housing and measurement of social impacts: steps towards a common toolkit. Valori E Valutazioni 19:11–40

Centro nazionale di documentazione e analisi per l'infanzia e l'adolescenza (2018) Inclusione dei giovani NEET, un aiuto dall'agricoltura sociale. Retrieved from https://www.minori.gov.it/it/not izia/inclusione-dei-giovani-neet-un-aiuto-dallagricoltura-sociale. Accessed on 3 July 2019

Cerreta M, Inglese P, Malangone V, Panaro S (2014) Complex values-based approach for multidimensional evaluation of landscape. In: Proceedings of the international conference on computational science and its applications, Guimarães, Portugal, 30 June–3 July 2014. Springer, Cham, Switzerland, pp 382–397

Council of Europe (2005) Council of Europe framework convention on the value of cultural heritage for society. Retrieved from https://rm.coe.int/1680083746. Accessed on 15 Nov 2019

Coryn CLS, Noakes LA, Westine CD, Schröter DC (2011) A systematic review of theory-driven evaluation practice from 1990 to 2009. Am J Eval 32(2):199–226

Coscia C, Chiaravalloti T (2018) Urban voids and public historical-artistic heritage: a road map for the Carlo Alberto complex in Acqui Terme. Vuoti urbani e patrimonio del demanio storico-artistico: Una road map per l'ex Carlo Alberto di Acqui Terme. Territorio, 84, 128–142

Coscia C, Curto R (2017) Valorising in the absence of public resources and weak markets: the case of "Ivrea, the 20th century industrial city". In: Stanghellini S, Morano P, Bottero M, Oppio A (eds) Appraisal: from theory to practice. Springer, Cham, Switzerland, pp 79–99

Coscia C, Rubino I (2021) Fostering new value chains and social impact-oriented strategies in urban regeneration processes: what challenges for the evaluation discipline? In: Bevilacqua C, Calabrò F, Della Spina L (eds) New metropolitan perspectives. knowledge dynamics and innovation-driven policies towards urban and regional transition volume 2. Smart innovation, systems and technologies, vol 178. Springer, Cham, pp 983–992

Coscia C, Russo V (2018) The valorization of economic assets and social capacities of the historic farmhouse system in peri-urban allocation: a sample of application of the corporate social responsible (CSR) approach. In: Bisello A, Vettorato D, Laconte P, Costa S (eds) Smart and sustainable planning for cities and regions, pp 615–634

Coscia C, Zanetta M (2018) Perchè mappare gli stakeholders? L'importanza dello stakeholder engagement nei processi decisionali. In Occasioni di Dialogo. Progetto di recupero a Vinovo: la Piccola Casa della Divina Provvidenza. Cult Herit 9:110–122. ISBN: 978-88-85629-31-8

Coscia C, Lazzari G, Rubino I (2018) Values, memory and the role of exploratory methods for policy-design processes and the sustainable redevelopment of waterfront contexts: the case of Officine Piaggio (Italy). Sustainability 10(9):2989, 1–22. https://doi.org/10.3390/su10092989

Curto R, Brigato MV, Coscia C, Fregonara E (2014) Assessing strategies for developing sustainable tourism in the Iglesias area, Sardinia. Valutazioni per strategie di sviluppo turistico sostenibile dell'iglesiente. Territorio 69:123–133

Eurostat (2019) Statistics on young people neither in employment nor in education or training. Retrieved from https://ec.europa.eu/eurostat/statistics-explained/index.php/Statistics_on_y oung_people_neither_in_employment_nor_in_education_or_training#NEETs:_analysis_by_ degree_of_urbanisation. Accessed on 20 Nov 2019

Fagnani MG (2017) Garanzia giovani inefficace, i "Neet" costano allo Stato 36 miliardi. *Buone Notizie*, 16 December 2017. Retrieved from https://www.corriere.it/buone-notizie/17_dicembre_ 14/generazione-neet-costo-totale-lo-stato-36-miliardi-3c5a43ea-e0e9-11e7-acec-8b1cf54b0 d3e.shtml. Accessed on 20 Oct 2019

Finzi MCV, Romero Aranda EC (2016) The NEEP and GLEAN project: guidelines and recommendations to a new approach for employability and entrepreneurship through agriculture. Retrieved from https://www.glean-project.eu/docs/GLEAN_O5_%20EN_Guidelines_Recomm endations.pdf. Accessed on 3 July 2019

Foster G (2020) Circular economy strategies for adaptive reuse of cultural heritage buildings to reduce environmental impacts. Resour Conserv Recycl 152:104507. https://doi.org/10.1016/j.res conrec.2019.104507

Fregonara E, Giordano R, Rolando D, Tulliani JM (2016) Integrating environmental and economic sustainability in new building construction and retrofits. J Urban Technol 26:3–28

Funnell SC, Rogers PJ (2011) Purposeful program theory: effective use of theories of change and logic models. Jossey-Bass, San Francisco

Fusco Girard L, Gravagnuolo A (2018) Circular economy and cultural heritage/landscape regeneration. Circular business, financing and governance models for a competitive Europe. BDC. Bollettino Del Centro Calza Bini 17(1):35–52

ICOMOS Australia (1999) Charter for the Conservation of places of cultural significance (The Burra Charter). Retrieved from https://australia.icomos.org/wp-content/uploads/BURRA_CHA RTER.pdf. Accessed on 12 June 2019

Jones S (2017) Wrestling with the social value of heritage: problems, dilemmas and opportunities. J Community Archaeol Heritage 4(1):21–37

Kee T (2019) Sustainable adaptive reuse—economic impact of cultural heritage. J Cult Heritage Manage Sustain Devel 9(2):165–183

Kohon J (2018) Social inclusion in the sustainable neighborhood? Idealism of urban social sustainability theory complicated by realities of community planning practice. City Cult Soc 15:14–22. https://doi.org/10.1016/j.ccs.2018.08.005

Korkmaz C, Balaban O (2019) Sustainability of urban regeneration in Turkey: assessing the performance of the North Ankara urban regeneration project. Habitat Int 102081. https://doi.org/10. 1016/j.habitatint.2019.102081

Lucchi E, D'Alonzo V, Exner D, Zambelli P, Garegnani G (2019) A density-based spatial cluster analysis supporting the building stock analysis in historical towns. In: Proceedings of the 16th international building performance simulation association conference. Rome, 2–4 Sept 2019, 3831–3838. https://doi.org/10.26868/25222708.2019.210346

Manganelli B (2007) Valutazioni economico-estimative nella valorizzazione di edifici storico-architettonici. Aestimum 51:21–42

Mohaddes Khorassani S, Ferrari AM, Pini M, Settembre Blundo D, García Muiña FE, García JF (2019) Environmental and social impact assessment of cultural heritage restoration and its application to the Uncastillo Fortress. Int J Life Cycle Assess 24(7):1297–1318. https://doi.org/ 10.1007/s11367-018-1493-1

Morra-Imas LG, Rist RC (2009) The road to results: designing and conducting effective development evaluations. The World Bank, Washington, DC

Murzyn-Kupisz M (2013) The socio-economic impact of built heritage projects conducted by private investors. J Cult Heritage 14(2):156–162. https://doi.org/10.1016/j.culher.2012.04.009

Nocca F (2017) The role of cultural heritage in sustainable development: multidimensional indicators as decision-making tool. Sustainability 9(10):1882. https://doi.org/10.3390/su9101882

Pendlebury J, Townshend T, Gilroy R (2004) The conservation of English cultural built heritage: a force for social inclusion? Int J Heritage Stud 10(1):11–31. https://doi.org/10.1080/135272503 2000194222

Plevoets B, Sowińska-Heim J (2018) Community initiatives as a catalyst for regeneration of heritage sites: vernacular transformation and its influence on the formal adaptive reuse practice. Cities 78:128–139. https://doi.org/10.1016/j.cities.2018.02.007

Roberti F, Oberegger UF, Lucchi E, Gasparella A (2015) Energy retrofit and conservation of built heritage using multi-objective optimization: demonstration on a medieval building, building simulation applications (BSA), 4–6 Feb 2015, pp 189–197

Rosato P, Rotaris L, Breil M, Zanatta V (2008) Do we care about built cultural heritage? The empirical evidence based on the Veneto house market. Research paper, Fondazione Eni Enrico Mattei. Retrieved from https://www.feem.it/m/publications_pages/NDL2008-064.pdf. Accessed on 12 June 2019

Testù F, Machiorletti P (2016) Le cascine volpianesi. Conoscere per valorizzare: l'analisi delle cascine volpianesi come strumento di promozione del territorio. L'Artistica Savigliano, Cuneo

Throsby D (2012) Heritage economics: a conceptual framework. In: Licciardi G, Amirtahmasebi R (eds) The economics of uniqueness. The World Bank, Washington DC, pp 45–72

W.K. Kellogg Foundation (2017) The step-by-step guide to evaluation. How to become savvy evaluation consumers. Retrieved from https://www.wkkf.org/resource-directory/resource/2017/ 11/wk-kellogg-foundation-step-by-step-guide-to-evaluation. Accessed on 10 June 2019

Yung EHK, Chan EHW (2015) Evaluation of the social values and willingness to pay for conserving built heritage in Hong Kong. Facilities 33:76–98

Zamagni S, Venturi P, Rago S (2015) Valutare l'impatto sociale. La questione della misurazione nelle imprese sociali, Impresa Sociale, p 6

Websites

https://ahfund.org.uk/
https://www.clicproject.eu/
https://www.visitchurches.org.uk/

Renewable Energy Communities: Business Models of Multi-family Housing Buildings

Valeria Casalicchio, Giampaolo Manzolini, Matteo Giacomo Prina, and David Moser

Abstract The new European directive on renewable sources (RED II), which entered into force in December 2018, has opened new perspectives on the consumers and the decentralization of energy production. The purpose of this work is to analyze the inclusion of the energy communities in the Italian regulatory framework. The analysis focuses on the strategies that can be adopted by tenants to share rooftop photovoltaic module production and stored electricity and the consequent economic impact on their electricity bills. We have created a program that simulates various business models to be proposed to prosumers of multi-family housing buildings through which the economic return of every participant in the energy community is evaluated. The program receives as input the data of the renewable energy community, and it returns as output the electricity bill of each tenant, while considering the cost of the energy purchased from the grid and the economic revenues from self-consumption and from the sale of the excess production for each of the business model adopted. Therefore, the advantages and disadvantages of the models are highlighted. The net metering is profitable, but excluding it, energy sharing within communities would be the best scenario: In the considered case study, with 10 kW of PV installed power for a community, the reduction in costs is equal to 15%. Moreover, the convenience of a heterogeneous set of electricity demand profiles of the members is clearly evident (in the considered case study, this entails a 10% reduction in costs). Finally, the most proper business model must be selected, to assure the benefit for each energy community participant: The examined models result in a bill differential between −20 and 36% for each participant.

Keywords Smart community · Business and value model · Challenge common spaces for the sharing economy · Prosumer · Energy community

V. Casalicchio (✉) · M. G. Prina · D. Moser
Institute for Renewable Energy, EURAC Research, Viale Druso 1, 39100 Bolzano, Italy
e-mail: valeria.casalicchio@eurac.edu

V. Casalicchio · G. Manzolini
Dipartimento Di Energia, Politecnico Di Milano, Via Lambruschini 4, 20156 Milano, MI, Italy

© The Author(s), under exclusive license to Springer Nature Switzerland AG 2021
A. Bisello et al. (eds.), *Smart and Sustainable Planning for Cities and Regions*,
Green Energy and Technology, https://doi.org/10.1007/978-3-030-57332-4_19

1   Introduction

The new European directive on renewable sources RED II, which entered into effect on December 24, 2018, as part of the Clean Energy Package for all Europeans, has opened new discussions about the concept of Self-Consumption of Electricity from Renewable Sources.

This directive has remarkable relevance because it introduces the "one-to-many" self-consumption solution in which the final consumer is not unique. So far, focusing on Italy, it has not been possible to realize a "one-to-many" self-consumption solution, but from now on collective self-consumption solutions will be investigated. Specifically, the RED II defines the "renewable energy communities":

> [...] Renewable energy community means a legal entity: (a) which [...] is based on open and voluntary participation, is autonomous, and is effectively controlled by shareholders or members that are located in the proximity of the renewable energy projects that are owned and developed by that legal entity; (b) the shareholders or members of which are natural persons, SMEs or local authorities, including municipalities; (c) the primary purpose of which is to provide environmental, economic or social community benefits for its shareholders or members or for the local areas where it operates, rather than financial profits (DIRECTIVE (EU) 2018/2001 (2018)).

The energy community (EC) topic has already been addressed in previous studies. Moroni et al. (2019) suggest an energy community taxonomy, distinguishing between geographical characteristics and purposes. Moura and Centeno Brito (2019) and (Brummer 2018) provide an overview of EC literature, compare different country experiences enabling the sharing of generation systems and investigate business model case studies. They highlight the importance of implementing ad hoc policies and regulations to support prosumer aggregation and projects already in place as Clean Energy collective (USA) (2010), SOM Energia (Spain) (2010), Sonnen (Germany) (2010) and Powerpeers (Netherlands) (2016) and provide conditions under which an EC movement can emerge. Moreover, Brummer (2018) describes EC benefits and barriers which are not only economic.

Research projects aim at deepening the knowledge around energy communities and at developing proper models such as those briefly presented in Appendix. As pointed out in this table, this study differs from the others because it performs the profitability evaluation with respect to the scenarios without communities and by comparing several regulatory frameworks, both with and without subsidies. Moreover, at the same time, it simulates more cases in which several EC participants take over the technology investment and use various pricing schemes to allocate the costs among them.

Therefore, the aim of this work is to perform an analysis of the inclusion of the EC in the Italian regulatory framework. Moreover, the role of prosumers and consumers participating in the EC and their relationships in terms of shared energy are evaluated, and forms of collective self-consumption have been simulated. The analysis mainly focuses on three issues. The first regards the assessment of the profitability of an EC according to the specific regulatory framework considered. The second issue is the evaluation of how the electricity flows are managed inside the EC: How the self-consumption of the photovoltaic module production, the energy-sharing (virtual), the storage and the grid trading are performed. The last consists in the definition of the best strategy that can be adopted by the EC, thus focusing also on the consequent economic impact on the electricity bill and annual costs to each EC member. To perform this study, a program that is suitable to simulate different systems composed of consumers and prosumers was developed. Several regulatory frameworks can be analyzed and compared, paying attention to the differences in the annual costs to EC members. Moreover, this program reproduces various possible business models (BM) to be proposed to prosumers and consumers of multi-family housing buildings participating in an EC. Through these models, the economic return of every participant is evaluated.

The plan of the paper is organized as follows: In Sect. 2, the methodology adopted to develop this work is presented. The methodology consists of two steps: The first focuses on modelling the electricity flows and costs in an EC system, in different regulatory frameworks; the second focuses on the definition of the business models that can be adopted inside the EC.

In Sect. 3, a few analyzed cases are presented in detail by paying attention to two main aspects: the EC profitability and the adopted business models. Lastly, in Sect. 4 the results are discussed and some conclusions drawn.

2 Methodology

A program developed in Python was implemented, and it executes several built-in functions that enable to perform a two-step simulation. The model consists of two steps: The modeling of the electricity flows according to whether an EC is available or not; the modeling of different regulatory frameworks and the assessment of the involved costs.

2.1 Modeling of Electricity Flows

The modeling and definition of the electricity flows involving the users of an EC were determined through a series of functions which are suitable to any multi-family housing building.

The input data include the composition of the EC members, their yearly electric demand profiles with hourly time-step and the information about their investment on shared PV and battery systems.

The data include also the details about the electricity bill component and about the installed rooftop photovoltaic and battery systems. The functions assess the electricity flows by setting priorities. This evaluation is performed both in case an EC does not exist and in case an EC is constituted. In case there is an EC and the members can share energy among them, we have defined few functions to assess how each user's electricity demand is met. These functions, listed according to their priority, evaluate:

- the self-consumption of each member;
- the excess of photovoltaic (PV) production or the shortage of electricity;
- the renewable electricity shared among the users of the EC;
- the electricity exchanged with the battery;
- the electricity sold or purchased from the grid.

Thus, in the EC the priority is given to self-consumption, followed by energy-sharing (ES), storage usage and grid electricity exchanges. Otherwise, if an EC has not been constituted, the function that evaluates the energy sharing will be skipped. In this latter situation, the program considers that each user (consumer or prosumer) has installed a photovoltaic panel and a storage batter. To simplify the comparison, PV and battery capacities are the same as they would be in case the tenants participated in the EC. Clearly, if a PV panel is not installed, the function still evaluates the electricity flows by performing only the last task of the previous list.

2.1.1 Storage Model in EC

One-year, hourly time-step input data on power generation from PV $P_i(t)$ and electricity demand $D_i(t)$ for each member i are available, and the electricity shared is evaluated $S_i(t)$. A storage system (a lithium-ion battery) with capacity $H^{\max}$ is limited so to analyze its role in the EC. Moreover, a storage system (a lithium-ion battery) with a limited capacity Hmax was modeled so to analyze its role in the EC. To implement the battery modelling, I took into consideration the approach illustrated in Weitemeyer et al. (2015). The storage device modeling approach developed in Weitemeyer et al. (2015) makes it possible to analyze the role of storage devices for the integration process of RES depending on two key parameters: the round-trip efficiency and the storage size. The same approach was adopted also in Prina et al. (2016) to study the impact of storage systems in the case of high renewable energy penetration.

The storage model of this work is described as follows:

$\Delta_i(t)$ is the mismatch in PV production, energy-sharing and energy demand at time t for each EC member:

$$\Delta_i(t) = P_i(t) - D_i(t) - S_i(t)$$

A general mismatch in PV production, energy-sharing and energy demand at time t was also evaluated for the entire EC and was defined as:

$$\Delta(t) = \sum_i P_i(t) - \sum_i D_i(t)$$

The storage battery time series $H(t)$ definition follows:

Table 1 Performance characteristics of lithium-ion batteries	η_c	0.9
	η_d	0.89
	η_{sd}	0.3% / month

$$H(t) = \begin{cases} \min[H \max, \Delta(t)\eta_c + H(t-1)(1 - \eta_{sd})] & \text{if} \Delta(t) \geq 0 \\ \max[0, H(t-1)(1 - \eta_{sd}) - \Delta(t)/\eta_d] & \text{if} \Delta(t) < 0 \end{cases}$$

with the storage considered as fully flexible and η_c being the charging efficiency of the storage, η_d being the discharging efficiency and η_{sd} being the self-discharge. The initial charging level of the battery $H(t = 0)$ has been specified and settled equal to zero. Table 1 shows the performance characteristics of lithium-ion batteries (Harding Energy 2004).

2.2 User Bill Definition and Regulatory Frameworks

Additional functions evaluate the costs associated with the electric flows of the defined energy system. Therefore, the yearly bill of each user is defined, and it mainly depends on the considered regulatory framework. Also, the LCOE and LCOS are evaluated and used to calculate the yearly costs associated with the PV and the battery for each user.

2.2.1 Bill Components

The first element considered to assess the EC profitability is the members' electricity bills:

> [...] Household consumers and communities engaging in renewables self-consumption should maintain their rights, [...] including the rights to have a contract with a supplier of their choice (Article 72 of Directive (EU) 2018/2001).

Therefore, the specific function that evaluates the electricity bill accounts for all the Italian bill components and allows the user to set the prices according to the selected supplier. Components include the energy prices, network services, general system charges and taxes, which are further subdivided according to fixed quota [€/year], power quota per power used [€/kW] and a variable quota for withdrawn energy [€/kWh].

2.2.2 LCOE and LCOS

LCOE is calculated by using the following formula (Fraunhofer ISE 2015):

Table 2 Rooftop PV and battery data for LCOE and LCOS evaluation

	Rooftop PV	Battery	Source
CAPEX 2020	1198 €/kW	587 €/kWh	Prina et al. (2019)
CAPEX 2050	753 €/kW	178 €/kWh	Prina et al. (2019)
OPEX	1.5%$_{CAPEX}$	5%$_{CAPEX}$	Veronese et al. (2019)
Discount rate	5%		Veronese et al. (2019)
Lifetime	30 years	15 years	Veronese et al. (2019)

$$\mathrm{LCOE} = \frac{I_0 + \sum_{t=1}^{n} \frac{\mathrm{O\&M}_t}{(1+i)^t}}{\sum_{t=1}^{n} \frac{P_{t,el}}{(1+i)^t}}$$

where I_0 is the PV capex, O&M is the opex and $P_{t,\,el}$ is the produced quantity of electricity in the respective year, i is the discount rate, n is the economic operational lifetime in years and t is the year.

LCOS is calculated by using the following formula (Sing Lai and McCulloch 2017):

$$\mathrm{LCOS} = \frac{I_0 + \sum_{t=1}^{n} \frac{C_{\mathrm{EES}t}}{(1+i)^t}}{\sum_{t=1}^{n} \frac{E_{\mathrm{EES}t}}{(1+i)^t}}$$

where I_0 is the battery initial investment cost, $C_{\mathrm{EES}t}$ is the total cost in year t and $E_{\mathrm{EES}t}$ is the total energy output in year t. Main data is provided in Table 2.

2.2.3 Regulatory Framework Definition

The profitability of an EC can be evaluated by making a comparison among several cases characterized by different regulatory frameworks. In this work, we have chosen six different frameworks:

- no PV or battery is installed. The households are consumers purchasing the electricity from the grid;
- PV panels are installed as a "one-to-one" self-consumption solution and no incentive is considered. The households can be both prosumers and consumers;
- as above, but the "net-metering" incentive (NM) is considered;
- ECs are established and the PV panel is a "one-to-many" self-consumption solution. Incentives are not considered. All the households participate in the EC;
- as above, and the "net-metering" incentive is considered separately for each member;
- as above, and the "net-metering" is applied, but the EC is considered as a unique user.

Table 3 summarizes the frameworks introduced above.

Table 3 Regulatory frameworks

	Framework 1	Framework 2	Framework 3	Framework 4	Framework 5	Framework 6
PV	✗	✓	✓	✓	✓	✓
Net-metering	–	✗	✓	✗	✓	✓*
Energy-sharing	–	✗	✗	✓	✓	✓

*EC considered for "net-metering" as a single user

2.2.4 Business Model Definition

More cases, where an EC is established, are simulated to assess what situation is expected by June 2021, the deadline to incorporate the RED II directive.

A further distinction is made according to which business model is adopted in the EC, to determine how the economic benefits resulting from energy-sharing are divided among EC participants.

The business models vary from each other because a different economic value is assigned to the self-consumed electricity.

Moreover, the benefits resulting from self-consumption can be allocated among the EC members, thus guaranteeing a more proper and fairer distribution and avoiding penalizing those users who have invested in the EC PV, but whose demand profiles cannot match with the PV production profiles. The benefit allocation has been performed according to the PV investment percentage of every member of the EC or according to the PV self-consumption percentage. Anyway, other allocation approaches can be introduced and adopted (Elemens and Kantar 2019). Every defined BM does not affect the overall EC community benefit that is generated by the self-consumed PV production, but it affects only the benefit of each member of the community.

Table 4 summarizes the BMs adopted for the evaluation and comparison. The upper section of the table defines what economic value has been assigned to the self-consumed electricity. The bottom section describes how the profit-pool arising from self-consumption tariffs is allocated to every EC member.

Table 4 Business models

		BM A (tariff)	BM B (tariff)	BM C (tariff)
Self-consumption tariff	Free	✓ (0 €/kWh)	–	–
	EC tariff	–	✓ (0.1 €/kWh)	–
	Grid tariff	–	–	✓ (0.1161 €/kWh)
Benefit allocation	PV investment	–	✓	–
	PV self-consumption	–	–	✓

The application simulates how much electricity each EC member self-consumes each hour and applies common tariffs to this consumption. The cash pool formed by the above tariffs is then redistributed among the members according to a specified approach.

We have chosen the following approaches: (i) redistribution according to the investment percentage of each member and (ii) redistribution according to the PV self-consumption percentage of each member. This latter approach aims at promoting self-consumption, therefore reducing grid usage and encouraging the application of various demand-side management techniques. The cost results derived from the combination of the just-mentioned approaches are then compared, and an analysis of the impact of every business model on the EC members is performed.

3 Case Studies

A few cases have been analyzed at a multi-family apartment building level, and through them we have attempted to answer the questions this work is based on: whether an EC is profitable according to the considered regulatory framework; what business model should be adopted in order to fairly distribute the economic benefit from PV consumption.

The analyzed case studies include several households, characterized by various compositions and by different habits. So, different electricity demand profiles have been considered and have been obtained through LoadProfileGenerator[1], a tool suitable to generate load profiles for residential electricity consumption.

The first case considered was a multi-family housing building made of five housing units, where a dual-earner couple, a one-earner couple, a couple over 65 years old, a family of a dual-earner couple and a child, and a dual-earner couple with home help live. This case study is interesting as it helped become more confident with the EC concept and identify the main factors for this sort of analysis. Table 5 presents the input data of installed PV in each of the six considered frameworks. The battery system capacity is set equal to zero because the year considered is 2020, and the storage system costs are still too high (Figs. 1, 2, 3, 4 and 5).

We considered also a second case: a multi-family housing building made of five housing units, as in the previous case study. Table 2 is still valid, but in this second case five dual-earner couples have the same electricity demand profiles, the sum of which is kept constant for both the case studies. They make different investments in PV and none in the storage system.

Global consumptions and profiles of the households in both case studies are reported in Appendix Tables 6 and 7, Appendix Figs. 6 and 7. The PV production profile is shown in Appendix Fig. 8.

---

[1]N.Pflugradt, available at: https://www.loadprofilegenerator.de/

Table 5 First case study data

	Household	Framework 1	Framework 2, 3	Framework 4, 5, 6	
					Investment (%)
PV (kW)	1st	Consumer -	Prosumer 4 kW	EC Members 10 kW	40
	2nd	Consumer -	Prosumer 3 kW		30
	3rd	Consumer -	Prosumer 2 kW		20
	4th	Consumer -	Prosumer 1 kW		10
	5th	Consumer -	Consumer -		0

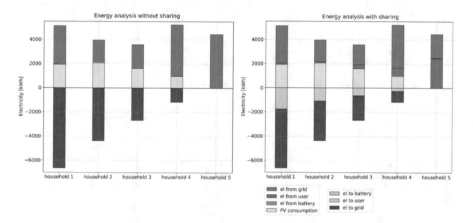

Fig. 1 Energy flows without energy-sharing (left), with energy-sharing (right), case study #1

4 Results

4.1 Case Study #1

Figure 1 shows the energy flow distribution evaluated by the developed application in the case without EC and therefore without energy-sharing among prosumers and in the case of an EC establishment.

After running the program, the results are summarized in Fig. 2 and include the costs of the overall multi-family housing building in different scenarios; Table 3 shows instead the different regulatory frameworks which have been taken into account.

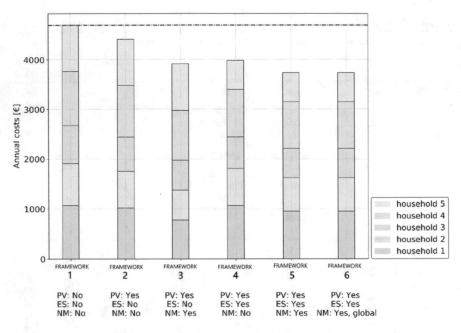

Fig. 2 Annual costs of EC members in different regulatory frameworks, case study #1

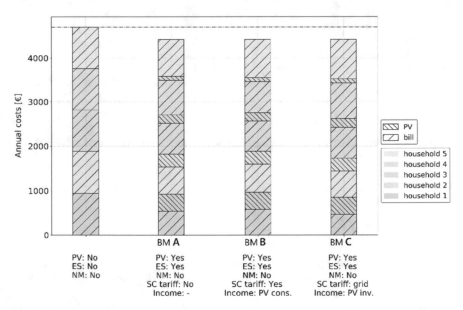

Fig. 3 Distributions of the EC benefit according to different business models, case study #1

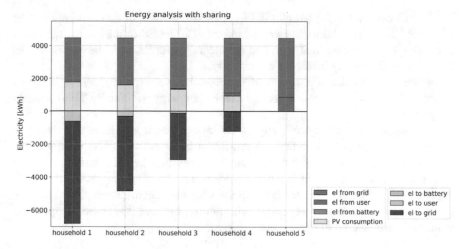

Fig. 4 Energy flows in case with energy-sharing, case study #2

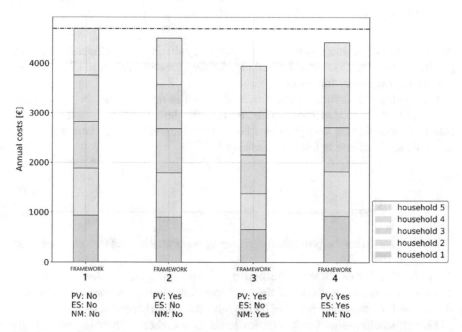

Fig. 5 Annual costs of EC members in different regulatory frameworks, case study #2

Focusing on the first four bars of Fig. 2 the third bar, which shows the framework with "net metering" (#3), is the lowest and represents the most profitable condition where an EC is not encouraged. However, in case there were no "net metering" incentive, the fourth bar would be the most convenient.

The last two bars of the bar plot in Fig. 2 show the effect of maintaining the "net metering" incentive even when establishing an EC. The EC of the last bar case (#6) is conceived as a unique final user, and the incentive impact could be higher thanks to an enhanced synergy inside the community.

Figure 3 shows the application of different business models, which bring to different distributions of the cost savings inside the energy community. These models are presented in Table 4.

Looking at Fig. 3 and specifically at the second bar, it is evident that the most favored household is the one that did not invest in the common PV panel. On the contrary, in the last bar the most favored one is the first member who made the biggest investment. However, also the last households had profits by participating in the EC. As already mentioned, there are fairer models for the benefit distribution: This is a key aspect in the formation of energy communities because it is based on open and voluntary participation of consumers or producers.

4.2   Case Study #2

Figure 4 shows the electricity flows in case of energy sharing, same electricity demand profiles and different investments, and Fig. 5 shows the annual costs of EC members in different regulatory frameworks.

Comparing Fig. 2 with Fig. 5, it is evident that in the first case the savings are higher even if the total electricity demand remains the same. This difference is the result of more various load profiles which lead to a greater synergistic effect and therefore to a decrease in bills by installing PV.

5   Conclusions

This paper addresses the profitability of energy communities and the recommended business model.

A case with the "net metering" incentive is the most profitable, and energy communities are not encouraged. But if the «net metering» incentive disappears, energy sharing within communities would be the most profitable scenario.

Energy sharing within communities leads to a higher renewable energy self-consumption and new incentives could push in this direction. Moreover, other profits must not be forgotten, such as the economic aspects, environmental and social and those related to network benefits which have not been examined in this work, but which could be studied in further developments.

The importance of different electricity demand profiles has been noted. The reason why this is so important lies in the necessity to identify the best clustering of EC members especially when passing from the multi-family apartment-building level to a district or more extensive level of detail.

Business models for the distribution of the economic benefit produced by the energy sharing are very significant. In fact, the participation is on a voluntary basis, which means that, if the benefit distribution is not fair for every member of the community and results into a disadvantage, citizen presumably won't be interested in it.

Therefore, this work aided in identifying the two main aspects to be considered: the global benefit for the entire energy community, as well as the profit of each participant of the EC. Because these two aspects do not always progress at the same pace.

Appendix

See Tables 6 and 7, Figs. 6 and 7.

Table 6 Scheme of the analyzed studies

Reference	Case study	EC shared technology	EC profitability evaluation	Regulatory frameworks	Internal BM (more participants take over the PV investment)
Bernadette et al. (2019)	Ten buildings (Austria)	• PV • Battery storage	Yes	• One scheme	-
Fleischhacker et al. (2018)	San Antonio, Corpus Christi (Texas)	• PV • Battery storage	No	• One scheme	• Three pricing schemes
Terlouw et al. (2019)	22 household community, Cernier (Switzerland)	• Battery storage	No	• Two schemes with subsidies	–
Riveros et al. (2019)	District of Schwyz (Switzerland)	• PV • Battery storage	No	• Four schemes with subsidies	–
This work model	Multi-family apartment building (Italy)	• PV • Battery storage	Yes	• Six schemes with subsidies	• Three pricing schemes

Table 7 Global electricity demand of the EC households

	Household 1l (kWh)	Household 2l (kWh)	Household 3l (kWh)	Household 4l (kWh)	Household 5 kWh	Total (kWh)
Case study #1	5129.5	3961.2	3579.1	5221.3	4454.7	22,346
Case study #2	4469.2	4469.2	4469.2	4469.2	4469.2	22,346

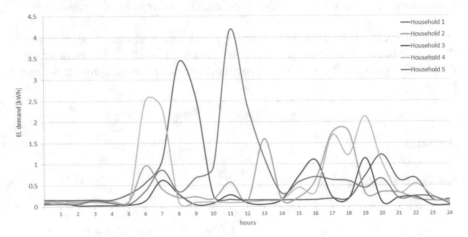

Fig. 6 One-week households' electricity demand profiles (case study #1)

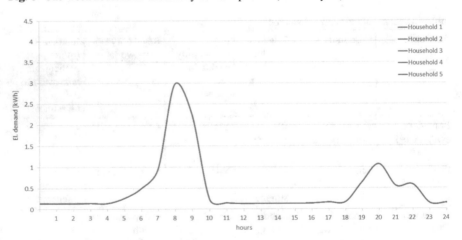

Fig. 7 One-week households' electricity demand profiles (case study #2)

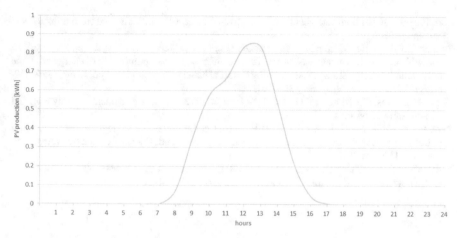

Fig. 8 PV production profile

References

Bernadette F, Auer H, Friedl W (2019) Profitability of PV sharing in energy communities: use cases for different settlement patterns. Energy. https://doi.org/10.1016/j.energy.2019.116148

Brummer V (2018) Community energy—benefits and barriers: a comparative literature review of community energy in the UK, Germany and the USA, the benefits it provides for society and the barriers it faces. Renew Sustain Energy Rev, 187–196. https://doi.org/10.1016/j.rser.2018.06.013

Clean Energy Collective (2010) Clean Energy Co. Retrieved from https://www.cleanenergyco.com/

DIRECTIVE (EU) 2018/2001 (2018, December 11) Official Journal of the European Union (on the promotion of the use of energy from renewable sources). Retrieved May 10, 2019, from https://eur-lex.europa.eu/eli/dir/2018/2001/oj

Elemens RSE, Kantar (2019). I prosumer condominiali. Associazione Energy@Home

Fleischhacker A, Auer H, Lettner G, Botterud A (2018) Sharing solar PV and energy storage in apartment buildings: resource allocation and pricing. IEEE Trans Smart Grid. https://doi.org/10.1109/TSG.2018.2844877

Fraunhofer ISE (2015) Current and future cost of photovoltaics. Long-term scenarios for market development, system prices and lcoe of utility-scale PV systems. Agora Energiewende

Harding Energy. Inc. (2004) Harding Battery Handbook for Quest® Rechargeable Cells and Battery Packs. Retrieved from https://nic.vajn.icu/PDF/battery/Rechargeable_Batteries.pdf

Moroni S, Alberti V, Antoniucci V, Bisello A (2019) Energy communities in the transition to a low-carbon future: a taxonomical approach and some policy dilemmas. J Environ Manage 45–53. https://doi.org/10.1016/j.jenvman.2019.01.095

Moura R, Centeno Brito M (2019) Prosumer aggregation policies, country experience and business models. Energy Policy 820–830. https://doi.org/10.1016/j.enpol.2019.06.053

Powerpeers power to the people (2016) Retrieved from https://www.powerpeers.nl/

Prina MG, Lionetti M, Manzolini G, Moser D, Sparber W (2019) Transition pathways optimization methodology through EnergyPLAN software for long-term energy planning. Appl Energy 235:356–368

Prina M, Garegnani G, Kleinhans D, Manzolini G, Vaccaro R, Weitemeyer S, Moser D (2016) Large-scale integration of renewable energy sources: technical and economical analysis for the Italian case. In: 32nd European photovoltaic solar energy conference exhibition, pp 2670–2674.

Sing Lai C, McCulloch MD (2017). Levelized cost of electricity for solar photovoltaic and electrical energy storage. Appl Energy 191–203.

Som Energia (2010) Som Energia | La Cooperativa d'Energia Verda. Retrieved from https://www.
    somenergia.coop/
Sonnen (2010) Retrieved from https://sonnen.de/sonnencommunity/
Terlouw T, AlSkaif T, Bauer C, van Sark W (2019) Multi-objective optimization of energy arbitrage
    in community energy storage systems using different battery technologies. Appl Energy 356–372.
    https://doi.org/10.1016/j.apenergy.2019.01.227
Veronese E, Moser D, Manzolini G (2019) PV lcoe for different market segments in Italy with and
    without storage systems
Riveros JZ, Kubli M, Ulli-Beer S (2019) Prosumer communities as strategic allies for electric
    utilities: exploring future decentralization trends in Switzerland. Energy Res Soc Sci. https://doi.
    org/10.1016/j.erss.2019.101219
Weitemeyer S, Kleinhans D, Vogt T, Agert C (2015) Integration of renewable energy sources in
    future power systems: the role of storage. Renew Energy 75:14–20

Relevance of Cultural Features in Contingent Valuation: A Literature Review of Environmental Goods Assessments

Valentina Antoniucci, Giuliano Marella, Roberto Raga, and Shinya Suzuki

Abstract This contribution is a literature review of contingent valuation (CV) applied to the assessment of environmental goods and municipal waste management (MWM), in particular. MWM activities and works do not guarantee private entrepreneurs a profit, so they are only conducted by public authorities. The positive externalities resulting from MWM can nonetheless be assessed and assigned a monetary value. The most often used approach in the environmental field is the contingent valuation method (CVM), which enables an estimation of the willingness to pay (WTP) for MWM. Given the paucity of empirical research on the economic sustainability of MWM, the aim of the present study is to review published CVM case studies in the field of waste management (municipal solid waste recycling, landfill mining). We collected 50 surveys on the WTP for MWM and established a set of statistically significant variables influencing the feasibility and/or enhancing waste treatment at the municipal level. The review underscores the prominent role of cultural factors, rather than strictly economic influences on the WTP for a given service. High levels of education and awareness of environmental issues and of the impact of waste management on the environment encourage people to pay more for enhancing MWM and to support new MWM policies. These data may be helpful in the design of further empirical research on other environmental activities, such as landfill mining, based on the benefit transfer (BT) methodology, given the lack of case studies and empirical research on these issues in southern Europe.

Keywords Circular economy · Municipal waste management · WTP · Contingent valuation · Non-market goods

V. Antoniucci (✉) · G. Marella · R. Raga
Department of Civil, Architectural and Environmental Engineering, ICEA, University of Padova, 35131 Padova, Italy
e-mail: valentina.antoniucci@unipd.it

S. Suzuki
Department of Civil Engineering, Fukuoka University, Fukuoka 814-0180, Japan

© The Author(s), under exclusive license to Springer Nature Switzerland AG 2021
A. Bisello et al. (eds.), *Smart and Sustainable Planning for Cities and Regions*,
Green Energy and Technology, https://doi.org/10.1007/978-3-030-57332-4_20

277

1 Introduction

The consumption of resources, their scarcity, and growth in the human population are increasing the pressure on policymakers and public authorities to focus on soil recovery and reducing pollution in urban areas worldwide. This situation has led to a rapidly growing demand to enhance ecosystem services and waste management and to consider the economic values involved (Giusti 2009). Waste is even one of the main focal issues in the "climate action, environment, resource efficiency, and raw materials challenge" in Horizon 2020. The European Union (EU) is working to convert the paradigm of waste management from just a problem into an opportunity to promote a virtuous circular economy, where the waste generated by one industry becomes a secondary raw material for another. Transition to a circular economy should reduce the EU's dependence on raw materials and help to safeguard the environment. The EU's commitment to waste management and a circular economy is demonstrated by the 150 billion euros it reported it allocated to these issues from 2014 to 2020 (European Commission 2020). Municipal waste management (MWM), and waste management in general, is crucial to the recovery and protection of the urban and regional environment, but their economic feasibility is still poorly understood. From a traditional economic perspective, MWM does not generate profits for private entrepreneurs, so only public authorities engage in these activities (Marella and Raga 2014). The positive externalities resulting from MWM can nonetheless be assessed and attributed a monetary value by focusing on the population's utility function. There are several valuation methods for assessing non-market goods (Caprino et al. 2018; Leeabai et al. 2019; Mangialardo and Micelli 2018; Nesticò et al. 2018), but the most often used, and most effective in the environmental field, is the contingent valuation method (CVM), which enables an estimation of the willingness to pay (WTP) for MWM.

The aim of the present contribution is to review CVM case studies conducted in the field of waste management (municipal solid waste recycling, landfill mining), given the paucity of empirical research on the economic sustainability of MWM and landfill mining (LM). Our main goal was to collect an original database on the WTP for MWM and to establish a set of statistically significant variables influencing the feasibility and/enhancing solid waste treatment at the urban level. These data may facilitate the design of further empirical research on MWM and other environmental recovery activities, such as LM, based on the benefit transfer (BT) methodology, which enables value estimates to be transferred from existing multiple "study sites" to a hitherto unstudied "policy site" with similar characteristics (Liu et al. 2010). The collection of such a set of variables, drawn from actual case studies, should be useful in assessing the value of non-market goods like LM and solid waste management (SWM) at the urban level. A second aim of our work is to design a BT model for estimating the positive externalities associated with LM in northern Italy, where this matter has yet to be investigated.

The remainder of this paper is organized as follows: the next section provides an overview of the methodology and describes the sample considered; Sect. 3 critically discusses the findings of our literature review; and Sect. 4 outlines the novelty of the present contribution and proposes steps for further research.

2 Methodology and Sample

WTP is "the maximum amount an individual would pay to gain an improvement in her/his quality of life or to avoid an undesirable change" (Damigos et al. 2016). The negative externalities of waste management are usually measured in terms of WTP to avoid or reduce, for instance, amenity losses, health risks, or a decline in property values (Antoniucci and Marella 2017; Bisello et al. 2020). The CVM is one of the methods most often used to assess the stated preferences of households and firms from an economic perspective. It relies on a survey that generates information concerning the distribution of respondents' WTP for a proposed change in an environmental good (Hanley and Barbier 2009; Ferreira and Marques 2015; Champ et al. 2017), such as a better MWM or the closure of a landfill site. Although CVM has been widely adopted, its application to empirical environmental research has been limited. This is partly because it is time-consuming and expensive when applied to the variety of features of environmental goods and the multiple benefits of environmental improvement works (Moroni et al. 2019).

The CVM necessitates a robust set of data on tested variables, so we conducted a systematic literature review in the Scopus database to collect and classify the most common and significant variables considered when the CVM has been applied to SWM. Scopus is the largest international database for scientific literature, including almost all the publications also indexed in the Web of Science, the other major research databases. We considered the research updates in the review performed by Damigos et al. (2016), where only the amount of WTP and the data needed to define the case study were taken into account (as explained subsequently in detail). Our chosen search string, "*landfill* AND *municipal* AND *solid* AND *waste* AND *contingent* AND *valuation*" (in the title, abstract and keywords), identified 64 articles. Fourteen of these articles were excluded because they related to research conducted using methods other than the CVM, such as choice experiments (Giannelli et al. 2018; Grilli et al. 2018). Our choice of keywords was intended to include a broad array of articles on MWM and SWM—an important line of research in the valuation of the environmental goods—and to identify any empirical analyses on WTP for LM at the municipal level.

Our sample consisted of 50 articles describing empirical measures of WTP for enhancing waste management (one concerned a quarry rehabilitation, and one was about LM). The review by Damigos et al. (2016) considered articles up until 2015, while our search conducted on the years from 1996 to 2020 identified a further 15 more recent case studies. We collected the core data needed to describe each case and the result, as listed in Table 1, i.e.,

- authors and year of publication;

Table 1 Core data from the surveys collected

Authors	Region/City	Country	WTP format	Type of survey	Sample size	Value	WTP (local currency)	WTP (USD) 2019
Aadland and Caplan (2006)	40 cities with > 50,000 population	USA	DBDC	Telephone	4000	USD	35.64 calibrated for bias	45.20
							67.32 uncalibrated for bias	85.37
Aadland and Caplan (1999)	Odgen, Utah	USA	OI	Telephone	401	USD	24.60	31.20
Afroz et al. (2009)	Dhaka City	Bangladesh	DBDC	Face-to-face	480	Taka	156	2.19
Afroz and Masud (2011)	Kuala Lumpur	Malaysia	DBDC	Face-to-face	467	MYR	264.00	70.63
Alhassan and Mohammed (2013)	New Juaben Municipality	Ghana	DC	Face-to-face	200	GHC	44.00	16.53
Arekere (2004)	Bryan, Texas	USA	DC + FU	Mail	618	USD	29.16	39.47
Ayalon et al. (1999)		Israel	OE	Telephone	600	NIS	170	73.33
Babaei et al. (2015)	Abadan	Iran	MBDC	Face-to-face	2400	Iranian lira	higher education > WTP; government employees > WTP	0.14
Banga et al. (2011)	Kampala	Uganda	DBDC	Face-to-face	381	USHS	29,268.00	10.57
Basili et al. (2006)	Siena Province	Italy	DBDC	Face-to-face	713	€	15.89	11.00
Begum et al. (2007)		Malaysia	OE	Face-to-face	130	RM	69.88	20.91
Berglund (2006)	Pitea	Sweden	OE	Mail	603	USD	56.25	71.33
Blaine et al. (2005)	Lake County	USA	DC	Mail	721	USD	28.20	36.92
Blaine et al. (2005)	Lake County	USA	PC	Mail	737	USD	18.48	24.19

(continued)

Table 1 (continued)

Authors	Region/City	Country	WTP format	Type of survey	Sample size	Value	WTP (local currency)	WTP (USD) 2019
Bluffstone and DeShazo (2003)		Lithuania	DBDC	Face-to-face	460	Litas	32.74	23.29
Bohara et al. (2007)		USA	DC	Face-to-face	458	USD	67.68	83.45
Cai et al. (2020)	Zhuhai	China	DC	Face-to-face	474	RMB	26.8	13.14
Chakrabarti et al. (2009)	Kolkata Metropolitan City	India	MBDC	Face-to-face	767	RS	2159	0.22
Challcharoenwattana and Pharino (2016)	Greater Phan Khon area; Hua Hin; Bangkok	Thailand	PC	Face-to-face	1064	USD	0.73 least urbanized areas; 1.96 urbanized areas; 1.65 most urbanized areas	0.78; 2.09; 1.76
Chung and Yeung (2019)	Hong Kong	China	DC–OE	Telephone	753	HKD	38.4	17.52
Damigos and Kaliampakos (2003)	Athens	Greece	DC	Face-to-face	200	GRD	10.477 Plan 1 16.856 Plan 2; 19.833 Plan 3	22.02; 36.62; 43.08
Danso et al. (2006)	Accra, Kumasi, Tamale	Ghana	DC + FU	Face-to-face	700	USD	Accra (urban) = 2.14; Kumasi (rural) = 1.63; Tamale (suburban) = 1.95	2.71; 2.07; 2.47
Ezebilo and Animasaun (2011)		Nigeria	PC	Face-to-face	224	Naira	4676.00	29.20

(continued)

Table 1 (continued)

Authors	Region/City	Country	WTP format	Type of survey	Sample size	Value	WTP (local currency)	WTP (USD) 2019
Ezebilo (2013)		Nigeria	DC	Face-to-face	236	Naira	3660.00	15.93
Ferreira and Marques (2015)		Portugal	DC	E-mail	1186	€	33.6	20.68
Fonta et al. (2007)	Enugu State	Nigeria	DC + FU	Face-to-face	200	Naira	2764.00	27.32
Gaglias et al. (2016)	Ikaria	Greece	OE	Face-to-face	150	€	6.5–6.7	3.70–3.81
Hagos et al. (2012)	Mekelle City	Ethiopia	DC	Face-to-face	226	ETB	142.68	28.02
Hagos et al. (2012)	Mekelle City	Ethiopia	OE	Face-to-face	226	ETB	94.60	18.60
Jin et al. (2006)		Macao	DBDC	Face-to-face	252	MOP	799.29	247.33
Jones et al. (2010)	Mytilene	Greece	OE	Telephone	140	€	0.50	0.28
Khattak et al. (2009)	Peshawar District	Pakistan	OE	Face-to-face	216	RS	1800.00	131.11
Ko et al. (2020)	Seoul, Incheon, Gyeonggi-do	South Korea	DBDC	E-mail	550	KRW	41,234	47.92
Koford et al. (2012)	Lexington, Kentucky	USA	DC	Mail	600	USD	27.48	30.60
Lake et al. (1996)	Hethersett	UK	OE-DC	Mail	285	GBP	35.69	40.07
Mulat et al. (2019)	Injibara	Ethiopia	DBDC	Face-to-face	903	ETB	29.7	2.98
Murad et al. (2007)	Kuala Lumpur	Malaysia	PC	Face-to-face	300	MYR	156.00	46.67
Rahji and Oloruntoba (2009)	Ibadan	Nigeria	DC	n.a	546	N	150–450	1.18–3.54
Roy and Deb (2013)	Silchar Municipal Area	India	OE	Face-to-face	378	Rs	127.47	8.91

(continued)

Table 1 (continued)

Authors	Region/City	Country	WTP format	Type of survey	Sample size	Value	WTP (local currency)	WTP (USD) 2019
Sarkhel and Banerjee (2010)	Bally	India	DBDC	Face-to-face	570	Rs	228.00	21.011
Song et al. (2016)	Macau	China	DC	Face-to-face	250	MOP	38.5	7.33
Tiller et al. (1997)	Williamson County	USA	DC + FU	Face-to-face	481	USD	48.00	76.46
Trang et al. (2017)	Thu Dau Mot	Vietnam	DC	Face-to-face	330	VND	24	0.0034
Vassanadumrongdee and Kittipongvises (2018)	Bangkok	Thailand	n.a	Self-administered survey + face-to-face	1076	Thai Bhat	71.6	5.90
Wang et al. (2014)	Yunnan	China	MBDC	Face-to-face	223	Yuan	205.20	63.36
Yuan and Yabe (2014)	Haidian and Dongcheng districts of Beijing city	China	DC	Face-to-face	391	Yuan	107.14	33.08
Zabala et al. (2019)	Mursia	Spain	OE–CE	n.a	352	€	5,26	3.33
Zen et al. (2014)	Kuala Lumpur	Malaysia	OE	Face-to-face	460	MYR	88.80	20.41
Zhang et al. (2015)	30 provinces	China	MBDC	Face-to-face	4638	Yuan	23.41	7.10

- city or region where the survey was performed (where available);
- country;
- WTP format;
- type of survey;
- sample size;
- currency in which WTP was expressed;
- amount of WTP.

These data were needed to make the results comparable in further steps of the research. The city and country variables enabled us to cluster the data by homogeneous economies and cultural habits (Del Giudice et al. 2019). For 14% of the sample (7/50 studies), the city (or cities) where the surveys were conducted was not stated. We included these studies nonetheless because the size of the samples analyzed was consistent with other surveys conducted at municipal and regional levels. The year of publication of the survey and the currency adopted were needed to convert the WTP into a single format (the currency was usually the euro or US dollar). The type of survey and sample size enabled us to test the robustness of the findings. For instance, face-to-face surveys achieve higher response rates, while "preferences do not seem to be significantly different or biased compared to face-to-face interviews" (Lindhjem and Navrud 2011: 1628). The WTP format concerns the various strategies adopted in the survey.

Some of the questionnaires used in the surveys adopted a dichotomous choice (DC) format (see Table 2), where the answer may be yes or no (Mitchell and Carson 1989). Depending on this answer, there may be further closed-ended questions, i.e., double-bounded or multiple-bounded dichotomous choices (DBDC, MBDC; see Table 2), or else an open-ended (OE) follow-up question (DC + FU; also see Table 2). Alternatively, the questions could simply be in an OE or closed-ended (CE) format (Table 2). Finally, the payment card (PC) approach estimates the WTP above the amount stated by the respondent and below the next higher one, if any (Venkatachalam 2004).

Table 2 Legend to the WTP formats

Acronyms	Definition
DC	Dichotomous choices
DBC	Double-bounded dichotomous choices
MBDC	Multiple-bounded dichotomous choices
DC + FU	Dichotomous choice + follow-up question
OE	Open-ended
CE	Closed-ended
PC	Payment card

3 Results of the Review

On the general topic of SWM and MWM, the WTP in the surveys collected related to:

- enhancing the existing waste management system;
- providing for curbside waste collection;
- choice of MWM policy (e.g., curbside collection, landfilling);
- shutting down of landfills and increased use of incineration;
- other policies to minimize environmental degradation.

The WTP was expressed as a total amount per year for a given service, or as the amount of increase in the charge per year or month.

The countries conducting most surveys were the USA and China, which each accounted for 7% of the whole sample. The distribution of the remainder of the surveys was quite homogenous, with a prevalence of the emerging economies (mostly in the Far East) and the most dynamic African states. The constant growth in their populations and economies, associated with their relatively recent development, probably induced these countries to pay more attention to the topic of waste management. European countries thus made up a minority of the sample, with seven countries together representing just 7% of the total sample. Asia is the continent most represented, followed by Africa and Europe (Fig. 1).

The average sample size was 652 respondents. The questions concerned households in all but one case (which dealt with businesses), and concerned from 130–4638 people.

We identified 29 explanatory variables for the WTP, which we divided into three groups (see Fig. 2): (i) socioeconomic variables that help to describe the composition of the sample; (ii) variables relating to present or expected SWM issues, the perception and understanding of recycling methods and respondents' own recycling habits;

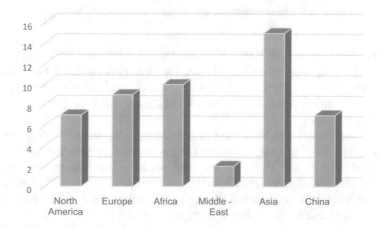

Fig. 1 Distribution of surveys by region

Category of variables	Explanatory Variables	Frequency (no.)
Socio - economic features	Gender	23
	Age	28
	Income	40
	Education level	37
	Marital status	4
	Religion	1
	Occupation	19
	Type of firm	1
	Dimension of firm	1
	No. Employees	1
	Racial group	4
	Travel cost/time	1
Information about SWM	Social pressure	4
	Distance drop-off	9
	MSW generation kg per capita	8
	Use of more recycleble materials	1
	Knowledge of governament funding	6
	Satisfaction with current MSW service	14
	Knowledge of recycling	20
	Concerns about MSW crisis/Recycling habit	22
Housing features	Location (urban-periurban)	5
	Household members	17
	Length of stay (in the home)	2
	House Ownership	6
	Home size (sqm)	7
	Home typology (house/flat)	4
	Frequency of sanitary inspection	1
	Soil input	1

Fig. 2 Frequency of explanatory variables in the sample

and (iii) housing features that were associated with households' behavior regarding MWM and their housing conditions. The bid and further bids were excluded from the explanatory variables, although they were the core elements of the surveys, because the aim of the present work was not meant to estimate the WTP, but to classify its explanatory variables.

The socioeconomic variables used to characterize the sample were those most commonly adopted. "Income" naturally correlates strongly with the WTP (as did "occupation"), so it was the most frequent variable. The "occupation" variable included variables like "occupied/unemployed," "occupied/retired," and "no. of women occupied in the household." The second most frequent variable in the sample was the "educational level," followed by "age" and "gender".

The WTP often correlated with the household's attitude to recycling and concern about the operational sustainability of the existing MWM system. Both these aspects were more relevant in developing countries, where awareness of the impact of waste on the environment was modest on average and MWM often poses severe problems. Understanding of recycling, the household's recycling habits and concern about existing MWM methods correlated strongly with the level of education and occupation. Awareness of the importance of recycling and of the efficiency of the recycling system for the environment was greater among respondents with a higher level of education. This attitude also showed a strong positive correlation with a greater willingness to pay for SWM and its improvement. The dimensions of households showed a negative correlation with the WTP, especially in the developing countries (no. 17 surveys, see Fig. 2): the larger the family, the less it was willing to spend money on expanding waste management. On the other hand, the cost of travel and distance to waste drop-off sites correlated positively with the improvement of MWM services. The per capita amount of waste showed an analogical correlation. In some cases, the size of the home and ownership of it correlated positively with the WTP; in other words, households were more willing to pay for WM when they owned their home and when their home was larger.

4 Conclusion: The Influence of Cultural Factors on the WTP

Along with the well-known importance of the economic features affecting the WTP for improvements to a given service, our literature review sheds light on the influence of cultural factors and formal education on the willingness to pay more for waste management. The educational-level variable emerged as being as significant as "Income," and it was mostly associated with other variables representing the degree of awareness of the impact of waste, and the effect on the environment of improving MWM. The habit of recycling often revealed a positive correlation with the WTP as well, going to show that the ordinary practice of recycling raises people's awareness and makes them more likely to be willing to pay more for a better service.

Our review of the state of the art on contingent valuation concerning SWM and MWM clearly indicates that younger people, women and the better educated are more willing to pay for additional services and innovative policies on waste management. A specific understanding of the effects on the environment of recycling materials also made people more willing to pay an extra charge for the necessary services. These aspects were not analyzed in existing studies on the topic, not even in the study by Damigos et al. (2016) taken as a reference, because all the empirical research conducted to date focused mainly on the amount of WTP and its variance rather than on the societal implications of its explanatory variables.

From an economic perspective, most of the literature concentrates on the amount of money that may guarantee the feasibility of improving MWM and LM services, which largely coincides with the amount that people are willing to pay. Little attention has been paid to the factors that justify paying more for waste management services. Counterintuitively, the cultural features of households seem to be more relevant than strictly economic aspects, such as income, occupation, or other traditional economic indicators (Nesticò and Galante 2015).

The present review analyzed 50 articles on the application of the CVM to measuring the WTP for MWM services in an effort to identify the most robust explanatory variables for the WTP for environmental goods. The study generated a set of indicators that can be helpful for assessing not only MWM services but also other kinds of environmental goods, such as landfills and landfill mining—even in areas scarcely investigated in the past, such as Italy and southern Europe in general.

Acknowledgements The present contribution was partially funded by the Department of Civil, Architectural and Environmental Engineering at the University of Padova, Italy (Research fellowship contract No. 205/2019, prot. 2982, dated November 14, 2019).

References

Aadland D, Caplan AJ (2006) Curbside recycling: waste resource or waste of resources? J Policy Anal Manage 25(4):855–874. https://doi.org/10.1002/pam.20211

Aadland DM, Caplan AJ (1999) Household valuation of curbside recycling. J Environ Planning Manage 42(6):781–799. https://doi.org/10.1080/09640569910821

Afroz R, Hanaki K, Hasegawa-Kurisu K (2009) Willingness to pay for waste management improvement in Dhaka city, Bangladesh. J Environ Manage 90(1):492–503. https://doi.org/10.1016/j.jenvman.2007.12.012

Afroz R, Masud MM (2011) Using a contingent valuation approach for improved solid waste management facility: evidence from Kuala Lumpur, Malaysia. Waste Manage 31(4):800–808. https://doi.org/10.1016/j.wasman.2010.10.028

Alhassan M, Mohammed J (2013) Households demand for better solid waste disposal services: case study of four communities in the New Juaben Municipality, Ghana. J Sustain Dev 6(11):16–25. https://doi.org/10.5539/jsd.v6n11p16

Antoniucci V, Marella G (2017) The influence of building typology on the economic feasibility of urban developments. Int J Appl Eng Res 12(15)

Arekere DM (2004) Analysis of recycling behavior. Waste Manage Environ II:397–406

Ayalon O, Avnimelech Y, Shechter M (1999) Issues in designing an effective solid waste policy—the Israeli experience. In: Sterner T (Ed.) The market and the environment: the effectiveness of market based instruments for environmental reform. Edward Elgar, UK

Babaei AA, Alavi N, Goudarzi G, Teymouri P, Ahmadi K, Rafiee M (2015) Household recycling knowledge, attitudes and practices towards solid waste management. Resour Conserv Recycl 102:94–100. https://doi.org/10.1016/J.RESCONREC.2015.06.014

Banga M, Lokina RB, Mkenda AF (2011) Households' willingness to pay for improved solid waste collection services in Kampala city, Uganda. J Environ Develop 20(4):428–448. https://doi.org/10.1177/1070496511426779

Basili M, Di Matteo M, Ferrini S (2006) Analysing demand for environmental quality: a willingness to pay/accept study in the province of Siena (Italy). Waste Manage 26(3):209–219. https://doi.org/10.1016/j.wasman.2004.12.027

Begum RA, Siwar C, Pereira JJ, Jaafar AH (2007) Factors and values of willingness to pay for improved construction waste management—a perspective of Malaysian contractors. Waste Manage 27(12):1902–1909. https://doi.org/10.1016/j.wasman.2006.08.013

Berglund C (2006) The assessment of households' recycling costs: the role of personal motives. Ecol Econ 56(4):560–569. https://doi.org/10.1016/j.ecolecon.2005.03.005

Bisello A, Antoniucci V, Marella G (2020) Measuring the price premium of energy efficiency: a two-step analysis in the Italian housing market. Energy Build 208:109670. https://doi.org/10.1016/j.enbuild.2019.109670

Blaine TW, Lichtkoppler FR, Jones KR, Zondag RH (2005) An assessment of household willingness to pay for curbside recycling: a comparison of payment card and referendum approaches. J Environ Manage 76(1):15–22. https://doi.org/10.1016/j.jenvman.2005.01.004

Bluffstone R, DeShazo JR (2003) Upgrading municipal environmental services to European Union levels: a case study of household willingness to pay in Lithuania. Environ Dev Econ 8(4):637–654. https://doi.org/10.1017/S1355770X0300342

Bohara AK, Caplan AJ, Grijalva T (2007) The effect of experience and quantity-based pricing on the valuation of a curbside recycling program. Ecol Econ 64(2):433–443. https://doi.org/10.1016/j.ecolecon.2007.02.033

Cai K, Song Q, Peng S, Yuan W, Liang Y, Li J (2020) Uncovering residents' behaviors, attitudes, and WTP for recycling e-waste: a case study of Zhuhai city, China. Environ Sci Pollut Res 27:2386–2399. https://doi.org/10.1007/s11356-019-06917-x

Caprino RM, De Mare G, Nesticò A (2018) Economic evaluation and urban regeneration: a new bottom-up approach to local development policies. In: Mondini G, Fattinnanzi E, Oppio A, Botter M, Stanghellini S (Eds.). Integrated evaluation for the management of contemporary cities SIEV 2016 green energy and technology. Springer, Cham, pp 379–390. https://doi.org/10.1007/978-3-319-78271-3_30

Chakrabarti S, Majumder A, Chakrabarti S (2009) Public-community participation in household waste management in India: an operational approach. Habitat Int 33(1):125–130. https://doi.org/10.1016/j.habitatint.2008.05.009

Challcharoenwattana A, Pharino C (2016) Wishing to finance a recycling program? Willingness-to-pay study for enhancing municipal solid waste recycling in urban settlements in Thailand. Habitat Int 51:23–30. https://doi.org/10.1016/j.habitatint.2015.10.008

Champ PA, Boyle KJ, Brown TC (2017) The economics of non-market goods and resources. In: A primer on nonmarket valuation, Springer, Dordrecht. https://doi.org/10.1007/978-94-007-7104-8

Chung W, Yeung IMH (2019) Analysis of residents' choice of waste charge methods and willingness to pay amount for solid waste management in Hong Kong. Waste Manage Pergamon 96:136–148. https://doi.org/10.1016/J.WASMAN.2019.07.020

Damigos D, Kaliampakos D (2003) Environmental economics and the mining industry: monetary benefits of an abandoned quarry rehabilitation in Greece. Environ Geol 44(3):356–362. https://doi.org/10.1007/s00254-003-0774-5

Damigos D, Kaliampakos D, Menegaki M (2016) How much are people willing to pay for efficient waste management schemes? A benefit transfer application. Waste Manage Res 34(4):345–355. https://doi.org/10.1177/0734242X16633518

Danso G, Drechsel P, Fialor S, Giordano M (2006) Estimating the demand for municipal waste compost via farmers' willingness-to-pay in Ghana. Waste Manage 26(12):1400–1409. https://doi.org/10.1016/j.wasman.2005.09.021

Del Giudice V, Massimo DE, De Paola P, Forte F, Musolino M, Malerba A (2019) Post carbon city and real estate market: testing the dataset of Reggio Calabria market using spline smoothing semiparametric method. In: Calabrò F, Della Spina L, Bevilacqua C (Eds.). New metropolitan perspectives ISHT 2018 smart innovation, systems and technologies Springer, Cham, pp 206–214. https://doi.org/10.1007/978-3-319-92099-3_25

European Commission online. https://ec.europa.eu/regional_policy/en/policy/themes/environment/circular_economy. Accessed on April 30, 2020

Ezebilo EE, Animasaun ED (2011) Economic valuation of private sector waste management services. J Sustain Dev 4(4):38–46. https://doi.org/10.5539/jsd.v4n4p38

Ezebilo EE (2013) Willingness to pay for improved residential waste management in a developing country. Int J Environ Sci Technol 10(3):413–422. https://doi.org/10.1007/s13762-012-0171-2

Ferreira S, Marques RC (2015) Contingent valuation method applied to waste management. Resour Conserv Recycl 99:111–117. https://doi.org/10.1016/j.resconrec.2015.02.013

Fonta WM, Ichoku HE, Ogujiuba KK, Chukwu JO (2007) Using a contingent valuation approach for improved solid waste management facility: evidence from Enugu State, Nigeria. J Afr Econ 17(2):277–304. https://doi.org/10.1093/jae/ejm020

Gaglias A, Mirasgedis S, Tourkolias C, Georgopoulou E (2016) Implementing the contingent valuation method for supporting decision making in the waste management sector. Waste Manage 53:237–244. https://doi.org/10.1016/j.wasman.2016.04.012

Giannelli A, Giuffrida S, Trovato MR (2018) Madrid Río Park symbolic values and contingent valuation. Valori Valutazioni 21:75–85

Giusti L (2009) A review of waste management practices and their impact on human health. Waste Manage 2227–2239. https://doi.org/10.1016/j.wasman.2009.03.028

Grilli G, Tomasi S, Bisello A (2018) Assessing preferences for attributes of city information points: results from a choice experiment. In: Green energy and technology, Springer International Publishing, pp 197–209. doi: https://doi.org/10.1007/978-3-319-75774-2_14

Hagos D, Mekonnen A, Gebreegziabher Z (2012) Households' willingness to pay for improved urban waste management in Mekelle City, Ethiopia. Environ Dev Discuss Pap Ser 1–25

Hanley N, Barbier EB (2009) Pricing nature: cost-benefit analysis and environmental policy. Edward Elgar, Cheltenham, UK

Jin J, Wang Z, Ran S (2006) Comparison of contingent valuation and choice experiment in solid waste management programs in Macao. Ecol Econ 57(3):430–441. https://doi.org/10.1016/j.ecolecon.2005.04.020

Jones N, Evangelinos K, Halvadakis CP, Iosifides T, Sophoulis CM (2010) Social factors influencing perceptions and willingness to pay for a market-based policy aiming on solid waste management. Resour Conserv Recycl 54(9):533–540. https://doi.org/10.1016/j.resconrec.2009.10.010

Khattak R, Khan J, Ahmad I (2009) An analysis of willingness to pay for better solid waste management services in urban areas of District Peshawar. Sarhad J Agri 25(3):529–536

Ko S, Kim W, Shin SC, Shin J (2020) The economic value of sustainable recycling and waste management policies: the case of a waste management crisis in South Korea. Waste Manage 104:220–227. https://doi.org/10.1016/j.wasman.2020.01.020

Koford BC, Blomquist GC, Hardesty DM, Troske KR (2012) Estimating consumer willingness to supply and willingness to pay for curbside recycling. Land Econ 88(4):745–763. https://doi.org/10.3368/le.88.4.745

Lake IR, Bateman IJ, Parfitt JP (1996) Assessing a kerbside recycling scheme: a quantitative and willingness to pay case study. J Environ Manage 46(3):239–254. https://doi.org/10.1006/jema.1996.0019

Leeabai N, Suzuki S, Jiang Q, Dilixiati D, Takahashi F (2019) The effects of setting conditions of trash bins on waste collection performance and waste separation behaviors; distance from walking path, separated setting, and arrangements. Waste Manage 94:58–67. https://doi.org/10.1016/j.wasman.2019.05.039

Lindhjem H, Navrud S (2011) Are Internet surveys an alternative to face-to-face interviews in contingent valuation? Ecol Econ 70(9):1628–1637. https://doi.org/10.1016/j.ecolecon.2011.04.002

Liu S, Costanza R, Troy A, D'Aagostino J, Mates W (2010) Valuing New Jersey's ecosystem services and natural capital: a spatially explicit benefit transfer approach. Environ Manage 45(6):1271–1285. https://doi.org/10.1007/s00267-010-9483-5

Mangialardo A, Micelli E (2018) Rethinking the construction industry under the circular economy: principles and case studies. In: Green energy and technology, Springer Verlag, pp 333–344. https://doi.org/10.1007/978-3-319-75774-2_23

Marella G, Raga R (2014) Use of the contingent valuation method in the assessment of a landfill mining project. Waste Manage 34(7):1199–1205. https://doi.org/10.1016/j.wasman.2014.03.018

Mitchell RC, Carson RT (1989) Using surveys to value public goods: the contingent valuation method. Resrces for the future, New York, London. https://doi.org/10.5860/choice.27-0417

Moroni S, Alberti V, Antoniucci V, Bisello A (2019) Energy communities in the transition to a low-carbon future: a taxonomical approach and some policy dilemmas. J Environ Manage 236:45–53. https://doi.org/10.1016/J.JENVMAN.2019.01.095

Mulat S, Worku W, Minyihun A (2019) Willingness to pay for improved solid waste management and associated factors among households in Injibara town, Northwest Ethiopia. BMC Res Notes BioMed Central 12(1):401. https://doi.org/10.1186/s13104-019-4433-7

Murad WM, Raquib AM, Siwar C (2007) Willingness of the poor to pay for improved access to solid waste collection and disposal services. J Environ Dev 16(1):84–101. https://doi.org/10.1177/107 0496506297006

Nesticò A, Galante M (2015) An estimate model for the equalisation of real estate tax: a case study. Int J Bus Intell Data Min 10(1):19–32. https://doi.org/10.1504/IJBIDM.2015.069038

Nesticò A, He S, De Mare G, Benintendi R, Maselli G (2018) The ALARP principle in the cost-benefit analysis for the acceptability of investment risk. Sustainability 10(12):4668. https://doi.org/10.3390/su10124668

Rahji MAY, Oloruntoba EO (2009) Determinants of households' willingness-to-pay for private solid waste management services in Ibadan, Nigeria. Waste Manage Res 27(10):961–965. https://doi.org/10.1177/0734242X09103824

Roy AT, Deb U (2013) Households' willingness to pay for improved waste management in silchar municipal area: a case study in Cachar district, Assam. IOSR J Humanit Soc Sci 6(5):21–31. https://doi.org/10.9790/0837-0652131

Sarkhel P, Banerjee S (2010) Municipal solid waste management, source-separated waste and stake-holder's attitude: a contingent valuation study. Environ Dev Sustain 12(5):611–630. https://doi.org/10.1007/s10668-009-9215-2

Song Q, Wang Z, Li J (2016) Exploring residents' attitudes and willingness to pay for solid waste management in Macau. Environ Sci Pollut Res 23(16):16456–16462. https://doi.org/10.1007/s11 356-016-6590-8

Tiller KH, Jakus PM, Park WM (1997) Household's willingness to pay for dropoff recycling. J Agric Resour Econ, Western Agricultural Economics Association, pp 310–320. https://doi.org/10.2307/40986950

Trang PTT, Toan DQ, Hanh NTX (2017) 'Estimating household willingness to pay for improved solid waste management: a case study of Thu Dau Mot City, Binh Duong. MATEC Web Conf 95:1–4. https://doi.org/10.1051/matecconf/20179518004

Vassanadumrongdee S, Kittipongvises S (2018) Factors influencing source separation intention and willingness to pay for improving waste management in Bangkok, Thailand. Sustain Environ Res 28(2):90–99. https://doi.org/10.1016/j.serj.2017.11.003

Venkatachalam L (2004) The contingent valuation method: a review. Environ Impact Assess Rev 24(1):89–124. https://doi.org/10.1016/S0195-9255(03)00138-0

Wang H, He J, Kim Y, Kamata T (2014) Municipal solid waste management in rural areas and small counties: an economic analysis using contingent valuation to estimate willingness to pay for Yunnan, China. Waste Manage Res 32(8):695–706. https://doi.org/10.1177/0734242X1453 9720

Yuan Y, Yabe M (2014) Residents' willingness to pay for household kitchen waste separation services in Haidian and Dongcheng districts, Beijing City. Environment 1(2):190–207. https://doi.org/10.3390/environments1020190

Zabala JA, Dolores de Miguel M, Martínez-Paz JM, Alcon F (2019) Perception welfare assessment of water reuse in competitive categories. Water Supply 19(5):1525–1532. https://doi.org/10.2166/ws.2019.019

Zen IS, Noor ZZ, Yusuf RO (2014) The profiles of household solid waste recyclers and non-recyclers in Kuala Lumpur, Malaysia. Habitat Int 42:83–89. https://doi.org/10.1016/j.habitatint. 2013.10.010

Zhang X, Wang R, Wu T, Song H, Liu C (2015) Would rural residents will to pay for environmental project? An evidence in China. Mod Econ 06(05):511–519. https://doi.org/10.4236/me.2015. 65050

Circular Economy Meets the Fashion Industry: Challenges and Opportunities in New York City

Younghyun Kim and Savannah Wu

Abstract With increasing awareness and urgency to address the risks of climate change, resource depletion, and waste accumulation, the concept of the circular economy is gaining momentum as a guiding framework to design and implement system-level strategies and policies to enable sustainable economic development. Today's fashion industry is one of the sectors that has the greatest opportunity to transform from a linear take-make-waste approach to a circular model that minimizes non-renewable resource consumption and landfilling of textile waste. The key is to extend the fashion product life cycle and incentivize value-creation opportunities. With the principles of making the fashion industry circular in mind, this paper focuses on New York City to examine the various stakeholders and their collaborations that seek novel ways of closing the loop of the fashion value chain. The stakeholders' current strategies and main bottlenecks are analyzed to assess opportunities for technological and policy interventions. We argue that, while technologies such as the Internet of things and digital identification have great potential to increase feedback loops and transparency throughout the fashion product value chain, thereby addressing existing challenges in making the circular transition, it is crucial to understand the context in which to apply these technologies and assess policy and regulatory incentives that need to accompany them. This paper provides an explorative overview of the experiences in New York City, which serves as a useful case study for policymakers, organizations, and individuals who are interested in knowing the actual practice and challenges in applying circular economy principles to the fashion industry.

Keywords Circular economy · Circular fashion · New York City · Waste management · Internet of things

Y. Kim · S. Wu (✉)
Department of Urban Planning, Graduate School of Architecture, Planning, and Preservation,
Columbia University, 1172 Amsterdam Avenue, New York, NY 10027, USA
e-mail: sw3318@columbia.edu

Y. Kim
e-mail: yk2767@columbia.edu

© The Author(s), under exclusive license to Springer Nature Switzerland AG 2021   293
A. Bisello et al. (eds.), *Smart and Sustainable Planning for Cities and Regions*,
Green Energy and Technology, https://doi.org/10.1007/978-3-030-57332-4_21

1   Introduction

Today's fashion industry poses many sustainability threats along its value chain, spanning from the extraction of non-renewable resources to the landfilling of textile wastes. In the USA alone, in 2017, almost 15.4 million metric tons of textile wastes were generated, of which 66% was landfilled. The global discussion around the circular economy, however, challenges this linear, take-make-dispose practice of the industry. As one of the fashion capitals in the world market, New York City (NYC) is a useful case to examine the various stakeholders and their collaborations that seek novel ways of closing the loop of the fashion value chain. These actors include not just upcycling designers and second-hand clothes shops but also digital platform providers and the Internet of things (IoT) technology developers. The insertion of circular economy in the City's OneNYC 2050 agenda and the Department of Sanitation's (DSNY) Zero Waste goals serves as an anchoring discourse that brings together the various initiatives and stakeholders, both existing and new, to rethink the entire life cycle of fashion products, beginning from the design process all the way to how used items are collected at their end-of-life.

However, disrupting the existing system to make the transition toward a circular one has many accompanying challenges. From a policy perspective, it is crucial to take a close look at these challenges so that the city's infrastructural and regulatory interventions account for these issues and parallel its sustainability vision and plans. As a starting point to document and build knowledge on circular fashion practices that are executed on the grounds, this paper takes the case study of NYC. NYC hosts diverse economic industries and entrepreneurial activities, including its historic Garment District and tech start-ups, hence choosing it as a case study provides the advantage to examine circular fashion practices and challenges from various angles. Increasing awareness around "greening" the fashion industry by both consumers and producers also makes NYC a suitable research site on the topic of circular fashion.

This paper addresses the following questions: (1) What are the current circular fashion initiatives taking place in NYC; (2) Who are the actors; (3) What challenges do the actors face; and (4) How would emerging technologies such as the IoT and policies address their challenges? By taking the case study of NYC, we use secondary data and semi-structured interviews with key stakeholders in the city's circular fashion practices to address the research questions. The paper first explores the background and prevailing sustainability challenges in the fashion industry, followed by a description of the methodology. It then presents the analysis of key stakeholders involved in activating NYC's circular economy in fashion. The paper then discusses major challenges faced by the stakeholders and opportunities for technological solutions, such as IoT and policy interventions.

2 Background

2.1 Sustainability Challenges in the Fashion Industry

The current fashion industry faces many sustainability challenges. First, the current system operates in an almost linear way: 98 million metric tons of non-renewable resources are extracted to produce clothes that are often utilized for a short period, after which materials are largely lost to landfill or incineration (EMF 2017a). Secondly, accompanied by falling costs, streamlined operations, and rising consumer spending, worldwide clothing utilization has decreased by 36%, while clothing production has doubled from 2000 to 2014 (McKinsey and Company 2017). Third, the increasing amount of textile waste that ends up in landfills and incinerators produce vast amounts of hazardous chemicals and greenhouse gases such as methane (Greenpeace International 2017). As fabric decomposes in landfills, dyes and chemicals can leach into the soil and contaminate local water systems. Last but not least, poor labor standards and working conditions across the fashion value chain are an ongoing societal challenge as companies strive to prioritize Corporate Social Responsibility in their business models.

In the USA, 85% of clothing ends up in landfills, and only 13.6% of clothing and footwear is recycled (EPA 2019). EPA estimates that textile generation in 2017 was 15.4 million metric tons, but only 2.4 million metric tons were recycled, while 2.9 million metric tons were combusted for energy recovery, and 10.2 million tons were landfilled (EPA 2019). NYC residents currently dispose of 181,436 metric tons of clothing, shoes, linens, and accessories into landfills each year, which makes up 6% of NYC's residential curbside waste stream (New York City Refashion Week 2020). Of these textiles, 1.1% are shoes, rubber, and leather, 2.9% are clothing textiles, and 2.3% are non-clothing textiles (DSNY 2017). These statistics are available as DSNY has been conducting waste-characterization studies since the 1980s to understand the composition of the city's residential waste stream and to inform planning for sustainable management (DSNY 2019). This annual textile waste is equivalent to the height of the Empire State Building or 4500 subway cars (New York City Refashion Week 2020).

Residential textile waste increased by a third between 2005 and 2013 (50 kg per household or 4.8% of aggregate discards by material category in 2005 and 56 kg per household in 2013 or 6.2%) and stayed relatively steady between 2013 and 2017 (57 kg per household or 6.3%) (DSNY 2017). The amount of improperly disposed textiles in residential curbside refuse has remained around 26 kg from 2013 to 2017 (DSNY 2017). This is substantial considering the existence of an array of programs to encourage donation or recycling offered to residents. Reusing textiles not only decreases the amount of material sent to landfills, it also saves natural resources such as water and petroleum, reduces disposal costs for the local government, businesses and residents, and creates jobs. According to the Department of Environmental Conservation, the number of jobs would be created state-wide if each New York resident recycled one additional pound (0.5 kg) of textiles per week is over 6700 (NYS DEC n.d.).

2.2 Circular Economy Meets the Fashion Industry

In the face of the unsustainable path that the fashion industry has evolved along, the circular economy (CE) model is discussed as a much-needed paradigm shift that the fashion industry must strive for. A circular economy for the fashion industry is an industrial economy where "clothes, textiles, and fibers are kept at their highest value during use and re-enter the economy afterwards, never ending up as waste" (EMF 2017b). Along with the systematic literature reviews that consolidated the definitions and principles of a CE (Kirchherr et al. 2017; Geissdoerfer et al. 2017; Ghisellini et al. 2016), scholars and practitioners generally agree that the CE is an industrial economy that is restorative and regenerative by intention and design, where biological components return to the biosphere and technical components are collected for reuse, recycle, repair, and remanufacture (EMF 2017a). As Stahel (2016) states, "reuse what you can, recycle what cannot be reused, repair what is broken, remanufacture what cannot be repaired". The CE, hence, aims to achieve a paradigm shift at the micro-, meso-, and macrolevels of society to reduce the demand for virgin materials in production and consumption cycles and decouple economic activity from environmental pressures.

New business missions and technological development that employ principles of CE in their operations are potential drivers of the CE transition at various scales of implementation. Specifically, addressing the relationship between Industry 4.0 technologies and the CE, there is a general consensus that IoT is an enabler of the CE (Rosa et al. 2019; Eon 2020; EMF 2016; Rusinek et al. 2018). As product traceability and supply-chain transparency are key to the enforcement of a circular economy, IoT innovations are discussed to play a fundamental role in establishing data on the location, condition, and availability of the products in question (EMF 2016). This data infrastructure would lie at the center of governing supply chains and opportunities for value creation based on the circular economy model, which enables businesses to become more efficient as well (Askoxylakis 2018). In fact, businesses are already seeking markets at the intersection between the circular economy and IoT, especially in the context of cities (EMF 2016).

With such advancement in technological innovation, considering the CE transition encourages new ways of rethinking the organization of the fashion value chain. First, the CE model incentivizes reverse logistics such that the values of the materials and components at their end-of-life are fully recovered. Second, new business models such as sharing or product as service systems could be developed. Third, processes of designing, making, and using fashion products would encourage maximization of value retention across the supply chain, thereby extending the products' lifetime. Fourth, end-of-use options are optimized by allowing for novel end-of-life processes and knowledge formalization on product disassembly and upgrading.

Acknowledging these theoretical advantages of applying circular economy principles in the fashion value chain and deploying IoT, what are the main challenges experienced by circular fashion's stakeholders and in what ways could they be overcome? The existing and emerging actors, business models, and technology solutions in the case of NYC provides a useful lens to address such questions.

3 Methodology

In order to understand how circular fashion strategies unfold in place and the potential role of technological innovation such as the IoT, this paper employs a case study research method to make an in-depth examination of the contemporary practices and actors involved (Yin 2013). Taking the case study of NYC, we ask the following research questions: (1) What are the current circular fashion initiatives taking place in NYC; (2) Who are the actors; (3) What challenges do the actors face; and (4) How would emerging technologies such as the IoT and policies address their challenges?

To answer these questions, we examined secondary data from consultancy reports, working papers, and academic journal articles that apply the principles of circular economy to the fashion industry. To assess how these theoretical principles unfold in practice, we conducted 15 semi-structured interviews with city government officers, apparel producers and designers, collectors and resellers of post-consumer apparel waste, and academics and businesses working on the IoT for circular fashion (Table 1). The interviewed stakeholders were located in NYC and were identified at circular economy events, panel discussions, online news on the topic of circular fashion, and snowball sampling. Through the interviews, the primary focus was to understand the stakeholders' activities and strategies in the circular economy and their challenges.

Table 1 Categories of interviewees

Sector	Number
City government officers	6
IoT practitioners	2
Academia	3
Collectors and resellers	1
Apparel producers and designers	2
Consultants	3
Total	15

4 NYC's Circular Fashion Stakeholders

Based on the consolidation of the gathered data, the various stakeholders who are driving circular fashion practices in NYC can be illustrated as in Fig. 1. Actors placed above the value chain arrow are considered to be in the industrial or manufacturing side of circular fashion. Upcycling designers take pre- or post-consumer textile waste and make higher-value fashion products with novel designs. Post-consumer waste could also be recycled by textile recyclers to become resources for fiber production. Innovation is also seen in the introduction of new, biodegradable raw materials. Actors placed below the value chain arrow provide services in circular fashion. Clothing swaps and repair shops extend the life cycle of clothes among consumers and communities. Second-hand retail brands collect, sort, and resell used clothes. Product as service or sharing platforms rent clothes. Fabric scraps are also collected from fashion manufacturers and resold by actors in resource recovery in Fig. 1. Local government actors and technology providers are crucial in the entire process of making fashion circular. The following sections discuss these actors in more detail by providing specific examples as seen in NYC. Table 2 summarizes a wider range of strategies and case study examples from NYC and beyond.

4.1 Government in Setting Vision and Collection Infrastructure

The City of New York's sustainability plan, OneNYC 2050, outlines goals and strategies for building a strong and fair city. One goal is to achieve a nearly 30% additional reduction in greenhouse emissions and to reach Zero Waste by 2030, while creating quality jobs (City of New York 2019a). Additional goals as part of joining the C40

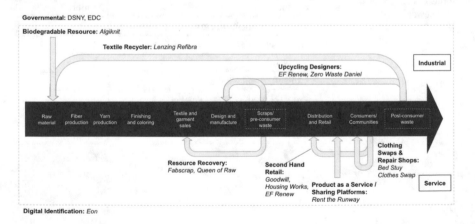

Fig. 1 Stakeholders in NYC and key strategies practiced in making fashion circular

Table 2 Strategies and case study examples of actors participating in making fashion circular

Actors	Strategies	Case study examples
Government: City, State, National, and International	Incentivize and establish minimum state recycling and diversion rates. Set forth minimum recycled-content requirements for a wide range of clothing products	**UN Alliance for Sustainable Fashion**: platform that works with member organizations and businesses to promote sustainable manufacturing practices (Fashion United Group 2019)
	Establish procurement requirements consistent with circular economy and deep decarbonization goals, such as banning the disposal and landfilling of textiles through extended producer responsibility, life-cycle assessment, and disclosure requirements for a range of products	**French Extended Producer Responsibility policy for textiles**: All legal entities putting garments, footwear, and household linens on the French market are held responsible for proper disposal of their products (refer to Bukhari et al. 2018)
	Raise awareness and education on reuse	**Department of Sanitation's Refashion NYC week, WearNext campaign**: tracks attendance, social media analytics of circular fashion initiatives
Industry: Businesses and R&D	Increase transparency in the supply chain	**Eon**: CircularID Protocol Pilot, a global protocol for digital identification of products in the circular economy
	Make effective use of biodegradable inputs for clothing production and phase out substances of concern (such as those that release microfibers)	**Algiknit**: R&D on biodegradable yarn made of kelp
	Offer take-back or extended producer responsibility programs	**Eileen Fisher Renew:** Take-back program for Eileen Fisher clothing of any condition. Customers receive a $5 Renew Rewards card for each item
	Share knowledge with other organizations	**Closed Loop Foundation and Closed Loop Fund:** Research and investment on film, electronics, glass, and plastic recycling and food waste innovations

(continued)

Table 2 (continued)

Actors	Strategies	Case study examples
	Transition to Product as a Service or Sharing Platform-based circular business models	**Rent the Runway:** A subscription fashion service that powers women to rent unlimited designer styles for everyday and occasion **For Days**: Organic cotton T-shirt brand that allows for a lifetime of swap-outs
	Design using upcycled fabric and create clothing items for repair, reuse, disassembly, and recycling	**The Renewal Workshop:** Takes discarded apparel/textiles and turns them into renewed apparel, upcycled materials or recycling feedstock **Zero Waste Daniel:** Clothing designer who uses pre-consumer waste and other hard-to-recycle materials, to create his line of genderless clothing and accessories
	Apply for certificates and standardization	**Cradle-to-Cradle Fashion Positive Materials Collection**: Assessment and verification for circular materials for apparel (Cradle to Cradle Products Institute 2020)
	Develop technology for chemical and mechanized sorting	**Lenzing Refibra**: cotton recycling technology **Valvan Baling Systems Fibersort**: fiber sorting machine
	Conduct a circle scan, establish baseline and map out circular economy opportunities and targets consistent with city, state, and national requirements	**Circle Economy:** Collaboration to launch Fibersort, which automatically sorts large volumes of mixed post-consumer textiles by fiber type. Based in Amsterdam **EnelX**: Circularity Report for Milan-based fashion brand Es'given to create the ES957 Coat using 70% recycled input materials and recommendations for energy efficiency interventions on entire production chain to improve circularity score from 3 to 4 (EnelX 2019)

(continued)

Table 2 (continued)

Actors	Strategies	Case study examples
Consumers and nonprofit community	Demand brands to implement circular economy principles, increase clothing utilization, learn to repair garments	**Citizens Committee of NYC Reduce, Reuse, Repair Grantee projects**: Provides local support for community projects. One example is Bed-Stuy Clothes Swap, which is a group that leads workshops to teach participants how to declutter their closets, providing reused business clothes for interviews, and showcasing artwork made with upcycled fabrics and materials
	Contribute to resource recovery efforts by urging brands to donate and buy deadstock fabric	**Fabscrap:** Nonprofit that provides convenient pickup of unwanted textiles from NYC businesses. Collaborates with Queen of Raw, a marketplace to buy and sell sustainable and deadstock fabrics
	Choose to shop second-hand and support the local economy	**Goodwill:** Nonprofit that aims to uplift local communities of New York and Northern New Jersey by using funds generated in its retail thrift stores and donation centers to support employment opportunities for people with disabilities and those faced with obstacles to employment **Housing Works**: Nonprofit fighting AIDS and homelessness through operation of a chain of thrift stores
	Sell or donate clothing, request a collection bin for your building, ask your company's building management what waste carter they use and if they have a textile recycling partner	**RefashionNYC:** NYC's official clothing-reuse program in partnership with NYC Department of Sanitation and Housing Works to make textile donations as easy **Wearables Collection:** Collects from commercial and residential buildings across NYC, partners with schools and organizations for clothing drives, hosts collection spots at 31 of GrowNYC's weekly Greenmarkets throughout the city, creates custom, sustainable solutions for keeping overstock, damages, and returns out-of-the-landfill for apparel brands

climate action pledge on achieving Zero Waste include reducing per capita waste generation 15% by 2030, reducing disposal to both landfills and waste to energy at least 50% by 2030, and increasing diversion rates to at least 70% (Rosengren 2019). The OneNYC sustainability plan goals that are related to the fashion industry are the following: (1) Stabilize and expand the city's industrial sector; (2) continue to support the city's creative sector; and (3) achieve carbon neutrality and 100% clean electricity (City of New York 2019b).

As the primary governmental actor in the circular fashion ecosystem, DSNY regulates businesses, collects post-consumer residential textile waste and supports research and development for implementing circular economy principles. Currently, if textiles comprise 10% or more of a business's waste, the organization is required by law to recycle it. However, enforcement is challenging because there are no audit requirements. DSNY is also responsible for picking up residential waste and conducting waste-characterization studies every three years. The department has a robust reuse program that tracks 72 nonprofit reuse organizations through their DonateNYC partnership, in which nonprofit partners are asked to submit materials-reuse data on a biannual basis. The nonprofits are divided into categories, such as retail (thrift stores and cooperative retailers such as flea markets), repair (shoe or clothing repair at dry cleaners), clothing rental, and social services (clothing or community center drop-offs) (DSNY 2019). A majority of the 72 nonprofits receive food and clothing donations. In fact, DSNY's recent study revealed 24% of NYC's rental reuse sector is comprised of clothing rental entities, 11% of which are online and virtual entities (DSNY 2019).

Another service that DSNY offers is the RefashionNYC clothing-collection program, available in residential buildings with ten or more units, institutions, schools, and businesses. Currently, the program covers 2,000 buildings out of NYC's total of 40,000 buildings, according to DSNY (2019). This is relatively low compared to the 14,000 buildings enrolled in the electronics recycling program, as the New York Disposal Ban in 2015 made it illegal to dispose of certain types of electronic equipment in landfills, waste-to-energy facilities, trash, or at curbside for trash pickup (DSNY 2019). RefashionNYC is a collaboration between DSNY and the nonprofit thrift stores in Housing Works, Goodwill, and The Salvation Army. The #WearNext Campaign from March 4 to June 9, 2019, was a successful and low-budget campaign created by the Ellen MacArthur Foundation and the New York City Economic Development Corporation (EDC) that presented a textile drop-off map with 1100 clothing reuse and recycling locations, which coincided with the launch of about two dozen additional drop-off points (Fig. 2). It is insightful to note that the majority of thrift stores and private textile drop-off locations are in Manhattan, Brooklyn, and Queens, while publicly accessible donation bins and textile recycling at farmers' markets are well distributed throughout New York City.

EDC's Brooklyn Army Terminal and the Brooklyn Navy Yard's Made In New York City Campus are also contributing to circular fashion business opportunities. There are multiple fashion industry companies, such as Fabscrap and Algiknit, that are pioneering circular initiatives to reuse and recycle scrap fabric and develop biodegradable materials respectively in the Brooklyn Army Terminal. EDC aims to

Fig. 2 NYC's 1100 textile drop-off locations. Data Source: New York Department of Information Technology & Telecommunications and DSNY

provide an environment to incubate and pilot ideas for circular practices in these hubs. At the World Circular Economy Forum held in June 2019, Lindsay Clinton, Executive Vice President of EDC shared that their circular economy industry strategy is to: (1) encourage business model innovation, (2) support technology, and (3) catalyze collaboration (2019).

4.2   Nonprofits in Resource Recovery and Second-Hand Retail

Nonprofits in the circular fashion ecosystem primarily focus on resource recovery. Several previous studies conducted by DSNY revealed that reuse entities divert thousands of tons of reusable materials from the waste stream and contribute significantly to the sustainability of NYC by generating local jobs and avoiding emissions associated with transporting materials across large distances (DSNY 2017). DSNY has supported reuse activities since 1997 through the creation of a materials exchange (DSNY 2017). Fabscrap is one such NYC-based nonprofit organization founded in 2017 (represents resource recovery in Fig. 1). Fabscrap collects pre-consumer or industrial textile waste directly from the brands that they work with, which are then sorted manually by volunteers. Industrial textile waste is estimated to be 40 times that of residential waste, and Fabscrap has only started working with design and sampling houses, which represent 255 brands in NYC (Fabscrap 2018). Fabscrap's 2018 annual report states that in 2017 and 2018, 97% of incoming fabric waste has been diverted from landfill. Fabscrap recently opened a shop in Chelsea, where they resell textiles and showcase clothes that are designed with the reused textiles. The accessible location in Manhattan contributes to raising awareness of circular fashion initiatives.

Goodwill, Housing Works, and the The Salvation Army are some of the main nonprofit organizations that collect post-consumer textile waste and sell second-hand items in their retail stores at various locations across NYC (represents second-hand retail in Fig. 1). In 2018, Goodwill collected 38 million pounds of clothing, of which around a third were resold (Goodwill 2018). In collaboration with DSNY's RefashionNYC program, these organizations enhance consumer awareness of second-hand shopping, thrifting, and the mindset that fashion does not always have to involve new items. Items that do not sell in the retail stores of these nonprofit organizations are sold to brokers or graders who sell to overseas markets, where used clothing and textiles supply local enterprises with materials to repair and sell (NYS DEC n.d.).

4.3   Businesses in Extended Producer Responsibility and Innovation

Private businesses are drivers of new product development and innovation that adhere to the circular economy principles. Eileen Fisher (EF) is a US-based luxury fashion brand targeted at middle-aged female customers (represents upcycling designers and second-hand retail in Fig. 1). Its line Renew was first launched as Green Eileen, a take-back program for EF clothing, representing the role of businesses in implementing extended producer responsibility (EPR). In the EPR model, producers bear the cost of collecting, treating, and recycling their end-of-life products. EPR for the textile industry is not required in the USA, hence EF's initiative is voluntary. In fact, the implementation of EPR policy for the fashion industry is uncommon, with the exception of France where the first legal framework for managing textile waste using the EPR policy was established in 2007 (Bukhari et al. 2018). In EF's EPR scheme, customers are encouraged to bring back old EF clothes and receive a $5 reward card for each returned item. The used items are sent to their warehouse in Irvington, NY, where staff members are involved in manually sorting, cleaning, repairing, and selling EF's returned clothes at a discounted price, which are also available online. Items that cannot be resold are turned into one-of-a-kind pieces in the Resewn Collection, or saved for recycling. The warehouse also has its "R&D Lab", where new product design ideas get tested for disassembly and remanufacturing.

One business example of creating biodegradable inputs in a circular supply chain is AlgiKnit (represents biodegradable resource in Fig. 1). Founded in 2016 and based in the Brooklyn Army Terminal, Algiknit is a biomaterials company that creates highly durable, yet rapidly degradable, yarns derived from a seaweed called kelp (AlgiKnit Press Kit 2018). This technology is currently undergoing research and development. It has the potential to tackle the textile-waste challenge because biodegradable textiles can be broken down to nutrients for the next generation of AlgiKnit's materials instead of contributing to harmful waste accumulation in landfills (AlgiKnit Press Kit 2018). The material can also absorb pigments using sustainable dye methods, reducing the need for excess chemicals and water usage. Kelp also does not require pesticides, fertilizers, and sequesters carbon, while filtering the surrounding water (Algiknit Press Kit 2018).

4.4 Internet of Things (IoT) to Advance a Transparent Circular Supply Chain

The fashion sector has been proactive in finding ways to close the industry's supply-chain loop with the aid of technologies. IoT makes possible the development of data infrastructure on the fashion product's material composition, transaction history, and environmental impacts along the supply chain (Rusinek et al. 2018; Eon 2020). A traceable resource unit could be assigned with unique identifiers such as universal product codes, 2-D barcodes, and radio-frequency identifiers (RFIDs) (Franco 2017; Eon 2020). As these tags get connected to sensors, stakeholders along the fashion value chain can have access to information about the product's condition, location, and availability, which facilitates reverse logistics, product services systems, value retention, and efficient end-of-life processes. When combined with separation and reprocessing technologies, these tags could also inform recycling or product-recovery partners about the specific material compositions and production processes that were used for the product (Franco 2017). Lastly, as Rusinek et al. (2018) explain, blockchain platforms could be developed to include information on the product's legal and corporate social responsibility requirements. Table 3 summarizes IoT and policy levers to address the circular fashion economy's main challenges.

Table 3 Summary of opportunities for IoT and policy levers to address challenges in making fashion circular

Main challenges	Opportunities for IoT	Opportunities for policy
Lack of awareness	Increases transparency in the supply chain for both producers and consumers Enables impact assessment of greenhouse gas emission curbed and water saved throughout the product supply chain	Raise awareness and education on reuse Increase accessibility of second-hand shops and textile drop-off points
Regulatory challenges	Develops data infrastructure to facilitate implementation of extended producer responsibility, life-cycle assessment, design standardization, and disclosure requirements Facilitates supply-chain traceability to monitor legal compliances and corporate social responsibility in the fashion industry	Incentivize and establish minimum state recycling and diversion rates Set forth minimum recycled-content requirements for a wide range of clothing products Establish procurement requirements consistent with circular economy and deep decarbonization goals, such as banning the disposal and landfilling of textiles through extended producer responsibility, life-cycle assessment, and disclosure requirements for a range of products
Technological challenges for sorting and recycling	Protects information on material composition, production processes, and ways to repair, disassemble, remanufacture, and recycle, which makes possible increased efficiency in the sorting process of textile waste	Support R&D in new sorting technologies and knowledge exchange among circular economy stakeholders Support innovation incubation and acceleration in biodegradable inputs for clothing production and in phasing out substances of concern (such as those that release microfibers)
Market and financial challenges	Provides sales and circular transaction history data to maximize value creation and material recovery Provides data infrastructure to develop new markets and circular supply chains Enables forecasting of quantity, quality, or timing of product take-back	Ease reverse logistics and inventory-management burden on budget-constrained stakeholders through transportation and zoning regulations

A key player in advancing IoT in the fashion industry is Natasha Franck, founder and CEO of Eon. Founded in 2015 in NYC, Eon provides a digital identity platform, in partnership with Microsoft, to generate and share data on fashion production, consumption, and reuse across the life cycle of fashion products. Eon's Circular ID Protocol, which includes a product's material passport, has great potential to enhance the management of a product's material components (e.g., material content, dye process, and thread type), commercial identification (categorizing brand, size, and retail price), and its end-of-life processes (with instructions on disassembly and recycling). Retailers and brands already have much of this data, but it is often lost after point-of-sale. This material passport enables that a product's details remain with it through its whole life cycle, to enable brands to automate resale and create a buyback price based on the manufacturer suggested retail price and condition of the product. Customer data would be protected because the product data follows the General Data Protection Regulation, and merging customer data with product data would be a violation. Eon's Circular ID Pilot Version was launched in January 2020, and the Circular ID Protocol Version 1.0 is set to be launched in January 2021.

5 Challenges and Opportunities in IoT and Policy Intervention

5.1 Lack of Awareness

The actual implementation of circular economy principles in NYC is accompanied by many challenges. Table 3 provides a summary of the challenges and opportunities listed in this section. First, stakeholders often raised the challenge of consumers and even clients and industry professionals' lack of awareness about the pollution impact of textile waste. On the production side, businesses are often unaware of the economic benefits of making the transition to a circular business model, or how much environmental impact the circular transition would bring about. On the consumption side, similarly, consumers are often unaware of the clothing-reuse sector, especially when it is much easier to find a new item of clothing that can be purchased at low price. Hence, due to such lack of predictability and knowledge, it is difficult for businesses or consumers (especially those under budget constraints) to make a new shift in their business model or consumer behavior.

Innovation such as the IoT data infrastructure could increase transparency in the supply chain by, for instance, allowing for impact assessment of greenhouse gas emissions curbed and water saved throughout the product value chain. In this way, producers and consumers would be able to observe outcomes of their behavioral change toward circularity, which makes the transition process more predictable. However, as one of our interviewees explained, it is much harder for established companies to change their processes, and they may not see the integration of technological solutions, such as the IoT as a feasible investment for future practices.

Hence, technological innovation is not sufficient for increasing awareness on textile waste and the unsustainable practices of fashion because deployment capacity and financial feasibility vary across stakeholders. This signifies opportunities and the necessity for local government involvement as well, just as DSNY has been active in raising awareness and education on textile reuse through events programming, by supporting local nonprofits, and increasing accessible infrastructure, such as textile drop-off points and second-hand retail shops.

5.2 Regulatory Challenges

The second challenge that stakeholders faced is lack of regulatory and enforcement power in the textile industry. As previously mentioned, EPR policy for the textile industry is not in place in the USA, while in France it has been enforced since 2006 and contributed to a threefold increase in the collection and recycling rates of post-consumer textiles (Bukhari et al. 2018). Hence, EPR policy could act as an effective policy incentive to strengthen the fashion reuse sector. Interviewees also voiced that the lack of design standardization and related regulation increases the variety of material blends and designs of fashion products, making it more difficult to implement the circular economy principles at scale. Blends of synthetic and natural fibers, together with various different chemicals, make the sorting, separation, reprocessing, and reuse of textile products extremely difficult, for which having eco-requirements implemented at the beginning of the product's design stage would be useful (Franco 2017).

The data infrastructure developed by IoT could facilitate implementation of regulations such as EPR, design standardization, and disclosure requirements. IoT not only provides benefits for the businesses in terms of cost savings and efficient supply-chain management, but it also facilitates monitoring of legal compliances and corporate social responsibility in the fashion industry. However, a point that some interviewees raised, the absence of data requirements for technologies such as IoT and material passports could increase the chance of greenwashing practices because there are no regulations on what information should or should not be included. Hence, stricter regulations and government policies are needed. Potential regulations to consider include establishing minimum state recycling and diversion rates of textile, setting forth minimum recycled-content requirements for government procurement products, and establishing design and data requirements.

5.3 Challenges in Sorting and Recycling Processes

The lack of technology for textile sorting is one of the biggest challenges for stakeholders that are directly involved in collecting and sorting textile wastes (such as actors in resource recovery and second-hand retail in Fig. 1). Both pre- and post-consumer textile sorting are manual, which is extremely inefficient since labels are often not attached to the garment, and the worker must visually and physically inspect every item to determine the fiber content of clothing and cutting-room scraps. With a lack of appropriate sorting technology, as the rate of incoming textile waste supersede those being sorted, inventory increases that requires additional storage area. Materials that are not reused are often downcycled by being shredded into rough fibers and re-spun into low quality yarns for insulation, carpet padding, or as moving-truck blankets. However, the existing technology for downcycling is such that Spandex, Lycra, and elastane fibers melt during the shredding process, contaminating other fibers and complicating sorting and reducing diversion options. The growing popularity and production of athleisure lines that emphasize fabric stretch is a mounting challenge to recycling options.

Technology such as the IoT provides information on the product's material composition, production processes, and ways to repair, disassemble, remanufacture, and recycle, all of which increase the efficiency of the sorting process. If incoming textiles are sorted fast, the economic costs that come from renting storage space and staff hiring could be resolved as well. However, some interviewees also noted that information and data requirements are different for every stakeholder along the value chain of the fashion industry, which makes it challenging to agree on a standardization protocol that would allow for building a robust data infrastructure system for analysis. Hence, it remains to be seen whether IoT data and digital platforms indeed could support the sorting, reuse, and recycling processes across different actors. In the meantime, local government actors could support R&D in developing new sorting technologies and knowledge exchange among the circular economy stakeholders. In order to phase out substances of concern, innovation could also be supported in developing biodegradable inputs for clothing production.

5.4   Market and Financial Challenges

Non-profit organizations such as Goodwill and Housing Works fund the operational costs with the revenues from clothing resale in their retail thrift stores. However, with the rise of fast fashion, the quality of donated clothing declines and so does the price for individual items. Thus, nonprofits must focus on procuring and sorting through donations to recover clothing of the highest quality for the current season, which is a manual, time-intensive process. Mixed-fiber textile waste has little or no resale value unless it is sorted by fiber content, but many nonprofits find it challenging to adapt to more advanced sorting methods because of added costs and a preference to reject pre-consumer textile waste (Schrieber 2017). Another cost burden comes from the lack of supportive infrastructure and markets for textile reuse. Our research indicates that stakeholders face obstacles in terms of rental cost for storage space and developing their own reverse logistics, from hiring drivers to establishing textile drop-off and pickup points.

Local government policies have opportunities to address these challenges through transportation and land-use regulations, which would ease reverse logistics and inventory-management burdens on budget-constrained stakeholders. IoT development could also be promising; for the sorting and recycling partners, the product's sales and transaction history data allows for maximum value recovery of the materials. By enabling forecasting of quantity, quality, or timing of product take-back strategies, stakeholders could find opportunities for novel business models and market demands. Still, the implementation and organizational change around IoT could be challenging for small-sized establishments under budget constraints. Re-training of staff, re-hiring, and investing in digital infrastructure development all contribute to additional cost burdens.

6 Conclusion

Conversations around Zero Waste and the circular economy are gaining traction in Europe and increasingly in the USA across many sectors. The fashion industry has great potential to adapt a circular approach, as $500 billion is lost each year due to clothing under-utilization, while less than 1% of clothing is recycled back into garments today (EMF 2017b). NYC's circular fashion ecosystem is growing with consumer and corporate awareness of the magnitude of textile waste. However, mainstreaming regenerative practices, increasing transparency across the supply chain, creating a more robust take-back market and infrastructure for sorting textile waste will take tremendous consumer, corporate and political willpower and coordination. Incremental steps taken by the various stakeholders introduced in this paper represent improvements in closing the fashion value chain and operationalizing sustainable economic development.

As demonstrated by NYC's experience, there are many open challenges to transition the fashion industry to more sustainable practices. The challenges and opportunities highlighted in this paper demonstrate that IoT technologies and digital identification have great potential to increase feedback loops and transparency in helping consumers and businesses responsibly handle fashion products throughout their value chain. Nevertheless, it is important to understand the context in which to apply these technologies and assess policy and regulatory incentives that need to accompany them. Continuing from this explorative research, future research will document additional practices and challenges for circular fashion stakeholders and their collaboration efforts. Comparative case study with cities outside of NYC, such as Milan and London that also have a high presence of fashion producers and consumer base, would be insightful as well.

References

Algiknit (2018). Algiknit Press Kit. Algiknit. Retrieved from: https://static1.squarespace.com/static/592f02709de4bb141cd80a50/t/5b46140570a6ad2664060439/1531319318750/AlgiKnit_Press_Summary. Accessed January 31, 2020.

Askoxylakis I (2018) A framework for pairing circular economy and the internet of things. In: Chapter from the 2018 IEEE international conference on communications (ICC) proceedings, Kansas City, MO, USA, 20–24 May 2018

Bukhari MA, Carrasco-Gallego R, Ponce-Cueto E (2018) Developing a national programme for textiles and clothing recovery. Waste Manage Res 36(4):321–331

City of New York (2019a) OneNew York City 2050 building a strong and fair city. Retrieved from https://1w3f31pzvdm485dou3dppkcq.wpengine.netdna-cdn.com/wpcontent/uploads/2019/11/OneNewYorkCity-2050-Full-Report-11.7.pdf. Accessed January 23, 2020.

City of New York (2019b) OneNew York City 2019 progress report. Retrieved from: https://1w3f31pzvdm485dou3dppkcq.wpengine.netdna-cdn.com/wp-content/uploads/2019/04/OneNewYorkCity-2019-Progress-Report.pdf. Accessed January 23, 2020

Cradle to Cradle Products Institute (2020) Textiles and apparel materials collection. Retrieved from: https://www.c2ccertified.org/fashionpositivematerials/. Accessed January 23, 2020

Department of Sanitation New York (DSNY) (2017) 2017 New York City residential, school, and New York City waste characterization study. New York City Department of Sanitation. Retrieved from: https://dsny.cityofnewyork.us/wp-content/uploads/2018/04/2017-Waste-Characterization-Study.pdf. Accessed January 31, 2020

Department of Sanitation New York (DSNY) (2019) 2010 New York City reuse sector report. New York City Department of Sanitation. Retrieved from: https://dsny.cityofnewyork.us/wp-content/uploads/2019/06/2019_Reuse_Sector_Report.pdf. Accessed January 31, 2020

Ellen MacArthur Foundation (2016) Intelligent assets: unlocking the circular economy potential. Retrieved from: https://www.ellenmacarthurfoundation.org/publications/intelligent-assets. Accessed January 31, 2020.

Ellen MacArthur Foundation (2017a) The circular economy in detail. Retrieved from: https://www.ellenmacarthurfoundation.org/explore/the-circular-economy-indetail. Accessed January 31, 2020

Ellen MacArthur Foundation (2017b) A new textile economy: redesigning fashion's future. Retrieved from: https://www.ellenmacarthurfoundation.org/publications. Accessed January 31, 2020

Ellen MacArthur Foundation (2019) Circular economy overview. Retrieved from: https://www.ellenmacarthurfoundation.org/circular-economy/concept. Accessed January 31, 2020

EnelX (2019) Enel X and Es'givien, beautiful and sustainable fashion thanks to the circular economy model. EnelX. Retrieved from: https://www.enelx.com/it/en/resources/stories/2019/12/esgivien-model-of-circular-economy. Accessed January 23, 2020

Environmental Protection Agency (EPA) (2019) Advancing sustainable management: 2017 fact sheet. Retrieved from: https://www.epa.gov/sites/production/files/2019-11/documents/2017_facts_and_figures_fact_sheet_final.pdf. Accessed January 31, 2020

Eon (2020) The connected products economy—powering fashion and retail's circular future, New York: circular ID initiative. Retrieved from: https://www.eongroup.co/connected-products-economy. Accessed February 13, 2020

Fabscrap (2018) The Fabscrap annual report, Fabscrap. Retrieved from: https://fabscrap.org/news/2018-annual-report. Accessed January 31, 2020

Fashion United Group (2019) UN launches alliance for sustainable fashion, Fashion United Group. Retrieved from: https://fashionunited.com/news/business/un-launches-alliance-for-sustainable-fashion/2019031826753. Accessed January 31, 2020

Franco MA (2017) Circular economy at the micro level: a dynamic view of incumbents' struggles and challenges in the textile industry. J Cleaner Prod 168:833–845

Geissdoerfer M, Savaget P, Bocken NMP, Hultink EJ (2017) The circular economy—a new sustainability paradigm? J Cleaner Prod 143:757–768

Ghisellini P, Cialani C, Ulgiati S (2016) A review on circular economy: the expected transition to a balanced interplay of environmental and economic systems. J Cleaner Prod 114:11–32

Goodwill (2018) 2018 annual report—Goodwill NY/NJ, Goodwill. Retrieved from: https://www.goodwillnynj.org/sites/default/files/reports/2018Annual%20Report_Final.pdf. Accessed January 31, 2020

Greenpeace International (2017) Fashion at the crossroads. Retrieved from: https://www.greenpeace.org/international/publication/6969/fashion-at-the-crossroads/. Accessed January 31, 2020

Kirchherr J, Reike D, Hekkert M (2017) Conceptualizing the circular economy: an analysis of 114 definitions. Resour Conserv Recycl 127:221–232

McKinsey & Company (2017) Mapping the benefits of a circular economy. Retrieved from: https://www.mckinsey.com/business-functions/sustainability/our-insights/mapping-the-benefitsof-a-circular-economy. Accessed January 31, 2020

New York State Department of Environmental Conservation (NYS DEC) (n.d.). Textile reuse and recycling, New York State Department of Environmental Conservation. Retrieved from: https://www.dec.ny.gov/chemical/100141.html. Accessed January 23, 2020

Refashion Week NYC (2020) Why refashion? Department of Sanitation New York. Retrieved from: https://www.refashionnyc.org/why-refashion-1. Accessed February 13, 2020

Rosa P et al. (2019) Assessing relations between circular economy and industry 4.0: a systematic literature review. Int J Prod Res, 1–26

Rosengren C (2019) Is New York's 2030 'zero waste' goal receding from reach? Waste dive. Retrieved from: https://www.wastedive.com/news/2030-zero-waste-goal-new-york-city/544155/. Accessed January 23, 2020

Ruisinek MJ, Hao Z, Radziwill N (2018) Blockchain for a traceable, circular textile supply chain: a requirements approach. American Society for Quality ASQ, 21(1)

Schrieber J (2017) Recycling fashion's remnants: residential and commercial textile waste. Cooper Hewitt Des J. Retrieved from: https://medium.com/design-journal/recycling-fashions-remnants-residential-and-commercial-textile-waste-e58a5f0c91a8. Accessed January 23, 2020

Stahel WR (2016) Comment: circular economy. Nature 531:435–438

Yin RK (2013) Case study research: design and methods, 5th edn. SAGE, Thousand Oaks, CA

Dissolving Borders: Towards Integrated Territorial Approaches, from Smart Cities to Smart Regions

Beyond the City Limits—Smart Suburban Regions in Austria

Nina Svanda and Petra Hirschler

Abstract The majority of the population in Austria lives and interacts in (sub)urban regions. Their areas of activity cross the borders of cities, municipalities, and countries. The urban regions consist of the core city and the catchment areas, and the balance and the equilibrium between the city and its surroundings is very important for sustainable spatial development. The size of the urban regions in Austria ranges from small- and medium-sized regions to polycentric agglomerations and the metropolis of Vienna. Nevertheless, urban regions are currently not sufficiently defined by type of region nor established at the planning or action level among political and administrative stakeholders. One of the reasons is that urban regions as functional spaces with flexible borders extend beyond political and administrative borders. Due to this lack of clarity with respect to political responsibility for urban regions, it is often difficult to find support for urban–regional cooperation on the political level. The "Agenda for Urban Regions in Austria" is a first milestone to implement an Austrian policy for its urban regions. It identifies measures for regional actors and especially policymakers on different administrative levels to remain urban regions sustainable in the future and to encourage and support collaboration. Urban regions have to master challenges in many different fields covering the entire spectrum of spatial development. They have to take actions to improve mobility and accessibility across city borders, to secure free space for everyone by a prudent use of free space and natural resources and to support diversity and cohesion to provide space for the diversity of lifestyles. A very important action is the sustainable development of settlements and business locations through improved interaction of cities and municipalities within urban regions to achieve more for less investment. One of the key points is that urban regions practice governance to support cooperation among their actors. The steering and coordination of the spatial development in urban regions (i.e., the governance) affects not only various actors in urban regions but also the coordination between federal government, provinces, municipalities, and

N. Svanda (✉) · P. Hirschler
Institute of Spatial Planning, Vienna University of Technology, Karlsplatz 13, 1040 Wien, Austria
e-mail: nina.svanda@tuwien.ac.at

P. Hirschler
e-mail: petra.hirschler@tuwien.ac.at

© The Author(s), under exclusive license to Springer Nature Switzerland AG 2021
A. Bisello et al. (eds.), *Smart and Sustainable Planning for Cities and Regions*,
Green Energy and Technology, https://doi.org/10.1007/978-3-030-57332-4_22

cities. With the establishment of governance structures in urban regions, steering and coordination areas will be adjusted to functional areas. In this way, it is possible to solve spatial challenges jointly, to bundle resources and to raise the willingness to cooperate among the actors in urban regions.

Keywords Urban regions · Governance · Collaboration · Smart regions · Integrated approach

1   Introduction and Method

Urban agglomerations are key drivers of regional development. In the year 2010, the global level of urbanization reached more than 50%. In Europe more than 70% of the populations live in urban regions. As a result, urban agglomerations face great challenges such as, for example, urban sprawl, high consumption of resources, and dense traffic flows. Imbalances within and between urban agglomerations occur or even grow, e.g., in the field of employment and/or demographic structure (inter- and intraregional polarization). These trends often result in disparate fiscal developments between the core cities and their surrounding municipalities.

While the forces impacting the growth of cities have changed dramatically in many parts of the world, planning systems have changed only very little. Steering those growing or shrinking agglomerations with the existing planning tools needs to be revisited for urban and regional planning, especially in the field of shopping malls, subsidized housing, or cooperation among provinces.

The spectrum of Austrian agglomerations is broad. It reaches from small- to medium-sized urban agglomerations with differing economic structures (e.g., urban services, industrial structure, and touristic structure) to polycentric agglomerations, as well as the metropolitan region Vienna (see Fig. 1).

The study "Spatial Development in Austrian Urban Agglomerations: Need for Action and Steering Options" (Hamedinger et al. 2009) dealt with selected Austrian regions. It was conducted by the Department of Spatial Development, Infrastructure and Environmental Planning of the Vienna University of Technology in 2007–2008. Based on analysis of the spatial development in functional urban areas, the work focused on various aspects like, e.g., innovative approaches for steering spatial development in urban agglomerations and recommendations for actions to meet the different challenges of city-regions.

The study followed the approach of interactive research including participation, grounded in experience and action-oriented approaches. Central features are the collaborative dialogue between researchers and research participants and a common understanding of research as a joint learning process (Astleitner and Hamedinger 2003). The aim was to develop practical knowledge (requirements for action) and recommendations for action. A broad range of procedures and structures for the steering and coordination of the development in urban agglomerations have been analyzed, leading with the result that—compared to classical direct "hard"

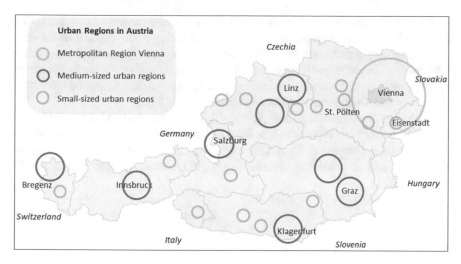

Fig. 1 Urban agglomerations in Austria

steering instruments—the significance (effectiveness) of more communicative and cooperative targeted "soft" steering instruments is evident.

Over recent years, these soft instruments especially have gained more and more importance—one guiding principle related to these cooperation models in Austria is voluntariness. The increased demand for coordination in urban agglomerations especially requires cooperative methods. Regional governance is based on concepts of cooperation and negotiation (Pütz 2004), aiming at a specific combination of various steering instruments (direct and indirect), as well as steering mechanisms (e.g., hierarchy, networks, and markets). The crucial point is always to come to a common answer to the competing interests of the involved actors. Therefore, incentive schemes are essential elements of regional governance to enable and support regional cooperation.

With the adoption of the Austrian Spatial Development Concept (ÖREK 2011),[1] the Austrian Conference on Spatial Planning (ÖROK) created a strategic framework for action for spatial development in Austria. In the ÖREK 2011, the "Development of an Agglomeration Policy for Austria" is defined as one of 14 actions. For the first time, spatial development of agglomerations made it to the front stage, while earlier the most important fields of intervention were lagging regions.

The majority of urban regions are growing and significant settlement entities, but nevertheless they still are not political or legal entities. Therefore, it is important to think about how urban regions can be demarcated, how they can take joint stances and gain legal status, as well as how the functional distribution of tasks within an urban region can be optimized.

---

[1]The Austrian Spatial Development Concept has been elaborated since 1981 every ten years. As the federal level has no spatial planning competence, it is a self-binding document adopted by the provinces, ministries and social partners.

The new positioning of urban regions embedded in new models of regional governance and of revenue sharing has the goal of achieving a fair balance of interests between core city and catchment areas and includes: rule-based and cooperative developments in traffic and settlements; securing and strengthening the functionality of core cities; complementary upgrading and functional enrichment of catchment areas (Austrian Conference on Spatial Planning 2011: 84).

In addition the potential of new partnerships between rural and urban regions is highlighted. Cities and regions will be involved in new forms of cooperation for integrated spatial development. Thus, they should achieve social and economic development, as well as sustainable economic growth (Austrian Conference on Spatial Planning 2011).

2 Findings

In general urban agglomerations with their dynamic development do not necessarily need other planning instruments for effective control, but certainly the instruments must be used. Integrated regional development focuses on interdisciplinary and intercommunal cooperation and steering of the development of agglomerations. While regional planning in lagging regions concentrates on the activation and initiation of development processes, the dynamic city-regions are in need of control and coordination of ongoing development processes.

Despite some progress in intercommunal cooperation in recent decades (often pushed by European Union funding), there is still great need for action. The metaphor of David and Goliath is often used for the cooperation in agglomerations. On the one hand, there is the core city with a high density of services, power but also costs, while, on the other hand, the smaller communities in the agglomeration benefit from the services, contribute only little financially and have less political power. Because of that, the starting point of the cooperation is not equal. Furthermore, the communities are strong legal entities in Austria, and one of their key competences is to design the spatial development on their territory. Financially, there is also no need to cooperate as they gain enough money through local taxes and do not rely on EU or other funds. The study tried to answer the following questions in depth and outlined possible pathways for cooperation in urban agglomerations:

- What prevents cities and towns from coordinating their development planning and working out common solutions?
- Why is the win–win situation not recognized?
- Why is development not controlled at the city-regional level?
- Why must every community offer all service tasks?

The observations and conclusions presented in this paper will hopefully contribute to improved cooperation in urban agglomerations.

2.1 Differentiated Strategies

Austria has nine different planning laws. The main paradigm of spatial development was to achieve and secure equal opportunities for all. Therefore, lagging region was the main focus for decades. Until now spatial planning instruments are equally designed for all regions and they are often used in the same way although the spatial challenges and requirements are different. In the future, the planning laws should recognize the difference between rural and central areas. While in agglomerations the development power is significant and the pressure on undeveloped land is constantly growing, instruments for protection and densification are desperately needed.

2.2 Cooperation Supported by Politics

The change in the European Union's regional funding policy in 2007 offered new perspectives and funding possibilities to urban agglomerations. With the option to enlarge the eligible areas to the entire country (including economically strong regions), the European Union co-financed projects can be realized in urban agglomerations.

Thus, it is up to the province to focus on the core areas. The funding programs of the European Union—through financial incentives—contributed significantly to the affiliation and advancement of cooperation in Austria. As a guiding principle, grants are not awarded to individual communities, but projects must be regionally coordinated and implemented.

Since one of the goals of politics is to exploit all funding, perhaps the focus will increasingly be on dynamic regions in the future. However, it is the responsibility of the province how to use Objective 2 (according to the European Union's regulations). For example, in the area of Graz–Graz Umgebung, the Operational Program "Competitiveness Styria 2007–2013 (objective 2)" has its own priority "URBAN plus: agglomeration development". Although the available funding of 5.5 million € is relatively low, compared to other programs, nonetheless the program offers the possibility to develop projects on sensitive cooperation topics, e.g., densification of residential areas. The pilot-like cooperation projects could be used successfully as a laboratory for the spatial development in the functional urban region of Graz and a further intensification of the collaboration between regional cooperation, surrounding communities and the city of Graz (Stadt Graz 2013).

2.3 Framework for Cooperation

Cooperation is a useful tool to promote coordination and integrated development in urban agglomerations, but especially in dynamic regions a defined framework is needed. The strength of cooperative instruments is the flexible and informal use rather than regulatory policies and measures. Moreover, the provinces are encouraged to set up a regulatory framework with development prospects and restrictions. Even now, the provinces have the appropriate planning instruments, but they are not used in practice.

For example, a prescribed strategy paper could be issued, focused on crucial development areas, including priorities and objectives to be broken down in detail by the communities (and not only a vague formulation of goals). Especially when "hard" cooperation issues (settlement and business development, social infrastructure) are addressed, a detailed framework is necessary. This requires a set of regulations on a higher level guiding the local and regional planners.

Successful impulses for regional cooperation are financial resources to be exclusively used for regional initiatives and projects. In addition, the community representatives can blame the province authority for unpopular measures. Certainly this contributes to cooperation of the involved stakeholders and shows added value of an integrated regional development. Generally, planning aims to enable processes with set targets and not prevent development. It is not about excluding competition, but controlling and steering how and where it should happen.

2.4 Top-Down Versus Bottom-Up

Based on the analysis of the case studies, the tension between top-down and bottom-up created partnerships is obvious. In general, the analyzed cooperation models generally follow a top-down driven approach. Cooperation is always a long-term process. The so-called soft cooperation issues are particularly suitable for the initial phase of cooperation.

For the involved parties, even small, visible successes are important. Especially for the politics but also for the public, the necessity and usefulness of regional cooperation must be noticeable. Moreover, cooperation is based on mutual confidence. Accordingly, the shift of persons (e.g., after public votes) always has an impact— either promoting or inhibiting. Often the knowledge and experience in cooperation is lost with a change of the involved persons.

Perhaps a common vision of urban agglomeration development is necessary that is especially based on bottom-up structures. Experience shows that, particularly in agglomerations, there is a high potential for intercommunal cooperation. Also at the regional level—particularly when the agglomeration is located in more than one province—steering is urgently needed.

2.5   Open Up "Hard" Cooperation Areas

The easiest way to cooperation is in so-called win–win situations. All partners benefit from the collaboration, and its successes are easily visible. Classic examples of this are regional cycling trails and sewage or waste associations. In more complex topics, like an intercommunal business location, it is not easy to promote the added value. Yet the politicians still aim to locate the companies within their community. Often communities overlook the fact that an intercommunal business location with a higher quality, as well as corresponding cost and revenue agreements, may bring higher profit for the municipal budget. At present, politicians are still in favor of "get jobs in the community," as a result of Austrian taxation, where the communities gain direct payments for each workplace within their territory.

In urban agglomerations, the competition primarily focuses on the surrounding communities and not on the core city. The central city generally has a differentiated profile and thus a unique selling position. Therefore, there is no direct competition with the surrounding communities, which differ little from each other in general.

Voluntary partnerships are typically based on the "give-and-take" principle between the actors. Accordingly, the integration of "weak" partners with nothing to offer but in who share the desperate need of cooperation is a challenge. In this case, the limit of voluntary cooperation is reached. Therefore, cooperation in win–win situations is easy to start and to handle, but, if some communities are affected negatively, clear guidelines are needed.

As a guiding principle, the dynamic development processes have to be increasingly interdisciplinary and integrated, as for example traffic planning and landscape protection have a massive impact on the settlement development not only locally but also regionally.

2.6   Participation and Collaboration with Individuals

Currently, cooperation in agglomerations is often just a matter for the local authorities. International examples show success in cooperation with industry and regional initiatives. The aim must be clearly defined for the cooperation with stakeholders. Possible areas of cooperation are health, education, or brownfields. Interested citizens, too, can be potential partners in urban–regional development processes. Especially when it comes to regional identities or development scenarios, the participation of civil society is essential. The inhabitants live an urban–regional life, and they do not care about administrative borders in their everyday lives. They choose to live in an urban surrounding with a high density of services, mobility, and life quality.

2.7 Professionalization and Management

Even though currently active organizations seek more "binding" cooperation instruments and regulatory solutions, there is also the chance for them to achieve a higher quality standard. Unfortunately there is not a single, complete formula for efficient cooperation. Each form of organization has its strengths and weaknesses and must continue to develop with the involved actors. Of course, it is not necessary to reinvent the wheel constantly, but rather to exchange and learn from other management units in urban agglomerations.

Urban agglomeration has to overcome the chicken-and-egg syndrome—they are still searching for the perfect form for cooperation as an excuse not to start with their coordination that will shape the form of cooperation by content. Even in informal forms of collaboration, methods or processes are developed to optimize the flow of information. For example, a mandatory "invitation principle" (for events and meetings on issues of regional interest) secures the transparent information flow and keeps all involved partners informed. Then it is up to them to follow the invitation. The organizational form of the management unit is not crucial for the success of the cooperation, but should be in line with its tasks and profile. In principle, the cooperation sets the frame for the organizational form and not vice versa. If it turns out that the existing legal framework is not adequate it can be adjusted easily. The choice of organizational form must correspond to the degree of willingness to cooperate. With upcoming and new challenges in urban agglomeration, new cooperation structures might be necessary and developed.

Participatory processes are important, not only because of the democratic political anchor, but to push and put pressure on the political level. They can give further impetus to regional cooperation and increase liability.

2.8 Awareness Raising and Long-Term Development

In general, the appropriateness and effectiveness of regional cooperation is not seriously questioned. There are clear benefits for the agglomeration, as well as for the involved parties, if the agglomeration development is coordinated. For example, financial resources are used more efficiently or information is more easily accessible and transparent.

The crucial factor for sustainable and successful cooperation is not in the theoretically available potential, but the willingness of the partners to work together and tackle sensible fields of development. Yet, there is no agglomeration awareness among the involved partners and the public. The knowledge, that an agglomeration is a common organism that must function as a whole and not partially, must be spread to a wider public (Hamedinger et al. 2009).

The survey showed the wide field of possible interactions, and the need for action became evident. In the next step, policies for urban agglomerations were elaborated in a broad process: Agenda for Urban Regions in Austria.

3 Need for Action

The "Agenda for Urban Regions in Austria" is the first milestone to implement an Austrian policy for urban regions. It identifies measures for regional actors and especially policymakers on various administrative levels to remain urban regions that are sustainable in the future and to encourage and support collaboration. It was approved by the political decision-making body of the Austrian Conference on Spatial Planning[2] in the year 2017.

3.1 Key Points for an Austrian Policy for Urban Regions and Agglomerations

According to the survey, urban regions have to master challenges in many different fields covering the entire spectrum of spatial development. Nevertheless, currently urban regions are neither sufficiently defined by type of region nor established as planning or action level among political and administrative stakeholders. One of the reasons is that urban regions as functional spaces with flexible borders extend beyond political and administrative borders. This lack of clarity regarding the political responsibility for urban regions often makes it difficult to find support for urban–regional cooperation on the political level.

The complexity of the challenges leads to a double policy: dealing with topics for cooperation and key points of an Austrian agglomeration policy, which are (see Fig. 2).

3.1.1 Urban Regions Have Instruments—To Plan and Develop

At the moment, urban regions have to plan within the legal framework of spatial planning laws. But, nevertheless, they are free to adjust the formal instruments according to their needs. Especially regarding informal instruments, they are flexible. If spatial planning laws are revised, provincial parliaments are advised to react to the spatial needs of agglomerations and design successful instruments. The establishment of

---

[2]The political decision-making body of the Austrian Conference on Spatial Planning is chaired by the Federal Chancellor and its members include all federal ministers and heads of the Länder, the presidents of the Austrian Association of Cities and Towns and the Austrian Association of Municipalities, as well as the heads of the social and economic partners with a consulting vote.

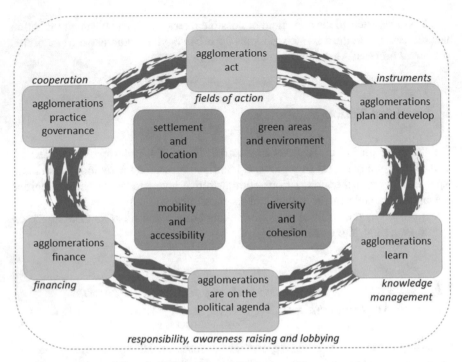

Fig. 2 Agenda for urban regions in Austria—key points

urban agglomerations as a relevant planning and action level can be achieved by, e.g., strengthening bottom-up processes, developing cross-border cooperation models for urban agglomerations, overcoming province and community borders by setting up common projects or intensifying use of public-relations instruments to communicate urban challenges.

3.1.2 Urban Regions Have Governance Structures—To Support Cooperation Among Actors

The experiences showed that cooperation structures support the integrated development. It again depends on the agglomerations as to which organization structure is suitable for the involved actors, the fields of action and the cooperation status. The survey provided an overview of successful forms of cooperation to be used as good practice examples and to be adjusted to agglomerations and their challenges.

3.1.3 Urban Regions Learn—And Engage in Systematic Knowledge Management to Achieve This

In the year 2011, the ÖROK established a platform, which had basically two tasks: to conduct surveys and to organize the yearly conference "Urban Regions Day." Because the ÖROK support for the platform ended in 2018, it should be supported by the involved actors in the future to secure a knowledge exchange and information platform.

3.1.4 Urban Regions Are Funded—And Have the Means to Create Incentives

This is one of the most fundamental but also most challenging fields of action. Since years, the authorities debate changes in the Austrian revenue-sharing system. It is a never ending story, but of course urban regions can find solutions on their own on a private basis. Maybe also in the future, the EU co-financing will be an important financial player in European agglomerations.

3.1.5 Urban Regions Are on the Political Agenda—And Engage in Awareness Raising and Lobbying to Position Themselves and Achieve Defined Areas of Competence

Finally, also the politicians get more active on urban regions as they learn about the benefits of intensified cooperation.

3.1.6 Urban Regions Take Action—In Their Own Fields of Action and Define Priorities

Each urban region in Austria faces different regional challenges, and the fields of action have to be in line with the local circumstances. There is no blue print to be unrolled to have successful cooperation in urban agglomerations. Because of that, each agglomeration has to identify its fields of action and define priorities.

In the center of the figure, the cooperation topics based on the results of the study are mentioned. To secure a high quality of living, the coordination of the following fields of spatial planning is desperately needed:

- Settlement and location: social housing, density, affordability, accessibility, high-quality internal development, and harmonized development within urban regions
- Mobility and accessibility: mobility partnerships, demand-side-oriented standards for public transport, public transport quality classes, high-speed infrastructure, and slow mobility

- Green areas and environment: landscape accounts in urban regions, regional parks, regional harmonization of building densities, renewable energy concepts and quantitative and qualitative soil protection
- Diversity and cohesion: integrated schemes for urban regions, spaces for encounters, affordable housing, open society, life-long learning, inclusion and participation (Schaffer et al. 2015)

When the policies were adopted by the ÖROK, the question arose where to start and what to do next. Therefore, a roadmap was elaborated in the next step.

3.2   Roadmap to Cooperation in Urban Regions

One of the key points of the "Agenda for Urban Regions in Austria" is that urban regions practice governance to support cooperation among their actors. "The establishment of governance structures in urban regions enables steering and coordination to adapt to the needs of functional areas" (Zech et al. 2016: 13).

Governance, the steering and coordination of spatial development in urban regions is a differentiated and extensive task that does not only concern different actors (which means horizontal coordination) but also the coordination between federal government, federal provinces, municipalities and communities.

In the practice of cooperation in urban regions, the breaks in the facilities with planning resources become apparent, especially between urban core and the surrounding communities. These general conditions complicate coordinated planning "at the urban level". In planning and administration associations, e.g., common building authority, building yards, professional expertise can be bundled, and at the same time service closeness can be offered. In the case of cross-border planning tasks in urban regions, the support of the federal provinces is necessary. In this, way spatial challenges can be solved jointly, resources can be bundled and the willingness of actors in urban regions to cooperate can be raised.

Different or unclear legal provisions complicate the cooperation. The reduction of existing legal and tax barriers, e.g., liability for turnover tax at cooperation businesses between communities shall support and promote the cooperation (Zech et al. 2016).

The roadmap suggests to start with adjustment at the local authorities and continues with the successful already existing cooperations and platforms. The actors are called to bring more formality and binding actions into the steering of the agglomerations.

4   Conclusion

In conclusion, there is no single model for successful governance in urban agglomerations. The study showed that there are various innovative approaches, but the transferability of these methods is limited as they are linked to the involved partners (politics and administration).

Despite the fact that the style of planning involving variable geometries in urban regions is especially critical, as well as the flexible and convertible forms of cooperation in planning processes, the future of politics for urban regions needs commitment, structure, and standards. The political representation of urban regions is often unclear or less visible, and the cooperation in various forums and processes partly overlapping in time and space is time-consuming. Nevertheless, urban–regional planning tasks emerge onto the political stage (Austrian Conference on Spatial Planning 2017). Summing up, there is still a long and winding road to go, but there is a light at the end of the tunnel.

References

Astleithner F, Hamedinger A (2003) The analysis of sustainability indicators as socially constructed policy instruments: benefits and challenges of 'interactive research.' Local Environ 8(6):627–640

Austrian Conference on Spatial Planning (ÖROK) (2011) Austrian spatial development concept ÖREK 2011. *Geschäftsstelle der Österreichischen Raumordnungskonferenz.* Wien

Austrian Conference on Spatial Planning (ÖROK) (2017) ÖROK-Empfehlung Nr. 55: „Für eine Stadtregionspolitik in Österreich" Ausgangslage, Empfehlungen & Beispiele. *Geschäftsstelle der Österreichischen Raumordnungskonferenz.* Wien

Hamedinger A, Bröthaler J, Dangschat J, Giffinger R, Gutheil-Knopp-Kirchwald G, Hauger G, Hirschler P, Kanonier A, Klamer M, Kramar H, Svanda N (2009) Räumliche Entwicklungen in österreichischen Stadtregionen. In: Handlungsbedarf und Steuerungsmöglichkeiten, Wien: Geschäftsstelle der Österreichischen Raumordnungskonferenz (ÖROK)

Pütz M (2004) Regional Governance—Theoretisch-konzeptionelle Grundlagen und eine Analyse nachhaltiger Siedlungsentwicklung in der Metropolregion München *(Hochschulschriften zur Nachhaltigkeit, 17).* Oekom Verlag, München

Schaffer H, Zech S, Svanda N, Hirschler P, Kolerovic R, Hamedinger A, Plakolm M, Gutheil-Knopp-Kirchwald G, Bröthaler J (2015) Für eine österreichische Stadtregionspolitik, Agenda "Stadtregionen in Österreich", Empfehlungen der ÖREK-Partnerschaft "Kooperationsplattform Stadtregion." Geschäftsstelle der Österreichische Raumordnungskonferenz (ÖROK), Wien

Stadt Graz (2013) URBAN PLUS—Integrierte, EU geförderte Stadt-Umland-Entwicklung im Süden von Graz. https://www.graz.at/cms/dokumente/10027232_8033447/c8e6fd0b/Präsentat ionURBANPLUS_131008.pdfs. Accessed February 13, 2020

Zech S, Schaffer H, Svanda N, Hirschler P, Kolerovic R, Hamedinger A, Plakolm M, Gutheil-Knopp-Kirchwald G, Bröthaler J (2016) Agenda Stadtregionen in Österreich, Empfehlungen der ÖREK-Partnerschaft "Kooperationsplattform Stadtregion" und Materialienband. *Geschäftsstelle der Österreichische Raumordnungskonferenz (ÖROK) (ed.).* Wien. ISBN: 978-3-9503875-7-5

Rural Areas as an Opportunity for a New Development Path

Stefano Aragona

Abstract The paper proposes an integrated view of the territory in which there is no conflict between rural areas and urban areas. This is in line with the indications of the 2007 EU *Charter of Leipzig* which calls for integrated planning strategies between rural and urban, small, medium, large, and metropolitan areas. Therefore, an ecological multidisciplinary approach to the territory, using the concept of ecology proposed by the 1950s by Dioxiadis, is suggested, concept then taken up by Appold and Kasarda in the early 1990s. And it is the key word of the Encyclical *Laudato Sii for the Care of the Common House* of Pope Francis 2015 which refers to the principles of the 1992 Rio Conference based on the centrality of "human ecology" and on the alliance between man and nature, which in 1995 Scandurra had requested in *L'ambiente dell'uomo* (*The Environment of Man*). The phenomenological, from the Greek world *phenomenon,* is the methodological keystone of the paper. The well-being of the person and of the communities must be the objective of those who deal with the development of the territory. Rural areas in this sense can offer great opportunities, generated today by the many technological innovations, both tangible and intangible, potentially available. However, this should not overshadow the need for links that facilitate access to these areas. One of the important positive effects of the presence in rural areas is the reduction of hydrogeological risk thanks to the presence and daily maintenance of them. Among the aspects that raise questions that are difficult to answer is the risk that the more accessible an area becomes, the more there is a threat of its loss of uniqueness. Last but not least, it should be pointed out that the rural areas are a sort of territorial "reserve" for populations that have to abandon their lands due to climate change or who seek a quality of life not possible in large metropolitan areas. All this means that some geographical areas, primarily Calabria, are potentially territories where scenarios can be hypothesized for a different modes of anthropizations indispensable to move toward the objectives of the *UN 2020–2030 Charter* on sustainable development. In light of the COVID 19 emergency, these indications are even more relevant.

S. Aragona (✉)
Department of Heritage, Architecture, Urban Planning, University Mediterranea of Reggio Calabria, Salita Melissari, 89124 Reggio Calabria, Italy
e-mail: saragona@unirc.it; stefano.aragona@gmail.com

© The Author(s), under exclusive license to Springer Nature Switzerland AG 2021
A. Bisello et al. (eds.), *Smart and Sustainable Planning for Cities and Regions*,
Green Energy and Technology, https://doi.org/10.1007/978-3-030-57332-4_23

329

Keywords Opportunity · Multidisciplinarity · Sustainability · New
anthropization · Scenarios

1 Rural and More

"Rurality" characterizes the internal areas, often defined as "minor", but is also
becoming a significant element of many urban suburbs. Emanuel already in the early
90s coined the expressions "urbanization of the countryside" and "ruralization of
the city" (Emanuel 1990).[1] This twofold phenomenon, if managed and addressed
by territorial and local strategies, can give further consistency to sustainable and
more useful methods of anthropization for the well-being of people, in line with the
objectives of the UN 2020–2030 Charter, inhabitants of the territory or citizens they
are. Phenomenon that characterizes Italy because of its ancient and recent history,
the reason for a landscape rich in social, architectural, and environmental diversity.
Phenomenon can be measured only partially, being largely qualitative[2] and with
indirect effects and externalities difficult to measure. Phenomenon linked in a relevant
way to the political choices to which technicians can only suggest mostly qualitative
scenarios.

The main conceptual reference is the resident of a place, even if this characteristic
no longer has the pervasiveness, the "hardness," which was present in the anthropization processes of the past. Historical settlements were often called the "stone cities,"
the more "stable" they were, the safer they were, the more they could grow. They
were in defense of agricultural areas, one of the three territorial invariants suggested
in 1987 by Raffestin.[3] Of the natural resources they needed, the most essential was
water. Earth, water, and stability were the elements that guaranteed the continuity
and prosperity of the local communities. The concept of "civic use," invented by
the Romans, recognized the community of residents and the lands burdened by this
denomination belonged to it, which is neither public nor private but collective.

The economic activities that have emerged and affirmed in recent years are often
"temporary," that is, precarious and uncertain, almost volatile. And therefore, the
settlements associated with them have often been, and are, equally unstable and
unable to build a "place".[4]

---

[1] These characteristics were related to the emerging of the so-called widespread city, of the dispersed
city, but they can also be useful regarding the issues that are being discussed here.

[2] ISTAT with CNEL since 2013 have proposed the BES—Fair and Sustainable Wellness, a multi-
criteria indicator made up of 134 indicators divided into thematic areas. Initially tested on 15 cities,
it is currently being extended to the various provinces of the Nation.

[3] This author considers the areas, nodes, and networks as "territorial invariants" that change in
importance in relation to the type of society, that is, the agricultural/rural one, the industrial, then
the contemporary or post-industrial.

[4] Situations that have in the speed, that increasingly is characterizing the contemporaneity, and
of which Augè (1993) places "the excess of time" as one of the elements of what he calls
"supermodernity" and which, together with "the excess of space and ego," creates the "non-places."

Settlements and rural activities have opposite characteristics: They need stability, continuity, presence, and control. So they participate significantly in the construction of the landscape, as the Charter of the same name defines it, i.e., the outcome of the relationship between human activities and nature. Thus, over the millennia, Italy has become "the country of 100 bell towers" (Coscetta et al. 2014). This occurred not with industrialized agricultural activities but with "human" relationship with the land, that is, that which is at the basis of the construction of local communities that used or built networks, canals, mills, etc., which were often seeds of villages and then often cities.

2 Transformation and Spatial Shapes

Innovation, technological, and technical transformations must be used, but this must be done in a "cultured" way (Del Nord 1991), that is, at the service of man and not vice versa. In social, economic, and spatial changes, rurality has been neglected both on a territorial and urban scale. The industrialist development model has overwhelmed rural landscapes, it has been indifferent to the territorial context. The post-World War II agrarian reform created many small or medium-sized owners, but immediately afterward the policies pointed at industrialization. So where there were inadequate conditions for it, the attempt was to build the elements for industries to arise. Then, a working class grew up, a wage, and finally, a "spending power" and "demand" related to industrial/modern goods.

With this mechanism, efforts were made to bridge the gap between the wealthiest and the least wealthy areas, both in relation to infrastructure and in relation to wage differences. And partially, they succeeded in. Because the most disadvantaged areas, from the unification of Italy onward, were mainly in the South, the Cassa per il Mezzogiorno—CASMEZ was born.[5] Situation worsened also following of the Second World War which had many battle theaters in the south and center of Italy. This has meant the widespread destruction of infrastructure networks and nodes as well as urban centers, towns, which represented military garrisons, defenses, of which the case of the Montecassino Monastery is a striking case but that small rural

---

[5]The *Cassa for extraordinary works of public interest in Southern Italy* was born in 1950 and ends in 1984 (instead of the planned 1960). It was intended to create the infrastructural conditions and primary urbanizations to facilitate the establishment of industrial activities. In the internal areas and especially in the South, from the Unification of Italy onward, a growing gap had arisen in terms of connection networks, energy, telephony, etc., i.e., what was called "The Southern Question." The cultural, political, and operational reference was the Keynesian vision, the basis of the New Deal USA, and the experimentation of the *Tennessee Valley Authority*. In 1986, the CASMEZ was replaced by the *Agency for the Promotion and Development of the South* (AgenSud) until 1992 when it was suppressed with the law 488 which came into force in 1993. From that year, the coordination and programming of the intervention in the depressed areas of the national territory passes into the hands of the Ministry of Economy and Finance. In the meantime, since 1986, areas whose development is lagging behind in Europe become the subject of *European cohesion policy*.

Fig. 1 Castelforte (FR), City Hall, painting representing the destruction of World War II. *Source* Aragona (2019)

settlements such as Castelforte (FR), in the valley of the Liri, are sadly exemplary of the almost total destruction to which they have been subjected (Fig. 1).

Since the beginning of the 90 s the internal areas, often defined as "minor," those where "rurality" is more present, undergo a progressive deconstruction first of all because the devastating "cutting of the dry branches of the railways" begins, which significantly worsens the accessibility (Aragona 1993).

Added to this are the privatizations and liberalizations of many services. Services that were people's rights and that were transformed into "products" to buy. The "excuse," the reason for this is that their price dropped thanks to the competition that was created. Thus, more and more the citizen, that is the cum-cives that shares the civitas, is transformed into a consumer. Since supply goes where demand is more dense and with higher spending capacity, the "smaller" areas, the internal ones, mostly rural, have seen many basic services disappear. These choices took place while, at the same time, the state made a drastic decrease in its presence: historical identity places, railway stations, barracks, etc., disappeared along with the closure of pharmacies, post offices, bank branches, and even ATMs. All this has deconstructed, and is continuing to deconstruct, these areas.

A different idea of rural and/or internal areas, we can say an attention to them, has been making its way only for a few years with the *National Strategy for Internal Areas*—SNAI (Lucatelli 2015, 2016) and then the Law n. 158/2017 *Measures for the support and enhancement of small municipalities*, as well as provisions for the redevelopment and recovery of the historic centers of the same municipalities or municipalities under 5000 inhabitants, although the former is being tested only in a few realities and the latter it has a budget of just 150 million Euros.

The National Union of Mountain Municipalities—UNCEM, which groups these small or very small centers, is requiring for essential services to be guaranteed or

Fig. 2 ADSL network in Italy. *Source* AGICOM (2017)

to return to these places. Among the many requests, there is also that linked to the presence of digital connections that allow the use of telematics services, so that this "competitive disadvantage"[6] can be reduced (Fig. 2). It should be noted that until the end of the 70s in Italy, the public companies that created and managed the basic services were obliged to make substantial investments—ca. 40% of the new ones— in areas whose development is lagging behind (Aragona and Pietrobelli 1989). This allowed a progressive rapprochement between the richer and less wealthy areas, bringing electricity, water, and telephony.

André Torre, President of the European Regional Science Association and professor of Agriculture, at the ERSA 2019 Congress, highlighted the huge opportunities that are offered in the rural area in having the presence of networks that allow the use of "online" activities. However, innovation should not be limited to technological aspects and those related to the production process but also which, in large part, involve social ones. That means that is necessary that it concerns innovation in the organization, in the social sphere, and in the institutional one. All this is with a vision, with transdisciplinary strategies and "integrated" policies. With this philosophy, the rural environment can be a relevant actor of the "glocal" (global and local) suggested byRobertson (1995), the unification of global and local expresses a concept that potentially allows the integration and contemporaneity of the different territorial scales. In order to do this, it is necessary to govern the network, the information, and the exchanges, tangible and intangible, in a way that is able to defend the

---

[6]Goddard and Gillespie as early as the mid-1980s called it "Digital devide," highlighting that virtual or material networks were constructed where there was a higher density of demand and this was richer (1986).

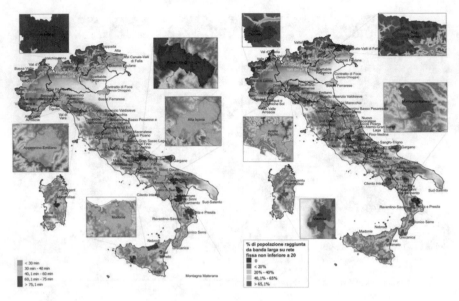

Fig. 3 From left, average distance of the areas, weighted for population, from its reference pole and wide band. *Source* Internal Areas Technical Committee—SNAI (2015)

complexities,[7] i.e., thinking globally but acting locally. But, accessibility to networks is needed.

The *Forum Inequalities and Diversity* in the Charter of the population reached by broadband (2018) shows the strong marginalization that the internal areas have as identified by SNAI: *"In looking at the rural/urban fault, the "distance" can be read with an indicator of physical/material access (average distance of the areas, weighted for the population, from its reference pole... and with an access indicator and intangible connectivity (digital divide...). In the latter case, it is a pre-condition strongly correlated with citizenship services (think of telemedicine, the development of e-learning approaches, the experiences of "remote school" or "online schools") and with the widest access of people and markets (knowledge, openness/attractiveness, production systems)"* (Fig. 3).

The "rural" components, both of areas defined as such and also in urban areas, can be a very useful "tool" to increase the resilience of the territory. Maintaining its presence will increase safety as it means lowering the level of risk as well as improving the level of local and "downstream" well-being. While the abandonment of internal and rural areas increases the fragility of the areas. So with a multicriteria approach, as well as multiscale, both negative and positive externalities must

---

[7]Together with the ease of access to information even in difficult regions, such as Calabria, for some time now, a technical and technological core of knowledge has started to form (see Aragona in the AISRe contributions 2001, 2003, 2004), a literacy process has started which is in fact already telematics, therefore beyond mere computerization, the basis of what Zeleny (1985) calls "higher technology knowledge."

enter the set that political decision makers use to choose strategies. Obviously, this means overcoming the microeconomic philosophy, previously mentioned, based on the (presumed) benefit of the consumer in terms of the purchase prices of the goods by introducing a wider, multidisciplinary dimension with short-, medium-, and long-term assessments. To all this must be added that maintaining bio-diversity means not decreasing the ecological resilience useful to the biosystem understood in a broader sense.

The utility mentioned above is particularly relevant considering the changed climatological conditions: from "water bombs" to periods of particular heat in seasons that are traditionally cold. In fact, much of what has been built, and the places where it has been built, are not prepared, designed to withstand these unexpected events. Therefore, every element, both locational and constructive, which increases its resilience must be valorized.

The elements of "rurality" offer a very significant opportunity to increase local resilience, both material and social resilience. Regarding the first, returning to the "ruralization of the city," the theme of "green" is emerging in an increasingly important way. It, together with the "water" element, can be a great resource of local resilience. Following this logic and taking note of the relevance of the theme, in 2017, the *Committee for the development of public green spaces* of the Ministry of the Environment and the Protection of the Territory and the Sea elaborates the *National Urban Green Strategy. Resilient and heterogeneous urban forests for the health and well-being of citizens* and *Guidelines for the management of urban greenery and first indications for sustainable planning* are the first structural, systemic indications on a territorial scale that go far beyond the law n. 10 of 2013 *Rules for the development of urban green spaces* that made it mandatory for municipalities over 50,000 inhabitants to adopt a "green plant that had a project and management of this, however, the first organic intervention.

With such attention to "rurality," there can also be a great social resilience as shown by the management of abandoned agricultural areas in Rome. In 2014, the municipal administration made a call to entrust the management of them to cooperatives of young farmers who proposed a mainly agricultural and didactic use (Municipality of Rome 2014). It is useful to note that many of the outskirts of the capital, but also of the other large and medium Italian cities, are in the same conditions. So, this method of governing the relationship between "rural" and "urban" can be particularly emblematic.

Alongside both material and social aspects, there are local experiences of which the "urban gardens" are emblematic situations. Even if not defined in this way, already in the late 60s and mid 70s, they appear in small to medium cities. When inhabitants move to new buildings, often economic and popular construction, in regions such as Umbria—strongly characterized by natural and agricultural components—various local administrations to maintain the historical relationship between man and nature, deeply felt by the elderly population, together with the apartments they assign agricultural plots. Over the years, this theme of the plot of land to be managed has taken the name of "urban gardens," as cited significant example is Città di Castello (PG) ca. 36,000 inhabitants, where the mayor says, speaking of the urban

Fig. 4 Città di Castello (PG), geography, territorial morphology, and urban gardens. *Source* Left and middle google map, right City Journal (2018)

gardens dedicated to the memory of Gualtiero Angelini, (Fig. 4) *"... The municipal administration cares a lot about this experience, which allows us to preserve and nurture fundamental social relationships for our community, offer the opportunity to maintain the relationship with nature and preserve the rural culture from which we all come, important elements of the quality of life that we try to promote daily in our city ... Ideal testimony of the benefits of outdoor life in the vegetable gardens of the municipality is Olga Capecci, with her 91 years is the oldest farmer among all pensioners holding a plot of land in the Municipality."* (Redazionale 2018).

The abandoned and uncultivated spaces grow, and many citizens take them into management with the intent to start participatory gardening initiatives. For this reason, several municipal administrations have drawn up "ad hoc" regulations. In the last 5 years, the record of over 1.9 million square meters of land owned by the municipalities for domestic use has been reached, with a growth of 36.4%.[8] Northern Italy is the area where the highest growth of urban gardens is recorded, with Emilia Romagna in the lead: 37% of public gardens cultivated nationally are located here. Lombardy (10.2%), Tuscany (9%), Veneto (8.5%), and Piedmont (7.6%) also fall into the "top 5" of the trendiest regions in urban gardens. Among the regions of Central Italy, the leadership lies with the Marche, with Umbria and Lazio behind them. Campania leads the ranking in the South, followed by Sicily, Sardinia, and Calabria. The passion for the land, the search for greenery, and the desire to recover more natural rhythms in daily life seem to be the reasons behind the boom in urban gardens.

Calabria is in the last places as urban gardens, most likely since the vast majority of its urban centers are very small, between a few tens, or hundreds of inhabitants to a few thousand. There are only 15 cities over 15,000 residents, Reggio Calabria, the most populous city, has approx. 189,000. So in many cases, there is already a close relationship with nature. In the face of the very low current settlement density, however, there is a widespread real-estate heritage. Not considering the illegal and unsafe one, however, there are resources, in terms of considerable potentialities. Urban, historic center, albeit small, and nature was functional to its construction and social recovery: urban waste collection made with donkeys, as in some municipalities of the metropolitan area of Rome, bees as indicators of air quality, repopulation of

---

[8]Coldiretti analysis (2018) conducted on Istat (2017) data.

the historic center and revaluation of ancient crafts, with international tourism. Landscape, therefore, which is built with new ways of anthropizing, of the relationship between rural and urban, conjugation between local identity and modernity.

3 Some Closing Remarks

The topic we are talking about belongs to, is part of, the change in the "industrialist" paradigm that has emerged from the first industrial revolution until the early 1970s, that is, for over 300 years. The unsustainability to continue with this development model emerged thanks to *The Limits to Growth* report, elaborated by the Meadows research group commissioned by Aurelio Peccei, President of the Club of Rome, in 1972. So references, theories, phenomenology, etc., they must be viewed in a historical key, that is, in the long term and not by reference to the short and medium term.

Thus, they are "periodizations" for experimentation and/or verification of steps that are being developed in the construction of different methods of anthropization, that is, of socially and environmentally sustainable scenarios. And of which rural areas, or urban areas with predominant rural features—in the contemporary and/or in the future—are an important part. So, continuing Emanuel's reflections, with a vision very close to that of the *Society of Territorialists*, what is proposed is a different concept of rurality, but also of city, which combine one and the other, "perhaps one within the other one" (Emanuel 1990).

And as the President of the National Union of Mountain Community Communities (UNCEM) said at the Assembly of the National BioArchitecture Institute (2019), the rural areas, the inner areas, may be "land of reserve" since more researches show that in large urban areas the quality of life is worsening. But they are also reserve areas linked to the climatological changes that are making the sea level rise. Thus, progressively, hundreds of millions of people currently living in marine cities will have to migrate to other safe places. However, the cause-effect relationship is not new as Bonardi (2004) recalls in the study of the relationship between the movements of people and climate. That is even more relevant considering that, due to the Coronavirus, there has been a blockage of many work activities and movements which have resulted in a large decrease in pollution. *Factum*, evidence not only in the more industrialized areas as shown by the satellite photographs but also in the natural ones where the fauna and flora have returned in good conditions (Fig. 5).

A positive signal comes from the will of the present government (2020) to demand that almost 40% of new investments be made in areas with a delay in development, thus returning to those policies which, after the Second World War, had managed, even if partially, in diminishing the gap between territories, as previously written. All these consider the territorial, social, economic, and spatial transformations, that is the landscape, as public space (Bonesio 2006).

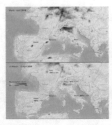

Fig. 5 Europe, level of NO2 in 2019 and during the lockdown, clean water in Venice and Sele river (SA). *Source* From left, ESA—KNMI (2020), © Andrea Pattaro Vision, Massimo Gugliuccello

References

AGCOM (2027) Sistema di mappatura nazionale delle reti di accesso Internet. In https://maps.agc om.it

Aragona S (1993). Infrastrutture di comunicazione, trasformazioni urbane e pianificazione: opzioni di modelli territoriali o scelte di microeconomia? In Proceedings of the XIV conference of Italian regional sciences association Per un nuovo regionalismo. Istituzioni, politiche regionali e locali, modelli di analisi e decisione. Bologna, 6–8 October

Aragona S, Pietrobelli M (1989) Innovazione tecnologica e trasformazioni territoriali. Il caso italiano: politiche, strategie, sviluppi. Pubblicazione del Dipartimento di Tecnica Edilizia e Controllo Ambientale, Facoltà di Ingegneria, Università La Sapienza: Roma.

Augè, M. (1993). Non luoghi. Introduzione a una antropolgia della surmodernità. Milano: elèuthera.

Bonardi L (a cura di) (2004) Che tempo faceva? Variazioni del clima e conseguenze sul popolamento umano. Fonti, metodologie e prospettive. Franco Angeli: Milano

Bonesio L, Università di Pavia, (2006). Il paesaggio come spazio pubblico. Dalle politiche del conflitto al patrimonio condiviso. Relazione presentata al Convegno living landscape: prospettive per una governance democratica del paesaggio. Cuneo.

City Journal. Il tuo quotidiano umbro (2018) Città di Castello, orti urbani dedicati alla memoria di Gualtiero Angelini. https://cityjournal.it/2018/07/citta-di-castello-orti-urbani-dedicati-alla-mem oria-di-gualtiero-angelini/

Coldiretti (2018) Crescono gli orti pubblici (+36%), ecco il decalogo. https://www.coldiretti.it/eco nomia/67599

Comune di Roma (2014). Terre pubbliche, assegnate le prime tre aree. Nasceranno nuove aziende agricole. https://www.comune.roma.it/pcr/it/newsview.page?contentId=NEW748950.

Comunicazione, Ministro per il Sud e la Coesione Territoriale, Sviluppo, Provenzano: Divario nordsud da colmare, ma anche centro e periferie, città e campagne deindustrializzate, aree urbane e aree interne. https://www.ministroperilsud.gov.it/it/comunicazione/notizie/aree-interne-1/

Coscetta P, Emiliani V, Sanfilippo M (2014) Mille borghi Cento città Un Paese: Libro Bianco sull'Italia delle origini. Minerva Edizioni, Bologna

Emanuel C (1990) L'organizzazione reticolare intermetropolitana: alcuni elementi per l'analisi e il progetto. In: Curti F, Diappi L (eds) Gerarchie e Reti di Città. Franco Angeli, Milano

Forum Diseguaglianze e Diversità (2018) Aree interne e il problema delle distanze: le proposte della SNAI. https://www.forumdisuguaglianzediversita.org/aree-interne-distanze-proposte-snai/

Goddard JB, Gillespie AE (1986) Advanced telecommunications and regional economic development. Geogr J (152)

ISTAT—CNEL (2013) Bes 2013 Il Benessere Equo e Sostenibile in Italia. Roma, Tipolitografia CSR

Legge n.10/2013 Norme per lo sviluppo degli spazi verdi urbani

Legge 6 ottobre 2017, n. 158 Misure per il sostegno e la valorizzazione dei piccoli comuni, nonche' disposizioni per la riqualificazione e il recupero dei centri storici dei medesimi comuni. First signatory Ermete Realacci, current honorary Presient of Legambiente.

Lettera Enciclica Laudato Sii del Santo Padre Francesco sulla Cura della Casa Comune (2015.05.24). Tipografia Vaticana: Città del Vaticano

Lucatelli S (2015) La strategia nazionale, il riconoscimento delle aree interne. Milano, Franco Angeli

Lucatelli S (2016) Strategia Nazionale per le Aree Interne: un punto a due anni dal lancio della Strategia. Agriregionieuropa, anno 12, n.45, Giugno. https://agriregionieuropa.univpm.it/it/content/article/31/45/strategia-nazionale-le-aree-interne-un-punto-due-anni-dal-lancio-della

Meadows HD et al (1972a) I limiti dello sviluppo. Milano, Mondadori

Meadows DL et al (1972b) The limits to growth. Universe Books, New York

Ministero dell'ambiente e della tutela del territorio e del mare, Comitato per lo sviluppo del verde pubblico (2017) Strategia nazionale del verde urbano. Foreste urbane resilienti ed eterogenee per la salute e il benessere dei cittadini. https://www.minambiente.it/sites/default/files/archivio/allegati/comitato%20verde%20pubblico/strategia_verde_urbano.pdf.

Ministero dell'ambiente e della tutela del territorio e del mare, Comitato per lo sviluppo del verde pubblico (2017) Linee guida per la gestione del verde urbano e prime indicazioni per una pianificazione sostenibile. https://www.minambiente.it/sites/default/files/archivio/allegati/comitato%20verde%20pubblico/lineeguida_finale_25_maggio_17.pdf

Raffestin C (1987) Repers pour une theorie de la territorialite' humaine. In Cahier n. 7, Groupe Reseaux: Parigi

Redazionale (2018) Città di Castello, gli orti urbani dedicati alla memoria di Gualtiero Angelini. https://www.umbriadomani.it/umbria-in-pillole/citta-di-castello-gli-orti-urbani-dedicati-alla-memoria-di-gualtiero-angelini-204655/

Robertson R (1995) Globalization: social theory and global culture. Sage, Newcastle upon Tyne, United Kingdom

UE (2000). Convenzione Europea del Paesaggio, Firenze. https://www.beap.beniculturali.it/opencms/export/BASAE/index.html

UE (2007) Carta di Lipsia sulle Città Europee Sostenibili. https://www.sinanet.isprambiente.it/gelso/files/leipzig-charter-it.pdf

UN (2015) The sustainable development agenda.17 goals to transform our world. https://www.un.org/sustainabledevelopment/development-agenda/

Scandurra E (1995) L'ambiente dell'uomo. Verso il progetto della città sostenibile. Milano, Etas Libri

The European Space Agency—ESA, Royal Netherlands Meteorological Institute—KNMI (2020). Coronavirus lockdown leading to drop in pollution across Europe. https://www.esa.int/Applications/Observing_the_Earth/Copernicus/Sentinel-5P/Coronavirus_lockdown_leading_to_drop_in_pollution_across_Europe

Torre A (2019) PIRs keynote Is there a smart development for rural areas? In: 59th ERSA Congress cities, regions and digital transformations, opportunities, risks and challenges, Lyon, 27–30 August

Zeleny M (1985) La Gestione a Tecnologia Superiore e la Gestione della Tecnologia Superiore. In: Bocchi G, Ceruti M (eds) La Sfida della complessità. Mondadori, Milano

The Impact of Action Planning on the Development of Peripheral Rural Villages: An Empirical Analysis of Rural Construction in Yanhe Village, China

Qiuyin Xu and Tianjie Zhang

Abstract As a central factor for building, smart and sustainable regions innovation requires the effective combination of local knowledge with expert knowledge, as well as the support of an extensive network. However, due to geographic distance, less agglomeration, and insufficient capacity, rural regions have fewer opportunities to actively communicate with the outside world, which limits their ability to learn and absorb external knowledge. Quite a lot of spatial planning has greatly improved ICT infrastructure in rural regions, but cannot guarantee the improvement of the knowledge level and innovation ability there. How can we effectively enhance knowledge learning and innovation networks in the construction of rural regions? This paper takes the rural construction of Yanhe Village in Hubei Province, China as a case, and discusses the issues mentioned above. Located in the southwest mountainous area of Hubei Province, Yanhe Village is relatively remote. It faces many problems, such as serious environmental pollution, economic backwardness. Since 2003, with the help of Green Cross NGO, the village committee led the villagers to carry out a large number of construction projects. The environment and economy of the village have been rapidly improved. More significantly, after the withdrawal of the NGO, the economy of Yanhe Village has maintained steady growth and the environment has been well maintained. Via literature research, fieldwork, and in-depth interviews, this research finds that besides the infrastructure improvement, the curriculum training conducted by Green Cross NGO considerably helps to enhance knowledge learning. In this process, local elites were effectively driven by external knowledge disseminators and became local knowledge disseminators. The seminars and collective construction actions carried out by the village collective itself further digested the new knowledge. As the initiative of village committee members, township and village enterprises and migrant workers grow, and the formal and informal relationships between a multitude of actors became richer. This article emphasizes that, besides ICT infrastructure, it is very important to provide training and guidance to villagers, especially the localelites. Moreover, diverse subjects should be encouraged to participate in rural construction. The resulting innovative environment makes the introduction and implementation of innovations possible.

Q. Xu (✉) · T. Zhang
School of Architecture, Tianjin University, No. 92, Weijin Road, Nankai District, Tianjin, China
e-mail: xu_qiuyin@163.com

© The Author(s), under exclusive license to Springer Nature Switzerland AG 2021   341
A. Bisello et al. (eds.), *Smart and Sustainable Planning for Cities and Regions*,
Green Energy and Technology, https://doi.org/10.1007/978-3-030-57332-4_24

Keywords Social practice · Smart region · Spatial planning · Socio-ecological transformation · Citizen participation · Local-based initiative approaches

1 Introduction

In the process of rapid urban development, rural planning is an important means of promoting rural governance. At present, most of the planning is based on spatial planning and economic benefits (Meng et al. 2015; Qiao and Hong 2017). Quite a lot of spatial planning has greatly improved the ICT infrastructure in rural regions, but it cannot guarantee the improvement of knowledge level and innovation ability there. Some studies show that the lack of social capital is an important factor restricting the development of rural society. Through the case study of Yanhe village, this paper analyzes how to cultivate social capital in the planning implementation to effectively strengthen the local knowledge learning and innovation network.

1.1 Cultivation of Social Capital as the Main Point of Rural Planning

Good rural governance depends on abundant social capital. Social capital takes "trust", "normative", "relationship network" as core concepts and includes the cooperation and mutual benefit formed by individuals and organizations within rural communities and internal or external objects in long-term interaction, and the historical traditions, mechanism concepts, beliefs, and behavioral paradigms accumulated behind the relationships. Social capital replaces and supplements the formal regulations to a certain extent, which can provide common capital support for group members (see Table 1), and has a positive relationship with villagers' health, rural economy, and social development (Pronyk et al. 2008).

Table 1 Role of social capital in rural development (Wilson 1997)

Social capital	Main content	Function
Trust	Intersubjective expectations	Lubricant for group operation, improve work efficiency
Reciprocity norm	Customs ethics system	Restricting members of society and helping people consciously abide by laws and disciplines
Relationship network	Vertical relational network Horizontal relational network	Link unequal or equal actors strengthen social cooperation and resolve social contradictions and differences internally

The lack of rural social capital is an important obstacle currently restricting rural construction. In traditional Chinese villages, social capital is reflected in the clan-based organizational structure: On the one hand, rural society mobilizes and uses resources through a set of strict institutional norms; on the other hand, the state can effectively and efficiently with minimal institutional costs manage rural society. In the 1950s and 1960s, the state comprehensively controlled the villages, causing rural social capital to be largely damaged and weakened. After the reform and opening up, great changes have taken place in the family system, the state governance model, and the rural industrial economy. The acquaintance society has become a half-person society, and the basis for generating the original rural social capital no longer exists. Moreover, traditional rural social capital is more manifested as "relation-ships" characterized by special trust, interpersonal trust, and mechanical solidarity, which is biased toward self-interest and cannot meet the needs of the development and development of rural modernization. New ways need to be explored to cultivate modern social capital characterized by universal trust, institutional trust, and organic solidarity.

Therefore, rural construction and social capital reconstruction should comple-ment each other and promote each other. As an important means to promote rural construction, rural planning not only pays attention to the construction of the phys-ical environment, but also to changes in the logic of rural resource allocation (Hong and Qiao 2016), breaks through the traditional compilation method, strengthens trust, strengthens the governance network, and improves policy specifications in construction.

1.2 Impact of Action Planning on Social Capital

Action planning is defined as a planning process that deals with issues and decision making at the local level. It emphasizes that planning and implementation should be considered at the same time, and the pursuit of "dynamic goals" rather than "ulti-mate blueprints," which is pragmatic, interactive, coordinated, flexible, and operable (Wang 2005).

Action planning can change the external environment of the village and promote the generation of social capital. Generally speaking, the generation of social capital has two methods of internal cultivation and external promotion. The social foundation for the production of rural social capital in China has been broken. In the face of the diminishing and severely inadequate social capital, the input of "external power" is urgently needed (Wei et al. 2016). Cultivating public spirit, strengthening villagers' moral construction, legal system construction, and developing non-governmental organizations are considered as alternate paths to cultivate rural social capital. In the preparation of the plan, scholars further put forward strategic suggestions, such as "increasing technological trust, strengthening governance networks, and improving policy specifications". Compared with "blueprint planning", rural construction under the concept of action planning places more emphasis on demand orientation, operability, and public participation. In the construction action, planners fully interact with the government and villagers to: tap "local knowledge" and enhance technical trust; coordinate and organize activities between different stakeholders to build a "life community" and strengthen the governance network; and provide diverse paths for participation to form policy norms and cultivate "public spirit".

The formulation and implementation of action plannings mainly include four links:

(1) Sort out the development intent. Combining the current status of resources, government control needs, and public development demands, it proposes reasonable development directions and formulates detailed targets;
(2) Planning for recent construction. The overall arrangement of the construction sequence will usually determine a series of catalyst projects to promote the overall project;
(3) Formulate action planning. Choose the appropriate action mode;
(4) Assess the effectiveness of implementation. In the implementation, dynamic benefit evaluation of resources, space, economy, society, environment. And culture is also needed.

2   Method

This article is an empirical study, and the selected case is Yanhe Village, which is located in the mountainous area in the southwest of Hubei Province (Fig. 1). Affected by the urban development of the Han River Basin, the ecological environment in the area has deteriorated sharply, and it faces environmental pollution and backward industries. In December 2003, Green Cross began a survey of Yanhe Village and found that its main problems are:

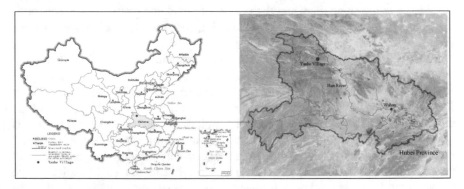

Fig. 1 Location of Yanhe Village. *Source* Draw by the author

(1) The environment is polluted seriously. All the trees on the mountain were cut down, causing serious soil erosion; lack of infrastructure such as garbage collection and sewage.
(2) The development of the industry is lagging. The villagers are mainly agricultural producers, and their per capita annual income is only 2000 yuan. Many villagers have to go out to work.
(3) The decline of rural civilization. The villagers mostly enjoy playing mahjong, the quality of school education is poor, and the overall spiritual life is empty.
(4) Rural governance is disorderly. Ordinary villagers and village committees are in a "split" state; supporting regulations are lacking; and the villagers' concepts of democracy and autonomy are weak.

Since August 2003, with the help of non-governmental organizations such as the Green Cross,[1] Yanhe Village has used action planning methods to strengthen multilateral cooperation and citizen participation. Village committees and villagers acquired much new knowledge in this process and began to actively carry out innovative activities to make their homes better. We think that this is a successful case, and analyzing and summarizing its experience can enrich our theory of improving welfare in rural areas.

The main methods used in this article are historical data collection, actual research, and interviews. First, we collected relevant materials of Yanhe Village from 2003 to the present from the Internet and related books. This mainly includes historical photos, economic development data, population changes, etc., so that we can quantify the changes that have occurred in Yanhe Village after implementing the action planning. Second, we conducted an on-site survey of the village, focusing on its ecological environment, industry and culture, where Yanhe Village performed poorly before. Third, narrative interviews were conducted with the villagers to understand their gains from participating in the construction process. A total of ten villagers were interviewed, including village committees, village capable people, and ordinary villagers. We conducted an in-depth qualitative interview with Green Cross, an important participant in this project. We learned about the strategies, methods, and difficulties they used in planning. At the same time, we discussed the changes in village names and their knowledge learning and innovation networks.

3 Results and Findings

Although the overall situation of Yanhe Village is relatively backward, there are still some resources that can be tapped and used: (1) There are few mountains, and

---

[1]Green Cross: Beijing Green Cross was founded in 2003 and registered with the Civil Affairs Bureau of Yanqing County in Beijing in June 2004 as a formal non-governmental organization. With environmental protection as its starting point, it has carried out multiple models of new rural construction for more than a decade and has now grown into a more professional non-governmental organization for rural construction in China.

the climate is suitable for the growth of tea; (2) the villagers are simple and kind and eager for knowledge. The Green Cross believes that with the support and help of the two village committees, rural construction can be partly successful. Starting from the village's tea resources, an ecological circle between life, production, and ecology is established, which is easily accepted by villagers, supported by higher-level governments, and has incentives for tea merchants.

3.1 The Main Contents and Methods of Yanhe Village Action Planning

The construction of Yanhe Village uses the method of action planning to learn, gradually advance, and continuously adjust. The construction mainly includes three processes: Start with the easiest and most effective environmental improvement, initiate villager actions and improve the village's appearance; second, use local natural resources to develop the industrial economy and improve the quality of people's lives; and third, then carry out cultural activities around folk characteristics to improve local identity. The action planning involves multiple interests (Fig. 2). Village committees and villagers have made great contributions to village construction. NGOs have promoted the improvement of the village environment, rural cooperatives have promoted economic development, and local enterprises have further accelerated village development (Wang et al. 2017).

The original action planning of Yanhe Village revolved around renovating the home environment, and the methods implemented mainly included seminars and training. When the plan was formulated, there were no clear goals, and the determination of the goals depended on the consensus reached between residents and organizations (Sun and Wang 2006). Depending on the task, the seminar has different sizes

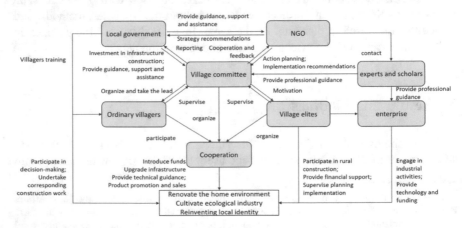

Fig. 2 Main participants of Yanhe Village action planning. *Source* Draw by the author

Fig. 3 Comparison of village environment before and after implementing garbage recycling. *Source* Provided by Green Cross

and discussion topics. The village committee accompanied the Green Cross to carry out in-depth visits to the villagers from house to house, to understand the villagers' psychological thoughts and neighborhood relations. After comprehensively considering the needs of villagers and the effect of implementation, Yanhe decided to adopt the construction mode of "living first and then producing," and started construction with garbage recycling as the catalyst (Fig. 3). The villagers also participated in setting a more detailed goal. To this end, the village's public hall is also built using the village's public space for assembly and exchange of opinions on village construction. In these processes, consultations go further than the informative public participation that was previously shown or distributed through planning exhibition halls (Jiang and Chen 2011). In addition, Green Cross has provided training courses for the village to popularize the knowledge of waste separation and recycling among the villagers and increase the villagers' own construction capacity. They encourage each household to use its own barrels, bags, etc., to prepare recycling bins for dry, wet, and hazardous waste for waste sorting; organize village volunteer teams to participate in voluntary labor for environmental remediation; school children will be transformed into envoys for ecological protection to help promote environmental protection knowledge and supervise the classification of household waste. Individual villagers are connected in this process, and their sense of collectiveness is strengthened.

To cultivation ecological industry, village cadres take the lead, while labor, crowdfunding, and the establishment of industrial cooperatives and other methods play a vital role in promoting. The beautiful ecological environment and tea gardens have greatly increased the appeal of the village, and many foreigners have come to see it since 2005. Village party members and cadres took the lead in trying to run farmhouses, and after that they passed on the experience to the villagers. Villagers learn the three-dimensional planting technology of tea gardens, as they optimize their own business, start farmhouses, and engage in folk skills such as paper cutting, which has enriched the service content of rural tourism. Yanhe Village gradually formed a "tea garden–farmhouse–hotel" tourism industry chain, but the government's finances are not enough to build the infrastructure. The village has built the village's main roads and riverbanks by means of labor planning, co-construction, and sharing (Fig. 4). In October 2007, after two years of planning, Yanhe Village officially established an eco-tourism professional cooperative. Cooperatives are responsible for introducing

Fig. 4 Tea Garden in Yanhe Village; villagers paving roads. *Source* Provided by Green Cross

funds, guiding villagers in tourism construction, investing in upgrading village facilities, and promoting local products. At present, more than 180 households in the cooperative have taken shares, and the dividend rate has reached 50%, which has directly led to the employment of more than 100 people.

The "ceremonial feeling" is the emphasis of the Green Cross in reshaping local culture. To enhance local cultural characteristics, the Green Cross not only helped design a series of rural buildings (Fig. 5), but also planned and organized a variety of cultural activities. In March 2004, the artisans in the village together built a tea altar, took the soil from the top of Wushan and the water at the foot of the five mountains, taking the soil at the right time and opening the altar at nine o'clock. Everyone worked hard and paid respect. Important construction details including time, people, assembly, and transportation were recorded (Fig. 6). On the opening day of the tea altar, the locals elected two of the most highly regarded tea people to unveil the tea altar. The construction of the tea altar is like establishing a belief among the villagers, setting a moral benchmark for loving the collective, self-discipline, peace of mind, and mutual dedication.

Fig. 5 Use of local stone and wood in the design of residential buildings. *Source* Provided by Green Cross

Fig. 6 Video recording, water extraction at the foot of the mountain, and the results during the construction of the tea altar. *Source* Provided by Green Cross

The process of action planning, which includes preparation, discussion, and construction, is not one-way. Many times, there are still various "emergency" problems in the follow-up tasks. Action participants need to rethink the planning concept and action design and then adjust the action content. For example, during the construction of the tea altar, the idea of building a structure was originally proposed by the village committee. After the survey, Green Cross put forward the construction idea and a specific design scheme for the tea altar by using the terrain, environment, and local materials. However, in the construction, it is difficult to unify the design and construction opinions, and we can only discuss them together and revise the tea altar design draft again. Each revision will cause quarrels between investors, tea farm teachers, and construction teams. It is also in these quarrels that everyone interacts with each other and weighs each other's interests.

3.2 The Cultivation of Rural Social Capital in Action Planning

The Yanhe Village Action planning promotes interaction among diverse subjects through discussions, training, and crowdfunding, and it cultivates social capital centered on trust, reciprocity, and relationships.

3.2.1 Promote Social Trust with "Effectiveness" and "Locality"

In the action planning, the trust among the main participants has increased significantly. The village committee has certain authority among the villagers, and has the characteristics of being in town, leading the crowd, and spirit. However, in the past, ordinary villagers and village committees formed a "double-layered structure" and were in a "split" state. "The cadres should be shown to the masses and carried by the masses" is the slogan of Yanhe Village cadres, emphasizing that cadres should lead by example. In addition to the role of organizer, the village committee members are also solid participants in waste sorting and road construction in the village. "Party members take the lead, demonstrations lead." For some new construction actions and development ideas, the villagers are afraid of risks and dare not try. Party members take the lead in practice. Only when it is effective, it can be persuasive. When the follow-up villagers encounter difficulties in the construction process, they will ask the village committee for help, and the village committee will use its own social resources to solve the difficulties for everyone.

With the participation of village committees and repeated visits, exchanges, and promotion by the Green Cross, villagers' doubts about the work of the Green Cross gradually lessened and participation increased greatly. At the same time, the Green Cross team respects the particularity of rural society in the design, adds local knowledge and social context, taps traditional rural culture and institutional resources, and embeds "local knowledge." Villagers have shown greater acceptance and increased villagers' trust in planning technology.

3.2.2   Forming Norms Through "Virtues" and "Township Treaties"

"Village-style postmodernism" believes that the rural world should reflect community and moral and cultural values, and the Green Cross and the village committees also recognize that "belief is a solution to eternal problems." In the rural construction of Yanhe Village, there are several key strategies:

- Strengthen traditional social virtues, moral education should be carried out, and the village morality should be formed.
- Emphasize that the village committee should lead by example, set an example, and drive everyone to follow suit.
- Promote tea culture in schools, teach tea arts, and teach environmental protection courses, making residents feel the Chinese traditional etiquette.
- Carry out the selection and publicity of "Top Ten Residents" and "Civilized Models" to inspire advanced and promote righteousness.
- Respect villagers' suggestions and improve the township rules and regulations.

As a contractual norm, the village covenant is a stabilizer and regulator of social relations. Starting from the environmental improvement, Yanhe Village has formulated village regulations for both production and living, providing a model for everyone to follow. At the beginning of rural construction, the village committee convened a meeting of the villagers, discussed and adopted the village covenant, and formed a consensus on caring for the environment and helping neighbors. The also improved the supervision mechanism, giving equal weight to rewards and penalties. Under the premise of the convention, a set of governance reward and punishment mechanisms has been established to ensure the effective implementation of actions. In the village, the health rating of the farmhouse is held once a month. From one star to five stars, each has its own standard, from reward to punishment. It is very strict. In rural construction, a number of organizational carriers, such as rural economic cooperatives, are also cultivated and developed, and each household is organized to cooperate in production cooperation.

3.2.3 Leverage "Foreign Aid" and "Interaction" to Strengthen the Relationship Network

The addition of the Green Cross has brought high-quality social resources to the weir, and the relationship network has become more complex and richer, providing new development concepts, models, and opportunities. The training course introduces advanced experience to villagers and enterprises and broadens their horizons. The tea farm owner You Bangli broadened his horizons and strengthened his confidence in education. He invested 100,000 yuan to establish a tea wholesale market in the township. Subsequently, more than 400 tea brokers appeared in the local area to promote high-quality tea to the outside world and promoted the development of tea industrialization. Under the introduction of the Green Cross, the village committee of Yanhe Village met more technical experts and obtained many valuable suggestions. For example, in the sewage-treatment problem, through the introduction of Green Cross, the village committee became acquainted with the experts of the environmental protection bureau. Experts advised the installation of a water-recovery device, so that the water can be subjected to pollution-free treatment.

The interaction between family and village constitutes an independent and interdependent organic life. Villagers from different social circles, different interest subjects, and different governance bodies are organized to carry out collective activities, promote internal exchanges and contacts among villagers, and increase trust among farmers. The village construction did not have sufficient financial support. The village committee proposed the method of "planning labor and labor co-construction and sharing," and adopting the method of mutual replacement during the "one construction and four reforms" process. Improved communication opportunities now exist between neighbors.

4 Conclusion and Discussion

In the context of rural space resource allocation, changing the logic from "relationship" to "social capital" and "blueprint planning" faces many difficulties in implementation, and it is necessary to adopt action planning to actively cultivate social capital. The village revitalization of Yanhe Village did not rely on government funds and special resource advantages. It was led by the village committee. The Green Cross provided advice and guidance. Villagers and enterprises participated extensively and started a series of construction operations. They started with a trivial matter related to villagers 'lives, waste sorting, mobilizing extensive villagers' participation so as to realize the first step in rural construction. Relying on local resources, a three-dimensional and comprehensive tea plantation model was established, and tourism development, handicraft production, and export of agricultural products are introduced to enrich industrial forms, increase villagers' income, and realize the second step of rural development. The third step of development was to tap the local culture, launch diverse activities, shape local identity, and enrich the spiritual world of the villagers. During the operation, interviews, seminars, trainings, and public-welfare organizations were organized to fully mobilize the enthusiasm of participants, enhance social trust, create a norm, and strengthen the network of relationships, effectively cultivate social capital, and achieve rural revitalization and sustainable development.

Action planning is not a simple policy formulation. It requires the planners' patient guidance and the full cooperation of the village committee. On the one hand, planners should conduct two-way interactions, learn from the countryside, and understand the complex relationships of geographic resources, cultural history, and blood interests and rights. Planners also need to play multiple roles such as organizer and coordinator to provide diversified paths for villagers to participate. In the process, the practice–reflection–action has become a catalyst for social capital (Wilson 1997). On the other hand, with a lack of coordination and support at the organizational level in the action, decentralized solutions cannot fully implement the construction content (Melish 2014). In the 1990s, community action planning in the United States lacked actual control over the community, and community organizations tended to target local governments, resulting in fragmented cities. Ensuring the dominant position of village committees in rural construction and cultivating their leadership is the key to rural construction and sustainable development.

Acknowledgements This research was sponsored by the National Natural Science Foundation of China (Nos. 51778403, 51778406) and the Discipline Innovation and Attraction Program of Universities (No. B13011).

References

Hong L, Qiao J (2016) 规划视角下乡村认知的逻辑与框架 [Logic and framework of rural cognition from the perspective of planning]. 城市发展研究 23(1):4–2

Jiang L, Chen J (2011) 作为行动过程的社区规划:目标与方法 [Community planning as a process of social action: goals and methods]. 城市发展研究 18(6):13–17

Melish TJ (2014) Maximum feasible participation of the poor: new governance, new accountability, and a 21st century war on the sources of poverty. Soc Sci Electron Publishing 13:1–125

Meng Y, Dai S, Wen X (2015) 当前我国我国乡村规划实践面临的问题与对策 [Problems and countermeasures of rural planning practices in china]. 规划师 2(31):143–147

Pronyk PM, Harpham T, Busza J, Phetla G, Morison LA, Hargreaves JR, Kim JC, Watts CH, Porter JD (2008) Can social capital be intentionally generated? A randomized trial from rural south Africa. Soc Sci Med 67(10):1559–1570

Qiao J, Hong L (2017) 从"关系"到"社会资本":论我国乡村规划的理论困境与出路 [From "relation" to "social capital": on the theoretical dilemma and outlet of China's rural planning]. 城市规划学刊 236(4):81–89

Sun J, Wang F (2006) 五山模式——一个建设社会主义新农村的典型标本 [The five mountains model: a typical specimen for the construction of a new socialist countryside]. People's Publishing House, Beijing

Wang H (2005) 引入行动规划 改进规划实施效果 [Introduce action plans to improve the effectiveness of plan implementation]. 城市规划 29(4):41–46

Wang X, Wang X, Wu J, Zhao G (2017) Social network analysis of actors in rural development: A case study of Yanhe Village, Hubei Province China. Growth Change 48(4):869–882

Wei C, Wei L, Deng H, Wei L (2016) 社会资本视角下的乡村规划与宜居建设 [Rural planning and livable construction from the perspective of social capital]. 规划师 32(245):124–130

Wilson PA (1997) Building social capital: a learning agenda for the twenty-first century. Urban Stud 5(34):745–760

Sustainability of Cultural Diversity and the Failure of Cohesion Policy in the EU: The Case of Szeklerland

Attila Dabis

Abstract The study focuses on the issue of preserving and maintaining the cultural and linguistic diversity of the EU and presents a possible way to utilize the tools of regional development planning for such purposes. The issue is of relevance for many reasons. Most importantly, the founding Treaties of the EU contain obligations of the Union to uphold its linguistic and cultural diversity. Additionally, maintaining this diversity is a crucial part of the human rights-based approach to sustainable development and as such forms an integral part of the implementation of the Sustainable Development Goals of the United Nations. The study first discusses why linguistic and cultural diversity is important from the point of view of sustainable development. It then presents the case study of Szeklerland in Romania as an illustrative example of how such diversity is under threat and concludes with a discussion on a possible way the EU can foster the preservation of its rich linguistic diversity through its regional development policy planning.

Keywords Cultural diversity · Social cohesion · Regional development policy · Human rights · Sustainable development · EU

1 Introduction

One of the main differences between the Millennium Development Goals (UN 2000) and the Sustainable Development Goals (SDGs) (UN 2015a) is that the latter exhibits an increased recognition of the fact that human rights are essential to achieve sustainable development.[1] Human rights principles and standards are thus strongly reflected in the 2030 Agenda for Sustainable Development (UN 2015b), as the precondition of the implementation of the Goals. This human rights-based approach to sustainable development informs us about the type of individuals and communities we as human beings need in order to tackle the challenges of climate change. It is about the capacity

---

[1] For a study on this evolution see e.g. (Nanda 2016: 398).

A. Dabis (✉)
Institute for the Protection of Minority Rights, Vármegye str. 7, Budapest 1052, Hungary
e-mail: dabis.attila@kji.hu

© The Author(s), under exclusive license to Springer Nature Switzerland AG 2021
A. Bisello et al. (eds.), *Smart and Sustainable Planning for Cities and Regions*,
Green Energy and Technology, https://doi.org/10.1007/978-3-030-57332-4_25

of human beings to transform laws, practices, public policies and social norms around them, as well as about empowering people with voice and agency to be active participants in designing their own futures and shaping sustainable-development solutions. The promotion and protection of human rights is crucial in empowering people to stand up for themselves and the communities they live in, as a means to accelerate the deep and transformative progress that is necessary in both economic, social and political practices for more sustainable development. This is why humanity needs: effective and accountable public institutions (SDG nr. 16); inclusive, resilient and safe societies, and human settlements that leave no one behind (SDG nr. 11); as well as education systems that provide inclusive and quality education for all, and at the same time help to preserve cultural and linguistic diversity (SDG nr. 04).[2] This latter finding follows the basic understanding that all cultures and civilizations of the globe can contribute to, and are crucial enablers of, sustainable development. Consequently, loss of linguistic diversity is not only a loss for humanity's heritage, but also a loss in human potential, and intellectual capacity that could be channeled to find viable solutions to the conundrum of climate change. Therefore, accommodating linguistic diversity is a hallmark of an inclusive society, and one of the keys to countering intolerance, discrimination and racism that undermine the sustainability of cultural and linguistic diversity.

In view of these considerations, one can state that cultural and linguistic diversity matters from the point of view of sustainable development. It is therefore worthy of scientific scrutiny how this diversity is diminishing due to various public-policy practices. To gauge such processes in relation to the cohesion policy of the EU, I will start by briefly introducing the evaluation system used to assess the spending of cohesion fund resources by the European Commission. I will then outline the "Szekler example", demonstrating the shortcomings of this evaluation system in assessing the overall impact of the allocation of development funds. Finally, the study will summarize the arguments of a grassroots initiative that was elaborated by the stakeholders affected by the said funding-allocation practices of Romania, with a view to presenting specific proposals to tackle the outlined complex socioeconomic challenge.

The results of the analysis will show that there is still room for improvement in terms of how the EU can implement her obligations pertaining to the maintenance of cultural and linguistic diversity and also in the way that EU institutions can monitor the spending of development funds on the Member-State level, to assure compliance with provisions of the founding Treaties, as well as to contribute to the realization of the Sustainable Development Goals. The subject matter is all the more topical given the ongoing debate whether EU funding should be utilized as an instrument to

---

[2]The goal to ensure inclusiveness in education for all is intrinsically connected with respect for all forms of diversity, including cultural and linguistic diversity: "Inclusive societies recognize and build development policies around the diversity of their members and enable everyone's full inclusion and participation, regardless of their status". See: OHCHR (2015: 4).

enforce the fundamental values of the Union, including respect for minority rights, by making it conditional for Member States that violate these values.[3]

2 Method

As far as the scientific literature is concerned, much more emphasis has been put on culture's consequences for economic development than examining the effects development programs might have on regional cultures.[4] This latter aspect lies in the core of this study, which belongs to the field of human rights research. As such, the paper assembles evidence from official documents, as well as findings of scientific literature, as a means to identify a systemic deficiency in EU development policy, which hinders the effective implementation of policy objectives, as well as the realization of the Sustainable Development Goals. Accordingly, this paper takes SDGs, as well as provisions of the founding Treaties of the EU and juxtaposes them with the case study of Szeklerland in Romania, to demonstrate one of the shortcomings of development policy planning in the EU. The study argues that due to a lack of targeted monitoring procedures the EU is not equipped to detect if one of its Member States uses the sources of development funds in a discriminatory manner, to the detriment of areas inhabited by national minorities. While similar phenomena can be seen, among others, in Greece or Slovakia, the case of Szeklerland provides a vivid example of how the distribution of EU funding for regional development can be used for exactly the opposite of the originally intended purpose.

To underpin my arguments, I will use: the findings of relevant scientific literature; the monitoring and evaluation guidelines of the European Commission; materials of the UN on the human rights-based approach to sustainable development; as well as specific legal documents of a lawsuit before the Court of Justice of the European Union, through which the mentioned grievance of Szeklerland was put forward by representatives of the Szekler community. Throughout the study, the economic indicator used to describe disparities in regional development, as well as to demonstrate the economic impact of the allocation of development funds, will be the Gross Domestic Product/capita measurement of the relevant counties of Romania, expressed in current price USD in terms of purchasing power parity (PPP).

---

[3] See e.g.: Ray (2019), Blauberger and van Hüllen (2020).
[4] See e.g.: Hofstede et al. (2010), Mindaugas and Matuzevičiūtė (2016), Fratesi and Wishlade (2017).

3 Diversity Under Threat

UNESCO's Atlas of the World's Languages in Danger (Moseley 2010) lists some 2,500 endangered languages worldwide.[5] More than 5% of these are located in Europe (see Fig. 1). UNESCO uses a six-tier system to categorize languages into: safe,[6] vulnerable (such as Basque),[7] definitely endangered (e.g., Scottish Gaelic or Irish),[8] severely endangered (e.g., Breton or Csángó Hungarian),[9] critically endangered (e.g., Cappadocian Greek),[10] all the way down to extinct languages, such as Dalmatian in Croatia or Akkala Saami (the last speaker of which died in 2003).

The vast majority of these languages are either definitely or severely endangered, counting altogether more than 100 endangered languages in the EU.[11] While the legal framework to protect these peculiar cultures is quite developed in the EU relative to other parts of the world, these numbers show that Member States and institutions of the Union have well-founded reasons to step up their efforts to prevent further erosion of cultural diversity. The least one could expect from all relevant actors is to refrain from practices that directly or indirectly eradicate cultural and linguistic diversity and to avoid EU money being utilized for such purposes. But is the current evaluation system of the EU ripe for detecting such practices?

Subsequent changes in the EU's cohesion policy system shows that decision makers were more preoccupied with addressing "macro-issues" (like balancing between diverging policy objectives of Member States, or choosing between methods of evaluation), rather than addressing "micro-issues", like the question of how the implementation of this policy impacts regional cultures.[12] Within the currently existing framework, the European Commission scrutinizes the use of cohesion policy

---

[5] For the interactive, online version of the Atlas visit: https://www.unesco.org/languages-atlas/. Accessed April 17, 2020.

[6] Meaning that the language is spoken by all generations and that intergenerational transmission is. uninterrupted.

[7] Most children speak the language, but it may be restricted to certain domains (e.g., the home).

[8] Children no longer learn the language as their mother tongue in their home environment.

[9] The language is spoken by grandparents and older generations, and, while the parent generation may understand it, they do not speak it to children or among themselves.

[10] The youngest speakers are grandparents and older, and they speak the language partially and infrequently.

[11] Acknowledging the urgency of the topic, the General Assembly of the United Nations declared 2019 as the International Year of Indigenous Languages with resolution No. 71/178. Furthermore, the Social, Humanitarian and Cultural Committee (Third Committee) of the General Assembly of the United Nations (UN-GA) approved a draft resolution on November 6, 2019 (at the 44th meeting of the 74th session of the UN-GA) on the rights of indigenous peoples. This stipulates among others that the UN proclaims the period 2022–2032 as the International Decade of Indigenous Languages to draw attention to the critical loss of indigenous languages and the urgent need to take steps to preserve, revitalize and promote such languages. Notwithstanding discrepancies between the terms "indigenous peoples" and "national minorities", this UN document, among others, shows the increased awareness of the international community that we live in a pivotal time for language revitalization.

[12] See e.g.: Bachtler and Wren (2006), Grazi (2012), Ferry (2013).

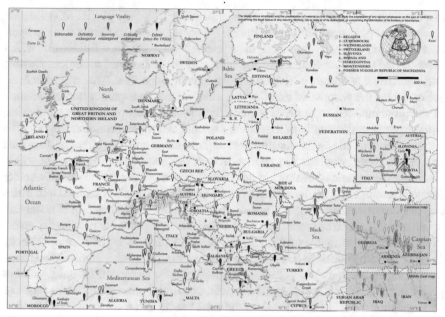

Europe

Alderney French (GBR)
Alemannic (AUT; CHE; DEU; FRA; ITA; LIE)
Algherese Catalan (ITA)
Alpine Provençal (FRA; ITA)
Aragonese (ESP)
Arbanasi (HRV)
Arbëresh (ITA)
Aromanian (ALB; BGR; GRC; MKD; SRB)
Arvanitika (GRC)
Asturian-Leonese (ESP; PRT)
Auvergnat (FRA)
Banat Bulgarian (ROU; SRB)
Basque (ESP; FRA)
Bavarian (AUT; CHE; CZE; DEU; HUN; ITA)
Belarusian (BRB; LTU; LVA; POL; RUS; UKR)
Breton (FRA)
Burgenland Croatian (AUT; HUN; SVK)
Burgundian (FRA)
Campidanese (ITA)
Cappadocian Greek (GRC)
Champenois (BEL; FRA)
Chuvash (RUS)
Cimbrian (ITA)
Corfiot Italkian (GRC)
Cornish* (GBR)
Corsican (FRA; ITA)
Crimean Tatar (2) (BGR; ROU; UKR)
Crimean Turkish (UKR)
Csángó Hungarian (ROU)
Dalecarlian (SWE)

Dalmatian (HRV)
East Franconian (CZE; DEU)
Eastern Mari (RUS)
Eastern Slovak (SVK; UKR)
Emilian-Romagnol (ITA; SMR)
Erzya (RUS)
Faetar (ITA)
Faroese (FRO)
Franc-Comtois (CHE; FRA)
Francoprovençal (CHE; FRA; ITA)
Friulian (ITA)
Gagauz (4) (BGR; GRC; MDA; MKD; ROU; TUR; UKR)
Gallo (FRA)
Gallo-Sicilian (ITA)
Gallurese (ITA)
Gardiol (ITA)
Gascon (ESP; FRA)
Ghomara (MAR)
Gottscheerish (SVN)
Griko (2) (ITA)
Guernsey French (GBR)
Gutnish (SWE)
Ingrian (RUS)
Irish (GBR; IRL)
Istriot (HRV)
Istro-Romanian (HRV)
Jersey French (GBR)
Judezmo (ALB; BGR; BIH; DZA; GRC; HRV; MAR; MKD; ROU; SRB; TUR)
Karagash (RUS)
Karaim (3) (LTU; UKR)
Karelian (4) (FIN; RUS)
Kashubian (POL)
Komi (RUS)

Ladin (ITA)
Languedocian (FRA)
Latgalian (LVA; RUS)
Ligurian (FRA; ITA; MCO)
Limburgian-Ripuarian (BEL; DEU; NLD)
Limousin (FRA)
Livonian* (LVA)
Logudorese (ITA)
Lombard (CHE; ITA)
Lorrain (BEL; FRA)
Low Saxon (DEU; DNK; NLD; POL; RUS)
Lude (RUS)
Manx* (GBR)
Mariupolitan Greek (UKR)
Megleno-Romanian (GRC; MKD)
Mòcheno (ITA)
Moksha (RUS)
Molise Croatian (ITA)
Moselle Franconian (BEL; DEU; FRA; LUX)
Nogay (2) (ROU; UKR)
Norman (FRA)
North Frisian (DEU)
Olonetsian (FIN; RUS)
Picard (BEL; FRA)
Piedmontese (ITA)
Plautdietsch (UKR)
Poitevin-Saintongeais (FRA)
Polesian (BRB; POL; UKR)
Provençal (FRA)
Resian (ITA)
Rhenish Franconian (DEU; FRA)

Romani (ALB; AUT; BGR; BIH; BRB; CHE; CZE; DEU; EST; FIN; FRA; GBR; GRC; HRV; HUN; ITA; LTU; LVA; MKD; MNE; NLD; POL; ROU; RUS; SRB; SVK; SVN; TUR; UKR)
Romansh (CHE)
Rusyn (HUN; POL; ROU; SVK; UKR)
Sanhaja of Srair (MAR)
Sassarese (ITA)
Saterlandic (DEU)
Scanian (DNK; SWE)
Scots (GBR)
Scottish Gaelic (GBR)
Sened (TUN)
Sicilian (ITA)
Slovincian (POL)
Sorbian (DEU)
South Italian (ITA)
South Jutish (DEU; DNK)
South Saami (NOR; SWE)
Tacenwit (DZA)
Tamazight (2) (DZA; TUN)
Tayurayt (DZA)
Töitschu (ITA)
Torlak (BGR; MKD; ROU; SRB)
Transylvanian Saxon (ROU)
Tsakonian (GRC)
Ubykh (TUR)
Urum (GEO; GRC; UKR)
Venetan (HRV; ITA; SVN)
Veps (RUS)
Vilamovian (POL)
Vojvodina Rusyn (HRV; SRB)
Võro-Seto (EST; RUS)
Vote (RUS)

Welsh (GBR)
West Flemish (BEL; FRA; NLD)
West Frisian (NLD)
Western Armenian (TUR)
Western Mari (RUS)
Yiddish (AUT; BEL; BRB; CHE; CZE; DEU; DNK; EST; FIN; FRA; GBR; HUN; ITA; LTU; LUX; LVA; MDA; NLD; NOR; POL; ROU; RUS; SVK; SWE; UKR)
Yurt Tatar (RUS)
Zenatiya (DZA)

Atlas of the World's Languages in Danger

Fig. 1 Endangered languages in Europe, based on the classification of UNESCO, https://openli bra.com/en/book/atlas-of-the-worlds-languages-danger, p. 183–184, Accessed June 11, 2020

resources, before, during and after their allocation.[13] Various phases of evaluation serve various purposes. The role of the ex-ante evaluation is to ensure that the operational programs clearly articulate their intervention logic, in a framework which identifies the social, environmental and economic needs and aims to achieve feasible and sustainable solutions in reducing development disparities. The mid-term evaluation serves the purpose of assuring the effective and transparent implementation of the programs that were given a green light by the decision-making authorities, while ex-post evaluation is to assess the regional and macroeconomic impacts the various programs had on the development of the country, and how effective the actual outcome of these programs was relative to their planned impact. At no point within this multi-tier system of evaluation does the European Commission gauge whether financial resources were allocated in a discriminatory way against national minorities and the territories they inhabit. Exacerbating the situation, the backlog in economic development that can be produced by such discrimination does not necessarily reveal itself statistically at first glance. Since the borders of NUTS regions follow already existing internal administrative borders,[14] in case the domestic administrative structure of a member state was set up, or aggregated into larger NUTS units, without giving due regard to reflect social, historical, or cultural circumstances (as required by Regulation (EC) No 1059/2003 on the establishment of a common classification of territorial units for statistics (hereinafter, NUTS Regulation)), then it can become quite difficult to detect the effects of economic discrimination targeted against areas inhabited by national minorities (as we will see in the following sections).

Despite these difficulties, the EU is required to prevent such discriminatory practices to implement its obligations under the founding Treaties. As the Fundamental Rights Agency of the European Union (EU-FRA) reminds us, the entry into force of the Lisbon Treaty confers a set of new horizontal obligations upon the EU (EU-FRA 2010: 23–24). Article 9 of the Treaty on the Functioning of the European Union (TFEU) obliges the EU to take various requirements into account when "defining and implementing its policies and activities", including "the fight against social exclusion". Additionally, in the context of the EU's overall values and objectives, Article 2 of the Treaty on the European Union (TEU) declares that the EU is founded, among others, on respect for human rights, including the rights of persons belonging to minorities. Article 3, paragraph 3, provisions two to four of the TEU, declares that the Union "shall combat social exclusion and discrimination, and shall promote social justice and protection", "promote [...] social cohesion" and "respect its rich cultural and linguistic diversity". This latter aspect is also reaffirmed by Article 167 TFEU, which stipulates that the EU shall "contribute to the flowering of the cultures of the Member States, while respecting their national and regional diversity". Cultural

---

[13]For a useful summary on the methodology applied during the evaluation of the regional policy, as well the shifts in these methods, see e.g., Nagy and Heil (2013: 155–200.), (European Commission 2014a, b).

[14]Nomenclature of Territorial Units for Statistics or NUTS is a geocode standard for referencing the administrative subdivisions of EU Member States for statistical purposes.

linguistic and regional diversity that exists "between and within Member States" (EU-FRA 2010: 10) is thus a fundamental value that the EU must respect and safeguard through affirmative action.[15]

Taking this further, the Court of Justice of the European Union concluded in its recent Judgment from September 24, 2019, in case T-391/17 (para 56), that there should be no obstacle for the European Commission to elaborate proposals for specific acts that are intended to complement the EU's action in the areas covered by its competences, to ensure compliance with the values set out in Article 2 TEU (including respect for the rights of persons belonging to minorities), and to ensure respect for the richness of the cultural and linguistic diversity of the EU, referred to in the fourth subparagraph of Article 3 (3) TEU. Put differently, this judgment substantiated the idea that it does not violate provisions of community law if the European Commission adopts specific measures related to the protection of national minorities, or the preservation of cultural and linguistic diversity within its own sphere of competence.

4 The Example of Szeklerland—Failure of the EU's Cohesion Policy

This is the point where the "Szekler experience" becomes relevant. Szeklerland is a national region located in south-east Transylvania, in the middle of Romania. This traditional area is around 13,000 km^2 large and is inhabited mostly by the Hungarian speaking Szekler community (Hungarian: *Székely*, Romanian: *Secui*). It is important to note that Szeklers are not just a linguistic minority but also a religious one. The vast majority of Szeklers belong to the denominations of western Christianity (mainly Catholic, Reformed and Unitary or other Protestant denominations), whereas most Romanians belong to the Orthodox Church. There are three counties with substantial Szekler–Hungarian populations (Hargita/Harghita—85.2%—257,707 people; Kovászna/Covasna—73.7%—150,468 people; and Maros/Mureș—38.1%—200.858 people),[16] and these are grouped together in the so-called Central Romanian NUTS II Development Region (hereinafter referred to as CDR) with overwhelmingly Romanian Alba, Sibiu and Brașov counties. Consequently, the proportion of Szekler–Hungarians in the CDR altogether is below 30%,[17] a grievance that

---

[15]Article 10 TFEU also contains new horizontal obligations for the EU to combat discrimination based on sex, racial or ethnic origin, religion or belief, disability, age or sexual orientation. As a matter of fact, this provision goes further than the mere prohibition of discrimination as set out in Article 21 of the Charter of Fundamental Rights of the EU. It enables and obliges the EU to proactively combat discrimination through affirmative action, as this clause builds on the enabling provision enshrined in Article 19 TFEU.

[16]For the number and percentage of ethnic Hungarians living in Romania by county, see the result of the last general census in Kapitány (2015: 230).

[17]This is particularly important given that earlier plans of the central government included vesting these eight NUTS II regions with administrative competencies.

Table 1 Proportion of GDP/capita in the CDR counties, relative to the CDR's GDP/capita average (in %)

County	2007	2008	2009	2010	2011
CDR	100	100	100	100	100
Alba/Fehér	108	103	102	99	107
Brasov/Brassó	122	124	118	122	126
Covasna/Kovászna	81	78	80	80	74
Harghita/Hargita	82	81	81	81	77
Mures/Maros	83	82	85	83	80
Sibiu/Szeben	109	115	117	119	116

Source Annex FB 3 of the Intervention of Kovászna County in Case T-529/13. https://nemzetiregiok.eu/dokumentumok/t-529-13.sz.per-kovaszna-megye-beavatkozasi-kerelme.doc, Accessed June 11, 2020

came to the fore in case T-529/13 (Izsák and Dabis v. Commission) of the General Court of the EU (concluded on 10 May 2016).[18] The intervention of the council of Kovászna County in this judicial procedure made references to the dramatic divergence in economic development by citing the GDP per capita data of the constituent counties of the CDR.[19] They presented data showing that, since Romania's EU accession in 2007, the average per capita GDP of the counties having Hungarian population (Maros, Hargita, Kovászna) has been only 80.5% of the CDR average, while the same proportion in the counties not having Hungarian inhabitants (Fehér, Brassó, Szeben) have remained well above 110% (see Table 1).

Furthermore, while the average per capita GDP in the two counties having the most substantial Hungarian majority (Hargita and Kovászna) has grown by only 4.6 and 1.7% between Romania's EU accession and 2013, counties with Romanian majority have increased multiple times faster. Brassó, for example, has experienced a growth more than eleven times greater in the same period (19.2%) than that of Kovászna (see Table 2; and Fig. 2).[20]

This general trend of economic divergence between Szekler and other Romanian counties continues to this day (see Fig. 3), forming, what the above stakeholder identified as indirect discrimination targeted against the Szekler community, given that there has been no reasonable justification whatsoever for these disproportionate effects.[21]

---

[18]The author of this study was one of the applicants of this lawsuit.

[19]For the intervention, see: https://nemzetiregiok.eu/dokumentumok/t-529-13.sz.per-kovaszna-megye-beavatkozasi-kerelme.doc. Accessed January 06, 2020.

[20]Due to similar issues encountered by the Hungarian community of Slovakia, the municipality of an overwhelmingly Hungarian speaking settlement, Debrőd/Debrad', located in Southern Slovakia, has also joined this lawsuit.

[21]The data enumerated in the Kovászna intervention is particularly relevant given that it can be quite difficult to identify the disproportionately prejudicial effects of public-policy measures that amount to indirect discrimination. On this topic see Barelli et al. (2011: 6–8).

Table 2 Annual increment of GDP/capita in the counties of the Central Development Region

County	2008 (Euro)	2013 (Euro)	Increment (%)
Alba/Fehér	6.318	7.111	12.5
Brasov/Brassó	7.285	8.681	19.2
Covasna/Kovászna	4.923	5.009	1.7
Harghita/Hargita	5.013	5.243	4.6%
Mures/Maros	5.269	5.953	12.9%
Sibiu/Szeben	7.25	8.011	10.9%

Source Annex FB 4 of the Intervention of Kovászna County in Case T-529/13. https://nemzetiregiok.eu/dokumentumok/t-529-13.sz.per-kovaszna-megye-beavatkozasi-kerelme.doc, Accessed June 11, 2020

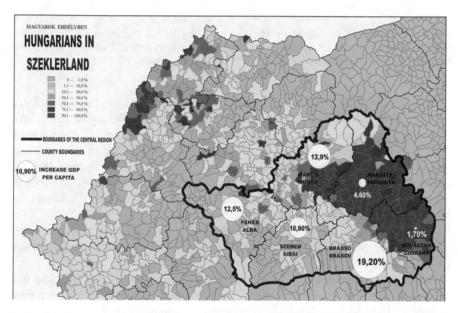

Fig. 2 Ethnic map of Transylvania, juxtaposed with borders of the administrative and the Central Romanian NUTS II region (with bold outlines), as well as the increment of their GDP per capita income (2007–2013). https://www.erdely.ma/a-eu-kohezios-politikajanak-kudarca-a-kozep-romaniai-fejlesztesi-regioban/, Accessed June 11, 2020

An additional aspect of this case was revealed by the expert report attached to the Kovászna Intervention. This material argued that the true economic backlog of the less developed Szekler regions remains hidden within the current NUTS II structure (see Fig. 4).

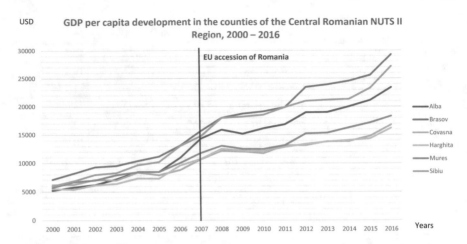

Fig. 3 GDP per capita development in the counties of the Central Romanian NUTS II Region, calculated in current price USD based on purchasing power parity terms (2000–2016). Own chart based on OECD statistics, available at: stats.oecd.org. Accessed January 07, 2020

As a matter of fact, measuring the GDP/inhabitant/region data, one can find that the CDR is the third best-performing region, preceded only by the capital (Bucuresti-Ilfov Region), and the westernmost NUTS II region (Vest region).[22] Even though the Romanian Development Plans (Government of Romania 2012, 2014) contain the strategic goal of stopping and preferably reversing the widening trend of regional development disparities, this expert report argues that the present borders of the CDR are economically not efficient to reach that goal due to the manifold differences between the Szekler and the Romanian regions. Furthermore, the boundaries of the CDR violate the provisions of the NUTS Regulation and statistically hide processes that contradict the main goal of the EU's cohesion policy and that of the Romanian Development Plans.

Viewing these statistics from the point of view of the SDGs, one can argue that the reason why it is sensible for development regions to be set up having regard to cultural, historical and linguistic circumstances is to build inclusive systems that foster subsidiarity as an effective way of making public-policy decisions that leave no one behind. The pledge to leave no person behind—as one of the defining features of the 2030 Agenda—is often hampered by the unequal distribution of wealth and decision-making power. In situations such as the Szekler example, where one deals with multiple identity factors (linguistic, cultural, religious) that can become the source of discrimination, it is particularly important for governments to "recognize and respond to multiple and intersecting deprivations and sources of discrimination that are compounded by one another and make it harder to escape poverty, live with dignity and enjoy human rights" (Secretary-General of the UN 2019: 35).

---

[22]For data, see e.g.: Surd et al. (2011: 23) and Oṭil et al. (2015: 42).

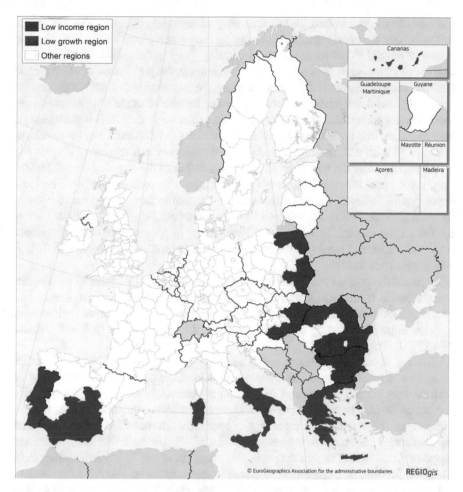

Fig. 4 Low income (red), low growth (blue) and other NUTS II regions (white). *Source* European Commission (2017), p. 1. https://ec.europa.eu/regional_policy/sources/docgener/studies/pdf/lagging_regions%20report_en.pdf, Accessed June 11, 2020

5 Safeguarding the Socioeconomic Cohesion of the EU Through Affirmative Action

Section 3 of this study discussed the obligations of the Union to combat discrimination and social exclusion, as well as to safeguard cultural and linguistic diversity. Accordingly, it is important to look at the proposals referred to in the mentioned lawsuit through which Kovászna County signaled its development issues before the Court of Justice of the European Union.

Case T-529/13, and the appeal procedure thereof (C-420/16 P), revolved around the registration of the European Citizens' Initiative entitled "Cohesion policy for the equality of regions and the sustainability of regional cultures". The aim of the initiative was to achieve special attention for national regions within the cohesion policy of the EU.[23] For these regions, including geographic areas with no administrative competencies (like Szeklerland, for example), the prevention of an economic backlog, the sustainment of development and the preservation of the conditions for economic, social and territorial cohesion should be carried out in a way that ensures their characteristics remain unchanged. In the organizers' view, this is the adequate way to ensure that the EU's overall and harmonious development can be sustained, and parallel with that, its cultural and linguistic diversity maintained.

As envisaged by the initiators, the implementation of the required "special attention" could entail the establishment of a separate funding scheme within the regional development policy of the EU that would be accessible directly and exclusively to national regions. The initiators also elaborated a material containing detailed information on the subject, objectives and background of the proposed citizens' initiative.[24] This material relates to the context of paragraph three of Article 174 TFEU, which stipulates that:

"Among the regions concerned, particular attention shall be paid to rural areas, areas affected by industrial transition, and regions which suffer from severe and permanent natural or demographic handicaps such as the northernmost regions with very low population density and island, cross-border and mountain regions." (Article 174 TFEU).

This list is exemplifying and not exhaustive, thus the possibility to add further categories to this list is open. It is the experience of the organizers of this initiative that the specific ethnic, cultural or linguistic characteristics of national minority regions could also be regarded as a demographic handicap that could make it more difficult for these regions to create the proper circumstances for economic, social and territorial cohesion. A separate legal act of the Union could thus widen the scope of Art. 174 and secure the particular attention of the EU to national regions. This new funding mechanism could create proper a financial incentive for states to value, cherish and preserve their own rich cultural diversity, and, at the same time, it would empower local communities in a subsidiary manner to invent and implement their own, smart and innovative regional development plans. Such a new financial framework would thus provide the leverage for local communities to tackle their respective development issues and prevent eventual demographic decline of their regions and the outflow of their population, which in itself erodes and endangers traditional minority cultures in the EU.

---

[23] National regions were defined by the organizers as territories that possess national, ethnic, cultural, religious or linguistic characteristics that are different from those of the surrounding regions.

[24] See: https://ec.europa.eu/citizens-initiative/public/documents/1764/annex.rtf, accessed: 12. December, 2019.

6 Conclusions

The assembled evidence outlines a systemic deficiency in properly monitoring and evaluating the impact that the state-level funding-allocation practices of EU development sources can have on regional cultures. If such practices are of a discriminatory nature, they hinder the effective implementation of policy objectives, such as reducing disparities among various regions. Additionally, they undermine the realization of the Sustainable Development Goals, especially those in connection with building inclusive societies that leave no one behind. Given the lack of an evaluation system that addresses this issue, there is a real risk that Member States could utilize EU sources for purposes that are contrary to the aims thereof and violate the fundamental values of the Union in the process.

The case of Szeklerland shows that there is a serious risk that even EU funds can be used for indirect discrimination, undetected by EU institutions. While there are numerous other variables that can determine the root cause of regional disparities in the level of development, the sources alluded to in the Szekler case are of importance due to their grassroots nature. The specific inputs mentioned under Sect. 4 came from local representatives of the affected community bottom-up and echo the community's own experience of discrimination, social exclusion and the demographic handicaps they suffer as a national region, on the grounds that the ethnic, cultural and linguistic characteristics of that area are different from those of surrounding regions. The EU is bound by the founding Treaties to avoid becoming an accessory to any form of discrimination, including indirect economic discrimination targeted against national minorities. That would be contrary to the goals and values of the Union and would compromise the goal of reaching the overall and harmonious development of the EU as a whole.

To realize this overarching goal, it is crucial to identify the sources of development divergence, which can be at least as difficult as finding the appropriate solutions to tackle it. The process of finding such a solution could also be hindered if the evaluation procedures disregard relevant factors of economic backlog, such as potentially discriminatory allocation of funding resources. The current monitoring and evaluation processes of the European Commission are not equipped to detect, let alone prevent, such policies. While it is first and foremost an obligation of the Member States to refrain from discriminatory practices targeted against members of national minority communities, given the fact that economic, social and territorial cohesion is a shared competence between the EU and its member states, it is reasonable to suggest that the European Commission should develop a monitoring capacity that is suitable for detecting discriminatory practices in the distribution of EU development funds.

Such a revision of the monitoring mechanisms of the European Commission could contribute to prevent EU funding being used to the detriment of national minorities, which in turn could help the EU to preserve its own rich cultural and linguistic diversity, in line with her obligations stemming from the founding treaties, as well as from the objectives of the Sustainable Development Goals. Development agencies, financial institutions and others involved in international cooperation and the distribution of development funds have a dual task vis-á-vis national minorities:

Firstly, to ensure that legitimate interests of minorities are not negatively affected by the measures implied in the cooperation envisaged; and secondly, to ensure that persons belonging to minorities can benefit as much as members of majorities from that cooperation. (UN Sub-Commission 2005: para. 74.).[25]

As the Office of the United Nations High Commissioner for Human Rights (OHCHR) warns, "development is a powerful tool, but it can also, be a tool of the powerful unless human rights for all, without discrimination, is part of its design" (OHCHR 2015: 2).

References

Bachtler J, Wren C (2006) Evaluation of European EU Cohesion policy: research questions and policy challenges. Regional Stud. 40(2):143–153

Barelli M, Guliyeva G, Errico S, Pentassuglia G (2011) Minority groups and litigation: a review of developments in international and regional jurisprudence. Minority Rights Group Guide Series (ed) Minority Rights Group International, London. https://www.minorityrights.org/733/guides/guides.html. Accessed April 28, 2020.

Blauberger M, van Hüllen V (2020) Conditionality of EU funds: an instrument to enforce EU fundamental values? J Eur Integr. https://www.tandfonline.com/doi/full/https://doi.org/10.1080/07036337.2019.1708337. Accessed 28 April 2020

European Commission (2014a) Guidance document on monitoring and evaluation for the European Cohesion Fund, and the European Regional Development Fund—concepts and recommendations for the programming period 2014–2020. European Commission, Directorate-General for Regional Policy, Brussels

European Commission (2014b) Monitoring and evaluation of European Cohesion Policy. Guidance document on ex-ante evaluation for the programming period 2015–2020. European Commission, Directorate-General for Regional Policy and Directorate-General for Employment, Social Affairs and Inclusion, Brussels

European Commission (2017) Competitiveness in low-income and low-growth regions. The lagging regions report. Commission staff working document. European Commission, Brussels

Ferry M (2013) Cohesion policy lessons from earlier EU/EC enlargements. Synthesis of case studies. In: GRINCOH Working Paper Series, Paper No. 8.02

Fratesi U, Wishlade FG (2017) The impact of European Cohesion Policy in different contexts. Regional Stud. 51(6):817–821

Fundamental Rights Agency of the European EU (2010) Respect for and protection of persons belonging to minorities 2008–2010. https://fra.europa.eu/sites/default/files/fra_uploads/1769-FRA-Report-Respect-protection-minorities-2011_EN.pdf. Accessed 03 Jan 2020

Government of Romania (2012) National strategic reference framework 2007–2013. Ministry of Economy and Finance, Bucharest

Government of Romania (2014) Romanian partnership agreement for the 2014–2020 programming period. Ministry of European Funds, Bucharest

---

[25] Moreover, as Henrard notes, the vision of integration and inclusion is mostly about ensuring that "minorities' integration in the wider society does not go hand in hand with forced assimilation" (Henrard 2015: 157–158).

Grazi L (2012) The long road to a cohesive Europe. The evolution of the EU regional policy and the impact of the enlargements the long road to a cohesive Europe. The evolution of the EU regional policy and the impact of the enlargement. Eurolimes 14:80–96

Henrard K (2015) The UN declaration on minorities' Vision on 'Integration'. In: Caruso U, Hofmann R (eds) The United Nations declaration on minorities: an academic account on the occasion of its 20th anniversary (1992–2012), p. 376. Brill–Nijhoff, Leiden

Hofstede G, Hofstede JG, Minkov M (2010) Cultures and organizations: software of the mind, 3rd edn. McGraw-Hill, New York

Kapitány B (2015) Ethnic Hungarians in the neighbouring countries. In: Monostori J, Őri P, Spéder Zs (eds) Demographic portrait of Hungary 2015. Hungarian Demographic Research Institute, Budapest

Mindaugas B, Matuzevičiūtė K (2016) Evaluation of Eu Cohesion Policy impact on regional convergence: do culture differences matter? Econ. Culture 13(1):41–52

Moseley C (ed) (2010) Atlas of the world's languages in danger, 3rd edn. United Nations Educational, Scientific and Cultural Organization, Paris

Nanda VP (2016) The journey from the millennium development goals to the sustainable development goals. Denver J. Int. Law Policy 44(3):389–412

Nagy SG, Heil P (eds) (2013) A kohéziós politika elmélete és gyakorlata. Akadémiai Kiadó, Budapest

Office of the High Commissioner for Human Rights (2015) Empowerment, inclusion, equality: accelerating sustainable development with human rights. https://www.ohchr.org/Documents/Iss ues/MDGs/Post2015/EIEPamphlet.pdf. Accessed 02 Jan 2020

Oțil MD, Miculescu A, Cismas LM (2015) Disparities in regional economic development in Romania. In: Scientific Annals of the "Alexandru Ioan Cuza" University of Iași. Economic Sciences 62 (SI), pp. 37–51

Ray, E. (2019). Enforcing the rule of law in the European EU, Quo Vadis EU? In: Harvard Human Rights Journal Online. https://harvardhrj.com/2019/11/enforcing-the-rule-of-law-in-the-european-EU-quo-vadis-eu/. Accessed 20 April 2020

Secretary-General of the United Nations (2019) Report of the Secretary-General on SDG Progress 2019—special edition. UN- Department of Economic and Social Affairs, New York. https://sustainabledevelopment.un.org/content/documents/24978Report_of_the_SG_on_SDG_Progress_2019.pdf. Accessed 2 Jan 2020

Surd V, Kassai J, Giurgiu L (2011) Romania disparities in regional development. In: Efe R, Ozturk M (eds) Procedia: social and behavioral sciences, vol 19, pp 21–30

United Nations (2000) G.A. Res. 55/2, U.N. Millennium Declaration, 8 Sept 2000. UN, New York

United Nations (2015a) G.A. Res. 70/1, U.N. Sustainable Development Goals, 25 Sept 2015. UN, New York

United Nations (2015b) Transforming our world: The 2030 agenda for sustainable development. UN, New York. https://sustainabledevelopment.un.org/post2015/transformingourworld/publication. Accessed 14 Jan 2020

UN Sub-Commission on the Promotion and Protection of Human Rights (2005) Commentary of the Working Group on Minorities to the United Nations Declaration on the Rights of Persons Belonging to National or Ethnic, Religious and Linguistic Minorities, 4 April 2005, E/CN.4/Sub.2/AC.5/2005/2. Available at: https://www.refworld.org/docid/43f30ac80.html. Accessed 21 April 2020

Thriving Governance and Citizenship in a Smart World: Environments and Approaches Fostering Engagement and Collaborative Action

Toward a Smart Urban Planning. The Co-production of Contemporary Citizenship in the Era of Digitalization

Enza Lissandrello

Abstract This paper investigates mediated negotiations in 'smart city' experimentalism. As often claimed, data can open pathways for innovative planning processes. However, the idea of planning underpinned by the interplay between citizens and data too often remains unquestioned. How might we move the idea of planning from data to provide (technical solutions) to data to transform (urban societal realities)? How can data empower citizens as true drivers of a transformative urban change? This paper argues for a planning perspective to enhance a new sense of citizenship in a future technology-driven urban democracy. The framework combines planning theory with theories of societal change under a critical pragmatism. The empirical research derives from Mobility Urban Values (MUV2020), a Horizon 2020 innovation and research project (2017–2020), with the ambition to change mobility endeavors toward a more participatory and sustainable urban policy. The paper synthesizes analysis of the 'practice stories' of professionals dealing with and facilitating the interplay between data and citizens in six European cities. It then discusses MUV's deliberative planning process in which citizens generate data (co-creation of values), interpret data (co-design of facts) and perform utterances to call for new urban policy (co-production of actions). The conclusions draw a possible pathway to enhance smart urban planning as a perspective to empower citizens *with* data for a progressive democracy in the era of digitalization. Change-oriented practitioners can potentially facilitate smart urban planning through: 1) technological devices that engage individual citizens (choices) with data practices in everyday life; 2) frames for the interpretation of data *with* citizens' and communities (practice) and 3) public conversations between citizens with other publics (system) for new street-level practices of urban democracy.

Keywords Planning theory · Data · Theories of change · Deliberative democracy · Critical pragmatism

E. Lissandrello (✉)
Aalborg University, Rendsburggade 14, Aalborg, Denmark
e-mail: enza@plan.aau.dk

1 Introduction

This paper aims to illuminate a new perspective to mediate the interplay between citizens and data by rethinking the nature of planning in an era of digitalization. Both data and citizen sciences have too often left unquestioned the very idea of planning as the very fundamental mindset for working data *with* citizens. The 'idea of planning' has been discussed in theory as a way of thinking 'prior to a particular set of practices or institutions, and provides a vantage point from which to judge the vagaries of regulatory or professional requirements, and hence the possibility for challenge and subversion' (Campbell 2012: 393). The planning perspective allows new insights into the contemporary interplay between citizens and data. Data can open up new pathways for innovative processes when they do not just remain anchored to evidence-based planning. Under an evidence-based paradigm, data often serve to provide 'solutions' among a close cycle of experts and professionals (rather than citizens): (a) proving scientific facts, (b) testing technology services and (c) creating evidence to present to policy actors for negotiating future strategies. However, the evidence provided by data remains too often distant from the real politics of planning. In other words, within an evidence-based paradigm, data generally fail to address the very question: What data do matter politically and which kinds of meanings and transformative potential do data represent for citizens and urban democracy? Thinking citizens as data-points reproduce and maintain (rather than transform) a technocratic idea of planning. A 'smart-mentality' focused on techno-scientific solutions risks to separate the city from its very politicization. Neglecting issues of citizens' accountability for participation and deliberative governance, the 'co-creation' with data and citizens—often claimed by smart cities experiments—risk remaining a pure 'exercise' in public engagement. A progressive idea of planning is at the base of rethinking future urban citizenship for contemporary change-oriented practitioners.

How might we move the idea of planning from data to provide (technical solutions) to data to transform (urban societal realities)? How can we open the idea of planning to empower citizens through data for a smarter and sustainable urban future? This paper advances the idea of smart urban planning. It draws on theories of change and a critical-pragmatism approach, and it elaborates the practical experience of Mobility Urban Values (MUV), an EU Horizon 2020 research and innovation project (2017–2020) aimed at change urban mobility and policy. MUV's change-oriented practitioners engage citizens through a gamified interaction (Di Dio et al. 2018), shape local communities and arrange new partnerships with local businesses, policymakers and Open Data enthusiasts in six EU cities neighborhoods (in Amsterdam, Helsinki, Barcelona, Palermo, Fundao and Ghent). Societal values related to mobility guide new visions for more sustainable, safer, inclusive and healthier future scenarios and urban innovation (Lissandrello et al. 2018) with an impact (Caroleo et al., 2019). This paper does not aim to assess the success of the MUV project in achieving more sustainable urban mobility in urban planning; instead, it focuses on the learning experience to elaborate further the idea of planning for a future technology-driven urban democracy.

The paper is structured as follows. The first part frames the current discussions on smart city experimentalism, questioning the planning idea underpinning such processes. A critical pragmatic perspective highlights the theories of change and advances a deliberative planning approach based on data values, facts and actions. The second part adopts this framework to examine the 'practice stories' of MUV's professionals dealing with the co-creation of values through data, the co-design of those data into a meaningful interpretation of facts and new citizens' utterances for conversations and calls of policy actions. It follows a pathway toward the idea of smart urban planning to orient and inspire change-oriented professionals to facilitate a future co-production of citizenship through data. Under the planning perspective, citizens—not just as data-points—become drivers of transformative urban change through new models of interaction and community building through data. The vision of smart urban planning in an era of digitalisation is all about underpinning the future sense of citizenship within the digital and physical ecosystem of knowledge and action.

2 Rethinking the Idea of Planning

Planning as the guide to future action is radically changing. The practical reason is that planning is deeply dependent on societal development. Therefore, every kind of change in society—as desired values of sustainability—creates pressure on the institutionalization of planning. Planning also changes in its very idea, therefore its purpose as the way of thinking the future, beyond particular regulatory and governance frameworks. For example, the ecological discourse on climate change and the transition to a low-carbon society has placed pressure on the production–consumption linearity within the growth paradigm. The technical and economic rationality in planning has, therefore, embroiled the process in uncertainty. The recent Covid-19 crisis has also accelerated awareness of the limit of planning in 'the risk society' (Beck 1992). Professionals need new methods, skills and attitudes for planning under conditions of risk and change, a change occurring suddenly without long-term warning and with significant consequences such as a recession and biodiversity collapse. Bauman (2007) argues that we are facing 'the passage from the 'solid' to the 'liquid' phase of modernity.

We are merely living in a time when social forms (structures that limit individual choices, institutions that guard repetitions of routines, patterns of acceptable behavior) can no longer (and are not expected) to keep their shape for long. These social forms 'decompose and melt faster than the time it takes to cast them, and once they are cast for them to set' (Bauman 2007: 1). Likewise, planning institutions and the way to think and govern the future are becoming unlikely to be given enough time to solidify. Liquid societal dynamics of transformation also entail smart city imagination as a flow of technological innovation (Cardullo and Kitchin 2019). While we still have not adopted routines to plan with and through data, the liquid smart-mentality and the digitalization place individual citizens at the center of future distributed urban transformations. The planning idea underpinning smart urban practice, however, often reproduces a citizenship of passive users. Finally disciplined by guidance on 'the correct' use of technology, the 'smart citizen' can assume 'the correct' behavior encapsulated through a multiplicity of digital devices and services, digital platforms, apps and wearables as pervasive technology-mediations. The smart citizen adopts a function as data provider.

A new technological urban imaginary (Vanolo 2016) develops the smart city's idea and big data production within an evidence-based idea of planning. Therefore, the latest phase of citizen-focused claims and language often just mirrors a one-way direction (Cowley et al. 2018; Saunders and Baeck 2015). Citizens providing data are a passive voice to inform, narrow, limit and control through the interplay between technology and participation. An interaction often facilitated by a particular entrepreneurial or pre-given design (Wilson et al. 2019; Baker et al. 2007; Kitchin 2015). The question is, therefore, on how technology and participation through data can co-produce a new type of citizenship, i.e., citizens as active and responsible voices. This paper argues for a focus on the idea of planning to think and to govern the future: an idea that can place at the center methods to produce new capacities for knowledge, communities of practice and common-hoods in the era of digitalization. Planning requires enabling skills and attitudes to navigate the risk society for an effective change of the role of citizens from data providers to data drivers of urban democracy (Lissandrello and Vesco 2020).

Planning as a process of change concerns of casual, emergent and co-evolving behaviors, social practices and systems that define and enhance diverse policy perspectives and drivers. In the urban context, these processes of change take form and reflect diverse temporalities. The city is a system of slow and fast dynamics of change. Sedimented historical layers of urban form and urban identity are resilient to change, while fast contemporary urban lifestyles, nowadays supported by technologies, transform urban dynamics and the sense of citizenship. Therefore, theories of change are important for planning because they offer a perspective to identify the process of transformation, the tension points (Flyvbjerg et al. 2016) and the policy angle that is already part of the system. For example, behavioral change-based policy on individual choice often offers a perspective that implies an external influencer that includes 'the different combinations of policy instruments—classically characterized as carrots, sticks, and sermons—to… facilitate choices such that individuals can make as a 'better' choices for themselves' (Shove et al. 2012). Data are often gathered from individual citizens, using, for example, techniques for rewarding behavior. In the field of urban mobility studies, low-carbon policy based on behavior change can consist of rewarding individual choices of biking or walking rather than using a car. This 'rewarding' can happen in forms of specific prizes, taxes and salaries. For urban change, the fact that individual citizens' choices produce behaviors, habits and routines is important. Indeed, when individual citizens consolidate their patterns of behavior, they also shape social practice. The perspective of social practice allows us to illuminate change through 'practice carrying'. In other words, choosing to bike rather than drive a car is not about an individual's choices alone but a patternation of practice and communities, for example, biking communities. In a perspective of change, policy-based social practice can consist of connecting individuals into communities. In the example of low-carbon policy based on urban mobility practice, a 'practice carrying' can be the car-sharing policy that connects individuals within a (digital) social context of communities of sharing. But changes in policy and planning occur within a complexity that includes behaviors (individual choices) and social practice (communities). This complexity can be understood as a system. A system is an ensemble or assemblage of multiple social practices as normalized behaviors and mechanisms of societal regulation that stabilize and maintain the system itself. In a policy perspective, a 'system change' consists of turning the existence of the system itself (Urry 2004), therefore the complexity of behavior, social practice and the holistic policy perspective represented by the system. Studies on socio-technical system change (Geels 2005a, b) show that that implies a long-term and complex transformation governed and maintained by both individual choices and social practices. A systemic change thus implies the alignment of innovations with 'turning points' or 'cracks' that might exist within the institutionalized and normalized behavior and practice under the flows or exogenous dynamics. These exogenous dynamics can be, for example, the climate change (landscape) that places pressure for a change of automobility (regime). Simultaneously, car-free neighborhoods (niches) can constitute an example of turning points or cracks in the current automobility practice and behaviors. From a system perspective, a change thus takes place through the alignment of multiple dynamics. Change-oriented practitioners and professionals cannot fully influence these alignments. However, the system approach is extremely relevant to change-oriented practitioners to identify 'turning points' or 'cracks' which might activate opportunities to co-construct, co-generate or co-produce systemic change. 'Things may look beak and hopeless, but for those who are nimble on their feet, the inevitable creaks and crevasses in the institutional structure always provide ever so many opportunities for positive action' (Krumholz and Forester 1990). Thinking the future requires the awareness that a system change will entail complex multi-level dynamics, unexpected consequences, risks and flows that require professionals to 'reflect in action' (Schön 1983). In other words, in the context of 'smart' digitalization and urban data, change-oriented practitioners need to rethink the fundamental idea of planning. The potential co-production of systemic change depends on the idea of thinking and governing the future as a way to enhance opportunities and political engagement and learning. The choices of individual citizens, communities and possible futures of urban citizenship need to be at the center of the system change. A critical-pragmatism framework to planning and public policy offers a pathway to pose questions about the interplay of citizens and data. Such an interplay can consist of dealing with creaks, crevasses and cracks in the current system. It might open possibilities for micro-politics 'in the trenches' (Forester 1999, 2013; Wagenaar 2011) in the deliberation about value, facts and actions (Forester 2017). In the remainder of this paper, this framework of theories of change and critical pragmatism contributes to exploring the 'practice stories' of MUV professionals engaged in the process of dealing with the interplay of data and citizens across the spheres of technology and participation. Smart urban planning includes the generation of data with citizens and the co-creation of values through individual choices, as well as the interpretation of data into a co-design of meanings with communities of practice, and the co-production of a collective redesign of policy actions. These stages aim to illuminate a planning perspective of co-production of citizenship for systemic urban change.

3 Shaping Citizenship *with* Citizens: MUV Mediated Negotiations

Mobility Urban Values (MUV) is a three-year Horizon 2020 project (2017–2020) in which an interdisciplinary team of EU academics and practitioners has envisioned the possibility of activating systemic urban change. The focus point of the change-oriented practitioners was bringing ethical urban mobility practices through technology-driven data devices. The theories of change, just mentioned, constitute the background for a vision that intertwines issues of individual choices, common social practice and urban ecosystem change. Citizens impact their environment through their choices and behavior, shape communities and transform their urban living system. MUV departs from a gamification strategy through an app that aims to influence the choices of individual citizens toward more sustainable mobility lifestyles. Uploading the app, citizens are transformed into MUVers, so they become active players of the digital world. By selecting their everyday active mobility choices (walking, cycling, public transport or car-sharing), citizens gain points connecting to local businesses that reward them with prizes when they become sustainable mobility champions. The MUV idea is that, through a motivational device (app) based on gamification and rewarding, citizens can produce data on their mobility choices. This approach to behavior change based on a policy of control exploits techniques of rewarding or nudging through technology. However, MUVers co-create sustainable mobility values as 'carriers' of practice in their everyday active mobility practice. These values are co-created when citizens engage in gaming communities and MUVerhoods. Sharing their sustainable mobility experience, citizens connect their journeys (points) to other MUVers, competing for the mobility challenge of winning points. MUVers connect to local businesses, as well as provide active mobility data to local planners and participating mobility managers. The next section synthesizes analysis and extracts some of the MUV practice stories of change-oriented practitioners (pilot coordinators in various cities), performing the participatory process of engagement of data and citizens for urban-policy innovation.

3.1 Generating Mobility Data with citizens—The Co-creation of Values

People think about mobility in terms of problems. When you approach citizens from this angle, they start to talk about frustrations: finding parking places, safety on roads and congestion. The turning point of the system is to transform the idea of mobility into something completely different from what citizens experience in everyday life: Let us talk about mobility as fun (MUV pilot coordinator 2017). When dowloading entering the digital device, citizens transform into MUVers. MUVers are digital individuals who, through a metaphor of sports narrative, play athletes to get rewards for their sustainable mobility choices, i.e., walking, biking, car-sharing, carpooling and traveling on public transportation. MUVers connect to public authorities that gather MUV mobility data and provide training sessions to coach–athletes to improve their sustainable mobility skills. MUVers also connect to local business communities that, as sponsors, have the opportunity to promote their brand and their products through the athletes' best achievements and provide prizes to them. The MUV app (Fig. 1), through gamification, collects and tracks spatio-temporal data on citizens' active mobility.

MUV gamification is therefore based on a 'nudging' policy, as depicted in Fig. 1, within the theories of behavioral change. However, MUV gamification is a means of mediation from the individual behavior of citizens to a common social practice of game communities. MUVers compete with each other, connect to local businesses and gain knowledge of their own impact on the urban environment. MUVers generate data and co-create values simultaneously when engaging in a more sustainable urban lifestyle. In everyday active mobility practice, connecting with other MUVers and shopping at local businesses, MUVers mobilize MUVerhoods. MUVerhoods are physical and digital environments that shape an urban context and provide the actors of urban transformations (citizens, local businesses, public authorities, active local communities) a sense of community with a playful vision of reality. The engagement and mobilization of MUVerhoods occur on-street level through playful

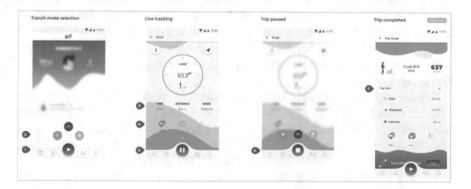

Fig. 1 MUV mobile app—on-screen visualization to generate data and co-create values

events with citizens as the MUV open-days. The design of these events aims to inform and diffuse the MUV game and shape game customizations (MUV pilot coordinator 2017).

MUVers champions are ambassadors who maintain strong individual ties within the MUVerhoods (community) and expand that community. MUVers target groups vary among cities. In Palermo, these groups include university students and tourists. In Fundao, the target is composed of workers in new enterprises (such as start-ups), in Helsinki residents already sensitive to traffic. In Ghent, families with children and schools' teachers are involved in the MUV project. In Barcelona, the target group is individual citizens already engaged with alternative modes of transportation and sports lovers; in Amsterdam, rather, participants are elderly communities and data hackers. Gamification on urban streets and the organization of events takes place when giving prizes to the MUV winners, establishing gamified competitions among cities and involving citizens in participation activities such as workshops during EU mobility weeks and other local festivals. MUVers have an active role in shaping their gaming communities. The effect of MUVhoods' motivations creates new urban values such as a healthy and cultural lifestyle, inclusive and safe shopping and 'smart' identity. MUV aims to inspire enjoyment of mobility to empower citizens' with their data measurement ('meten is weten'—'to measure is to know'—to quotes a famous Dutch sentence). This quote is a model of urban citizenship for the digital future (be the change he/she wants to see) (MUV pilot coordinator 2018).

3.2   Interpreting Facts—The Co-design of Meanings

The MUVers data on active mobility gathered from the app have been visualized in MUVmaps in each city. How do you ensure that people feel not only like data-points? The pilot coordinator in Amsterdam proposed this leading question when preparing workshops with citizens. The stake is the kind of difference that MUV will make for people when interpreting data into facts. MUV is not the only platform that creates mobility data. Nowadays, we have several route-planning and ridesharing platforms and other digital products related to mobility. Large flows of data also sometimes do not involve the users of these platforms. In MUV, the interpretation of data into facts—mobility tracks and journeys—has been the center of the co-design strategy. Data call for the design of meanings along with the citizens (MUV pilot coordinator 2019) (Fig. 2).

Besides mobility journeys, MUVMaps serve to visualize and interpret data collected from MUVers through the app. Citizens and policymakers need greater transparency of data. Data collected are often perceived as evaporating from the hands of those who generate them. How do the data collected by a cyclist help the cyclist? Specific questions have been a leading role of pilot coordinators (MUV pilot coordinator 2018). Interpreting the maps with participants, such as start-up companies, municipality workers and media agents, afforded the opportunity to visualize

Fig. 2 MUVmap (Ghent)

for the first time the data gathered by the MUV app. All the participants demon-strated real satisfaction in being directly involved in this process of mapping and visualization. Their contribution was active and productive (MUV pilot coordinator 2019). The interpretation of facts through maps has enabled translating data mean-ings into values. For example, maps of mobility practice in Fundao have deepened values for a more healthy lifestyle; in Ghent, safety-related data have pinpointed specific areas in MUVerhoods. Data interpretations have also emphasized the inter-twined importance of quantitative data on the diverse tracks with the qualitative perspectives and approaches of citizens (MUV pilot coordinator 2019). The 'MUV ambassadors' in Ghent, for example, have provided qualitative insights on the safety of bikers in their everyday mobility. The issue was to pinpoint specific critical areas in the neighborhood to engage the local knowledge. The 'citizen expert panel' in Helsinki, responding to surveys and tracking routes, has contributed direct interpre-tations of data on living MUVers experience. The collective understanding of the data on walking, for instance, reproduced and visualized on the maps, has led to insights on pedestrians' diverse safety issues in various cities and neighborhoods. In Palermo, walking issues highlighted by citizens identified specific safety needs for tourists or young citizens during the night across the historical center that would be improved with better lighting. In Barcelona, citizens interpreted maps to define issues with the timing of green lights for pedestrians on the crosswalks in peack-hours. The voice of the new green wave of an active citizens' movement in Barcelona proclaims alter-natives to car-mobility that are emerging but still require strategies of connectivity among, for example, existing bicycle lanes. The absence of data on the map raised citizens' safety issues or specific lacks of service such as an efficient public-transport ticket service in Barcelona that impedes easier hop-on and hop-off.

3.3   Calling for Policy Actions—The Co-production of Conversations

The interpretations of the maps produced by the citizens' journeys and their analysis with the visualization of tracks on maps have enabled highlighting particular problems. These problems have been utilized to shift from 'complaints' to 'policy action' with citizens. Every MUV pilot city in these three years of experience with MUVers and MUVerhoods also opens channels of co-policymaking between citizens and urban mobility planners, policymakers and other publics. The MUVers' active, playful activity has sparked positive energy to talk freely about new ideas for future policy actions (MUV pilot coordinator 2020). The calls for policy actions have proven to be more effective when designed in combination with festivals and other events in cities. The 'EU mobility week' has been the anchoring event to produce new conversations among citizens and various types of publics, for example, through temporary communication campaigns. In Palermo, a guerilla marketing campaign has raised the attention for citizens-policy interaction on mobility issues, mediated by MUV professionals. Posters produced after the interpretation of the maps with citizens have been placed on the street level to trigger several conversations (MUV pilot coordinator 2019). In Ghent, a campaign facilitated the information about the safer routes on the neighborhood and the crossroads that kids can use. MUVers' are equipped with fluorescent covers that show 'safety across the neighborhood'. A 'neighborhood house' has been established to provide more information on safety and the MUV app. Another campaign has been the to chalk-spray Emoji's conversations that visualize the bikers' experiences on 'hot spots' (Fig. 3). New ideas on how

Fig. 3 Living MUV emoticons (Ghent)

to improve urban mobility policy in MUVerhoods concerning citizens' everyday-mobility practice have produced conversations adapted to the diverse pilot contexts and urban identities. In Palermo, the safety of pedestrians during evening hours has created the idea of streetlights designed by artists that would activate as people pass by to reduce fear on the streets and criminality and encourage walking instead of taking the car (MUV pilot coordinator 2019).

Safety for tourists to enjoy the city and discover urban experiences has been advanced by the conversations between cyclists and the public administration as adopting some temporary obstacle-free bike lanes. Values of sustainability have forged ideas of carpooling among citizens by multiple people. In Ghent, conversations on the upcoming Sustainable Urban Mobility Plans (SUMP) have been enhanced targets for MUVers data (MUV pilot coordinator 2019). In Barcelona, the conversations have activated citizens' ideas on the safety needs of pedestrians on crosswalks and traffic adjustments such as the green light timing; also, emerging ideas have been facilitating hop-on-hop-off on public transport with the use of contact-less cards or smartphones (MUV pilot coordinator 2019). In Fundao, the idea of converting rural ways into bike lanes will facilitate a healthy lifestyle; promoting a bike lane to school will reduce car-dependency; and education can enhance sustainable mobility orientations. Peripheral car parking in the city and pedestrian routes crossing the whole town will facilitate walking instead of other modes of transport (MUV pilot coordinator 2019). In Helsinki, the conversations between citizens and other policy actors have also underlined the relationship and the role of citizens and data providers. Citizens providing data have pursued the idea to become immediately informed about the role and nature of the data provided (e.g., automated graphs generated in the response), increasing the motivation for data production (MUV pilot coordinator 2019).

4 Toward a Smart Urban Planning: Co-Producing Citizenship in the Era of Digitalization

The MUV project and the practice stories of the change-oriented practitioners in various cities have provided exciting lessons on the interplay between data and citizens to inspire the idea of planning: data shape not just pieces of evidence to point to specific solutions but can co-produce a diverse view of the role of the citizens and future citizenship. The MUV participatory process has developed toward deliberately meet inclusive, safe, resilient and sustainable urban mobility values with citizens. Six EU neighborhood communities have been transformed in MUVerhoods, living urban experiences based on mobility data and game communities. The idea of planning that emerges here is the shift from a mindset of thinking the future *for* citizens as data-points to imagining the future *with* citizens as active agents of transformative urban governance. Lessons from MUVs consist of the redesign of the deliberative stages through which the role of citizens change concerning the data for a transformative urban democracy. MUV contributes to illuminate a pragmatic pathway to re-imagine the idea of planning with data and citizens—for the generation of data (the co-creation of values), the interpretation of facts (the co-design of meanings) and the call for policy actions (the co-production of conversations) (Table 1). MUV inspires the idea of planning in which data open new pathways to transform urban societal realities and co-produce a new sense of future citizenship. Smart urban planning, under a critical-pragmatism perspective, emerges as a participatory process in which values are co-created with data citizens, meanings are co-designed by their interpretation and actions are co-produced by conversations at street-level urban democracy. The role of the professionals as change agents consists of mediating citizens' everyday practice to generate data and values in their daily life, facilitating citizens' interpretation of data through representation and negotiate citizens' ideas with policy actors. A key lesson from MUV is the continuous data-driven mediation to cultivate conversations among data, citizens and policy actors.

Acting as a smart urban planning process, MUV has opened a new mindset regarding the interplay of data and citizens in policymaking. Smart urban planning is not just about the final destination of data—if data will serve traffic planning, or urban development, or new sustainable mobility plans or the provision of new services—but the very way change-oriented professionals think in action future citizenships. A shift from an idea of planning *for* citizens to planning *with* citizens requires reflexive professionals to re-imagine the very co-production of urban ecological and digital

Table 1 Staging a critical pragmatic pathway for smart urban planning

Co-creation of values Generate data	Co-design of meanings Interpretation of data	Co-production of conversations Transformation of data into action
Professionals mediate citizens' everyday practice to generate data of value for individual and collective choices	Professionals facilitate citizens' representation and interpretation of data into facts	Professionals elicit citizens to negotiate policy actions
Lessons from MUV: the app and gamification strategy collect data and shape urban mobility values as an active and healthy lifestyle	Lessons from MUV: data aggregated are visualized and communicated through maps that citizens can interpret	Lessons from MUV: communication campaigns and temporary urbanism engage citizens' ideas at street-level practice
Citizens create data communities	Citizens design data meanings	Citizens produce urban citizenship
Lessons from MUV: game communities shape MUVerhoods where citizens connect to play active mobility together	Lessons from MUV: visualization of aggregated data on maps elicits local knowledge of MUVerhoods	Lessons from MUV: campaigns at street-level produce a new sense of citizens' ownership of MUVerhoods

ecosystems of knowledge through citizens and data. In MUV, this system of knowledge has created new positive energy for policy change toward a new culture of participation and deliberative democracy. In the era of digitalization, professionals urge rethinking the idea of planning through data for future urban democracy. Change-oriented practitioners can enhance a transformative potential by rethinking the role of citizens through data with: (1) technological dispositive that do not just 'gather' data but engage individual citizens' to make sense of those data in the practice of the everyday life (choices); (2) frame the means for the interpretation of data with citizens' to shape new communities of knowledge (practice) and (3) create public conversations between citizens and other publics for transformative utterances of urban societal realities (system), to be enhanced possibly by street-level practices. The citizens' practice in their everyday life is the essential setting for re-imagining and redesigning new digital and physical urban futures for planning the next city.

Acknowledgements This research has received funding from the European Union's Horizon 2020 research and innovation program, under grant agreement No 723521. Thanks to all the research team and especially to Nicola Morelli (AAU), Jesse Marsh (atelier.it), the coordinators of MUV 2020 Salvatore Di Dio (PUSH), Domenico Schillaci (PUSH), and all the rest of the smart urban planners Andrea Vesco (LINKS), Max Kortlander (Waag), Judith Veenkamp (Waag), Alessia Torre (PUSH), Rafel Nualart (i2cat), Emilia Pardi (BAG), Inge Ferwerda (LUCA), Heli Ponto (FVH). The paper derives and synthesizes several deliverables produced in the course of the MUV project, but the author has sole responsibility for the content of this publication. Thanks to the two anonymous reviewers for contributing to the improvement of the quality of the manuscript in its final form.

References

Baker M, Coaffee J, Sherriff G (2007) Achieving successful participation in the new UK spatial planning system. Planning Practice Res. 22(1):79–93. https://doi.org/10.1080/026974506011 73371

Bauman Z (2007) Liquid times: living in an age of uncertainty. Polity

Beck U (1992) From industrial to risk society. Theory Culture Soc 9:97–123

Campbell H (2012) 'Planning ethics' and rediscovering the idea of planning. Planning Theory 11(4):379–399

Cardullo P, Kitchin R (2019) Smart urbanism and smart citizenship: the neoliberal logic of 'citizen-focused' smart cities in Europe. Environ Planning C. Politics Space 37(5):813–830. https://doi.org/10.1177/0263774X18806508

Caroleo B, Morelli N, Lissandrello E, Vesco A, Di Dio S, Mauro S (2019) Measuring the change towards more sustainable mobility: MUV impact evaluation approach. Systems 7(2):30

Cowley R, Joss S, Dayot Y (2018) The smart city and its publics: Insights from across six UK cities. Urban Res Practice 11(1):53–77. https://doi.org/10.1080/17535069.2017.1293150

Di Dio S, Lissandrello E, Schillaci D, Caroleo B, Vesco A, D'Hespeel I (2018) MUV: a game to encourage sustainable mobility habits. In: International conference on games and learning alliance. Springer, Cham, pp 60–70

Flyvbjerg B, Landman T, Schram S (2016) Tension points: learning to make social science matter

Forester J (1999) The deliberative practitioner: encouraging participatory planning processes. MIT Press, Cambridge

Forester J (2013) On the theory and practice of critical pragmatism: deliberative practice and creative negotiations. Planning Theory 12(1):5–22

Forester J (2017). On the evolution of a critical pragmatism. In: Encounters with planning thought, pp 280–296

Geels FW (2005a) Processes and patterns in transitions and system innovations: refining the co-evolutionary multi-level perspective. Technol Forecast Soc Chang 72(6):681–696

Geels FW (2005b) The dynamics of transitions in socio-technical systems: a multi-level analysis of the transition pathway from horse-drawn carriages to automobiles (1860–1930). Technol Anal Strategic Manage 17(4):445–476

Kitchin R (2015) Making sense of smart cities: addressing present shortcomings. Cambr J Regions Econ Soc 8(1):131–136. https://doi.org/10.1093/cjres/rsu027

Lissandrello E, Morelli N, Schillaci D, Di Dio S (2018) Urban innovation through co-design scenarios. In: Conference on smart learning ecosystems and regional development. Springer, Cham, pp 110–122

Lissandrello E, Vesco A (2020) Editorial preface. Int J Urban Planning Smart Cities 1. https://www.igi-global.com/pdf.aspx?tid%3D244196%26ptid%3D228593%26ctid%3D15%26t%3DEditorial%20Prefaceandisxn=null

Saunders T, Baeck P (2015) Rethinking smart cities from the ground up. Retrieved from https://media.nesta.org.uk/documents/rethinking_smart_cities_from_the_ground_up_2015.pdf

Schön D (1983) The reflective practitioner: how professionals think in action. Basic Books, New York

Shove E, Pantzar M, Watson M (2012) The dynamics of social practice: everyday life and how it changes. Sage, London

Urry J (2004) The 'system' of automobility. Theory, Culture Soc 21(4–5):25–39

Vanolo A (2016) Is There Anybody Out There? The Place and Role of Citizens in Tomorrow's Smart Cities. Futures 82:26–36. https://doi.org/10.1016/j.futures.2016.05.010

Wagenaar H (2011) A beckon to the makings, workings, and doings of human beings: the critical pragmatism of John Forester. Public Administration Review 71(2):293–298

Wilson, A., Tewdwr-Jones, M., and Comber, R. (2019). Urban planning, public participation, and digital technology: App development as a method of generating citizen involvement in local planning processes. Environment and Planning B. Urban Analytics and City Science, 46(2), 286–302.

Digital Technologies for Community Engagement in Decision-Making and Planning Process

Antonella Galassi, Lucia Petríková, and Micaela Scacchi

Abstract The way that we describe and understand cities is radically transforming—just like the tools we use for designing and implementing them. The change is often seen only as a technological aspect, for example, in the concept of smart cities. Smart cities are believed to provide societies with a higher quality of life thanks to modern technologies. However, there is also a human factor that is needed to make these changes go smoothly: acceptance. For many, change and innovation cause fear and disrupt everyday habits. Public participation is crucial both for understanding citizens' needs and for adopting new programs. The ability to try, engage, or entertain with new technologies will move innovation from the abstract level to the level of understanding. A smart city can be a living laboratory that tests new technologies and services where citizens and urban communities are active actors in the process. Innovation can be used by the city to improve its services, mutual communication, and engage citizens in its activities and projects, co-creating urban space and city strategy through new participatory tools. Trends in European cities show that the use of modern digital technologies and interactive tools can be used to involve citizens in urban decision-making processes, e.g., when creating or revitalizing public spaces. Modern participatory technologies that enable citizens to explore, analyze, design, and evaluate spatial information on the basis of shared and open data that bring new challenges and new opportunities to cities, as well as for citizens. Our knowledge of the use of these new technologies, however, is still narrow and limited today. In the following research, the authors intend to explore the potential of digital technologies for community engagement in the decision-making process in smart cities by examining the specific settings upon which social innovation builds. We discuss

A. Galassi · M. Scacchi
Dipartimento di Pianificazione Design Tecnologia dell'Architettura, Università degli Studi di Roma "La Sapienza", Rome, Italy
e-mail: antonella.galassi@uniroma1.it

M. Scacchi
e-mail: mica.scacchi@tiscali.it

L. Petríková (✉)
Institute of Management, SPECTRA Centre of Excellence EU, Slovak University of Technology, Bratislava, Slovakia
e-mail: lucia.petrikova@stuba.sk

© The Author(s), under exclusive license to Springer Nature Switzerland AG 2021
A. Bisello et al. (eds.), *Smart and Sustainable Planning for Cities and Regions*,
Green Energy and Technology, https://doi.org/10.1007/978-3-030-57332-4_27

the potential of digital participation for community development and propose good-practice examples for facilitating the process of adopting and integrating digital technologies within such settings. Rather than conclusions, some final reflections are proposed, based on how digital technologies can play a crucial role in involving new groups of people, empowering citizens and building new relationships at the local level.

Keywords Innovation · Digital tools · Decision making · Smart cities

1 Introduction

The boom in the field of information and communication technologies (ICTs) has fundamentally affected the development of societies and the way they interact. The 2030 Agenda for Sustainable Development (United Nations 2015) and the Sustainable Development Goals (SDGs) (Griggs et al. 2013) support the concept of ICTs as the potential means to advance knowledge societies and ameliorate the digital divide—a gap between people with effective access to digital and information technology and those with poor access (Fleming et al. 2018). Governments, policymakers, and city authorities, as well as citizens, realize the power of ICTs and digital engagement as the way to improve communication between various urban stakeholders and public institutions and improve public service-delivery capacities. Digital engagement can play a vital role in building more effective, inclusive and accessible institutions to support policy-making and service delivery for the SDGs to all people and, at the same time, build public trust and ensure transparency, participation and collaboration in the planning process. A well-fitted participation process may prevent later dissatisfaction and better meet citizens' expectations. Even if public consultations are not mandatory, the process usually includes some type of public meetings or hearings. However, most of citizens do not attend such meetings for various reasons (inconvenient time, location, unfamiliar subject, technical language, etc.). Additionally, public meetings typically held at city halls are not very popular among generation Y[1] (MacKinnon 2008; Sloam 2012). To address these limitations, we explore the use of two digital technologies, *CvikerAr* and *InViTo*, interactive engagement tools that help citizens better understand various urban development projects and enable them to express their opinions through digital platforms. This paper shares lessons from our research[2] into some of the pioneering innovations in digital engagement that are taking place across Europe and beyond.

---

[1]The publication *Advertising Age—Ad Age* was one of the first to coin the term "Generation Y" also known as "Millennials," generally refers to the generation of people born between the early 1980s and 1990s to early 2000s as ending birth years (Advertising Age—Ad Age, 30 August 1993, p. 16).

[2]Common research topics among our departments and especially part of the process of doctoral thesis of PhD student Lucia Petríková.

2   Method

In the context of the rapid advancement in technologies relevant to community engagement, this paper attempts to explore the relationship between new ICTs and participation, by examining the role played by specific kinds of digital participatory tools, CvikerAr and InViTo, to engage wider community in the planning process. In this regard, the research into community engagement in decision-making and planning process is in the spotlight when we talk about smart cities, yet little research has been carried out in the area of digital-supported community participation. Addressing this gap, we study best practice on the use of digital technologies in communities to reveal the significance of digital participation for community-led development in smart cities and the role of communities in decision-making processes. We explore the role of interactive tools, as the digital engagement—'connectedness' in the technical sense; and as opportunities for their effective use by various communities in cities—'connectedness' in the social sense.[3] A summary of individual cases will be discussed.

3   Digital Tools for Community Engagement

The era of digitalization brings new forms of civic participation. The success of a community to deal with challenges that contemporary cities face is predicated by its community members feeling a sense of belonging and place attachment. "The highly profiled identification with the living space and deeply articulated place attachments" implies the rebirth of the civic sense and belonging, a desire for identity and participation (Jaššo and Petríková 2016). The new generation Y is more attentive and active, with predominant bottom-up movements of social, political, environmental, community interest. Community participation has moved from traditional approaches (such as town hall meetings, opinion polls, etc.) (Glass 1979) to more active and engaging approaches such as e-participation[4] (Macintosh 2004), supported using ICTs. To help address very complex urban challenges, public and private sectors have begun developing tools that use technology to make participation more informed, transparent, and relevant to citizens' daily lives. Modern ICTs enable citizens to create, share content, and participate in planning processes using a wide range of digital tools. Such tools can make easier for people to share their views with large groups of people, support greater education, enhance connections between institutions and citizens or small communities, give input to decision-making, connect like-minded

---

[3]People with "Connectedness" find meaning, purpose, and deeper relationships. They often feel personal responsibility to the connections they make, actively participating. A structural connectedness is based on the idea that policy and community engagement are made within a context of a network of actors and institutions.

[4]E-participation is the term referring to ICT-supported participation in processes involved in government and governance.

people to work with on a common goal and raise attention or money (Bolívar and Muñoz 2018). Modern community tools build up techniques that have been already used to engage communities (such as workshops, meetings, etc.); however, they are not meant to replace them, but rather to complement them. One of the main aspects of e-participation is to motivate and engage citizens in the decision-making process, promoting the following advantages: enabling broad participation; adjusting a range of tools to citizens' varied technical and communication abilities; providing relevant and up-to-date information to citizens to make more informed decisions and deliberation; and enabling analyzing data provided by citizens (Vito 2018).

3.1 Digital Social Innovation

Digital technologies are especially well fitted for civic action: from mobilizing various communities and sharing resources to spreading capabilities. Some of these are particularly aimed to deal with social challenges. These are, for example, online platforms for citizen participation in policymaking or new development projects or open data to promote better transparency around public spending (Kuriyan et al. 2011). This is what we call digital social innovation (DSI). DSI is defined as:

> a type of social and collaborative innovation in which innovators, users, and communities collaborate using digital technologies to co-create knowledge and solutions for a wide range of social needs and at a scale and speed that was unimaginable before the rise of the Internet. (Bria et al. 2015)

Particularly when the level of public participation increases, the tools become more interactive to foster ever advancing complex discussions. For example, tools based on *open knowledge*[5] refer to online platforms through which diverse groups of citizens can collectively create, analyze various scopes of issues, or crowdfund social projects. Popular types of digital social innovation include participatory platforms that enable citizens to crowd-map local problems (e.g., unsafe areas, broken roads, polluted zones, etc.), e-petitioning, e-budgeting, e-governance, and the like, while impacting local communities or the wider society. Another interesting example is the *open ministry* concept (also known as crowdsourcing legislation) that enables citizens to co-write and grant citizen-led policy proposals, e.g., this concept is implemented in Finland (Finnish Citizens Initiative Act 2012). Tools based on *open data*[6] refer to innovative ways to open, capture, use, analyze, and interpret data. This approach has been successfully tried and tested in the city of Vienna, Austria, which has set up over 160 databases to cover issues from budgeting to planning information (Homeier et al. 2019). This led to development of more than 109 open database apps for the city and its citizens. Similarly, the city of Barcelona created an open-data digital tool to

---

[5]Open knowledge is free to use, reuse, and redistribute without legal, social or technological restrictions.

[6]Open data are freely available to everyone to use and republish, without restrictions from copyright, patents, or other mechanisms of control.

keep citizens informed on processes and to receive their input, which has eventually become a global initiative (Peña-López 2017). The city of Bologna established the so-called "Office of Civic Imagination" with a specific purpose to advance greater participation by creation of "engagement laboratories" throughout the city and the use of interactive digital tools (d'Alena et al. 2018). It works as a customizable platform for communication between city planners and various communities. Some of those communities have later created their own tools, e.g., *YouthScore*, where youngsters rate their neighborhoods based on their youth-friendliness. We can assume that these examples present a promising trajectory toward more inclusive participation with a potential to engage various different stakeholders and enable people to build positive attitudes toward the places where they live, work, or study. The main idea is to allow communities to easily influence decisions that may impact them and where decision-making process becomes more reactive to community input. The premise is that, the more people feel empowered to shape their communities, the more they will participate—and the more they participate, the more inclusive decision-making process will be toward the community voice, with aspiration to motivate more community members to participate.

3.2 Good Practice

As previous examples showed, there are many ways by which interactive tools, along with traditional approaches, can advance civic participation. Based on the outcome of a collaboration between the Slovak University of Technology and the Slovak Smart City Cluster,[7] we chose *CvikerAr* tool as an example of a local best practice of innovative technology for improving transparency in planning and decision-making processes by enabling community members to better understand specific situations and encourage more collective decisions ahead of individual interests. *InViTo*, the second tool that we present was generated during the LUMAT[8] project in which the authors participated based on its positive outcomes in multiple case studies (Coppola et al. 2014). This tool is conceived as a toolbox for a visual analysis, exploration, and communication of spatial and non-spatial data to support policy and decision making. These tools are based on the open-data concept with a focus on the visualization of spatial data. A visual interface there is used as a new criterion to display both positive and negative impacts on territories while respecting the complexity of multiple stakeholders' choices. The interactive form enables users to analyze data themselves. Comparing various scenarios and modifying different features of the subject supports discussion of specific issues related to the community. The tools have shown to be effective, especially when evaluating various planning scenarios.

---

[7]Slovak Smart City Cluster is an association integrating representatives of business sector, public administration, academic environment, and technology innovators.

[8]LUMAT Project—Interreg Central Europe Programme 2019—Implementation of Sustainable Land Use in Integrated Environmental Management of Functional Urban Areas.

Fig. 1 Illustration of the use of CvikerAr to engage communities in the planning process. *Source* www.cvikerar.com (the illustration elaborated by the authors)

3.2.1 CvikerAr, Poprad, Slovakia

In 2016, the Municipality of Poprad, among other cities in Slovakia, adopted the first smart-development strategy with the aim to provide better services to its citizens and improve the quality of life in the urban environment by introducing smart solutions supported by modern technologies. This has caused the city to be ranked among the European Smart Municipalities.[9] In addition, the strategy covers topics of smart economy and smart governance. The city runs a pilot initiative in the development of digital technologies for community engagement in the planning process with the mobile app CvikerAr that enables visualizing the real world through 3D modeling in virtual reality. Thanks to its virtual feature, it makes it possible for citizens to view any proposal in its realistic environment before their realization. For example, people can visualize how new building plans or areas will look, how they will affect the urban fabric of the city or choose between different scenarios. This interactive tool, at first, will be tried and tested in a revitalization process for current brownfields (a former area of military barracks) located in the city. So far, the tool has been used in minor projects, e.g., for a proposal for a new pedestrian bridge connecting two residential neighborhoods (see Fig. 1) which allowed various stakeholders and residential communities to engage in the planning process. It enabled citizens to actively participate in decision-making process from the beginning, see different drafts in a "real" picture, analyze, and comment, as well as actively co-design.

3.2.2 InViTo, Torino, Italy

The second example of an interactive tool InViTo is an acronym for *Interactive Visualization Tool*. InViTo has been classified within the category of spatial decision-support system (sDSS) as a Web-based GIS tool (Geographic Information System). It was developed to deal with various spatial issues and disciplines with the aim of sharing the spatial information to visualize urban effects in real time and to improve the territorial decision-making process in general (Pensa and Masala 2014). The main purpose of the tool is to help people build their spatial knowledge by interacting with dynamic maps. Similarly to CvikerAr, it is able to display the relationship between an area and a proposed intervention in real time. Having been designed to encourage

---

[9]https://ec.europa.eu/info/eu-regional-and-urban-development/topics/cities-and-urban-development/city-initiatives/smart-cities_en [accessed October 19, 2019].

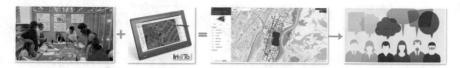

Fig. 2 Illustration of the use of the interactive map in the InViTo platform during a stakeholders' workshop. *Source* www.urbantoolbox.it (the illustration elaborated by the authors)

discussion among different actors, e.g., urban planners, stakeholders, non-experts, focus groups, and urban communities, it enables the exchange of knowledge in collaborative and participative activities. At the same time, it allows a full interaction between users and the information. For example, citizens can share a map of their neighborhood with access to local information and, by clicking on the map, see further details with the possibility to leave a notice (e.g., comment, complaint, suggestion, etc.). Moreover, it allows citizens to choose an area and receive statistical data associated with the chosen locality. Citizens can access, edit, or save their maps and open them later. It is flexible to use for various applications, purposes, and scales, with the possibility to actively manage and modify data (variables) and create dynamic scenarios. Because of this, it can showcase areas of interest and create immediate outputs. The outcomes can be presented in various forms according to the user expertise and used together with other community planning instruments, during collaborative working activities, meetings, and workshops (LUMAT Project 2018) (Fig. 2).

4 Results

It is clear that all spatial decisions involve a number of different actors, opinions, and interests, and we consider data communication as fundamental to achieve common agreement. With the boom in ICTs in the field of territorial development, the vast amount of complex information is not easily understandable through simple reasoning anymore. Based on the original study of Tang and Waters (2005) and the research carried out under the LUMAT project, we elaborated a new table on the effectiveness of different participation techniques, having evaluated them according the selected criteria—* poor; ** fair; *** good; **** outstanding (see Table 1). The table shows comparisons between the traditional and new approaches to citizen engagement in participatory planning practice.

Traditional methods of participation in the planning process (such as questionnaires, surveys, public meetings and public hearings) have shown to be relatively demanding in terms of time and money necessary for their collection. As we can see from the table, some traditional techniques prove to be more successful than others, particularly workshops and contacts in communities. Others, i.e., public hearings and questionnaires, seem to work less successfully. On the other hand, the field of new participation techniques has been widely introducing a new range of interactive

Table1 Evaluation of the effectiveness of selected participation techniques

	Participation technique	Providing information	Receiving information	Interaction with communities	Giving assurance to communities	Broad cross-section of opinions	Communication
Traditional	Public hearings, meetings	***	*	*	**	*	one-way
	Workshops, focus groups	****	*****	*****	**	***	two-way
	Presentations to clubs and groups	***	**	**	**		one-way
	Contacts with people in community	****	*****	*****	****		two-way
	Questionnaires, surveys	*/**	*****	*	*	***	one-way
Up to date	E-participation (e.g., online opinion surveys, e-budgeting)	***	***	***	***	***	one-way
	Digital participatory tools (e.g., CvikerAr, InVito, sDSS)	*****	*****	*****	****	***	two-way

tools for citizen involvement. Among the new techniques, digital participatory tools appear to be most effective in various aspects. The right-hand column shows the level of communication. This level involves two dimensions—one-way communication flow and two-way communication flow. As we can see, the one-way communication in e-participation (without the possibility for interaction) may eventually suffer a communication barrier, in comparison with digital participatory tools that provides the space for immediate feedback. The one-way-oriented approach does not allow such a degree of interaction and feedback on both sides. We assume that modern decision-support tools, such as InViTo and CvikerAr presented in this paper, are emerging as promising tools for solving complex urban issues, collaborative participatory planning and effective spatial analysis in territorial decision-making processes. Based on the outcomes of the research, these tools have proved to be able to make very complex information more comprehensible even to people who are not familiar with the technology by combining interactive maps with pictures and text information in a user-friendly visual interface. They can provide better interaction and mutual feedback, while encompassing a broad space for opinions, exchange of ideas and discussion. The basic knowledge of urban dynamics is essential for addressing specific community-related issues. We believe that the better is the knowledge in the planning process, the higher is the chance of community to make better decisions. Communities may have a better chance to actively influence issues that directly concern them with and come up with new social innovations.

5 Discussion

Of the many different technologies that support participatory approaches to community engagement in decision making, modern participatory technologies are gaining increased attention as a means of fostering more inclusive planning process. There are many good-practice examples of the successful application of digital participatory tools in urban communities; some are used to monitor a local quality of environment (quality of air, water contamination, dangerous pollutants); others use smart mapping based on local knowledge and information to reveal 'critical' areas where people do not feel safe, or suffer a lack of greenery, and spaces for new cycling routes, etc. There are, however, challenges that digital engagement will need to address in the future. In particular, this includes how to define a better understanding of what we mean by "participation," more flexible communication at the institutional level (open governance) and how to tackle the digital divide. As an important factor, we recognize the digital education at the level of communities with the emphasis on underrepresented groups, e.g., elderly people, the poor, minorities, etc., as well as the diversity and inclusion in the development and testing of new participatory techniques. We can assume that the demand for more participatory and more inclusive ways in decision making will continue to grow, either from city authorities, urban planners or citizens, and the technology will advance to bring more inclusive, cheaper, and easier ways to provide a greater participation and more transparency.

6   Conclusion

This paper provides an overview to help understand its wide scope of activities and the new emerging techniques. Rather than draw a conclusion, we wish to end with some reflections on the role that technological innovation can have in the decision-making process. Lessons from international case studies show that digital tools are being used to engage communities in more meaningful participation, while they are improving the quality and validity of decision making. Experimentations with implementation at the local level have shown that digital technologies can play a crucial role in engaging new groups of people, empowering citizens and building new relationships between cities and local communities, as well as local governments and citizens. As the meaning of community engagement is particularly relevant at the local level, local governments have begun initiating platforms to enable citizens to contribute with their ideas and local knowledge, evaluate priorities, and influence allocation of public resources. In this sense, the community is to be considered both actor and beneficiary. We assume that the biggest concern when using such tools is to overcome the citizens' lack of familiarity with digital technology. This is particularly important in territories with no or little experience in this field. Also, it is important to emphasize that the traditional participation techniques should not be forgotten, simply because even the modern digital methods are not a cure-all for all community-related issues. The 'smart' techniques should be integrated simultaneously with the traditional ones. Eventually, the digital engagement in the field of citizen participation is a multispectral concept that brings a set of challenges for modern cities that will require yet deeper research in the future.

References

Bolívar MPR, Muñoz LA (2018) E-participation in smart cities: technologies and models of governance for citizen engagement, vol 34. Springer, Berlin

Bria F, Gascó M, Kresin F (2015) Growing a digital social innovation ecosystem for Europe. DSI final report. European Commission, Brussels. 10.2759/448169

Coppola P, Pensa S, Masala E, Tabasso M, Papa E (2014) Visualising accessibility: an Interactive tool and two applications to empirical case studies of urban development and public engagement. https://doi.org/10.13140/RG.2.1.3459.0484.

CvikerAr Online. https://cvikerar.wixsite.com. Accessed 13 Sept 2019

d'Alena M, Beolchi S, Paolazzi S (2018) Civic imagination office as a platform to design a collaborative city. In: ServDes2018—service design proof of concept, pp 646–648. https://www.ep.liu.se/ecp/150/053/ecp18150053.pdf

Finnish Citizens Initiative Act 2012. Online. https://www.finlex.fi/fi/laki/alkup/2012/20120012. Accessed 24 Sept 2019

Fleming A, Mason C, Paxton G (2018) Discourses of technology, ageing and participation. Palgrave Commun 4:54. https://doi.org/10.1057/s41599-018-0107-7

Glass JJ (1979) Citizen participation in planning: the relationship between objectives and techniques. J Am Planning Assoc 45(2):180–189

Griggs D, Stafford-Smith M, Gaffney O, Rockström J, Öhman MC, Shyamsundar P, Noble I (2013) Policy: sustainable development goals for people and planet. Nature 495(7441):305

Homeier I, Pangerl E, Tollmann J, Daskalow K, Mückstein G (2019) Smart City Wien Rahmenstrategie 2019–2050. Magistrat der Stadt Wien, Wien

Jaššo M, Petríková D (2016) Towards creating place attachment and social communities in the SMART cCities. In: Smart city 360°: First EAI international summit. Bratislava, Slovakia and Toronto, Canada. Smart City 360°, 13–16 Oct

Kuriyan R, Bailur S, Gigler B-S, Park K (2011) Technologies for transparency and accountability: implications for ICT policy and implementation. https://doi.org/10.13140/RG.2.2.19320.24320

Longford G (2008) Community networking and civic participation: surveying the Canadian research landscape. J Commun Inform 4(2)

LUMAT Online. www.interreg-central.eu. Accessed 19 Oct 2019

Macintosh A (2004) Characterizing E-participation in policy-making. In: The proceedings of the thirty-seventh annual Hawaii international conference on system sciences

MacKinnon MP (2008) Talking politics, practicing citizenship. Educ Canada 48(1):64–66

Open Definition—Defining Open in Open Data, Open Content and Open Knowledge. Online. opendefinition.org. Accessed 19 Sept 2019

Open Ministry—Crowdsourcing Legislation Online. https://openministry.info. Accessed 20 Sept 2019

Peña-López I (2017) Decidim. Barcelona, Spain. IT for Change, Barcelona

Pensa S, Masala E (2014) InViTo: an interactive visualisation tool to support spatial decision processes. In: Pinto NN, Tenedorio JA, Antunes AP, Cladera JR (eds) Technologies for urban and spatial planning: virtual cities and territories. IGI Global Book, Hershey, PA, pp 135–153

Sloam J (2012) Introduction: youth, citizenship and politics. Parliamentary Affairs 65(1):4–12. https://doi.org/10.1093/pa/gsr048

Tang KX, Waters NM (2005) The internet, GIS and public participation in transportation planning. Progress Planning 64:7–62

United Nations (2015) Transforming our world: the 2030 agenda for sustainable development. UN Publishing, New York

Vito D. (2018) Enhancing participation through ICTs: how modern information technologies can improve participatory approaches fostering sustainable development. In: Petrillo A, Bellaviti P (eds) Sustainable urban development and globalization. Research for Development. Springer, Cham

YouthScore Online. https://resources.esri.ca/education-and-research/. Accessed 1 Oct 2019

Emerging Interpretation Models of Social and Institutional Innovation in the City. The Role of 'Intermediate Places' Between the USA and Italy

Bruno Monardo and Martina Massari

Abstract Distinguished schools of thought, within the increasing impact of the 'Innovation District' phenomenon, have highlighted the 'Innovation Center' idea and its interpretations as physical structures able to build a powerful nexus for social and institutional innovation in urban regeneration. These 'intermediate places' can be considered interactive playgrounds, triggering new horizons in urban policies toward shared, inclusionary solutions more likely to meet the needs of local communities. Social innovation is strictly path-dependent, enabled by 'opportunity windows' in which local actors get mutual engagement and advantage, addressing contextual needs, while creating virtuous cooperation and new governance arrangements. The thesis of the paper is that innovation centers—as 'intermediate places'—can be successful if they are built to recover direct relationships between the various stakeholders in the urban arena; consequently, the operative capacity of practices can expand the planning strategies providing the perspective of a long-term change. Through comparisons and drawbacks arising from case studies selected from different cultural and physical geographies (City of Boston, Massachusetts, USA, and Bologna, Italy), the paper emphasizes the recognition of the variety of 'intermediate places', encompassing the diversity of actors, within a strategy for an authentic urban innovation ecosystem.

Keywords Social innovation · Intermediate places · Institutional innovation

1 Introduction

Since the last decades of twentieth century, the issue of innovation has been exploding in the Western world discourse and public policies. Supported by the theoretical

B. Monardo (✉)
Department of PDTA, Sapienza University of Rome, Rome, Italy
e-mail: bruno.monardo@uniroma1.it

M. Massari
Department of Architecture, University of Bologna, Bologna, Italy
e-mail: m.massari@unibo.it

© The Author(s), under exclusive license to Springer Nature Switzerland AG 2021
A. Bisello et al. (eds.), *Smart and Sustainable Planning for Cities and Regions*,
Green Energy and Technology, https://doi.org/10.1007/978-3-030-57332-4_28

debate in academic networks, think tanks and research foundations, the governments of the advanced economies have promoted urban and regional innovation policies to tackle the recurring crisis cycles that had occurred during recent years.

Innovation is a polysemous concept, but the usual rhetoric of conventional policies often highlights and privileges its technological and economic dimensions. However, as recent research findings show (Moulaert et al. 2013; George et al. 2019), the most intriguing contemporary interpretation of innovation, referring to the urban realm, is mostly to be identified in the social (Gerometta et al. 2005; Garcia 2006; Hillier 2013; Massari 2018; Nyseth and Hamdouch 2019) and institutional dimensions (Gonzalez and Healey 2005; Ostanel and Attili 2018).

Social innovation as an investigation concept is not new: decision makers, researchers, professionals, economic lobbies, civic organizations and other stakeholders in the urban arena have always been looking for new solutions to pressing needs (Donolo 2003), especially in the season of 'public shrinking' and market crisis; however, nowadays the social innovation issue is becoming more and more '[…] fundamental for policy-makers, urban scholars, practitioners, and, especially, citizens.[…]' (Palermo and Ponzini 2015: 3). It is an 'umbrella idea', an evocative concept, alluding rather than explaining and, although one struggles to define it, it is socially and conceptually mobilizing. Its goals are aimed at increasing the quality of life in urban settlements (Drewe et al. 2008; Brandsen et al. 2016), with consequent effects on the well-being and empowerment of the local community (BEPA 2011). However, the extent of the initiatives and the spillovers they provoke are currently opening up original opportunities for urban attractiveness and regeneration.

European agendas and international agencies (OECD 2001) have long since established the need for cities to interact with the social innovation sphere and to define approaches that can hold together the dimensions of innovation, its tools, timing and constant change, with an institutional structure able to guide them. The 2030 Sustainable Development Agenda (United Nations 2015) and the new EU Cohesion Policy 2021–27 confirm that cities are expected to experiment with original urban solutions, fostering collective agency and exchange by promoting 'place' and 'people-based' approaches. Within this scenario emerges the mismatch between the intrinsic episodic actions of social innovation practices—with their short-term, immediate and constantly evolving results—and the long-term, regulated strategic institutional vision of urban planning, which is inherently oriented to locate such actions in spatially bounded frameworks.

Another crucial interpretation of innovation considers it as a trigger for institutional change. Modern reflections on 'institutional innovation' started in the late 1980s and caused the revision of regional economic development policies. Within this context, innovation acquired increasing importance as a cross-cutting process able to empower the potentials of local governmental agencies and, at the same time, provide space for new institutional arrangements to blossom (Galston and McElvein 2015).

In contemporary urban policies, the emerging trend of advanced local administrations seems extremely sensitive to the 'institutional innovation' approach, playing a sophisticated game in tailoring 'ad hoc' initiatives, specific tools and adaptive partnerships among anchor actors, higher-education institutions, advanced investors, venture capitals, non-profit organizations and local communities (Mieg and Topfer 2013). This model represents a 'virtuous hybridization' of the wisely mixed dimensions in the planning initiatives: from the overwhelming role of real-estate development to the increasing sensitiveness for local inclusion (George et al. 2019), from socioeconomic and environmental quality to changeable government/governance profiles.

2　Structure and Method

This paper, after the previous introduction and some synthetic notes about the investigation method (this section), concisely develops the profile of 'Innovation Centers' as 'intermediate places' (Sect. 3). The case studies selected in one of the most vibrant, thriving urban areas in USA, the municipality of Boston (Massachusetts), and in the city of Bologna (one of the most advanced local public administrations in Italy) are explored in Sects. 4 and 5, using the analytical lens of key actors and places. Finally (in Sect. 6), the authors highlight and discuss the lessons learned and the evolutionary perspectives, together with open issues, that the galaxy of 'intermediate places' can open toward policies supporting virtuous urban innovation ecosystems.

As already noted, the paper aims to explore how the emerging models of innovation center represent 'intermediate places', contemporary urban nexuses for both social and institutional innovation; these 'new agoras' engender dense social interactions and networks, reconnecting locally based practices and macro policies.

In order to address the question, an in-depth comparative analysis has been carried on by taking advantage of the synergy between two research projects run by the authors (see acknowledgements).

The research method follows the inductive approach (from the particular representative case to the general lessons) and the classic case-study interpretation key (Yin 1984) developed with a qualitative approach and supported by direct sources and interviews. In particular, the following sections draw the principal findings emerging from the empirical case analysis through two main keys:

- innovation centers as 'intermediate places', excellence civic poles, 'cognitive platforms' infilling revitalized districts or critical neighborhoods within urban strategies. What is the 'ontological sense', the essence, the emerging 'DNA' of 'intermediate places'? According to the 'anthropology of postmodernity', the 'place' is recognizable through three dimensions: history, identity and relationship (Augé 1992, 2007). Looking at the consolidated examples from all over the world, innovation centers embody at least the latter two;
- emerging heterogeneous actors and players on the urban stage; private, non-profit, hybrid stakeholders interacting in the place management and design with collective, general-interest objectives. They promote policies, conceive and implement strategies and manage schedules for activities and spaces.

3   Innovation Centers as 'intermediate Places'

As recent research contributions highlighted (Fransman 2018; Rissola et al. 2019), the possibility of pursuing authentic social and institutional innovation in urban ecosystems can find an intriguing opportunity in the rising phenomenon of 'Innovation Districts' (Katz and Wagner 2014; Baily and Montalbano 2018; Wagner et al. 2019) and related 'Innovation Centers' (ICs), catalysts for contemporary urban regeneration strategies in Europe and Northern America (Monardo 2018).

The on-going evolution process of innovation centers is not limited to represent the concentration of high-level economic opportunities, but nurture the ambition to become urban 'cognitive infrastructures for developing collective intelligence' (Mulgan 2019), identity interfaces explicitly pursuing the mission of decreasing the social gap between privileged and recessive actors, formal and informal stakeholders in the urban scene.

Across different latitudes, IC models as 'intermediate places' are promoted by significant partnerships of public, private and non-profit organizations. They represent powerful magnets, civic turbines and relational hubs spreading beneficial effects within single neighborhoods, districts or in the whole city. ICs favor cross-fertilization, catalyze the interweaving and positive stakeholder dialectics and promote opportunities for job creation, setting up training activities, relationship potential, spatial contiguity and virtuous serendipity.

ICs assume the role of places for interaction and mediation, 'intermediate places', structured 'windows of opportunity' (Deleuze 1994; Hillier 2013), where urban actors engage within the urban space.

'Intermediate places' have been defined as 'boundary-spanners' (Williams 2011), hybrid platforms (Bergvall-Kåreborn et al. 2015) linking internal networks (social innovation practices) to external players (business, high education, research) and public resources of the city (municipalities). 'Intermediate places' aggregate an intentional community, identified by the sense of belonging to the context and the system of values. The use of the place dimension refers to the *civitas* spatialization, which today is increasingly determined by the contextual, specific and niche conditions (Donzelot 2009).

The case studies presented in the next sections, both for physical places and key players, represent a significant 'fractal' of the emerging interpretation styles of ICs in urban innovation ecosystems. On the two sides of the Atlantic, the cultural style embedded in urban redevelopment policies in Boston and Bologna can be deemed emblematic in terms of relationships between local public administration and private and not-for-profit actors, proposing an intriguing blending of three main 'modes of governance': hierarchy, market and network. Emerging innovation models are open to variable geometry, flexibility, retrofitting and resilience, with municipal institutions ready to introduce new 'organizational architectures'. Boston and Bologna, despite their deep cultural and juridical diversities, represent exemplary cases, icons of a wider trend that involves other American and European realities. They are different in terms of physical dimension and socioeconomic ranking; however, they share some similarities:

- huge concentration of higher-education institutions, historically embedded in the city. The numeric ratio between students and residents is about 22% (Bologna hosts an average of 86,000 students every year, with 390,000 inhabitants; Boston has about 152,000 students with 685,000 inhabitants);
- innovation represents the main attraction driver for the city (Boston in knowledge, high-tech and business; Bologna in public/civic administration and education as well);
- explicit policies stressing innovation as contextually emerging from the local and territorial milieu.

4 The Boston Case

The Greater Boston area represents a paradigmatic case not only in terms of inno-vation and knowledge driven policy but also in terms of parallel urban planning and redevelopment initiatives. The density of high-profile education institutions and their relationship with the proliferation and vitality of new start-ups is impressive. Since 2010, the Menino administration has launched an explicit urban redevelop-ment policy based on some European urban regeneration experiences (Barcelona, London) opening the new season of Northern American 'Innovation Districts'.

4.1 Intermediate Key Places: District Hall and Roxbury Innovation Center

The Boston Innovation District (BID) in the Seaport neighborhood is a flagship example in that sense. In terms of economic 'critical mass', BID is the most signif-icant long-term planning initiative launched ten years ago by the Menino adminis-tration. The project aims to create a complex neighborhood following the example of 22@Barcelona, one of the most controversial cases of the 'Innovation District' phenomenon (Monardo 2018). The BID project has been conceived to redevelop the South Boston Waterfront, an underutilized area of 1000 acres which previously hosted industrial activities and parking, transforming it into a thriving hub of inno-vation and entrepreneurship together with new residential, commercial and retail spaces with a mixed-use configuration (Monardo and Trillo 2016).

The BID centerpiece is the District Hall, a large public space where innovators meet, exchange ideas, explore potential synergies, finalize their creativity and come to concrete agreements. The building, opened in 2013, is the result of the partnership between the City and private real-estate developers who funded its construction within the development of the Seaport Plan. Recently, District Hall hosted plenty of events ranging from hackathons and training sessions to start-up networking meetings and brainstorming sessions. More than 70 percent of District Hall's space rental value has been donated for community use and about US$1 million invested in the local start-up community.

The District Hall is the natural catalyst of a vast array of favorable conditions: the attraction of an educated and 'creative' workforce, 'local supply' of entrepreneurs, a high concentration of private venture capital and a diversified business environment. It is not only a space for debating, discussing and discovering, but also the ideal place for recovering the physical proximity dimension in new social relationships, entangling work with leisure and consumption, discussion and play.

However, the most significant idea for implementing an authentic social and institutional innovation policy is the Roxbury Innovation Center (RIC), located in one of the most distressed neighborhoods in Boston. Here, the local public administration is the inspirer and turbine for the conception and implementation of the idea of the 'Neighborhood Innovation District' redevelopment strategy.

Roxbury's urban fabric is anchored around Dudley Square, the vibrant heart of the neighborhood. Part of the investment plan in Dudley Square has included the opening in 2015 of the RIC. Located in the heart of the neighborhood, it was created through a public–private partnership with the City of Boston and the Venture Café Foundation. RIC is an intermediate non-profit center that supports the economic growth of the neighborhood by encouraging innovation and local entrepreneurship. Its mission is to support local economic development and job creation in Roxbury by empowering and guiding innovation and entrepreneurship as viable career options. RIC provides a diverse variety of resources for small business owners of all stages and industries, through instructional workshops and courses, networking events and office-hour mentorship.

After some years following the beginning of the implementation strategy, the attraction of private investors is relatively limited, but the Municipality is still giving priority to the initiative in which the refurbishment of Roxbury Center, entirely funded by the City of Boston, is one of the most important steps within the inclusive regeneration plan of the entire neighborhood.

4.2 Key Players: Venture Café Foundation and MONUM

The new horizon of ICs in the emerging Boston urban ecosystem cannot be assessed without investigating at least two key players, the Venture Café Foundation and the Mayor's Office of New Urban Mechanics (MONUM), respectively, representing the private/non-profit galaxy and the City government. The former is a privately driven innovative management actor that contributed to making the District Hall and the Roxbury Innovation Center vibrant and thriving places, bridging the mutual support of big firms and grassroot start-ups. The latter is an 'ad hoc' public agency created by the City of Boston in support of civic engagement and an entrepreneurial approach to local governance policy.

Venture Café is a global movement working to build stronger and more inclusive urban ecosystems enhancing the innovation process into cities around the world. In Seaport and Roxbury districts, Venture Café operates dedicated civic innovation centers, as physical gathering spaces to help foster and sustain innovation communities. The goal is the creation of a connection place for the whole civic arena, with no barriers to entry. In other words, the mission of Venture Café Foundation is to educate and connect anyone with an idea to the resources and relationships they need to successfully launch and grow businesses. With high-touch programming, spaces, storytelling and broad innovation engagement, Venture Café Foundation provides the connective framework to link various parts of the local innovation ecosystem, amplifying and convening the local community.

The Mayor's Office of New Urban Mechanics (MONUM) was created in 2010 as one of the initiatives of the Mayor Thomas Menino and serves as the City's R&D Lab. Currently, it continues its original mission under the leadership of Mayor Walsh, including civic engagement, racial equity, city infrastructure and education. The peculiarity of MONUM is that, although it is a public office operating as an emanation of the public administration, it shows common traits of a typical entrepreneurial actor.

In MONUM, experimentation and risk-taking are two major features, undertaking threats that traditional city departments do not usually take, by developing low-cost, small prototypes piloted to test their potential to be upscaled. Failures are considered as learning opportunities and shared across the larger community of public decision makers; successful experiments are ready to be implemented by the traditional departments and implemented on the city scale. Besides testing innovative ideas, MONUM also plays the role of 'front door' for start-ups, universities and residents who want to explore cooperation spaces with the City.

5 The Bologna Experience

Bologna is renewing its inspirational City Lab model with the aim to retrieve and create opportunities through interaction between *urbs* and *civitas*. Interaction is facilitated by specific places (existing innovation centers, hubs, 'ad hoc' initiatives) and guided by key actors. In this approach, social innovation is informing urban policies, while institutions are more responsible toward the local practices. This overcomes a subsidiarity logic: local engagement, entrepreneurial skills and local knowledge are newly created values. The Urban Innovation Plan is shaping this vision, with a discursive framework for co-design processes led by places.

5.1 Intermediate Key Places: Opificio Golinelli and COB Social Innovation

Opificio Golinelli was born thanks to its eponymous foundation and is defined as a palimpsest of interactive clusters. In this former industrial building, formal and informal, physical and virtual exchanges take place. The space is about 14,000 m², divided into pavilions and independent structures, connected by a system of paths, gardens and squares. The center hosts the training, educational and cultural activities of the Golinelli Foundation and its project areas: School of Ideas, Science in Practice, Business Garden, Science in the Square and Educating to Educate. The Opificio was conceived as a metaphor for the city in which all activities take the form of icons of representative buildings within the urban fabric, such as the City Hall, the School and the Building Site. It provides spaces for children, youngsters, teachers and citizens, following a 'learning-by-doing' approach. The long-term development plan of the Golinelli Foundation— 'Opus 2065'—is designed to support the young generations in their knowledge path, for the implementation of the accelerator of ideas G-Factor, for the formation of business culture in all sectors, and for the delivery of highly innovative, scientific and technological services. With this program, Opificio seeks to integrate the places of knowledge, experimentation and production in order to face the challenges and the contemporary dynamics of the city.

COB Social Innovation was born in 2013 in Bologna and is now located in Valsamoggia, a municipality in the Bologna metropolitan area. It was born with the aim to ensure a wide social impact from the produced goods, whether they were objects, processes or new organizational models. The name COB is born from 'corn-cob', the idea of creating a methodology of diffuse, clustered social innovation; it can also be associated with 'collaborative Bologna' or the contraction of 'co-business'. The municipality offered the spaces to the association in exchange of their reactivation for a public use. After a three-year investment by Philip Morris, the space was configured in its present use. This created a virtuous circle between great corporations, the municipality and a newborn social-purpose association. The space is variously used by many actors: the public administration for institutional meetings, informal groups to discuss and create shared projects for the whole territorial context, associations and freelance subjects for working activities; moreover, the relationships with the closest institutions (e.g., the municipal library) generate and distribute impacts providing new value for the local and extended territory.

The group of innovators works as a consultant for other innovation hubs, and the main idea is to expand and create further clusters in the regional context. Their role as intermediaries is visible in the ability to boost and connect different competences, answering to local necessities, in particular those related to the employment of young people and refugees, to synergize with the business companies and the public actors. Their mission is not to get immediate results but to produce impact and narrate it in a broad and clear manner.

5.2 Key Players: Foundation for Urban Innovation

As stated in the recent Urban Innovation Plan (Comune di Bologna 2017), Bologna aims at fostering the dimension of 'spaces and places' as relational capital, harnessed to create opportunities for communities to interact. The Plan is managed by the 'Office for Civic Imagination', born in 2016 and managed by the Foundation for Urban Innovation (FIU).

FIU is the current configuration of the previous 'Urban Center Bologna', an inter-mediary between the city and its citizens, where multiple resources and tools (skills, data, training and spaces) find a virtuous collision. As the primary experienced player, FIU takes part and enhances the debate about the re-thinking of participatory democracy and policymaking. It is a city agency acting as a broker between the local ecosystems of actors, the municipality and ultimately the European Union level. Its multidisciplinary staff is composed of project managers with different backgrounds in urban planning, architecture, economics, political science, art and communication. FIU's role is twofold: it is an 'access point' (Barbi and Ginocchini 2019) for urban challenge management and urban service supply and fruition, and it is a 'factory' that relates on global models, localized to tackle contemporary challenges, urging citizens' agency on a wide range of sectors and especially in prototyping new products and services.

FIU moves toward the role of triggering new territorial practices, going beyond mere dialogue by giving up some power and resources, in order to put transformative actions into practice. The case of FIU is paradigmatic in its ability to be a 'quadruple-helix' platform between public administration, university, entrepreneurs and citizens. The main success of FIU is the shift in the relationship between the public institution and the citizens: The local 'antennas' in the neighborhoods represent a concretization of the distribution of administrative power that is typical of Bologna's decentralization policies (Cafiero et al. 2014).

6 Lessons Learned and Final Remarks

Boston and Bologna suggest intriguing lessons to be learned on the role of ICs as 'intermediate places' both on social and institutional matrix.

In terms of social innovation, the most relevant issue is focused on the space matter and the role of bottom-up initiatives in cities. The spatial configuration of social innovation is the key element to generate a demand-driven externality (positive or negative).

In Boston, local authorities are recognizing the socioeconomic value of the built environment as one of the main enablers of cognitive networks by delivering mixed use-led master plans, emphasizing the value of public pedestrian-oriented spaces, and encouraging the rise of collective domains. This approach enables a voluntary community to be created and enhanced through proximity work triggering innovative ideas. In parallel, Bologna is expressing a clear urban vision intertwined with social innovation. Here, interaction is the key for a new planning scenario: urban space matters in its multiple and generative dimension as the playground where the interaction takes place. However, interaction still needs to overcome bureaucratic styles in order to foster a more critical, continuous and conscious approach; the cities also need to raise awareness of the innovation center's ability to be both an observatory of practices and a factory of innovative and more coherent policies. Moreover, 'intermediate spaces' in both cases are considered as enabling infrastructures for bottom-up initiatives. While in Boston, the policy is oriented toward the concentration of enabling spaces in specific areas of the city, in Bologna, the strategy aims at creating a hierarchical diffuse network of enabling spaces.

Concerning the emerging strategies of institutional innovation, a shared pattern in the governance structure of the two cities is to be highlighted. In Boston, the quadruple-helix model (i.e., government, university, enterprise and society) is properly embedded in the spatial structure of the city, in which higher-education institutions nurture the entrepreneurial context. Innovation centers are both feeding and applying spillovers from prestigious universities and research institutions, while small start-ups and international corporations prove to be synergistic between them and to interact within the urban fabric. In Bologna, on the other hand, the governance models try to add another element to the quadruple-helix configuration: the organized society (NGOs, associations, committees, social innovators) as a widespread network of urban agents. This is a key element in the value dissemination produced in 'intermediate places', including groups of people who usually are not involved in innovation processes.

In terms of government, institutional innovation is witnessed in both cities as transitioning toward more entrepreneurial models. In Boston, this tendency is eliciting risk-taking and civic participation in tackling key issues in the city (e.g., the unique experiment of MONUM). In Bologna, the urban government embraces the possibility to get feedback and evaluation, a 'trial-and-error' strategy of policymaking, supported by 'intermediate places'. Opificio is an example for the mechanism of mutual learning where the place has been modifying the user action and promoting institutions to foster widespread education by merging art and science. In both cases, a key role is played by the networking structure of intermediary actors supporting entrepreneurial collaboration and cross-fertilization on multiple scales. They enable positive spillovers from the strongest to the weakest contexts: the Venture Café network enabling international cooperation and cross-fertilization, the Foundation for Urban Innovation supporting the emergence and stabilization of the network of urban social innovation.

As a final remark, we sustain that innovation centers could be interesting socio-urban observatories (Nicolas-Le Strat 2015), local strongholds of the municipal administration and research and development units for social and institutional innovation. Despite their growth and success, they increasingly require recognition as real 'intermediate places' for both active citizenship and institutions. For this reason, innovation centers deserve to be structured by a set of policies promoted by an urban agenda at both the local and national level.

Innovation in its social and institutional dimension can be achieved if the urban ecosystem as a whole is successfully reorganized and reinforced, including its physical and socioeconomic features. Locating 'innovation centers' in distressed neighborhoods is one of the biggest challenges of the urban administrations involved in promoting specific place-based policies. Local and regional governments are expected to discard traditional strategies in favor of something truly innovative like disrupting the patterns of inequality. The regeneration vision is expected to recover and emphasize the 'consciousness of places' including the territorial 'DNA' of local communities. This approach takes time, culture and sensitivity. The 'social construction of spatial forms', fostered by Manuel Castells almost 50 years ago, is still waiting to be successfully implemented.

Acknowledgements This paper is related to the dissemination of the EU research project 'MAPS-LED' (*Multidisciplinary Approach to Plan Specialization Strategies for Local Economic Development*), Horizon 2020, Marie Sklodowska-Curie RISE, 2015-2019. The project was carried on by a European and USA university network. Bruno Monardo was the coordinator of 'Sapienza' University of Rome unit. This paper is also connected to the PhD research of Martina Massari in architecture of the University of Bologna—Architecture Department, with supervisor Valentina Orioli and Elena Ostanel.

Author Contributions This paper is the result of common reflections by the two researchers. However, Bruno Monardo developed Sects. 2 and 4, Martina Massari Sects. 3 and 5, while both authors wrote Sects. 1 and 6.

References

Augé M (1992) Non-lieux, Introduction à une anthropologie de la surmodernité. Seuil, La Librairie du XXe siècle, Paris
Augé M (2007) Tra i confini. Città, luoghi, interazioni. Mondadori, Milano
Baily NM, Montalbano N (2018) Clusters and innovation districts: lessons from the United States experience. The Brookings Institution, Washington D.C.
Barbi V, Ginocchini G (2019) Cities in change: urban agencies as a strategic player to face new challenges and needs. The case of the Foundation for Urban Innovation in Bologna. In: Topi C, Lucchini C (eds) The city agencies working papers. Urban Lab Torino, Torino, pp 55–63
BEPA, Bureau of European Policy Advisors (2011) Empowering people, driving change social innovation in the European Union. Publications Office of the European Union, Luxembourg
Bergvall-Kåreborn B, Eriksson CI, Ståhlbröst A (2015) Places and spaces within living labs. Technol Innov Manage Rev 5(12):37–47
Brandsen T, Cattacin S, Evers A, Zimmer A (eds) (2016) Social innovations in the urban context. Springer, Cham
Cafiero G, Calace F, Corchia I (2014) La rigenerazione urbana tra politiche economiche e innovazione istituzionale. Rivista Economica Del Mezzogiorno 3(2014):667–696
Comune di Bologna (2017) Piano per l'Innovazione Urbana di Bologna. www.comune.bologna.it/pianoinnovazioneurbana. Accessed 30 Nov 2020
Deleuze G (1994) Difference and repetition. Columbia University Press, New York
Donolo C (2003) Il distretto sostenibile. Governare i beni comuni per lo sviluppo. Franco Angeli, Milano
Donzelot J (2009) La ville à trois vitesses. Villette, Paris

Drewe P, Klein J-L, Hulsbergen E (eds) (2008). The Challenge of Social Innovation in Urban Revitalization. Techne Press, Delft

Fransman M (2018) Innovation ecosystems: increasing competitiveness. Cambridge University Press, Cambridge

Garcia M (2006) Citizenship practices and urban governance in European cities. Urban Stud 43(4):745–765

Galston WA, McElvein E (2015) Institutional innovation: how it happens and why it matters. The Brookings Institution, Washington D.C.

George G, Baker T, Tracey P, Joshi H (eds) (2019) Handbook of inclusive innovation. Edward Elgar Publishing, Cheltenham

Gerometta J, Hausserman H, Longo G (2005) Social innovation and civil society in urban governance: strategies for an inclusive city. Urban Stud 42(11):2007–2021

Gonzalez S, Healey P (2005) A sociological institutionalist approach to the study of innovation in governance capacity. Urban Stud 42(11):2055–2070

Hillier J (2013) Towards a Deleuzean-inspired methodology for social innovation research and practice. In: Moulaert F, MacCallum D, Mehmood A, Hamdouch A (eds) The international handbook of social innovation. Edward Elgar Publishing, Cheltenham, pp 169–180

Katz B, Wagner J (2014) The rise of innovation districts: a new geography of innovation in America. The Brookings Institution, Washington D.C.

Massari M (2018) The transformative power of social innovation for new development models. In: Calabrò F, Della Spina L, Bevilacqua C (eds) New metropolitan perspectives. Smart innovation, systems and technologies. Springer, Cham, pp 354–361

Mieg HA, Topfer K (2013) Institutional and social innovation for sustainable urban development. Routledge, London

Monardo B, Trillo C (2016) Innovation strategies and cities. Insights from the Boston Area. Urbanistica 157:151–155

Monardo B (2018) Innovation districts as turbines of smart strategy policies in US and EU. Boston and Barcelona Experience. In: Calabrò F, Della Spina L, Bevilacqua C (eds) New metropolitan perspectives. Smart innovation, systems and technologies. Springer, Cham, pp 322–335

Moulaert F, MacCullum D, Mehmood A, Hamdouch A (2013) The international handbook of social innovation. Edward Elgar Publishing, Cheltenham

Mulgan G (2019) Innovation districts: how cities speed up the circulation of ideas. www.nesta.org.uk/blog/innovation-districts. Accessed 30 Nov 2020

Nicolas-Le Strat P (2015) Une politique de la co-création, https://blog.le-commun.fr/?p=873. Accessed 30 Nov 2020

Nyseth T, Hamdouch A (2019) The transformative power of social innovation in urban planning and local development. Urban Planning 4(1):1–6

OECD (2001) OECD economic outlook. OECD Publishing, Paris

Ostanel E, Attili G (2018) (2018) Self-organization practices in cities: discussing the transformative potential. Tracce Urbane, Italian J Urban Stud 4:6–17

Palermo PC, Ponzini D (2015) Place-making and urban development: new challenges for contemporary planning and design. Routledge, London

Rissola G, Bevilacqua C, Monardo B, Trillo C (2019) Place-based innovation ecosystems: Boston-Cambridge innovation districts. Publications Office of the European Union, Luxembourg

United Nations (2015) Transforming our world. The 2030 agenda for sustainable development, www.sustainabledevelopment.org. Accessed 30 Nov 2020

Wagner J, Katz B, Osha T (2019) The evolution of innovation districts: the new geography of global innovation. GIID, https://www.giid.org. Accessed 30 Nov 2020

Williams P (2011) The life and times of the boundary spanner. J Integr Care 19(3):26–33

Yin RK (1984) Case study research and applications: design and methods, 1st ed. Sage Publications, Beverly Hills

Smart Creative Cities and Urban Regeneration Policy: Culture, Innovation, and Economy at Nexus. Learning from Lyon Metropolis

Maria Beatrice Andreucci

Abstract In 1995, the English planner Charles Landry and Franco Bianchini published "The Creative City", focusing on three intertwined topics: the cultural, social, and economic impact that arises from creativity in cities; the need to promote integrated urban planning levering on knowledge from other disciplines; and the active inclusion in urban planning processes of ordinary, often marginalized people. A few years later, Landry issues "The Creative City. A Toolkit for Urban Innovation", a book in which he challenges and further develops his ideas by proposing them as a "toolbox for urban renaissance." At the beginning of the twenty-first century, the American economist Richard Florida delivers what is considered a milestone on the subject of the creative city: "The Rise of the Creative Class", in which he emphasizes the characteristics of people performing creative activities in cities, as well as the conditions that cities must offer in order for the "creative class" to be attracted and settle in them. The Smart Creative City is a more recent concept. It grew out of economic science, especially the so-called Experience Economy. Regarding specifically the economic development of cities, creativity, art, and culture represent strategic assets in the urban regeneration process, and the socioeconomic feature of smart creative cities can be considered the most evident and critical one. This study thus springs from the recognition of the relevance of smart creative cities, and of an integrated and visionary planning approach to urban regeneration—itself creative. This analysis has been conducted focusing on selected experiences developed by Lyon metropolis, aiming at understanding whether and how the municipality is levering on creativity, art and culture within its urban regeneration programmes. This objective is addressed through a mixed-qualitative methodology that investigates the political discourse and adopts a descriptive case study approach to analyse policy processes, drivers, and obstacles that are fostering or limiting that vision in the local context of Lyon. The research responds to the questions posed, showing both the transformative capacity and the trade-offs of explicitly integrating cultural and artistic projects and events, as urban "innovative" regeneration devices, within the "common" planning and design practice of the municipality of Lyon.

M. B. Andreucci (✉)
Department of Planning, Design, Technology of Architecture, Sapienza University of Rome, Via Flaminia 72, 00196 Rome, Italy
e-mail: mbeatrice.andreucci@uniroma1.it

© The Author(s), under exclusive license to Springer Nature Switzerland AG 2021
A. Bisello et al. (eds.), *Smart and Sustainable Planning for Cities and Regions*,
Green Energy and Technology, https://doi.org/10.1007/978-3-030-57332-4_29

Keywords Creative clusters · Experience economy · Lyon Gerland · Lyon confluence · Urban art

1   Introduction

In 2010, the European Economic and Social Committee expressed the opinion on "The need to apply an integrated approach to urban regeneration" (EESC 2010), highlighting the inefficiency of ordinary planning measures in modern cities. Since then, the political debate about urban regeneration significantly increased in Europe and became embedded in the framework of the urban dimension of the EU Cohesion Policy. A large part of post-industrial cities realized, at that point, that the necessary momentum for economic revitalization had arrived, and started substantiating a significant number of implementation projects, at different scales.

The past two decades have been dominated by a considerable international debate around the creative city. Some researchers (Landry et al. 1996; Helbrecht 1998; Florida 2003, 2005; Hospers 2003; Scott 2006; Ponzini and Rossi 2010; Grodach 2017; Montalto et al. 2019) pointed to understanding creativity as an asset that can be levered and exploited to regenerate post-industrial sites and other degraded urban areas. Creative and cultural industries (CCIs)—normally developed out of marginalized areas, or on the urban fringe where convenient financial conditions and abandoned post-industrial buildings are available—can provide suitable situations for the establishment of work studios, art spaces, and start-ups (Bayliss 2004; Jensen 2007; Andreucci 2019). Moreover, re-inventing cities as places of *consumption* of attractive cultural events, such as arts festivals, exhibitions and other flagship projects, are also believed to magnetize investments, inhabitants, and labor opportunities (Bianchini 1993; Jensen 2007; World Economic Forum 2016).

In parallel, other authors (Zukin 1995; García 2004; Evans 2009; Gunay and Dokmeci 2012; Markusen 2014; Murdoch et al. 2016; Florida 2017) have been questioning or critically evaluating culture-led and creativity-based urban development strategies which, they say, tend to cater to the tastes of economically privileged and well-connected business people, leading to *cultural commodification*, high-cost projects, gentrification, and social exclusion based on ethnicity, wealth, and gender.

The interest in better understanding the controversial effects of CCIs on urban revitalization, briefly documented above, has led to the research objectives of this study, aiming: (i) to explore if and how cities are integrating creativity, art and culture within urban regeneration policies; (ii) to analyze the main factors supporting this integration, specifically referring to the socioeconomic context; (iii) to evaluate the transformative capacity and the trade-offs of explicitly integrating cultural and artistic projects and events, as urban "regeneration devices," within the "common" planning and design practice; and (iv) to assess how evidence speaks to the shared understanding of the relationship between CCIs and urban regeneration in the European context.

In order to achieve these objectives, the research has adopted a mixed-qualitative methodology. In the first phase, the key concepts of CCIs and integrated urban regeneration programmes, as well as their interplay, have been explored through a literature review. In the second phase, the work focused on Lyon metropolis, critically analyzing selected urban regeneration experiences of the French city, levering on creativity, culture, and art.

The chapter is consequently structured as follows: Sect. 2 explains the applied methodology; Sect. 3 presents the conceptual framework, in which the theoretical and practical interrelations between urban regeneration and creativity are highlighted; Sect. 4 introduces and develops the Lyon metropolis case study, summarizing and discussing selected urban regeneration experiences; Sect. 5 presents concluding remarks taking into account the limits of the conducted research.

2 Method

Both the multiple intertwined relations between creative strategies, spatial and financial policies, and the emblematic structural transformations of the economy that derive from urban regeneration plans can be explored, first conceptually and then empirically, through a qualitative multidimensional case study design (Yin 1984; Stake 2005; Creswell 2007; Baxter and Jack 2008).

Lyon metropolis was specifically selected to undertake this part of the work as its urban regeneration dynamics allow to investigate the phenomenon under study in relation with its diversified urban context, levering on different sources of evidence.

The development of the Lyon "descriptive" case study (Yin 1984) was based on a storyline analysis, as it "identifies assumptions and logics underlying the choice of particular policy directions over others" (Maccallum et al. 2019: 44), building on an inductive work based on a political discourse critique. The gaps of information identified in the public evidence were filled through semi-open interviews to public servants and local experts; while attending conferences further supplemented the local research.

3 Creative Industries, Clusters, and Cities in the Context of Urban Regeneration

We refer to creative cities as conurbations characterized by a high rate of individual, institutional and pervasive creativity, and to those cities that are able to use this resource as a tool for urban competition (Hospers 2003). In this framework, particularly interesting is the concept that promotes the combination of creative industries and culture (Cooke and Lazzeretti 2007; European Commission 2010; Grodach

2017), aimed at increasing the attractiveness of cities in terms of living, working, visiting and spending leisure time.

In 1995, the planner, Charles Landry and the expert in policy and cultural planning, Franco Bianchini, published "The Creative City", focusing on three intertwined topics while investigating the concept of the Creative City:

- the cultural, social, and economic impact that arises from creativity in cities;
- the need to promote integrated urban planning levering on knowledge from other disciplines (economy, sociology, ecology, psychology, etc.); and
- the active inclusion in the urban planning processes of ordinary, often marginalized persons or groups, such as minorities or migrants.

A few years later (2000), Landry publishes "The Creative City. A Toolkit for Urban Innovators", a book in which he challenges and further develops his ideas by proposing seven concepts, plus a series of techniques to help creative thinking and planning, as a toolbox for urban *renaissance*, where "the goal is to find interpretative 'keys' that improve our understanding of urban dynamics, and enable us to act on them." (Landry 2000: 165).

At the beginning of the twenty-first century, the American economist Richard Florida delivers what are also considered milestones on the subject of the creative city: "The Rise of the Creative Class" (2003), and "The Flight of the Creative Class" (2005) in which—relying on facts and figures—he emphasizes, on the one hand, the characteristics of people developing creative activities in cities, and on the other hand the environmental conditions that cities must offer in order for the *creative class* to be attracted and wishing to settle in them.

Richard Florida's books and the publication of John Howkins "The Creative Economy" (2001) gave a dramatic lift to the new planning paradigm, advocated by Landry. Florida's writings were particularly significant, as they connected three areas: a creative class – a novel idea –, the creative economy, and what conditions in cities attract the creative class (IPoP 2011).

Three years ago, in "The New Urban Crisis" (2017), Florida considered the shortcomings—such as, artist-led gentrification, and short-term duration of creative industries—of the last two decades of the type of urban renewal he has advocated. Flourishing cities—many of which developed along the lines of his theory—have become victims of their own success, as widespread inequality has risen alongside success and innovation, reaching its peaks, sadly, in the most open-minded and creative cities (Sussman 2017; Liang and Wang 2020), such as London, New York, and Los Angeles.

The Smart Creative City is also a twentieth-century concept. It grew out of economic sciences, especially the so-called Experience Economy (Pine and Gilmore 1998). For the past twenty-five years, the concept has been studied by a growing number of authors and researchers from different disciplines, so as to count nowadays on a diversified literature (Jensen 1996; Foley et al. 2012; Lehmann 2019). Under this paradigm, experiences are a distinct economic offering, as different from services as services are from goods (Pine and Gilmore 1998). Although the concept of the experience economy was initially focused on business, it rapidly crossed into tourism, architecture, nursing, urban planning, and other fields (Lonsway 2009; Liang and Wang 2020). For the economic development of cities, creative industries represent a strategic sector in the urban regeneration process, and the socioeconomic feature of smart creative cities can be considered the most evident, as well as the most critical one. Different spatial arrangements occur with the contribution of the creative and cultural sectors, as real driving forces of innovative urban development strategies (WEF 2016).

Creative quarters and clusters seem to be the key urban systems, i.e., powerful organizations aimed at advancing the creative economy.

Michael Porter argued—already thirty years ago—that competitive success tends to concentrate in particular industries and groups of interconnected industries (Porter 2009). Landry (2008), building on that, emphasized the role of clustering of talents, skills and support-infrastructure—central for the creative economy and the innovative *milieu*. The encompassing paradigm for urban development thus changed, from an urban manufacturing or infrastructure-based approach, to creative and innovative city-making. This is the art of making cities for people, including the connections between places and people, program and urban form, nature and the built environment, as well as the design and construction processes toward successful settlements (IPoP 2011).

The urgent need for urban regeneration occurs as an outcome of wider socioeconomic dynamics, like conversion into the post-industrial era of the Anthropocene, which results in empty cities and deserted post-industrial sites. An abandoned area *colonized* by a creative group soon becomes attractive for others, thus activating local revitalisation and wider regeneration (Scheffler 2016).

Urban policies are trying to stimulate the renovation of degraded areas into creative hubs in very different ways. One common way in European cities is through the implementation of so-called flagship projects (van Aalst and Boogaarts 2002), where innovative clusters make a connection between old and new, between large and small scale and between functions in, and close by, the new complex, attracting both locals and visitors alike. Examples include, to cite just a few, Dublin with the Temple Bar (1991), the Guggenheim museum in Bilbao (1994), the *Cultureplan* 2005–2008 of Rotterdam, and the Docks and old harbour requalification project in Marseille (2017). All of them are examples of integrated and visionary strategies, levering on culture, creative skills, and local identity, while delivering outstanding quality in architecture and urban design, with the rest of the actions under the public sphere just supposed to follow.

4   Lyon Creative City

In France, creativity has increasingly gained momentum within urban policies in deprived neighbourhoods, as a cultural-correlate of sustainability, notably within the "Politique de la Ville" (PdV) launched by the State already in the late 1970s, aiming at reducing territorial inequalities.

Ever since then, several challenging generations of the PdV have been set up addressing the specific domain of housing and urban economy, as well as more general issues in health, law and order, security and urban services and, lately, civic art. Over time, the cultural issue has been embedded as such in the fight against discrimination launched by the local agencies (*Agence nationale pour la cohésion sociale et l'égalité des chances,* that complements the *Agence nationale de rénovation urbaine*) by enhancing socioeconomic and professional integration, and supporting both cultural and artistic practices, as well as widespread access to cultural infrastructure (Palazzo 2013). Culture has become a major issue targeted by policy-makers, entering the PdV Agenda in a threefold approach:

- Access to cultural facilities;
- Organisation of cultural events and festivals;
- Support for artistic and cultural activities.

As a study site, Lyon, located in the region of Auvergne-Rhône-Alpes, and the second area of economic activity and distribution of wealth in France, has been selected as an emblematic case offering useful insights. Classified as the second French city by population, the agglomeration of Lyon today has all the attributes of a metropolis, i.e., an urban complex of great importance, which performs functions of governance, organization, and impetus in the political, economic, cultural, and innovation fields, all in a region that integrates it with the rest of the world. The Lyon metropolis, established since January 1, 2015, as an administrative entity, even represents the first metropolis generated in France by law (MAPTAM *Modernisation de l'Action Publique Territoriale et d'Affirmation des Métropoles*), before Paris and Aix-Marseille (Mollé 2019).

Lyon Métropole (formerly *Greater Lyon*) encompasses 59 municipalities and 1,262,000 inhabitants in an area of around 500 km^2 featuring a longstanding sense of strong inter-municipality responsible for the *Politique de l'Habitat* and the economic development.

Three are the reasons why Lyon Creative City (Fig. 1) is an excellent study case: It is the first among French creative cities (before Saint-Etienne, for Design;

Fig. 1 Lyon, France Image credits: UNESCO

Angoulême, for Literature; and Metz, for Music); the city levers on the brand *ONLY-LYON* to be exploited and broadcasted; and the creative development of the city is now 15 years old, and all the story-telling plan has already proved successful.

Lyon, a pioneering city in the field, has also put in place a true Smart City strategy to combine economic dynamism and sustainable development. The major urban projects carried by the city, i.e., Lyon Confluence, Lyon Part-Dieu, Lyon Gerland, Villeurbanne Carré de Soie, have become life-size areas of experimentation for imagining and developing new ways of living and working in the city (E&Y 2019). The urban projects Lyon Gerland and Lyon Confluence, described below, are particularly emblematic of the city's strong focus on innovation and entrepreneurship.

4.1 Lyon Gerland

In Lyon, Gerland (20,000 inhabitants, 700 ha) deserves specific attention. Located in the outskirts of the city center, the district has been marked by an imposing industrial history. Still characterized by big voids within post-industrial estates, Gerland is hosting 150 new social housing units/per year, a category which is expected to reach 25% of total housing by 2020.

In Gerland, urban regeneration has been carried out since the late 1990s by the agency, *Mission* Gerland, delivering a number of improvements addressing the public transport network, challenging re-development operations, high-quality infrastructure and green spaces, and aiming to create a culturally rich living environment, benefitting also from the presence of leading universities and private research centers, such as the *Grandes Ecoles*, and the *Biopôle* (Fig. 2).

It must be highlighted that at the turn of the century, in Gerland, main concerns were still raised by the share of inhabitants getting no benefits from the overall re-development, notably in the *Cités Sociales* estate, dating back to the 1930s. The district was suffering from massive concentration of social dwellings (over 50%) and increasing discrimination and precariousness, not to mention health problems, isolation and ageing, poor associative dynamism among inhabitants and tenants, low presence of local facilities and social infrastructure.

The *Contrat Urbain de Cohésion Sociale* (2007–2010; 2012–2014; 2015–2020), agreed by the *Mission* Gerland and the Greater Lyon since 2007, determined the turning point, envisaging the opportunity for disadvantaged people to conveniently move elsewhere, and conversely attracting middle class households for a more balanced *mixité sociale*.

Under the scheme of the *Contrat Urbain de Cohésion Sociale*, the district has been supported with investments in buildings' energy efficiency and new developments, while its open space has been thoroughly re-designed, with people increasingly reclaiming more qualitative open spaces and art works in an attempt to change the urban perspective.

Located on *Rue Georges-Gouy* in the 7th *arrondissement*, at the heart of Gerland, the *Espace* Diego Rivera is anchored in a working class area of Lyon, renowned for

Fig. 2 Emlyon Business School, Lyon Gerland, Image credits: PCA-STREAM Philippe Chiambaretta Architecte

its large concentration of housing for foreign workers. Three *trompe l'oeil*, on two buildings at the entrance to the street, represent Mexico throughout its history with the pre-Columbian civilizations Aztec and Maya, the Spanish military conquest with Cortès, the enslavement of the local populations, the political social upheavals and land reform, with windows paying tribute to the work of Mexican muralist Diego Rivera, one of the fathers of wall art. Created by the cooperative *Cité Création*, and inaugurated on December 4, 2007, this 450 m^2 fresco (Fig. 3) was intended by the Diego Rivera Foundation and his daughter Guadalupe, on the occasion of the 50th anniversary of the artist's death.

Fig. 3 Fresco by Diego Rivera. Image credits: France 3/Culturebox

At the time, the project was well received by the population, although some saw it as an apology for slavery. But, since then, the wall carrying the fresco has become an ideal hiding place for offenders. "What the residents were complaining about was mainly drug trafficking" declares Myriam Picot, mayor of the 7th *arrondissement*. As a result, since 2017, ten years after the inauguration of the *Espace* Diego Rivera in the working-class district of Lyon, only two walls remain painted out of the three. The heart of the structure has been destroyed. For the project supporters, the arguments advanced by the mayor do not justify its destruction, and a petition has been launched calling for the rehabilitation of the artwork.

4.2 Completing the Confluence to Redefine Lyon's Image

Lyon's race toward the future is personified in this reborn industrial district near the southern tip of Presqu'île, where the rivers Rhône and Saône converge. The land was reclaimed from the water between 1770 and 1850, and for a long time the area was used for industrial and logistics activities that made it less attractive: a postal sorting centre, wholesale structures, a natural gas plant, and prisons. The departure of those activities gradually created post-industrial brownfields and a land reserve at the heart of the urban area, a controversial landscape of exceptional quality at the confluence of the two rivers, featuring 5 km of riverbanks.

The decision taken in 1998 to transform the Confluence's 150 hectares of industrial brownfields was based not only on the desire to recover a prime location close to the city centre, but also to transform the area into a showcase of an ambitious city of the future (Genevois 2005), that is:

- A smart, sustainable city;
- A walkable city conducive to new forms of mobility;
- A city with bold architectural statements;
- A city for everyone, fostering social diversity.

Planned in two phases—Phase 1—2003/2018 and Phase 2—2010/2025—Lyon Confluence is one of the most ambitious city-center projects in Europe. The urban project characterizing 50 hectares of transformable land will be completed by 2025, with expected 1 million m^2 of newly built volumes (E&Y 2019). Lyon Confluence is already a smart district, with a level of architectural requirements and building processes unequalled in France. It is also Lyon's creative heart, whose key stakeholders in the creative economy include Lyon's French Tech. The district hosts 860 companies, i.e., large groups and many SMEs and start-ups from diverse sectors: communication and media, digital technologies, building and construction, energy and environment. Lyon's largest museum, *Musée des Confluences*, and Lyon's second largest shopping mall are also present in the quarter (E&Y 2019). A former factory dating from 1857, the newly renovated Halle Girard represents the last vestige of the industrial past of the Confluence. Today, the *halle* is strategically located in the new masterplan, as designed by Herzog and De Meuron and Michel Desvigne Paysagistes, as an interface between the development of the dense city and a large natural area which will join the southern tip of the site. Once a landscape of empty warehouses and urban blight, the newly styled Confluence, with its contemporary architecture and innovative re-design, truly embodies the city of the future. Upon completion, 16,000 inhabitants, 25,000 employees, and an office stock of 500,000 m^2 are expected to characterize and animate the Confluence area (E&Y 2019).

5   Concluding Remarks

The overall objective of the conducted study has been to grasp the essential nature and complexity of smart creative urban transformations, under the more general urban regeneration policy and conceptual frameworks, learning from Lyon.

The literature review informed that the interaction between culture and urban economy, and the influence of creativity on conventional economic activities have resulted, internationally, in significant as well as controversial expressions.

Creativity, art, and culture have, within the Smart City and Community concept, a much broader power than they are usually attributed (Matovic et al. 2018), and the illustration of successful, as well as controversial experiences and practices implemented in Lyon stimulated a profound reflection on: the different dimensions of urban creativity; the effectiveness of cultural and artistic projects and events as urban regeneration devices; and the role of national and supranational stakeholders in supporting culture, art, and economy at nexus.

The development of CCIs in Lyon has a long history intertwined with urban regeneration, i.e., the development of CCIs has been widely used as a tool in urban planning and economic development strategies, and especially in some district is closely related to industrial revamping and new urbanization. CCIs have adapted to a variety of spatial configurations in Lyon, while contributing to the transformation of the social dimension of its urban economy in many ways. Local governance characteristics include top-down approach, close relation with ongoing urbanization—especially in Gerland—, and relatively weak ties with local communities. Claims for the distributive effects of creative regeneration strategies (social, economic, and environmental) generally lack evidence of impacts and benefits. Like many other cities in Europe, gentrification and displacement are also characterizing CCIs development in Lyon (Evans 2005). Sometimes, intents of control from the local government have exacerbated social issues, generating vandalism, and lack of consent.

Social issues related to urban landscape democracy in Lyon mirror existing tradeoffs between different goals of urban regeneration policy that can be generalized as follows: to seek city branding, market interest, and entrepreneurialism through creativity-based placemaking strategies, on the one hand; and to support top-down governance legitimacy and mobilize consensus, on the other.

The main purpose of this study has also been to point out research gaps and stimulate future reflections on the potentialities of CCIs for urban regeneration at a European level. Not seeking to focus on benchmarking the economic effectiveness underlying the production of urban creativity, the study instead aimed to convey the importance of exploring and better understanding synergies and trade-offs of the urban creativity phenomenon, and of pointing out, through future research, more inclusive practices for its management, promotion, and regulation, learning from cities which have championed this approach. When planning urban regeneration, informal networks and synergies should be the central focus of decision-makers. Examples of good practice in this regard are rare and therefore deserve greater attention and nurturing.

Creative clusters and common spaces of production should be instrumental as groundwork for cultural and social bonding, which are based on collective rules, conventions, knowledge, and diversified forms of sociocultural identification (Evans 2004). Social ties are crucial for organic upspring and effective functioning of creative clusters. The process becomes self-organized and functions as a grassroots initiative, launched and driven by the citizens (Scott 2010). When the bottom-up initiative is encouraged and sustained by strategic and visionary urban policies, it can be considered an effective device towards urban regeneration (Ley and Dobson 2008; IPoP 2011; Scheffler 2016). Without social dimensions, cohesion, governance, and vision specifically addressed, clusters are in fact nothing more than a plain concentration of economic activity in a particular area.

References

Andreucci MB (2019) Progettare l'involucro urbano. Wolters Kluwer, Milano, Italy

Bayliss D (2004) Denmark's creative potential; the role of culture within Danish urban development strategies. Int J Cultural Policy 10(1):7–30

Baxter P, Jack S (2008) Qualitative case study methodology: study design and implementation for novice researchers. Qual Rep 13(4):544–559

Bianchini F (1993) Remaking European cities: the role of cultural policies. In: Bianchini F, Parkinson M (eds) Cultural policy and regeneration: the West European experience. Manchester University Press, Manchester

Cooke P, Lazzeretti L (2007) Creative cities, cultural clusters and local economic development. Edward Elgar Publishing, Cheltenham

Creswell JW (2007) Qualitative inquiry and research design: choosing among five approaches, 2nd edn. Sage, Thousand Oaks

E&Y (2019) Why invest in Lyon, the European city of the future? Available at: https://www.ey.com/Publication/vwLUAssets/ey-etude-why-invest-lyon-2019/$File/ey-etude-why-invest-lyon-2019.pdf. Accessed on 12 Feb 2020

European Economic and Social Committee (2010) Urban regeneration: integrated approach. ECO/273–2010–760, Brussels, 7 June 2010. Available at: https://www.eesc.europa.eu/en/our-work/opinions-information-reports/opinions/urban-regeneration-integrated-approach. Accessed on 20 May 2020

Evans G (2004) Cultural industry quarters: from pre-industrial to post-industrial production. In: Bell D, Jayne M (eds) City of quarters: urban villages in the contemporary city. Ashgate, Aldershot

Evans G (2005) Measure for measure: evaluating the evidence of culture's contribution to regeneration. Urban Studies 42:959–983

Evans G (2009) From cultural quarters to creative clusters: creative spaces in the new city economy. In: Legner M (ed) The sustainability and development of cultural quarters: international perspectives. Institute of Urban History, Stockholm, Sweden, pp 32–59

European Commission (2010) GREEN PAPER Unlocking the potential of cultural and creative industries. Brussels, 27 April 2010. Available at: https://op.europa.eu/en/publication-detail/-/pub lication/1cb6f484-074b-4913-87b3-344ccf020eef/language-en. Accessed on 25 May 2020

Florida R (2003) Cities and the creative class. City Commun 2(1):3–19

Florida R (2005) Cities and the creative class. Routledge, New York

Florida R (2017) The new urban crisis. Basic Books, New York

Foley M, McGillivray D, McPherson G (2012) Event policy: from theory to strategy. Routledge, London

García B (2004) Cultural policy and urban regeneration in Western European cities. Local Econ. 19:312–326

Grodach C (2017) Urban cultural policy and creative city making. Cities 68:82–91

Gunay Z, Dokmeci V (2012) Cultural-led regeneration of Istanbul waterfront: golden horn cultural valley project. Cities 29:213–222

Helbrecht I (1998) The creative metropolis: services, symbols and spaces. Int J Architect Theory 3(1):1–10

Hospers G-J (2003) Creative cities in europe: urban competitiveness in the knowledge economy. Intereconomics 38(5):260–269. Available at: https://hdl.handle.net/10419/41712. Accessed on 22 Feb 2020

Howkins J (2001) The creative economy: how people make money from ideas. Allen Lane, London

IPoP (2011) Creative cities. Institute for spatial policies, Ljubljana

Jensen R (1996) Dream society. Futurist 30(3):9–13

Jones P, Evans J (2008) Urban regeneration in the UK: theory and practice. SAGE Publishing, Newbury Park

Landry C, Bianchini F (1995) The creative city. Demos, London

Landry C, Greene L, Matarasso F, Bianchini F (1996) The art of regeneration: cultural development and urban regeneration. Comedia, in Association with Civic Trust Regeneration Unit, Nottingham City Council, London and Nottingham, UK

Landry C (2008) The creative city: a toolkit for urban innovators, 2nd edn. Comedia, London

Lehmann S (2019) Urban regeneration. a manifesto for transforming UK cities in the age of climate change. Palgrave Mc Millan, London

Ley D, Dobson C (2008) Are there limits to gentrification? The contexts of impeded gentrification in Vancouver. Urban Stud 45(12):2471–2498

Liang S, Wang Q (2020) Cultural and creative industries and urban (re)development in China. J Planning Liter 35(1):54–70

Genevois S (2005) La France: des territoires en mutation Lyon-Confluence, un exemple de réno-vation urbaine. GéoConfluence, 18 July 2005. Available online: https://geoconfluences.ens-lyon. fr/doc/territ/FranceMut/FranceMutDoc2.htm. Accessed on 10 Oct 2019

Lonsway B (2009) Making leisure work: architecture and the experience economy. Routledge Press, Oxford

Maccallum D, Babb C, Curtis C (2019) Doing research in urban and regional planning. lessons in practical methods. Routledge, London

Markusen A (2014) Creative cities: a 10-year research agenda. J Urban Affairs 36(2):567–589

Matovic M, Madariaga A, San Salvador del Valle R (2018) Creative cities: mapping creativity driven cities. 12 good practices from UNESCO Creative Cities Network. Cities Lab, Bilbao

Montalto V, Tacao Moura CJ, Langedijk S, Saisana M (2019) Culture counts: an empirical approach to measure the cultural and creative vitality of European cities. Cities 89:167–185

Mollé G (2019) Un changement de regard sur la verticalité urbaine, de nouvelles tours d'habitation dans le paysage de la métropole de Lyon. Géoconfluences. Avail-able at: https://geoconfluences.ens-lyon.fr/informations-scientifiques/dossiers-regionaux/lyon-metropole/articles-scientifiques/verticalite-urbaine-tours-d-habitation. Accessed on 15 Feb 2020

Murdoch J III, Grodach C, Foster N (2016) The importance of neighborhood context in arts-led development: community anchor or creative class magnet? J Planning Educ Res 36(1):32–48

Palazzo AL (2013) La Politique de la Ville nell'esperienza di Lione. In: De Matteis M, Marin A (eds) Nuove qualità del vivere in periferia. Percorsi di rigenerazione nei quartieri residenziali publici. EDICOM, Gorizia, pp 93–98

Pine BJ II, Gilmore JH (1998) Welcome to the experience economy. Harvard Bus Rev 76(4):97–105

Ponzini D, Rossi U (2010) Becoming a creative city: the entrepreneurial mayor, network politics and the promise of an urban renaissance. Urban Stud 47(5):1037–1057

Porter ME (2009) The competitive advantage of nations. Collier Macmillan, London

Scott AJ (2006) Entrepreneurship, innovation and industrial development: geography and the creative field revisited. Small Bus Econ 26(1):1–24

Scott AJ (2010) Cultural economy and the creative field of the city. Human Geogr 92(2):115–130

Scheffler N (2016) Waking up the sleeping giants. Activation of vacant buildings and building complexes for a sustainable urban development. In: URBACT III 2nd Chance—Baseline Study: Available online https://www.comune.genova.it/sites/default/files/2nd_chance_baseline_study_20160311_wp.pdf. Accessed on 14 May 2020

Stake RE (2005) Qualitative case studies. In: Denzin NK, Lincoln YS (eds) Handbook of qualitative research. Sage, Thousand Oaks, pp 443–467

Sussman AL (2017) Richard florida on why the most creative cities are the most unequal. Available at: https://www.artsy.net/article/artsy-editorial-creative-cities-unequal. Accessed on 14 Feb 2020

van Aalst I, Boogaarts I (2002) From museum to mass entertainment: the evolution of the role of museum in cities. Eur Urban Regional Stud 9:195–209

Yin RK (1984) Case study research: design and methods. Sage, Beverly Hills

World Economic Forum (2016) Factors for enabling the creative economy. Available at www.weforum.org. Accessed on 15 May 2020

Zukin S (1995) Cultures of cities. Blackwell Publishers, Oxford

Analysis of National Research Programs to Boost Urban Challenges in Transnational Cooperation

Gilda Massa

Abstract The paper analyzes of National Research and Innovation Programs related to urban topic, still ongoing in 2018; the main aim is in to highlight the possibility to align, around joint urban research priorities, the national programs of 16 European Member States (Austria, Belgium, Cyprus, Denmark, Finland, France, Germany, Italy, Latvia, Netherland, Norway, Poland, Romania, Slovenia, Sweden, and the United Kingdom). According to the GPC (High Level Group on Joint Programming), "alignment" is the strategic approach taken by Member States to modify their strategies, priorities, or activities as a consequence of the adoption of joint research priorities in the context of joint programming, with a view to implement changes to improve the efficiency of investment in research at the level of Member States and the European Research Area. The analysis is based on data collected with an online survey among funding agencies in the framework of the EXPAND project, the Coordination and Support Action to boost the Joint Programming Initiative Urban Europe. The main goal of the survey is to highlight how the research national programs are close to each other, in order to develop strategies for sustainable and liveable cities. The overall approach analyzes the key elements of each program and compare it with the five thematic priorities of Strategic Research and Innovation Agenda—S.R.I.A. The aim is to cause all the funding agencies involved in JPI Urban Europe to have a common vision of what is "aligned" beyond the different rules and applications at the national level. The information was provided and gathered through the Web tool and collected to identify similarities and differences among national programs relating to their main features: research topics, aims, eligibility, and funding criteria. Finally, the activity identifies suitable tracks and criteria to foster the alignment, thus providing a basis to build on a strategy for the sustainable, resilient, and liveable urban areas. The main outcomes of this analysis were: national program' aims, objectives, and results; national funders/management; programs name linked with research area or research topic; program relations to SRIA thematic priorities.

G. Massa (✉)
Environment and Sustainable Economic Development, ENEA National Agency for Energy, Piazzale Enrico Fermi n.1 80055, Portici (Na), Italy
e-mail: gilda.massa@enea.it

© The Author(s), under exclusive license to Springer Nature Switzerland AG 2021
A. Bisello et al. (eds.), *Smart and Sustainable Planning for Cities and Regions*,
Green Energy and Technology, https://doi.org/10.1007/978-3-030-57332-4_30

Keyword Public engagement · National coordination · Alignment · Research programs · Urban issue

1 Starting Point of the Analysis

This analysis has its starting point in the main lesson learnt in the BOOST project; during BOOST[1] the project partners defined two milestones: The first one is that the funding agency network and collaborative research programs are the main needs to be reached if we want to talk of "alignment" in Urban Europe; the second is that at the national level, there are structural funds and cohesion funds that address many research priority area related to societal challenges. Starting from these two results, ENEA has developed this study, within the framework of the EXPAND project,[2] focusing on national programs to gather the results of a two-step activity for the identification of a "common framework" among them. The aim is to enable all the funding agencies involved in Urban Europe to have a common vision of what is "aligned" beyond the different rules and applications at the national level.

In the BOOST project, an analysis of ongoing national programs related to urban Europe's main objective was carried out. The information was provided and gathered through the Web tool and has been collected to identify similarities and differences among national programs, relating to their main features: research topics, aims, eligibility, and funding criteria. The overall aim of the activity was to identify suitable tracks and criteria fostering the alignment, thus providing a basis to build on an alignment strategy for the JPI UE.

A report based on the desk analysis, illustrating the data on kind of research, funding management, eligible subjects, research results, etc., at an aggregated level, is provided as a framework for discussion, contributing to the reciprocal knowledge among funding agencies, in particular with respect to research areas, topics, and goals.

1.1 Constraints of the Analysis

The scenario stated during BOOST Project was completed 2012–2013 and needs to be updated. Analyzing the previous data collected, we identified a set of constraints at the context level and at the data level.

Related to context constraints, we highlighted that: (1) Member States organizations have different as to skills, internal management, and the coordination of national

---

[1] BOOST—Cooperation in Urban Science Technology and Policy, Funding from the European Union's FP7—Regions under grant agreement: 618,994.

[2] EXPAND—Enhancing Co-creation in JPI Urban Europe through widening Member State and stakeholder participation—Funding from the European Union's Horizon 2020 research and innovation program under grant agreement: 726,744.

programs; (2) Not all the national funding agencies have initiated the stage of data compilation; (3) The management of the individual programs at the national level does not enable secure scalability of the total budget data over a time period different from that of the specific program duration; (4) Some funding agencies have not quantified the annual program budget; and (5) Some Member States have thematic programs that others do not have.

On the other hand, some data need to be better defined: update of ongoing programs is mandatory; timelines need to be fixed for same programs; in-depth analysis of sub-programs for Italy, Slovenia, Romania, and Germany is necessary; and some data are lost or not clear.

A detailed analysis was carried out to have a common check list for each data type requested to avoiding having too many open answers and to cover and align all the possible national specific scenario. For each answer, we leave an open field to enter an answer not covered in our check list, and the few answers entered in this field made us confident that the check list adequately identified each field, as better explained in following section.

1.2　The Survey

The survey was developed using the Monkey Survey platform to be easy-to-use for all recipients and to reach most of them rapidly even if they belong to different countries and have different technical skills. The survey was dispatched to the mailing list representing 47 funding agencies of 19 Member States or Observers (o) of JPI Urban Europe. The countries involved in the survey were: Austria, Belgium, Cyprus, Denmark, Finland, France, Germany, Italy, Latvia, Netherland, Norway, Poland (o), Portugal (o), Romania (o), Slovenia, Spain (o), Sweden, Turkey (o), and the United Kingdom. We have collected answers in April–May 2018, and all the national programs in the references section were accessed in this period.

In the structure of the questionnaire, we tried to overcome the constraints that were presented in the previous version and 18 questions were been proposed. The first section aims to collect the information about the respondents in terms of contact information and Funding Agency represented; in the first section, we also added data related to programs that were still ongoing from the previous analysis to achieve an easy update.

Then, there is a specific session in which we analyze the elements of a research national program: what type of program is it?; if it is open or thematic, meaning a program that does not focus on a specific domain or a program with a specific domain and related budget, as well as the identification, if feasible, of an annual budget. For each program, the duration, aim, research area, and objectives are requested. In the aim section, with a multiple-choice option, we have eight possible response and an open field; some feasible answers could be: setting or implementing strategic development objectives; internationalization; and multilateral cooperation/international cooperation. We identified a list of 16 research area fitting the urban scenario and an equally long list of specific objectives, for example: new knowledge for urban and rural interrelations; resilient cities/districts; reduction of emissions/resource consumption; improvement of the quality of life; better living/working environments; improvement of building efficiency; providing solutions for innovative use of ICT on the urban scale; and so on.

Different from the previous analysis, closed lists were defined within which the various answers can be selected for uniformity of data. However, an open field was envisaged if the choices made were not considered appropriate by the interviewee.

In the last section, we questioned the eligible subjects regarding: the activities expected to be financed; results like new technologies, new city concepts/strategies, integrated business solutions, knowledge/collaborations, pilot solutions, etc.; the international openness of the program: and if it encourages and/or co-finances collaborations with foreign subjects clarified.

Finally, the Strategic Research and Innovation Agenda (SRIA) thematic areas were presented and respondents indicate the level of adherence of the program to each area. The SRIA has five priorities: vibrancy in changing economies; welfare and finance; environmental sustainability and resilience; accessibility and connectivity; and urban governance and participation.

2   Results

We have identified 24 national programs related to JPI Urban Europe that could be still ongoing judging from their starting date and durations declared in the past. In the new analysis, four national programs have been updated (Austrian…Mobility of…2018; Norwegian…Demos…2018; Norwegian… Environment…2018; Swedish…the Challenges…2018), while seven new programs have been added (Italian…Smart cities…2018; Finnish…Innovative 2018; Austrian…City of Tomorrow…2018; Austrian…City of the future…2018; Belgium…Research Project…2018; Cyprian…European Initiative.. 2018; Dutch… Smart Urban…2018) (see Fig. 1).

The main aims identified (see Fig. 2) are multilateral cooperation and interdisciplinary domain research, followed by sustainable development and implementing strategic development objects. Less relevant is the purchase or construction of research equipment. The interviewees have added two extra aims: funding excellent of science in general, a bottom-up approach, and a framework for the participation in calls for P2P initiatives.

The national programs' main research areas (Fig. 3) comprise 42% based on policy/governance and urban planning in the area of environment, housing/living and mobility, and 37% focused on sustainable system, urban communication and ICT, smart cities, and energy. There are fewer relevance related to natural science, aging society, and infrastructure. Others area identified by respondents are: "Bottom-up approach across all scientific domains" and "Water Challenges for a Changing World:

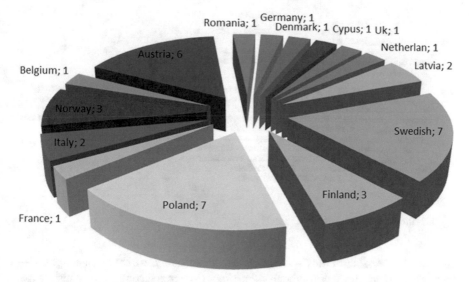

Fig. 1 Number of NPs analyzed for each country

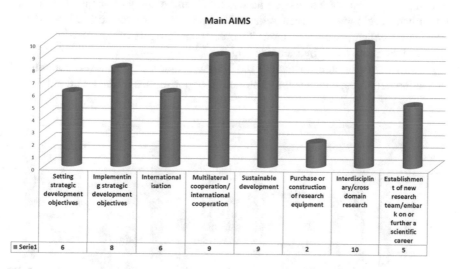

Fig. 2 National programs' main aims

Environment-Tourism, Agriculture, Food Security and Climate Change, Materials-Energy, Solar Energy, ICT-Health-Aging Population."

When we analyze the objects, we find a wider scenario where 15 of the 16 proposed objects having been selected and four more added. "To encourage cooperation between science, public authorities and civil society in the urban field " is the top one,

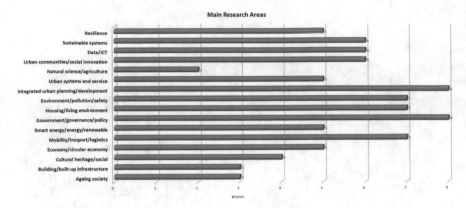

Fig. 3 National programs' research areas

followed by "resilient cities"—" reduction of emission/resource consumption"—
"better living/working environment"—"promote integrated solutions for sustainable
urban and societal development" selected for the 34% of the solutions (Fig. 4).

Others objectives added in open field are: "planning and governance for sustainable urban development", "reconciliation of interests between traffic route, human
living environment, and ecosystem," "overcome or mitigate of the country's weaknesses, such as its small market size and the lack of critical mass of stakeholders
and infrastructure in order to achieve the implementation of high-level research and
innovation activities; enhance the exploitation opportunities of the RTDI outputs."

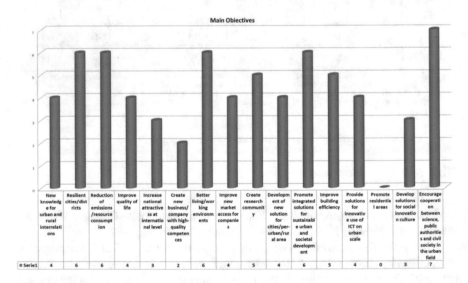

Fig. 4 National programs' main objectives

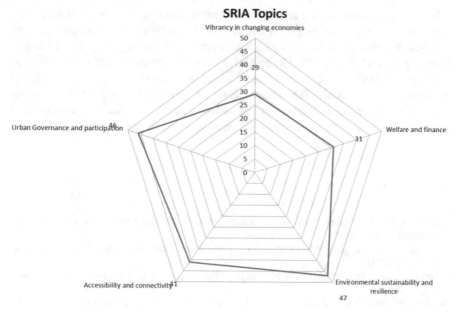

Fig. 5 SRIA topics and national programs' themes

Interesting data also for SRIA2.0 development is how similarly each funding agency evaluates its national program according to SRIA topics (2015).

The national programs relate for the 94% the priorities enounced in "urban governance and participation" and in "environmental sustainability and resilience," followed at 82% by "accessibility and connectivity." Less addressed are "welfare and finance "with 62%, and last is "vibrancy in changing economies" with 58% alignment (Fig. 5).

3 Conclusions

The first results include that the check lists in the survey setting have been well defined and are able the cover more than the 90% of the national program analyzed scenario. We can confirm that the national programs have been well aligned if we focus on aims and research areas. The approach chosen by several Member States is multidisciplinary research based on international cooperation, and this is a strategic opportunity for a joint programming initiative and specifically the JPI Urban Europe, focusing on a sustainable and liveable urban future, which is very well positioned in this scenario. Several research common area have been identified such as: integrated urban planning, governance, mobility, sustainability, open innovation, and ICT. In the final scenario, there are several different objectives of research that could be explored in the future to identify synergies for funding instrument among national calls and joint programming funding instruments. Looking at strategical collaborations among the national programs and the SRIA, the topics of "urban governance" and "environmental sustainability" are an opportunity to focus on with JPI Urban Europe partners while needing to better address the activity related to "vibrancy in changing economies" and "welfare and finance" because they do not fit the national programs aims and goals. The funding agencies' partners in the JPI Urban Europe have, at the end of this analysis, a shared vision of each other national programs and can build on this first data for future activities to speed research in the urban area.

References

Austrian Research Program "City of Tomorrow", https://www.hausderzukunft.at/results.html/id. Accessed April–May 2018

Austrian Research Program Smart Cities Demo https://www.smartcities.at/home-en-us. Accessed April- May 2018

Austrian Research Program Mobility of tomorrow https://www2.ffg.at/verkehr/. Accessed April–May 2018

Austrian Research Program Electric Mobility Flagship https://www.klimafonds.gv.at. Accessed April–May 2018

Austrian Research Program Energy Research Programme https://www.klimafonds.gv.at. Accessed April–May 2018

Austrian Research Program Urban Solution https://www.zit.co.at/en/funding/calls/call-ur. Accessed April–May 2018

Belgium-Brussels-Program RD for universities, separate program RD for enterprises. https://innoviris.brussels/get-funded. Accessed April–May 2018

Cyprian Research Program Urban and Built Environment https://www.research.org.cy/EN/index.html/Accessed. April–May 2018

Danish Research Program Transport and Infrastructure www.innovationsfonden.dk. Accessed April–May 2018

Dutch Research Program Smart Urban Regions of the future https://surf.verdus.nl/1345. Accessed April–May 2018

Finnish Research Program Innovative Cities Program https://www.tekes.fi/Accessed. April–May 2018

Finnish Research Program the Future Living and Housing https://www.aka.fi/en-GB/A/. Accessed April–May 2018

Finnish Research Program Vibrant and Integrated https://www.ymparistoministerio.fi/en-US. Accessed April–May 2018

French Research Program Sustainable Cities and Mobility Challenge https://www.agence-nationale-recherche.fr/fileadmin/aap/2016/ANR-Work-Programme-2016.pdf. Accessed April–May 2018

German Research Program Research for Sustainable Development https://www.bmbf.de/en/city-of-the-future-2321.html. Accessed April–May 2018

Italian Research Program Smart Cities and Communities https://www.researchitaly.it/en/tagNRP/Smart%20Communities. Accessed April–May 2018

Italian Research Program Social Innovation https://www.miur.it. Accessed April–May 2018

JPI Urban Europe—The Strategic Research and Innovation Agenda—Sept 2015. https://jpi-urbaneurope.eu/app/uploads/2016/05/JPI-Urban-Europe-SRIA-Strategic-Research-and-Innovation-Agenda.pdf

Latvian Research Program Support to International Programmes in Science and Technology https://viaa.gov.lv/lat/zinatnes_inovacijas_progr/. Accessed April–May 2018

Latvian Research program Applied Research https://www.cfla.gov.lv/lv/es-fondi-2014-2020/izsludinatas-atlases/1-1-1-1-k-1. Accessed April–May 2018

Norwegian Research Program Demos https://www.forskningsradet.no/prognett-demosr. Accessed April–May 2018

Norwegian Research Program Energix https://www.forskningsradet.no/servlet/Satelli. Accessed April–May 2018

Norwegian Research Program Demos Environment https://www.forskningsradet.no/prognett-miljo2. Accessed April–May 2018

Polish Research Program Opus https://ncn.gov.pl/finansowanie-nauki/konkursy/typy/1?language=en. Accessed April–May 2018

Polish Research Program Symfonia https://ncn.gov.pl/finansowanie-nauki/konkursy/typy/8?language=en. Accessed April–May 2018

Polish Research Program Maestro Poland Research Program https://ncn.gov.pl/finansowanie-nauki/konkursy/typy/5?language=en. Accessed April–May 2018

Polish Research Program Harmonia https://ncn.gov.pl/finansowanie-nauki/konkursy/typy/4?language=en. Accessed April–May 2018

Polish Research Program Sonata https://ncn.gov.pl/finansowanie-nauki/konkursy/typy/3?language=en. Accessed April–May 2018

Polish Research Program Preludium https://ncn.gov.pl/finansowanie-nauki/konkursy/typy/2?language=en. Accessed April–May 2018

Polish Research Program Sonata Bis https://ncn.gov.pl/finansowanie-nauki/konkursy/typy/7?language=en. Accessed April–May 2018

Romanian Research Program European and International cooperation https://uefiscdi.gov.ro/Public/cat/571/Cooperare-internationala.html. Accessed April–May 2018

Slovenian research Program Targeted Research Programme https://www.arrs.gov.si/sl/progproj/crp/. Accessed April–May 2018

Swedish Research Program Urban and rural development https://www.formas.se/en/Financing/Calls-For-Proposals/#current. Accessed April–May 2018

Swedish Research Program Sustainable Building and Planning https://www.formas.se/en/financing/calls-for-proposals/211-million-sek-to-a-strategic-call-within-the-area-sustainable-building-and-planning. Accessed April–May 2018

Swedish Research Program the challenges and opportunities of urbanization https://www.formas.se/urbanization-2016. Accessed April–May 2018

Swedish Research Program Well-functioning journeys https://www.trafikverket.se/Om-Trafikverket/Fo. Accessed April–May 2018

Swedish Research Program Cultural Heritage https://www.raa.se/om-riksantikvarieambetet/fo. Accessed April–May 2018

Swedish Research Program European regional development https://www.tillvaxtverket.se/euprogram/artigh. Accessed April–May 2018

Swedish Research Program Challenge Driven Innovation https://www.vinnova.se/en/Our-acitivities/Cros. Accessed April–May 2018

The Role of Stakeholders' Risk Perception in Water Management Policies. A Case-Study Comparison in Southern Italy

Stefania Santoro and Giulia Motta Zanin

Abstract Across Italy, water-related risks have affected communities, environmental systems, urban areas, and economic activities, due to the hydro-geomorphological characteristics of the country. Around 16.6% of the Italian territory is classified as being vulnerable to such risks, and the approximately 8,300 km length of the Italian coastline further increases the complexity of this system. Evidence demonstrates that it is not easy to determine the effectiveness of a risk-management policy to reduce water-related risks. The unsuccessful results of such policies, based on the traditional paradigm of operation research, led practitioners and policy-makers to consider stakeholders' risk perception, such as socioeconomic dynamics, interaction, previous experience, values, and cultural factors, facilitating bottom-up approaches. The literature highlights that the effectiveness of risks related to water management policies heavily depends on human behaviors, decisions, actions, and interactions that depend on the perspectives and frames of stakeholders involved. The perceived risk influences stakeholders' decisions and actions. Therefore, differences in risk perception could have a twofold implication. On the one hand, they could lead to conflicting situations. On the other hand, they can offer opportunities for the development of innovative solutions hampering the effectiveness of the risk-management policies. In order to understand the role of stakeholders' risk perception about natural hazards in urban contexts, a multistep methodology has been applied to two case studies in the Apulia region (Southern Italy). Specifically, analyses of flood risk in the city of Brindisi and the coastal erosion in the town of Margherita di Savoia have been conducted. This work is subdivided into two steps. The first part gives an overview on the traditional risk-management tools and the factors influencing stakeholders' risk perception. The second part tries to elicit stakeholders' risk perception through problem structuring methods. Finally, a comparison is carried out between

S. Santoro (✉) · G. Motta Zanin
Department of Civil, Land, Environmental, Building Engineering and Chemistry (DICATECh),
Polytechnic University of Bari, via Orabona 4, Bari 70126, Italy
e-mail: stefania.santoro@poliba.it

G. Motta Zanin
e-mail: giulia.mottazanin@poliba.it

© The Author(s), under exclusive license to Springer Nature Switzerland AG 2021
A. Bisello et al. (eds.), *Smart and Sustainable Planning for Cities and Regions*,
Green Energy and Technology, https://doi.org/10.1007/978-3-030-57332-4_31

the two case studies. It is aimed at highlighting the common points and the differences regarding the role of stakeholders' risk perceptions about water-related risks in management policies.

Keywords Water management policies · Water-related risks · Stakeholders' involvement · Fuzzy cognitive maps

1 Introduction

The concept of water-related risk (WRR), which integrates physical and social components, is increasing over time, giving rise to a new discipline called socio-hydrology (Di Baldassarre et al. 2013). Stakeholder collaboration and involvement, participation, and shared knowledge have become watchwords in the decision-support process for water-related risk-management strategies (WRRMS) (Wehn et al. 2015). Risk perception can influence stakeholder behavior and may lead to failures in WRRMS (Flynn et al. 1999; Bickerstaff 2004; Savadori et al. 2004; Harclerode et al. 2016).

A literature review in the field of water management to reduce flood risk (Raaij-makers et al. 2008; Terpstra 2011; Armas and Avram 2009; Heitz et al. 2009) shows that the most commonly used methodological approaches are based on the collection of information through interviews or surveys and the building of models as conceptual maps and indicators.

Similarly, in the case of coastal management, the analysis of risk perception, which supports for management policies and adaptation and mitigation actions, is becoming essential for understanding the social dynamics in such territories (Meur-Ferec et al. 2010; Kettle and Dow 2016).

The activities described in this work are in line with the results of the just-mentioned work, but nevertheless, it aims to take a step forward in the analysis of the interactions between risk perception and risk management by comparing two case studies.

Although some risk perception studies have been conducted (e.g., Birkholz et al. 2014; Chowdhooree et al. 2018), limited research has been performed on the identification and analysis of differences in risk perception (Santoro et al. 2019). To this aim, this paper shows a methodology based on two approaches to collect knowledge and fuzzy cognitive maps to structure it, through the analysis of two case studies. Subsequently, the analysis of fuzzy cognitive maps was conducted through the centrality index, which enables scenario building based on the perception of the actors involved. The comparison of two case studies tries to highlight the importance of considering the perception of risk whatever the type and the context, helping policy-makers to design more effective risk-management policies and to implement mitigation and adaptation actions.

After the present introduction, Sect. 2 briefly presents the two case studies. Section 3 describes the developed methodology, while Sect. 4 outlines and discusses the results derived from the case studies. Finally, some remarks and future developments close the paper (Sect. 5).

2 The Case Studies

Case studies as a research approach are usually adopted "as an empirical enquiry that investigates a contemporary phenomenon within its real-life context; when the boundaries between phenomenon and context are not clearly evident; and in which multiple sources of evidence are used" (Zainal 2007: 2).

This paper focuses on two case studies in the Apulia region (Southern Italy): the city of Brindisi and the town of Margherita di Savoia (Fig. 1). Both case studies are affected by natural hazards: Floods are particularly remarkable in the case of Brindisi and coastal erosion in the case of Margherita di Savoia.

The anthropogenic activities, such as soil lamination for the progressive sealing of surfaces, the removal of green areas, and the growth human settlements, make the landscape vulnerable to flooding. In addition, the morphology of Italian country, characterized by hydrographic network between mountainous areas and the sea, makes the landscape particularly exposed to flood events. Short and intense meteorological phenomena in these types of areas generate event known as flash floods.

Specifically, according to Trigila et al. (2018), 12,405 km^2 (4.1% of the national territory) is affected by significant hydraulic hazard, and the average danger zone amounts to 25,398 km^2 (8.4%) while those with lower hazard (maximum expected scenario) at 32,961 km^2 (10.9%).

Fig. 1 Brindisi and Margherita di Savoia within Apulia

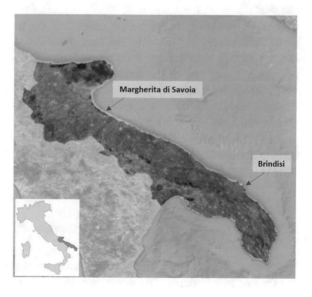

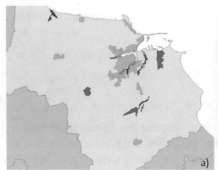

Fig. 2 a Yellow area represents the Municipality of Brindisi; the blue area represents hydraulic hazard map; the green area represents the residential area of Brindisi (*Data source* AdB; ISTAT), b Image (May 2005) of the coast of Margherita di Savoia and the coastline of 1952 (*Data source* Region Puglia and Polytechnic University of Bari, 2012)

Italy, with almost 7500 km of coastline, is also prone to several pressures and coastal hazards (ISPRA 2014). As a matter of fact, about 42% of the over 4000 km of beaches are subject to coastal erosion processes (Falco and Barbanente, in press).

2.1 The City of Brindisi

Brindisi is a town of 86,375, along the Adriatic coast. The surface hydrography of cities requires a high level of attention in relation to hydraulic hazard, due to the predominantly karst nature of highly permeable soils. The almost total absence of meager outflows on a ten-year scale has led to a misreading of the hydro-geomorphological grid, which had been considered inactive and thus becoming hostage to urban expansion. It therefore happens that, due to precipitative events of an exceptional nature, the sudden activation of the articulated river complex can transport large volumes of water and mud towards the sea. The effects are physical damages and tragic consequences for humans and the environment, as happened in October and November 2005 (AdBP 2012).

An application in a GIS environment, obtained thanks to the overlapping of the layer of floodplain areas (Hydrogeological Setting Plan, PAI) and the residential area, demonstrates the presence of buildings in areas at risk (Fig. 2a).

2.2 The Town of Margherita di Savoia

Margherita di Savoia is a small coastal town of almost 18,000 inhabitants, along the Gulf of Manfredonia. The urban conformation is long and narrow because it is enclosed between the sea and the saltworks. The history of this town is linked to the saltworks and their use and the relationship between the inhabitants, and sea-related economic activities are still strong.

The coastal area of Margherita di Savoia is 18-km long and made up of low sandy coast, and it is characterized by many different uses and activities, such as: (i) beach tourism and related activities, (ii) port activities, (iii) residential use, (iv) agriculture, and (v) saltworks.

Furthermore, this territory faces huge problems of coastal erosion and flooding due to the construction of the harbor which has interrupted the sediment flow, enhancing the related risks and damages which are impacting the economic activities and uses of the coastal area.

In order to manage the problem of coastal erosion in such an area, several mitigation measures have been built, shifting the erosive process to the west without any significant benefits. As it is possible to see from Fig. 2b, it has clearly emerged that Margherita di Savoia is one of the municipalities at greater risk of erosion and flooding (Saponieri et al. 2016).

3 Methodology

This section describes the methodology adopted in order to structure stakeholders' perception of WRR. Four different phases have been identified (Fig. 3):

3.1 Data Collecting

Two different participation techniques have been used: Semi-structured Interview for the case study of Brindisi and a Scenario Workshop in Margherita di Savoia. The two case studies are part of two distinct research projects originating from different

Fig. 3 Methodology applied

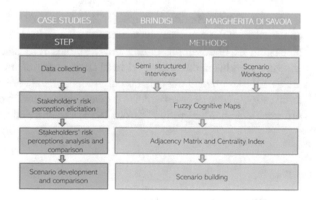

purposes. For this reason, two different participation techniques have been adopted depending on the context of reference according to participation techniques theory (see Voinov and Bousquet 2010).

Specifically, to analyze the degree of knowledge and to elicit stakeholders' risk perception in Brindisi city, a semi-structured interview was structured in two sections: (i) user profile and (ii) twelve questions. The user profile, in addition to collecting general information (role, affiliation, position, etc.), aims to gather insights regarding past experiences with flood risk; the second section aims to collect information on flood-risk causes, effects and potential actions to mitigate it, according to individual perception.

The semi-structured interviews have been submitted to technicians and councilors of the municipality, civil protection officers, local police, external technicians, and journalists. Stakeholder involvement has taken place through an approach known as "snowballing" (Prell et al. 2008; Reed et al. 2009). Each stakeholder suggested the involvement of other agents considering their role in the previous flood management: Twelve stakeholders have been invited, including councilors, municipal technicians, civil protection, and local police.

With the same purpose as the previous case study, in Margherita di Savoia, a Scenario Workshop was designed and organized, taking place on the May 9, 2019 from 10 am to 6 pm. It was an adapted version of the Future Workshop approach, aiming at changing or transforming the reality of a system through three main steps: (i) criticize the actual situation, (ii) theorize about a preferable future situation, and (iii) find ways to move from the actual situation to the preferable one. This approach is a particularly adaptable one that can be used in different forms depending on the research context, the issues to be investigated, and the results to be obtained. Fifteen representatives of stakeholders were invited to participate in the Scenario Workshop, to ensure the broadest representation of the interests involved (Jungk and Müllert 1987; Barbanente et al. 2002; Khakee et al. 2002; Barbanente and Khakee 2003).

3.2 Elicitation of the Stakeholders' Risk Perception

Individual fuzzy cognitive maps (FCM) have been built from the information collected from the previous phase. Graphically, an FCM is a feedback-oriented graph, consisting of nodes representing the variables (C_i) and weighted arcs (W_i) with assigned weights in the range [-1 to 1] (Papageorgiou and Kontogianni 2012). FCM are useful to represent the conceptual models of stakeholders. They make it possible to investigate how people perceive a given system and to compare the perceptions of different stakeholder groups (Kosko 1986; Özesmi and Özesmi 2004).

3.3 Analysis and Comparison of Stakeholders' Risk Perception

In order to identify differences and similarities in the two case studies, FCM variables have been analyzed and compared through the centrality index (CI).

According to graph theory, the developed Collective FCMs were transformed into adjacent matrices (Harary et al. 1965) from which it is possible to extrapolate a CI by summing the input and output connections (Harary et al. 1965; Eden 1992). The out-degree and in-degree indices describe the aggregate strengths of the connections respectively as sums of rows and columns of absolute values of matrix (Papageorgiou and Kontogianni 2012). The CI enables one to identify the most important vertices within a graph (Özesmi and Özesmi 2004).

3.4 Scenario Development and Comparison

The analysis of the previous step was completed through scenario development. The transition from concept maps to scenario building is often used in the literature (see Goodier et al. 2010). The construction of FCM can be a stepping-stone for the development of scenarios. Scenario aims to simulate a series of interconnected and uncertain future events and their possible consequences. Scenario building is useful to develop a plausible future. It helps to understand the relationship between a context and the repercussions of the stakeholders' actions (Harty et al. 2007).

The actions proposed by stakeholders are simulated to understand the effects on the urban water management process (van Vliet et al. 2010).

4 Results and Discussion

4.1 Elicitation of Stakeholders' Risk Perception

In the case study of Brindisi, the aggregation of individual FCMs built the Collective FCM (for details, see Santoro et al. 2019). The Collective FCM was subsequently validated by the same stakeholders involved (Fig. 4).

From the observation of the Collective FCM, it is possible to identify the direct and indirect impacts and the direct and indirect causes of flood risk in Brindisi as identified by stakeholders (Table 1).

The stakeholders involved in Brindisi recognize two main causes of flood risk: (i) the runoff due to the impervious surfaces caused by the excessive urbanization on the flood plain; (ii) the ineffectiveness of mitigation measures due to poor maintenance or creation of infrastructure due to lack of funding.

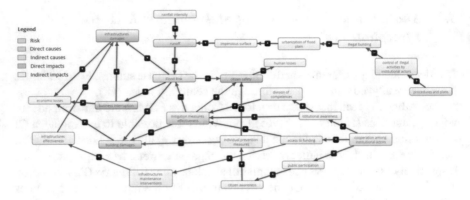

Fig. 4 Collective FCM for Brindisi

Table 1 Direct and indirect impacts and causes of flood risk in Brindisi

Direct impacts of flood	Indirect impacts of flood	Direct causes of flood	Indirect causes of flood
Economic damages	Economic losses	Runoff	Lack of cooperation among institutional actors
Building damages	Human losses	Lack of effective mitigation measures	Lack of procedures and plans
Infrastructures damages			
Citizen safety			

The failure to observe plans that led to an increase of illegal actions and the lack of cooperation between the institutions are considered indirect causes of flood risk. As a result of the flood event, the direct impacts are perceived damages in terms of interruption of economic activities, damage to infrastructure, buildings, and the security of citizens. The economic and human losses are recognized as a direct effect of direct impact and therefore classified as indirect impacts.

In the case study of Margherita di Savoia, the Collective FCM has been built as a result of the Scenario Workshop and represented in Fig. 5:

As in the case of Brindisi, from the Collective FCM, it is possible to identify the direct and indirect impacts, and the direct and indirect causes of coastal erosion identified by stakeholders.

As it is possible to notice from Table 2, the stakeholders of Margherita di Savoia identify the following impacts of coastal erosion: The width of the beach, which is reduced due to coastal erosion and the business interruption, in particular, regarding activities related to beach tourism, which may cause economic losses. Moreover, they recognize two main causes of such phenomenon: (i) Anthropic pressures that increase the stiffening of coastal areas due to urbanization and intensive use of the territory;

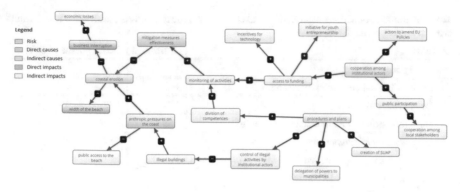

Fig. 5 Collective FCM of Margherita di Savoia

Table 2 Direct and Indirect impacts and causes of coastal erosion in Margherita di Savoia

Direct impacts of coastal erosion	Indirect impacts of coastal erosion	Direct causes of coastal erosion	Indirect causes of coastal erosion
Width of the beach	Economic losses	Anthropic pressures	Lack of cooperation among institutional actors
Business interruption		Lack of effective mitigation measures	Lack of procedures and plans

(ii) mitigation measures that are ineffective and that are worsening the situation of the coastal area of Margherita di Savoia. The long-lasting and complicated procedures and the inadequate plans are perceived by stakeholders as indirect causes which rise the exacerbating of coastal erosion.

4.2 CI Comparison

The detection of the most impacted causes was analyzed and compared. Specifically, each variable has been converted into action, and the value of the CI has been normalized. Table 3 shows a ranking of actions identified by stakeholders to reduce flood risk in Brindisi and coastal erosion in Margherita di Savoia.

As it is possible to see from Table 1, four are the common actions identified by stakeholders of Brindisi and of Margherita di Savoia. A comparison between CIs has been applied to understand the degree of importance related to stakeholders' perception.

In both case studies, the actions considered most important by stakeholders is related to access to funds. Splitting competences is also essential for proper management policies. Both recognize the role of monitoring controls of illegal activities and to increase participatory processes aimed at raising public awareness of risk.

Table 3 Ranking of actions identified by stakeholders to reduce flood risk in Brindisi and coastal erosion in Margherita di Savoia

Actions	CI Brindisi	CI Margherita di Savoia
Increase public participation	2	2
Increase individual prevention measures	3	–
Increase access to funding	2.71	3.42
Increase infrastructure maintenance intervention	1.59	–
Improve division of competences	2	1.54
Increase control of illegal activities by institutional actors	2	2
Increase action to amend EU policies	–	0.84
Creation of SUAP	–	0.78
Increase delegation of powers to municipalities	–	1
Increase monitoring of activities	–	2.62

4.3   Scenario Comparison

The scenarios were built by changing the value of actions in Table 3, according to the importance recognized by stakeholders through the CI. The values of actions were changed by a weight of $[-1]$ in the case of decreasing and of $[+1]$ in the case of increasing.

In order to understand the changes, two scenarios have been simulated for both case studies (Figs. 6 and 7), i.e., the business-as-usual (BAU) scenario and the shared scenario. On the one hand, the BAU scenario represents a future vision of the current situation without any implementation of actions to manage risks. On the other hand, the shared scenario has been built implementing the actions identified by the stake-holders to reduce flood in the case of Brindisi and coastal erosion in the case of Margherita di Savoia.

As shared scenario shows (Fig. 6b) when implementing the actions suggested by stakeholders, the value of flood risk and also direct and indirect damages (see Table 1) is decreased.

As it possible to see from the two scenarios for Margherita di Savoia (Fig. 7), the implementation of the actions suggested by stakeholders enables reduction of coastal erosion and the related negative impacts.

In light of the above, the scenario comparison highlights that the actions proposed by the stakeholders in both case studies reduce the risk of flood and coastal erosion.

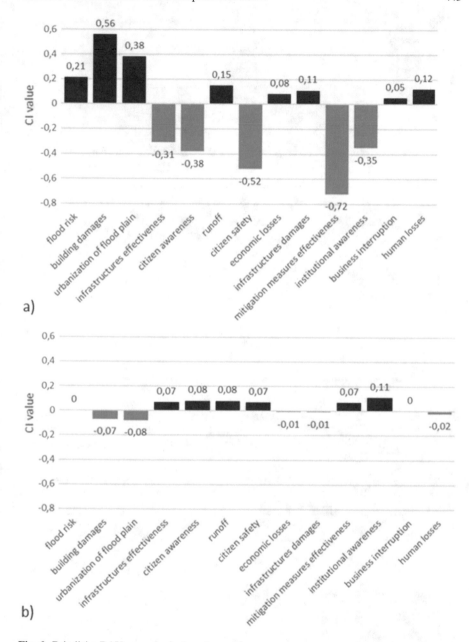

Fig. 6 Brindisi **a** BAU scenario, **b** shared scenario

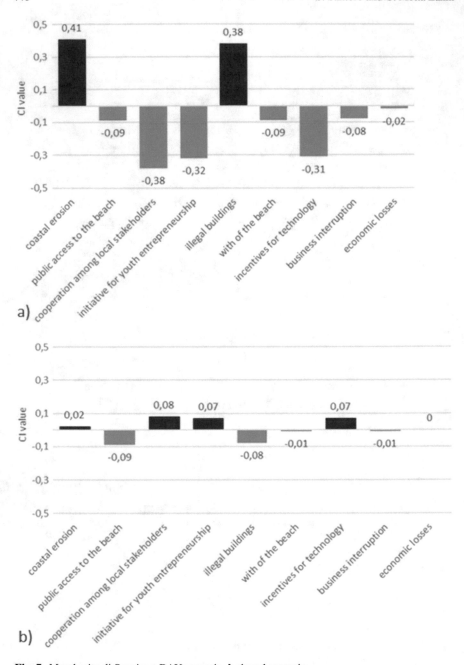

Fig. 7 Margherita di Savoia: **a** BAU scenario, **b** shared scenario

5 Conclusion

Water-related risks affect both human and environmental systems. They are connected through complex and uncertainty network. In addition, there could often be the coexistence of multiple risks that need to be managed in urban areas. In this context, it is therefore essential to analyze the complexity of the network managing the system to understand how to deal with risks.

This work tries to highlights, through a multi-step methodology, the importance of considering the perception of risk whatever the type and the context, helping policy-makers to design more effective risk-management policies and to implement mitigation and adaptation actions.

The complexity of the human system involved in risk situation and the effect of its actions on the management process to reduce risks in environmental system needs to be considered. A bottom-up approach makes it possible to gather stakeholder knowledge. Starting from their perception of the risk, the effectiveness of the proposed actions simulated the effects on the management process through the building of a scenario. A comparison between two case studies shows the possibility to pursue common strategies for two different risks aimed at enhancing management policies, adaptation, and mitigation actions.

In fact, if it is true that hard measures are necessary to deal with every type of risk, e.g., flood risk and coastal erosion (reducing the runoff or increasing the efficiency of the infrastructures), it is also true that actions, such as the involvement of stakeholders and local communities, are necessary in order to increase awareness of risk and therefore helping them to act properly for the risk management.

However, according to stakeholders' perception, access to funds and the fight against the illegal activities are the most useful actions to reduce the risks of flooding and coastal erosion. This highlights the lack of a broader view of the problems and of the complexity that characterizes urban areas.

Nevertheless, structuring the stakeholders' perception helps to deeply define the problem situation regarding the complexity of risk and can provide useful information to decision-makers in risk-management support system. Furthermore, it could be useful to help to develop models that integrate hard approaches with social aspects.

Acknowledgements The analysis of the case study of Margherita di Savoia is part of the STIMARE project, funded by the Italian Ministry for the Environment and Protection of the Territory and the Sea—CUP J56C18001240001.

The authors would like to thank Angela Barbanente for her support.

The authors would also like to thank Raffaele Giordano, Alessandro Pagano and Irene Pluchinotta for their insights and methodology.

References

AdBP (2012) Studio per la definizione delle opere necessarie alla messa in sicurezza del reticolo idraulico interessato dagli eventi alluvionali di ottobre e novembre 2005 nelle province di Bari e Brindisi, Autorità di Bacino della Puglia

Armas I, Avram E (2009) Perception of flood risk in Danube Delta Romania. Nat Hazards 50(2):269–287

Barbanente A, Khakee A, Puglisi M (2002) Scenario building for metropolitan Tunis. Futures 34:583–596

Barbanente A, Khakee A (2003) Influencing ideas and inspirations. Scenarios as an Instrument in Evaluation, Foresight 5(5):3–15

Bickerstaff K (2004) Risk perception research: Socio-cultural perspectives on the public experience of air pollution. Environ Int 30(6):827–840

Birkholz S, Muro M, Jeffrey P, Smith HM (2014) Rethinking the relationship between flood risk perception and flood management. Sci Total Environ 478:12–20

Chowdhooree I, Sloan M, Dawes L (2018) Community perceptions of flood resilience as represented in cognitive maps. J Flood Risk Manag

Di Baldassarre G, Viglione A, Carr G, Kuil L, Salinas JL, Blöschl G (2013) Socio-hydrology: conceptualising human-flood interactions. Hydrol Earth Syst Sci 17:3295–3303

Eden C (1992) On the nature of cognitive maps. J Manage Stud 29:261–265

Falco E, Barbanente A (in press) Chap 13: Italy. In MARENSTRUM Project

Flynn J, Slovic P, Mertz CK, Carlisle C (1999) Public support for earthquake risk mitigation in Portland. Oregon Risk Anal 2:205–216

Goodier C, Austin S, Soetanto R, Dainty A (2010) Causal mapping and scenario building with multiple organisations. Futures 42(3):219–229

Harary F, Norman R, Cartwright D (1965) Structural models: an introduction to the theory of directed graphs. Wiley, New York

Harclerode MA, Lal P, Vedwan N, Wolde B, Miller ME (2016) Evaluation of the role of risk perception in stakeholder engagement to prevent lead exposure in an urban setting. J Environ Manag 184:132–142

Harty C, Goodier CI, Soetanto R, Austin S, Dainty AR, Price AD (2007) The futures of construction: a critical review of construction future studies. Constr Manag Econ 25(5):477–493

Heitz C, Spaeter S, Auzet AV, Glatron S (2009) Local stakeholders' perception of muddy flood risk and implications for management approaches: a case study in Alsace (France). Land Use Policy 26(2):443–451

ISPRA (2014) Mare e ambiente costiero, in Annuario dei dati ambientali, pp 1–39

Jungk R, Müllert N (1987) Future workshops: how to create desirable futures. Institute for Social Inventions, London, UK

Khakee A, Barbanente A, Camarda D, Puglisi M (2002) With or without: comparative study of preparing participatory scenarios for Izmir with computer-based and traditional brainstorming. J Futures Stud 6(4):45–64

Kettle NP, Dow K (2016) The role of perceived risk, uncertainty, and trust on coastal climate change adaptation planning. Environ Behav 48(4):579–606

Kosko B (1986) Fuzzy knowledge combination. Int J Intell Syst 1(4):293–320

Meur-Ferec C, Flanquart H, Hellequin A-P, Rulleau B (2010) Risk perception, a key component of systemic vulnerability of coastal zones to erosion-submersion. Case study on the French Mediterranean coast, Littoral 2010—Adapting to Global Change at the Coast: Leadership, Innovation, and Investment 2011, UK

Özesmi and Özesmi (2003) A participatory approach to ecosystem conservation: fuzzy cognitive maps and stakeholder group analysis in Uluabat Lake Turkey. Environ Manag 31(4):518–531

Papageorgiou E, Kontogianni A (2012) Using fuzzy cognitive mapping in environmental decision making and management: a methodological primer and an application. INTECH Open Access Publisher

Prell C, Hubacek K, Quinn C, Reed M (2008) 'Who's in the Network?' when stakeholders influence data analysis. Syst Pract Action Res 21(6):443–458

Raaijmakers R, Krywkow J, van der Veen A (2008) Flood risk perceptions and spatial multi-criteria analysis: an exploratory research for hazard mitigation. Nat Hazards 46(3):307–322

Reed MS, Graves A, Dandy N, Posthumus H, Hubacek K, Morris J, Prell C, Quinn CH, Stringer LC (2009) Who's in and why? A typology of stakeholder analysis methods for natural resource management. J Environ Manag 90(5):1933–1949

Santoro S, Pluchinotta I, Pagano A, Pengal P, Cokan B, Giordano R (2019) Assessing stakeholders' risk perception to promote nature based solutions as flood protection strategies: the case of the Glinščica river (Slovenia). Sci Total Environ 655:188–201

Saponieri A, Damiani L, Bruno MF (2016) La valutazione del rischio in ambiente costiero: il caso studio della Puglia, pp 1–9

Savadori L, Savio S, Nicotra E, Rumiati R, Finucane M, Slovic P (2004) Expert and public perception of risk from biotechnology. Risk Anal 24:1289–1299

Terpstra T (2011) Emotions, trust and perceived risk: Affective and cognitive routes to flood preparedness behaviour. Risk Anal 31:1658–1675

Trigila A, Iadanza C, Bussettini M, Lastoria B (2018) Dissesto idrogeologico in Italia: pericolosità e indicatori di rischio - Edizione 2018. ISPRA, Rapporti 287/2018

Van Vliet M, Kok K, Veldkamp T (2010) Linking stakeholders and modellers in scenario studies: The use of Fuzzy Cognitive Maps as a communication and learning tool. Futures 42:1–14

Voinov AA, Bousquet F (2010) Modelling with stakeholders. Environ Modell Softw 25:1268–1281

Wehn U, Rusca M, Evers J, Lanfranchi V (2015) Participation in flood risk management and the potential of citizen observatories: a governance analysis. Environ Sci Policy 48:225–236

Zainal Z (2007) The case study as a research method. J Kemanusian

Devising a Socioeconomic Vulnerability Assessment Framework and Ensuring Community Participation for Disaster Risk Reduction: A Case-Study Post Kerala Floods of 2018

Fathimah Tayyiba Rasheed

Abstract In addition to planning for smart, sustainable, equitable and economically strong communities, physical planners must now simultaneously tackle the imminent ramifications of disasters like floods and cyclones to ensure that no individual is left behind from being a part of a resilient community. This study attempts to prove that, by developing and adapting an approach that understands, quantifies and maps vulnerability, we can help in curbing the adverse effects of disasters. The research is done in the context of the 2018 Kerala floods, in a case-study area specific to the coastal region of Ernakulam District. In August of 2018, the state experienced its worst-ever widespread calamity; floods, affecting more than 75% of the villages spread across its 14 districts and impacting the lives of around 5.4 million people. This humanitarian crisis exposed an array of hidden, as well as obvious, vulnerabilities of the many coastal communities in the small state. Therefore, the aim is to develop an approach-based framework for decision-makers and physical planners to understand and reduce vulnerability to disasters and ensure community participation; we strive to make the existing disaster management process and disaster risk reduction measures much more effective. The research began as an attempt to explore the concept of disaster risk and vulnerability while examining the past trends and methods of flood vulnerability assessments, with a primary focus on the parameters/indicators that are used to quantify it. Subsequently, an approach was devised to assess vulnerabilities, and the analysis that follows traces this approach in the context of the floods, across the various levels of jurisdiction in the state, from the state to the community level. Finally, the study examines the multiple dimensions of vulnerability and disaster management across state policies and district plans in Kerala to identify gaps. In the context of those findings, an analysis is done of the missing link of socioeconomic vulnerability, from district to local self-government to community level (top-down approach), mapping its entire process. Later, an identification of the issues and their implications is done through community participation, and the findings are added to the skeleton of the proposed approach to complete a disaster (flood) vulnerability-reduction framework and prove its applicability.

F. T. Rasheed (✉)
Department of Physical Planning, School of Planning and Architecture, Indraprastha Marg, IP Estate, New Delhi, Delhi 110002, India
e-mail: tayyibafathima@gmail.com

© The Author(s), under exclusive license to Springer Nature Switzerland AG 2021   451
A. Bisello et al. (eds.), *Smart and Sustainable Planning for Cities and Regions*,
Green Energy and Technology, https://doi.org/10.1007/978-3-030-57332-4_32

Keywords Disaster risk reduction · Socioeconomic vulnerability · Community participation · Governance · Urban planning

1 Introduction

Kerala is a small south Indian coastal state (see Fig. 1), internationally recognized for its impressive achievements in human development. With a population of around 34 million, the width of the state varies between 15 and 120 km, and it comprises three geographical regions: highlands, which slope down from the Western Ghats onto the midlands of undulating hills and valleys into an unbroken 580 km long coastline with picturesque backwaters interconnected with 44 streams (Government of Kerala 2018). In 2018, during 1–20 August, Kerala experienced incessant and abnormally high rainfall, and these rains triggered landslides and flash floods all over the state. The abnormal rainfall subsequently led to the release of excess water from 37 dams across the state, further aggravating the impact of the flash floods on 16 August 2018, causing 433 deaths (Government of Kerala 2018). When the risk in a vulnerable situation manifests along coastal areas in the form of a disaster, such as in this case, coastal insularity tends to delay a timely response with the needed resources, worsening the consequence of the disaster (Kelman and Lewis 2005). The interactions with the environment and the neighboring societies, as well as their remoteness, is what makes coastal areas highly vulnerable to some of the most devastating disasters involving natural hazards (Kelman and Lewis 2005). Understanding, quantifying and mapping vulnerabilities in such scenarios can perhaps help in curbing the adverse

Fig. 1 Kerala's location in India

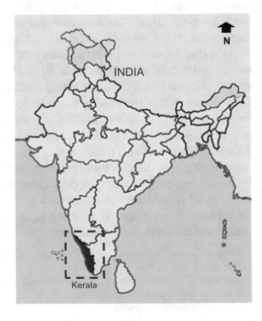

effects of flooding in coastal regions/cities and consequently help to build in more resilience and craft actionable solutions for the areas that most need it. This can save lives, livelihoods and investments.

1.1 Understanding Risk, Hazard and Vulnerability: In the Context of Disasters

The relationship between the risk of a disaster, hazard and vulnerability can be represented in a pseudo-equation as:

$$Risk = Hazard \times Vulnerability (R = H \times V).$$

The risk of disaster, in the literature, is seen as a compounded function of the natural hazard/phenomenon and the number of people involved, characterized by their varying degrees of vulnerability to that particular hazard (Davis 2004). Here, *hazard* refers to "The natural events that may affect different places singularly or in combination (on hillsides, rainforests, coastlines, earthquake faults, etc.) at different times (time of the day, season of the year or over periods of different durations) with varying degrees of severity or intensity" (Davis 2004: 131). In the context of hazards and disaster risk, the concept of *vulnerability* can be defined as "A combination of the susceptibility of a given population, system or place to harm from exposure to the hazard, that directly affects the ability to prepare for, respond to and recover from hazards and disasters." (Balica et al. 2012: 7). When a natural hazard impacts the region having inherently vulnerable conditions (which can be certain socioeconomic–cultural, political and structural formations that are independent of the hazard), it causes difficulty in accessing resources due to poverty, the political-economy system and/or low social protection, thereby aggravating the overall impact of disasters (Lewis 2009). This perception, which has been developing in recent times, helps overcome the misrepresentation that natural hazards are the sole cause of disasters.

Now more than ever, large populations are found in coastal areas where the exposure to floods is known to be high (Nicholls et al. 2008), and therefore, it is necessary that *vulnerability assessments* and other similar disaster management tools are mandated for coastal resilience planning and the policy-making processes.

1.2 Disaster Risk Reduction (DRR) and the Need for Socioeconomic Vulnerability Assessments

In theory, 'disaster risk reduction' is defined by the United Nation's office for Disaster Risk Reduction (UNDRR) as the practice and concept of managing residual, reducing existing and preventing new disaster risks. Its objectives are reduction of exposure to possible hazards, lessening the vulnerability of people and property, sustainable management of the environment and improving preparedness, as well as early warning for adverse events (UNISDR 2014). Prior to the twentieth century, disaster management was mostly considered as a post-disaster relief-centric process, but, following the 2004 Indian Ocean earthquake, the concept of DRR was formally introduced at the United Nation's World Conference on Disaster Risk Reduction (WCDR) in Kobe, Japan, in 2005. After that, it became increasingly recognized that disasters are not natural even when the associated hazard is natural. Since it is not possible for us to reduce the severity of the hazard itself, our main opportunity to reduce the disaster risk is by assessing and reducing the vulnerability and exposure to the hazard. It is a community's socioeconomic vulnerability that determines how well it responds and recovers. Social vulnerability refers to potential harm to people, involving a combination of factors that determine the degree to which someone's life and livelihood are put at risk by an identifiable event in nature or in society (Wisner 2002). Therefore, there exists a need to understand and assess how certain socioeconomic factors have an effect on vulnerability to disasters and why some groups of people are more susceptible to its brunt, so that we can ensure efficacy of more targeted solutions.

2 Methodology and Approach

The dimensions that influence vulnerability are usually physical, social, economic, environmental and institutional. The various indicators of these five dimensions are clearly incorporated in the 'Climate Disaster Resilience Index' (CDRI) developed by Kyoto University, Japan. The CDRI measures resilience of cities to climate change and disasters, with modifications, and the same can be used to measure the vulnerability of local self-government areas (such as a panchayat/municipality[1]) to disasters. Hence, with reference from these indexes (see Fig. 2), it was possible to derive and understand the indicators for measuring vulnerability for the purposes of this case study and perform a limited check/assessment of vulnerability at the state, district and community levels. It should be noted here that there is no definite method for assessing vulnerability: Methods are usually divided into those that consider the tangible dimensions (i.e., physical, environmental and institutional) and the intangible dimensions (i.e., socioeconomic). In this study, an approach is devised to assess socioeconomic vulnerability. The analysis that follows traces this approach in the context of the recent Kerala floods of 2018, across the various levels of jurisdiction (see Fig. 3), from which an identification of the issues, as well as their implications, can be arrived at. The findings of the assessments can then be added to the skeleton of the whole approach to complete the framework and prove its applicability. The idea is to incorporate and reflect on each of the principles at every stage of this framework.

---

[1] A panchayat/municipality in the grassroots governance level of the Indian local-self-government system.

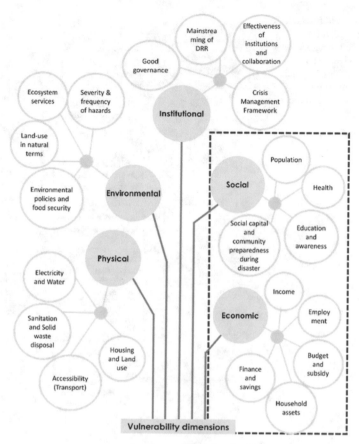

Fig. 2 Five dimensions of vulnerability and their respective indicators as adapted from the climate disaster resilience index (CDRI). *Source* Razafindrabe et al. (2009)

Since it is important to view DRR as part of sustainable development, it should be involved in every part of society, government/non-government, with a people-centric and multi-sectoral approach (UNISDR 2015). Therefore, a successful incorporation of DRR would result from the combination of top-down institutional strategies with bottom-up, local/community-based approaches. Recognizing the importance of community participation in DRR, the case-study analysis is performed on two levels: the macro (state–district) and the micro (local self-government–community). At the macro-level, data collected from secondary sources are used for the analysis and at the micro-level the data is acquired first-hand from primary sources: through one-on-one surveys of panchayat members (local leaders) and community consultation (via focus group discussions, community participatory mapping, transect walks and household surveys).

Fig. 3 Proposed research specific approach for top-down action agenda (problem-driven) and bottom-up strategic agenda (solution-driven) for assessing socioeconomic vulnerability

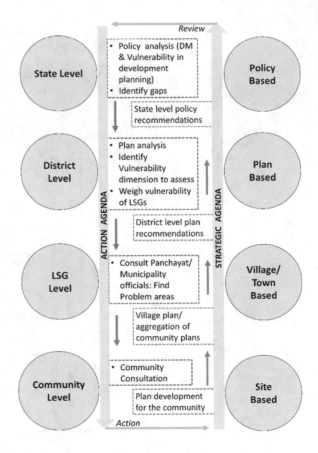

2.1   *Method Applied to Quantify Socioeconomic Vulnerability*

The method is adapted from the Social Vulnerability Index (SoVI) constructed by Sherbinin and Bardy (2016), which uses indexes for a synthetic measure of poverty and social exclusion to examine differential exposure to floods. In a developing country like India, there are other layers to be considered, and there exist location-specific factors that contribute to vulnerability, such as the existence of the caste system or of systematized gender discrimination (see Table 1). For the purposes of this research, to understand the hierarchy of socioeconomic vulnerability and find out the most vulnerable localities in the chosen Ernakulam district, a Socioeconomic Vulnerability Index (SEVI) has been developed. This SEVI analysis has been done for three reasons: (a) To introduce an additional layer of assessment into the existing hazard mapping system to provide disaster response and preparation in accordance with the socioeconomic standings of different settlements; (b) to narrow down the most socioeconomically vulnerable locations in the district for further analysis and c)

Table 1 Variable list for socioeconomic vulnerability assessment adopted and modified for the Indian context from (IPCC 2014) and (Sherbinin and Bardy 2016). The indicators finalized for the SEVI has been highlighted

Description of factors	Concept		Variable description
Demographic characteristics	Gender	*1*	*% main female working population*
	Race and ethnicity	2	% of Adivasi/tribal population
		3	*% of scheduled caste or scheduled tribe households*
	Migration	4	% of migrant workers (labor force)
Livelihood characteristics	Income status	5	% of population living in HH earning < Rs. 7000 per capita
		6	*% of irregular (marginal)/informal workers)*
	Employment loss/risk	*7*	*% population employed in agriculture/fishing/forestry*
		8	*% population unemployed*
	Renters	9	% population living in rented households
	Education	10	% of illiterate population over 15 years
Community development/identity	Family structure	11	Average number of people per household
		12	% of female headed households
	Social dependency	*13*	*% of population below 14 years of age and above 65*
Health	Medical services and access	14	% of labor force working in human health or social work
		15	*% of households more than three km from any medical facility*

to establish the relationship between the incidence of high vulnerability and coastal dependence/proximity.

2.1.1 Indicators for the Socioeconomic Vulnerability Index (SEVI)

From the indicators of socioeconomic vulnerability to floods that were added and modified according to the context of the case study (see Table 1), seven indicators were finalized for constituting the SEVI for this study. The indicators chosen are: the percentage of (age) dependent population, the percentage of backward population

Fig. 4 Case study location

(Scheduled Caste and Scheduled Tribe—SC and ST), the percentage of illiterate population, the percentage of household (HH) dependent on coastal livelihood, the percentage of marginal workers, the percentage of main[2] female working population and the percentage of unemployed population.

2.1.2   Assessment Tools for Calculating SEVI

After narrowing down and finalizing the indicators, the socioeconomic-indicator data is collected from 2011 District Census and Panchayat Level Statistics and standardized using principal component analysis (PCA) and Varimax rotation with Statistical Package for Social Science (SPSS) as a software tool. Then, the suitability of the data for factor analysis is confirmed using the Kaiser Meyer Olkin (KMO) test (which is available on the SPSS toolbox as well). The results of PCA are summed up and multiplied by the respective indicator weightages, using the Delphi method, to calculate numerical SEV scores for each gram panchayats/municipalities. Finally, mapping of the scores is done using the ArcMap 10.4 GIS to show their spatial variability toward socioeconomic vulnerability.

3   Case Study: Implementation of the Proposed Approach

3.1   Macro-Level Study: The State and the District

At the macro-level, analysis is done to identify gaps in the DRR efforts by the Kerala state and the Ernakulam district (see Fig. 4 for location). At the state level,

---

[2]Those workers who had worked for the majority of the year (i.e., 6 months or more) are called main workers.

while analyzing the Kerala State Disaster Management Policy (2010), it was found that the policy clearly asserts that *disaster management should be built into the development planning* (Government of Kerala 2010). But, a review of the state's development policies shows a lack of said incorporation. Among eight development policies reviewed, only two incorporates elements of disaster management into their contents; these were the State Policy for Persons with Disabilities (PwD) (2015) and the State Agriculture Development Policy (2015) (the former mandates that the evaluation of all infrastructure earmarked for emergency is accessible to PwD with involvement of PwD in disaster management training, while the latter has a separate chapter addressing the need for systematic disaster management and contingency planning at district, block and panchayat levels). Another important matter outlined in the states' disaster management policy is the setting up of 'community-based disaster management programs, plans and committees at each village level,' but, as of now, no such program, plan or committee exists across the state. It was possible to identify the socioeconomic losses to the floods from the Kerala State Post-Disaster Needs Assessment report (Government of Kerala 2018), which was prepared by assessing the damage in 1120 villages across ten districts. It brings to light some of the inherent socioeconomic vulnerabilities that could be addressed with the efficient incorporation of DRR measures in the state's policies and development plans. Some of the problems faced were of: (a) health: need felt for vulnerability assessment of all health facilities toward floods and training of health workers; (b) education and awareness: unsafe school buildings that needed repair, especially schools in low lying areas in the path of flood waters and landslides and unsafe access to schools with damaged roads/bridges; (c) exclusion and community preparedness: majority of SC and ST (80%) live in dilapidated houses, temporary huts and traditional mud houses along flood zones and suffered total damages; (d) livelihood: emerging need felt for support for farmers to restart cultivation and alternate sources of income needed for households with small plantations and coastal dependency.

3.1.1 District Level Socioeconomic Vulnerability Indexing (SEVI)

Among the worst-hit districts in Kerala, Ernakulam has been taken up for this case study since it serves as the economic hub of the state and suffered the highest loss in micro, small and medium enterprises (MSME) sector (Government of Kerala 2018). More than 70% of the district was inundated by the floods and landslides in August 2018, as can be seen in Fig. 5.

After calculating the Socioeconomic Vulnerability Index (SEVI) for Ernakulam (see Table 2), the weightage was the highest for dependence on coastal livelihood. When the SEVI was calculated for each panchayat in the district and mapped (see Fig. 6, in which the intensity of shading shows higher vulnerability), it was found that 68% of the villages in the high vulnerability range are coastal villages. This district SEVI map, when viewed in conjunction with the inundation map (Fig. 5), shows that almost all the district panchayats that are found to have higher inherent vulner-abilities (i.e., high SEVI) were also the ones inundated and located in high hazard

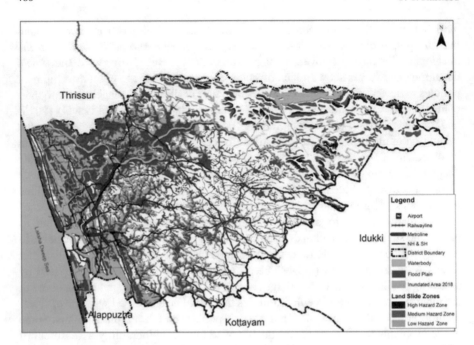

Fig. 5 2018 flood inundation and landslide map with the existing flood plains of Ernakulam District

Table 2 SPSS-generated variance and weightage allocation for SEVI in Ernakulam District

S. No	Indicator name	Mean	Std. deviation	Extracted communality	% Variance	Weightage
1	Age-dependent population	9.12	1.06	0.927	30.60	0.25
2	Marginalized population (SC and ST)	9.9	4.44	0.181	26.35	0.30
3	Illiterate population	14.51	2.43	0.895	16.07	0.40
4	HH dependent on coastal livelihood	19.95	18.06	0.791	13.92	0.50
5	Marginal workers	17.47	6.13	0.843	8.27	0.35
6	Main female working population	19.46	3.12	0.776	3.07	0.40
7	Unemployed population	43.45	2.116	0.699	1.69	0.30

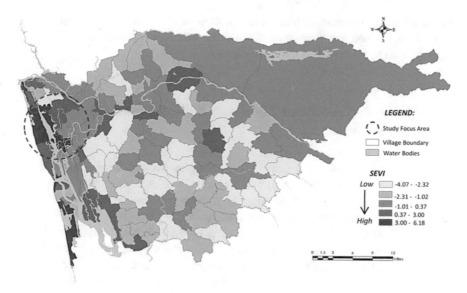

Fig. 6 Mapped Socioeconomic Vulnerability Indexing (SEVI) in Ernakulam District

zones of flooding/landslides. From mapping the immediate government/institutional interventions and relief measures after the floods (see Fig. 7), it is evident that they focused mainly on constructing relief camps and rebuilding schools in areas with

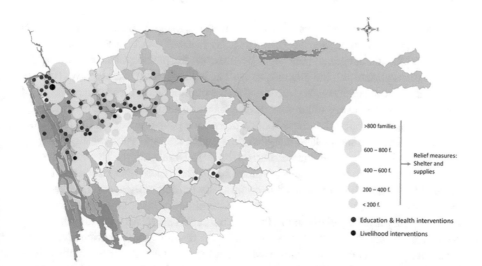

Fig. 7 Post-flood Interventions of main government departments after the Kerala floods 2018; the yellow circles indicate the varying intensity of numbers of families (f.) accommodated in each relief camp

higher reports of infrastructure damage, but there is a lack of livelihood reconstruction efforts. For some of the coastal island communities, the designated relief camps were difficult to access/reach during the floods.

3.2 Micro-level Analysis: The Local Self-government and the Community

Analysis at the micro-level is done by assessing the socioeconomic vulnerability, at the local self-government (LSG) and community scales, of populations such as the elderly, poor, racial and ethnic minorities, those with special medical needs and those living in remote and/or inaccessible areas. At the LSG level, it was found that the vulnerable households were without accessible/reliable transportation and faced challenges evacuating to safer areas. Consequently, many were forced to seek shelter in facilities that were unsafe and ill-equipped. Other residents were forced to flee to distant villages and towns. But all of them faced a difficult and prolonged recovery due to the dissolution of social networks, extended periods of unemployment, a complicated aid process and inadequate resources to repair or replace damaged homes/property. Some critical facilities, such as nursing homes and hospitals, did not have a plan for evacuating elderly or infirmed residents.

3.2.1 Selection of Community, Vulnerable Groups and Households

An island community was selected from the most inundated panchayat in Ernakulam to be surveyed in detail. *Cheriya-Paramb (C.P.) Thuruth* (vernacular for island) was chosen in Chendamangalam panchayat because, on the panchayat survey during the various interactions with the officials, as well as the local civil society, it was clearly evident that certain inland-island areas within the boundaries of the panchayaths faced flash floods almost every other year during the monsoon season (June–August). It was also communicated that these islands often lack proper access to the mainland. *C.P. Thuruth* is one such island that also falls under the high vulnerability category per SEVI and was heavily inundated during the 2018 Kerala floods. To understand the reasons for their high socioeconomic vulnerability and exposure to disasters, the residents of C.P. Thuruth were engaged in community participatory exercises (involving the members in focus group discussions and community mapping). The vulnerable groups mainly involved in the community participation exercises were the marginally poor, women, children and the elderly. The main result of these participatory exercises was a Community Vulnerability Map that the members of the community drew for their own Island, highlighting all the land-uses, buildings they considered important and problematic areas/aspects within the boundary. Figure 8 is the digitized and labeled version of the hand-drawn community vulnerability map.

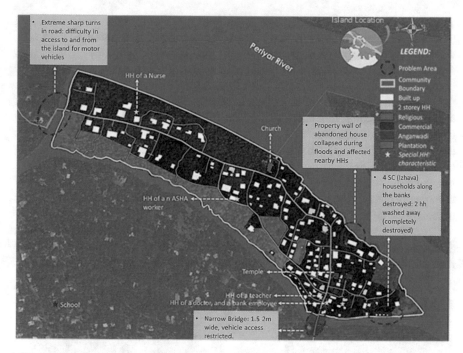

Fig. 8 Digitized community mapping of the island *C.P. Thuruth* in Chendamangalam panchayat

3.2.2 Participatory Community Mapping and Discussion

Through the participatory mapping, discussion and personal observations, it was possible to understand the main characteristics and issues on the island. *C.P. Thuruth* can only be entered through a bridge 1.5-m wide (hence, ambulances cannot enter in emergencies). The southern part of the island is mainly inhabited by Hindu households with a temple and anganwadi in the middle, and the northern part has Christian households starting from the church location, but these castes are interlinked, and there is no conflict. The main damage suffered in the community was to minority households, which were completely washed away in the floods. More than 40% of the working class in the community depend on the MGNREGA[3] scheme, and their social vulnerability is due to the heavy dependence on the neighboring town for social and physical amenities. Moreover, the economic vulnerability was mainly due to the high unemployment rates of women in the community and the heavy dependence on informal sectors. The residents also pointed out the lack of two-storied buildings on the island where they could immediately seek safety when the flood waters swept in (there were only five two-storied houses on the entire island).

---

[3]The Mahatma Gandhi National Rural Employment Guarantee Act of 2005, otherwise termed the MGNREGA for short, is an act of the Indian government that provides direct supplementary wage-employment to the rural poor (the underemployed and surplus labor force of the rural sector) in the country for 100 days, through public works.

These socioeconomic vulnerabilities do not stand alone: rather they are linked to the physical vulnerabilities that exist within various communities. Due to the weak transport linkages and access to such small island communities, the disaster warnings and subsequent evacuation were delayed, leading to weakened community preparedness. Here, this weak transport linkage continues to hinder the women in the community from seeking employment outside their community. The location of critical infrastructure, like hospitals, is either along the flood plains or further away toward the mainland. This lack of efficiency in land-use planning makes access to resources difficult and risky for the already vulnerable population residing in coastal communities. From studying the drivers of socioeconomic vulnerability at a micro-level, it was possible to understand the existence of high social vulnerability of community preparedness in remote islands, exhibiting the need for strong community preparedness of logistics, materials and management.

To arrive at recommendations for these issues stated, the research followed a process of informing, consulting, involving, collaborating and empowering the residents (see Fig. 9).

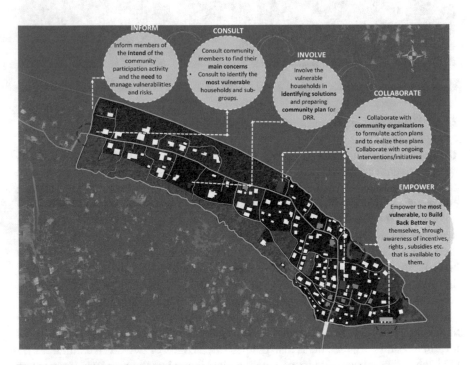

Fig. 9 Community level proposal identification through participatory exercises

4 Recommendations and Conclusion

Was It Ever Discussed in Your Community Meetings How Each One of You Would Safeguard Your Lives and Livelihoods During and After Disastrous Events?

We Are Known to Be a "thotti" (Scavenging) Community, Anything Could Happen Tomorrow, it's God's Will, Whatever Would Happen to Us is Our Fate. So, We don't Think About These Things and Worry.

—Kannan 47, Municipality Owned Laborer, Pipeline Colony, Ernakulam

The core objective of this research is in the spirit of leaving no one behind, and there are various processes to ensure that there is effective community participation as well as action. In this case study, the people of the community put forward two major recommendations, which seem very simple but address the core issues. The first one is to make the existing religious institutions and existing *Kudumbashree*[4] units in the community responsible for constituting a DRR task force. The task force would be entrusted with the primary responsibility of implementing and monitoring community disaster management and evacuation plans, as well as keeping records of the community. The second recommendation, as simple as it sounds, is that every community should have a two-storey community building. So, within the community the panchayat officials could construct a two-storied or higher anganwadi/lower-primary school, with disaster management incorporated into its curriculum and where awareness workshops could be conducted for the community members regarding sanitation, health, risk reduction measure, etc. Ideally, this facility should be maintained by the local government but run by the community members themselves to ensure effective participation, as well as to create additional employment opportunities.

From the evidence gained from the implementation of the proposed approach on the case-study area, the Ernakulam district officials should follow the below-mentioned strategies to reduce the district's vulnerability to disasters:

- Mandating participatory tools for the planning process on site;
- The repurposing of the local employment generation schemes for coordinated risk reduction efforts by the generation of employment in work relating to pre-disaster (employment in mitigation efforts and awareness campaigns), during (training in immediate relief operations) and post-disaster (skill development, reconstruction);
- Adding a layer of *socioeconomic vulnerability* mapping to the overall risk mapping at the district level.

In conclusion, this research revolved around four main objectives that were: to understand the concept of vulnerability to disasters; to examine disaster management being addressed by the government institutions; to assess the socioeconomic vulnerability in the case-study area; and to develop an approach-based framework for

---

[4]Kudumbashree is the poverty eradication and women empowerment program implemented by the State Poverty Eradication Mission (SPEM) of the Government of Kerala. The name *Kudumbashree* in Malayalam language means 'prosperity of the family'. It is essentially a woman-led community network that covers the entire State of Kerala.

decision-makers and physical planners to reduce disaster vulnerability. The first three objectives are satisfied in the analysis, and the fourth and final objective is fulfilled in the application process of this approach. In linking the issues and drawing recommendations, the logic is to move from action toward strategy. That being said, our desired overall output is to have reduced vulnerabilities: an effective reduction, from what we have seen so far, is only possible through the participation of the most vulnerable people. It is crucial that such vulnerability assessments are undertaken with a thorough input from local stakeholders, incorporating the experiential knowledge of people who have lived through these disasters and who understand how such disasters make them vulnerable. Involving the community in the preparation of vulnerability assessments can improve its effectiveness and ensure that such assessments are relevant to those who are the most at risk. This helps create awareness about the risks posed by certain hazards and motivates community members and organizations to take steps to become more prepared.

To reduce the impact through participatory action the most important aspect is that we "Hear out the Survivors," and the main focus of disaster risk reduction plans and policies being proposed should revolve around the idea of working 'with' them rather than 'for' them.

References

Balica SF, Wright NG, Meulen F (2012) A flood vulnerability index for coastal cities and its use in assessing climate change impacts. Natural Hazards, pp 1–33

Davis I (2004) Progress in Analysis of Social Vulnerability and Capacity. In: Greg B, Georg F, Dorothea H (eds) Mapping vulnerability: disasters, development and people. Earthscan, London, pp 128–144

Government of Kerala (2010) Kerala State Disaster Management Policy. Kerala State Disaster Management Authority, Thiruvananthapuram

Government of Kerala (2018) Kerala Post Disaster Needs Assessment, Floods and Landslides Report, August 2018. Jointly developed by Kerala State Disaster Management Authority, United Nations agencies, European Commission, European Union Civil Protection and Humanitarian Aid (ECHO). Thiruvananthapuram, UNDRR

IPCC (2014) Summary for policymakers. In: Climate change 2014: impacts, adaptation, and vulnerability. Cambridge University Press, New York. https://www.ipcc.ch/report/ar5/wg2/. Accessed 4 Feb 2019

Kelman I, Lewis J (2005) Ecology and vulnerability: islands and sustainable risk management. Int J Island Aff 14(2):4–12

Lewis J (2009) An Island characteristic: derivative vulnerabilities to indigenous and exogenous hazards. Int J Res Island Cult 3(1):3–15

Nicholls RJ, Hanson S, Herweijer C, Patmore N, Hallegatte S, Morlot J, Chateau J, Wood R (2008) Ranking port cities with high exposure and vulnerability to climate change: exposure estimates. OECD, New Delhi

Razafindrabe B, Parvin G, Surjan A, Takeuchi Y, Shaw R (2009) Climate disaster resilience: focus on coastal urban cities in Asia. Asian J Environ Disast Manag, p 1

Sherbinin DA, Bardy G (2016) Social vulnerability to floods in two coastal megacities: New York City and Mumbai. Vienna Yearbook Popul Res 1:131–165

UNISDR (United Nations International Strategy for Disaster Reduction) (2014) United Nations General Assembly Third United Nations World Conference on Disaster Risk Reduction: Outcome of Sixth Asian Ministerial Conference on Disaster Risk Reduction, 22–26 June 2014. UNDRR, Bangkok

UNISDR (United Nations International Strategy for Disaster Reduction) (2015) Sendai framework for disaster risk reduction 2015–2030. UNDRR, New York

Wisner B (2002) Who? What? Where? When? In an emergency: notes on possible indicators of vulnerability and resilience: by phase of the disaster management cycle and social actor. In: Plate E (ed) Environment and human security: contributions to a workshop in Bonn, pp 7–14

Towards Sustainable and Inclusive Cities: Discrimination Against Vulnerable and Marginalized Groups—A Review of a Hidden Barrier to Sustainable Urbanization

Vivien Benda

Abstract Implementing the United Nations' Sustainable Development Goals (or, in abbreviated form, SDGs) is a global obligation. SDG 11 focuses on the issues of sustainability in cities and other communities, which has a strong correlation with significant human rights, such as the right to a healthy environment or the right to water and sanitation. The most worrying barrier to the implementation of this SDG is the discrimination against vulnerable and marginalized groups of the society. This study aims to identify the various forms of discrimination in this context and propose a solution framework. In the study's context, sustainable urbanization is extended to the promotion of human rights and the improved meaning of security, which involves the respect of fundamental rights. This context is based on the principle of sustainable development and its three systems (ecological, economic, and social). The classical legal methods—the analysis of the international and national legal background and the related literature—are complemented by the qualitative case-study legal method. Analysis of the secondary and primary data, from Hungary and Transylvania (Romania), revealed the various, sometimes additional, discrimination against specific vulnerable groups, like passengers with special needs or socially, politically disadvantaged ethnic groups. The results indicate that sustainable urbanization must be based on the respect for human rights, and this shall be addressed at all levels of decision making and practical implementation. To achieve this, on the one hand, the exploration of the potential discrimination is essential, and this should be based on independent actors' reports. This also requires the extension of the SDGs' indicators, therefore the criteria of the evaluation. On the other hand, the state's active steps are also needed to eliminate the violations, such as specific measures concerning the special need of the vulnerable people of the particular community. Taking all this into consideration, it is possible to create human-centred cities and other human communities to bring about territorial balance and prevent human-rights violations.

Keywords SDG11 · Hungary · Romania · Discrimination · Passengers with special needs · Right to water and sanitation

V. Benda (✉)
PhD-student of the Pázmány Péter Catholic University's Faculty of Law and Political Sciences and lawyer at the Institute for the Protection of Minority Rights, Budapest, Hungary
e-mail: benda.vivien@kji.hu; benda.vivien18@gmail.com

© The Author(s), under exclusive license to Springer Nature Switzerland AG 2021
A. Bisello et al. (eds.), *Smart and Sustainable Planning for Cities and Regions*,
Green Energy and Technology, https://doi.org/10.1007/978-3-030-57332-4_33

1 Introduction

Sustainable urbanization is one of the significant aspects of sustainable development, especially, because most of the world's population—55% currently—is living in cities, a number that is estimated to reach 68% by 2050 (World Urbanization Prospects: The 2018 Revision 2019: 19). The United Nations' Sustainable Development Goals (or, in abbreviated form, SDGs) also consider this issue, as SDG 11 aims to create sustainable and inclusive cities and human settlements. The SDGs create an integrated and holistic network that focuses on the whole world, with special attention on the developing countries. This is in accordance with the concept of sustainable development, which meets the needs of the present generation without compromising the ability of future generations to meet their own needs (United Nations, Report of the World Commission on Environment and Development 1987). Sustainable development also means the coordinated operation of the ecological, economic and social systems (Barbier and Burges 2017: 3–6). This concept is related to the intra- and intergenerational equity described by Edith Brown Weiss. Intergenerational equity is the equity between generations of today and tomorrow, while intragenerational equity is the equity between those who are living today, within and among countries (Brown Weiss 2008: 615–619; 627). Moreover, the concept of "no one should be left behind" is also defining. This was reflected by the fact that the Secretary General addressed its importance in the latest comprehensive report on SDG progress. The report highlights that the concept of leaving no one behind is one of the eight areas where fundamental changes are needed. It is among the defining features of the 2030 Agenda, and it could help to realize human rights for all. This also indicates that states must advance economic, social, and cultural rights by ensuring access to the most vulnerable groups to high-quality essential services (Report of the Secretary-General on SDG Progress 2019: 35).

Several recent studies have confirmed that the human security and sustainability are dependent on the enjoyment of human rights (e.g. Fukunda-Parr and Messineo 2012) and that the relationship between cities and human rights has been stronger since the early 1990s (Oomen and Baumgärtel 2014). The consideration of these holistic and human rights-centred approaches highlights one of the main barriers to the implementation of SDG 11 and the creation of sustainable, safe, and inclusive cities and human settlements. This barrier is a violation against fundamental human rights, namely the discrimination against the vulnerable and marginalized groups of the society. According to Tomasevski, elimination of discrimination related to development is the core human-rights demand, and this requires the proper legal framework to access equally to economic opportunities and public services (Tomasevski 1992: 88–89). Bearing this in mind, the study presents a multifaceted selection of cases, each representing a particular form of discrimination, targeted against different social groups to answer the question, whether the discrimination against vulnerable and/or marginalized groups could be a barrier to the implementation of SDG 11 and the realization of sustainable urbanization. The aim is to outline an overview of the multiple challenges that need to be confronted to make progress in sustainable urbanization.

2   Method

To address the above topic, it's vital to clarify why it is important to examine human-rights issues in light of sustainable urbanization. To realize this, I combined the traditional legal methods with improved approaches that reflect the new legal challenges and complement the classical legal considerations. Some authors draw attention to these new kinds of legal research methods, like qualitative empirical legal research (Langbroek et al. 2017: 1–4; 7–8). The authors mention Argyrou, who combined the internal perspective and the external perspective, therefore, the doctrinal legal research and the empirical legal-research methods (Argyrou 2017: 97). In this light, the methodology of this paper can be divided into two parts. On the one hand, according to the traditional legal methods, I examined the theoretical background—the principle of sustainable development, the intra- and intergenerational equity, the concept of no one should be left behind—and the relevant international documents and domestic rules. On the other hand, from the external perspective, I used the so-called case study qualitative legal method to evaluate the implementation of the legal norms. The next section will assess three main types of discriminations, including:

a. Discrimination against passengers with special needs (case study no. 1).
b. Discrimination related to the right to water and sanitation (case study no. 2).
c. Environmental discrimination against ethnic Hungarian minorities living in the Carpathian Basin (case study no. 3).

The type and collection of data are different. In case-study no. 1, about the discrimination related to accessibility (para 3.1.), the data are secondary—data from the Ombudsman for the Future Generations' statement and from official statistics of the metro stations that are accessible to the passengers with special needs (Deputy Commissioner for Fundamental Rights, Ombudsman for Future Generations 2017). In case-study no. 2 about the discrimination related to water and sanitation (para 3.2.), the data are also secondary, originating from the Report of the European Roma Rights Centre. Finally, the data in case-study no. 3 about the environmental discrimination of the indigenous Hungarian people (para 3.3.) are fully primary, collected by the author as a legal researcher of the Institute for the Protection of Minority Rights (IPMR) by using surveys. Particular mention should be paid of the fact that the data have a significant point in common: their sources are independent of the Hungarian and Romanian Executive. Evaluating this method, it can be concluded that it is capable to reveal discrimination in practice, even if the legal background appears appropriate and, for this reason, it can help to identify the potential barriers to sustainable urbanization.

3   The Sustainable Development Goals and SDG 11

The Sustainable Development Goals are set in an international soft law document—United Nations' General Assembly resolution (United Nations 2015)—which determines the agenda between 2015 and 2030. SDG 11, which seeks to make cities and human settlements inclusive, safe, resilient, and sustainable, is the most relevant from the perspective of sustainable urbanization. The SDG addresses universal access to: housing and basic services; to an inclusive, safe, affordable, accessible, and sustainable transport system; and safe, inclusive, and accessible, green and public spaces, in particular for women and children, older persons and persons with disabilities. It also aspires to upgrade slums, protect the world's cultural and natural heritage, promote the importance of disaster management and the reduction of the cities' environmental impact (United Nations 2015: 21/35–22/36).

In this research work, the potential barriers, to the implementation of SGD 11 in Hungary and in neighbouring countries where ethnic Hungarians are living (in particular in Romania), are to be examined.

The cases have some common elements because the persons concerned belong to a minority, based on their defining attributes. These attributes are their different physical status, belonging to a social and ethnic minority or indigenous community. This fact involves that the discrimination, which is common in each case, violated various fundamental rights (i.e. the right to free movement and dignity (case study no. 1)); the right to water and sanitation (case study no. 2); and the right to bilingualism and access to environmental information (case study no. 3) and seriously infringed on the principle of equality and non-discrimination. A further common feature is that the state and local governments are required to prevent and eradicate these violations, but the involvement of independent actors, like civil society, could also effectively support this.

3.1   Case Study No. 1: Discrimination Against Passengers with Special Needs

In Hungary, the current Ombudsman for the Future Generations'[1] issued a statement about the implementation of the SDGs in 2017 and, in the light of SDG 11, he concluded that physical accessibility is one of the main indicators of SDG 11 (Deputy Commissioner for Fundamental Rights, Ombudsman for Future Generations 2017). He expressed his concern about the old infrastructures in the public transport system and mentioned a case before the Hungarian Constitutional Court (3023/2015. (II.9.) ABH.), which was an ex-post review of the related law's[2]—the law about passenger

---

[1] Prof. Dr. Gyula Bándi.

[2] Hungarian law Nr. XLI. 2012 about passenger transport services 51 § (4).

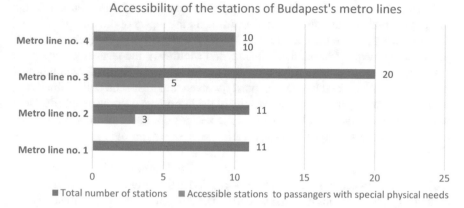

Fig. 1 Number of accessible stations to passengers with special physical needs in Budapestss

transport services—conformity with the Fundamental Law,[3] initiated by the former Commissioner for Fundamental Rights. According to the law, the state must provide equal physical accessibility in scheduled public transport vehicles, stations, and stops. The state must provide it gradually and, according to the initiator, this condition did not conform with the Hungarian Fundamental Law and the Convention on the Rights of Persons with Disabilities (2006).[4] The decision of the Constitutional Court was the following: providing equal accessibility is a constant obligation of the state and the term gradually does not mean exemption from the requirement. But, according to the Ombudsman for Future Generations, this issue remains worrying and points the national legislature's attention to this. To illustrate the gravity of this problem in Hungary, there are four metro lines in the capital city Budapest with a total of 52 stops, out of which only 18 are accessible to passengers with special physical needs, such as persons who need to use wheelchairs (see Fig. 1).[5]

3.2 Case Study No. 2: Discrimination Related to the Right to Water and Sanitation

On one hand, the issue is linked to SDG 6, which foresees ensuring the availability and sustainable management of water and sanitation for all; on the other hand, it aspires to promote these universal human rights, and there is also a strong connection with SDG 11; for example, in Budapest, the Budapest Waterworks is mostly municipality-owned.

---

[3] According to the initiator, it is against the provisions, which grant human dignity (Art. II;), free movement (Art. XXVII. (1); the requirement of equality (Art. XV. (4).

[4] Convention on the Rights of Persons with Disabilities (CRPD), New York, 13 December 2006.

[5] See more: https://bkk.hu/akadalymentes/kotottpalyas/ Accessed April 28, 2020.

Clean drinking water for all citizens is yet an unresolved issue even within the European Union. To give an example, according to the report of the European Roma Rights Centre, in more than half of the places visited (places in Hungary, Slovakia, France, Montenegro, Macedonia, Albania, and Moldova), the nearest water source was more than 150-m away. In some case, the interviewed Roma people reported needing to walk kms and that those, mostly women and girls who collect the water, are often attacked by animals. The majority of the interviewed Roma people declared that, even if piped water would be made available, they could not afford it. So, besides the physical accessibility, the unaffordable water costs are also causing concern (Thirsting for justice… 2017: 30–32).

It is worth mentioning that the first successful European Citizens' Initiative (ECI)[6]—the so-called Right2Water ECI[7]—reacted to this problem. The Right2Water led to a legislative proposal on the so-called Drinking-water Directive made by the Commission[8] which reflected the SDGs. After the release of the proposal, there was a successful vote in the European Parliament,[9] and a General Approach[10] of the Council of the EU was also released. However, the legislative process is currently ongoing, and the aspirations of the ECI's initiators were restricted, but the Right2Water could be considered as a success because the European Union finally addressed the right to water and sanitation, as an individual human right.

3.3   Case Study No. 3: The Environmental Discrimination against Ethnic Hungarian Minorities Living in the Carpathian Basin

Finally, the third aspect is a lesser known kind of discrimination against ethnic Hungarian minorities living in the Carpathian Basin, outside the borders of Hungary. A survey was carried out by the Budapest-based NGO the Institute for the Protection

---

[6]The ECI is a unique transnational tool of direct democracy that allows the European Citizens to put issues on the European agenda. It is technically an agenda initiative, and this tool's disadvantage is the lack of legislative obligation. The European Citizens' Initiative's details are regulated in the Regulation no 211/2011 of the European Parliament and Council, Regulation no. 211/2011 of the European Parliament and the Council of 16. February 2011 on the citizens' initiative. OJ L 65, 11.3.2011.

[7]The ECI's former name is the following: Water and sanitation are a human right! Water is a public good, not a commodity! See more: https://europa.eu/citizens-initiative/initiatives/details/2012/000 003_en. Accessed February 14, 2020.

[8]Proposal for a Directive of the European Parliament and of the Council on the quality of water intended for human consumption (recast). COM(2017) 753 final, Brussels, 1.2.2018.

[9]Quality of water intended for human consumption ***I(vote) 2017/0332(COD). https://www.eur oparl.europa.eu/doceo/document/PV-8-2018-10-23-ITM-007-13_EN.html. Accessed Jan. 6, 2020.

[10]Proposal for the Directive of the European Parliament and of the Council on the quality of water intended for human consumption (recast)—General approach. 6876/1/19 REV. Brussels, 27 February 2019.

of Minority Rights (IPMR)[11] with the help of its partners who are members of other NGOs and/or individual lawyers. The survey's intension was to reveal the following situations in regions where Hungarian minorities are living: installation, maintenance, or improper liquidation of facilities that endanger the human health or are pose a significant risk; violation or restriction of fundamental human rights, like the right to a healthy environment, the right to health; violation or restriction of the right to access information or public participation in decision making or access justice in environmental matters; refusal or restriction of environmentally valuable, useful innovations, aids, financial assistances, and tenders.

In general terms, we can draw the conclusions that most of the feedback described situations where environmental discrimination existed, but the environmental characteristic was often the result of the violation of other human rights. In the following, two cases will be presented from Transylvania, Romania, where the Hungarian minority represented 6.5% of the total population at the time of the last general census in 2011 (Veres 2013:30).

One case revealed a discriminatory practice of the local government of Kolozsvár (Cluj-Napoca) which launched a call for tenders related to education and other civil projects, such as educational projects on sustainable development or climate change or projects related to inclusive and sustainable urbanization. IPMR's responding partner pointed out that the tenders who attend to the Hungarian minorities or are in the Hungarian language are rejected because of this ethic circumstance.

In the other case, the right to access information on environmental matters was violated. The Romanian national environmental authority has lower-level authorities in counties, even in counties where ethnic Hungarian minorities are living. According to the relevant Romanian law about the public administrations[12] in the places where the proportion of the minority population is above 20%, the state must provide bilingualism. We need to mention that, if this proportion is below 20%, bilingualism is still allowed under this law; however, in practice, it is not implemented by local authorities. Despite the relevant law, the websites of the environmental authorities functioning in Szatmár, Maros, and Kovászna, were not available in the Hungarian language at all. This also applies to environmental reports, informative documents, decisions, etc.; however, in Hargita county, these data are at least partially available.[13] Due to this situation, members of the Hungarian community whose native language is Hungarian are discriminated against, and they cannot access information about the environmental issues of their city or community on the same terms as the majority of the population.

---

[11] See more about the IPMR: https://www.en.kji.hu/. Accessed January 6, 2020.

[12] Legea administrației publice locale nr. 215/2001.

[13] The percent of the Hungarian population is 34,7 in County Szatmár, 38,1 in County Maros, 73,7 in County Kovászna and 85,2 in County Hargita (Veres 2013, 31).

4 Results and Discussions

There are several possible solutions related to the issues just described. The proposed solution framework can be divided into two parts: possible solutions related to the exploration of the discrimination and solutions related to the elimination of these violations. On the one hand, to properly reveal the possible discriminatory situations, the need for independent actors, like ombudspersons and NGOs, is undeniable. Moreover, in particular, to effectively evaluate the implementations of SDG 11, the scope of the indicators should be broadened and more emphasis should be put on the implementation of human rights and human capabilities (see: Fukuda-Par et al. 2014: 3–4). The expansion of the indicators' scope has to be done at the local level, too. According to Kaufman, the SDGs provide the opportunity to advance human rights locally by developing the locally relevant human rights indicators, too (Kaufman 2017: 119–120).

On the other hand, to solve these violations and prevent other similar ones related to safe and inclusive urbanization, the extended concept of human security should be used as guidance for the policymakers. This extended concept includes all aspects of human rights, like meeting basic needs and the enjoyment of political and social freedom. It also involves the possibility to take into account the significant human rights related to safe urbanization (Fukuda-Parr et al. 2012: 3–4). Practical solutions are also needed, like strengthening the responsibilities of the local level related to the accessibility and promoting negotiations with representative bodies to identify the problems (e.g., some stations are not accessible by wheelchair). To this end, the local governments' source of financing should be increased (Gradwohl and Vámosi 2012: 41–42.). Promoting the right to water and sanitation involves not only the issue of physical affordability but also economic affordability. The latter requires the states to reconsider liberalization in this field, and they should take positive steps to identify the concerned people and provide special aid to them. Local governments should be partners in granting these persons an advantage. The solutions to the lingual discrimination against the ethnic Hungarian communities in Transylvania are different. First, advocacy steps should be taken, like an inquiry into the competent authority by a local NGO or engaged citizens. If this does not solve the issue, a complaint should be submitted to the competent bodies, like the National Council for Combating Discrimination in Romania.

The common characteristics of the proposed solutions to prevent these kinds of discriminations are the following: the need for ensuring close cooperation between the local and central level, providing more resources to the local level, and the importance of independent actors, like NGOs.

It is clear that, like the forms of discrimination, the possible solutions can come in many forms involving important actors from different levels to create safe and sustainable, human rights-centred cities and human settlements.

5 Conclusion

This study sought to answer the question of whether the discrimination against the vulnerable groups in a society can prevent the implementation of SDG 11 and the creation of inclusive, sustainable cities. Based on the qualitative case-study analysis, it can be concluded that the answer is yes because sustainable urbanization must be based on the concept of sustainable development that takes into account social and human rights issues, in line with the SDGs. Inclusive and sustainable cities and regions should be based on social justice and equality, and these require affirmative action by states to combat discrimination against the vulnerable and/or marginalized groups of the society. To achieve all this, the exploration of the potential discrimination by independent actors like ombudspersons and NGOs or international organizations is necessary because their reports can reveal the discrimination and demonstrate best practices, too. It should be added that the report should take into account the circumstance that a region or city is inhabited by ethnic, national, or other minorities. These considerations must be incorporated in the policy-making at the national, regional and local levels as previously described (see Sect. 4).

Implementing this human rights-based approach with special attention paid to the vulnerable groups of the society and the application of the extended concept of human security, we can create safe, inclusive, resilient, and sustainable cities and other communities and properly implement SDG 11 by 2030. In this light, this concept can help non-legal researchers and policy-makers to give greater priority to fundamental human rights during urban planning.

Acknowledgements Supported by the ÚNKP-19-3-I-PPKE-80 New National Excellence Program of the Ministry for Innovation and Technology.

References

Argyrou A (2017) Making the case for case studies in empirical legal research. Utrecht Law Rev 13(3)

Barbier EB, Burgess JC (2017) The sustainable development goals and the systems approach to sustainability. Econ OpenAccess, Open-Assess E-J, 11, 2017/28

Brown Weiss E (2008) Climate change, intergenerational equity, and international law. Vermont J Environ Law 3

The decision of the Hungarian Constitutional Court No. 3023/2015. (II.9.) ABH. https://public.mkab.hu/dev/dontesek.nsf/0/98ab3478425794ffc1257b3400212928/$FILE/3023_2015_hat%C3%A1rozat.pdf. Accessed 6 Jan 2020

Deputy Commissioner for Fundamental Rights, Ombudsman for Future Generations (2017) *Doctrinal statement about the Hungarian implementation of the United Nation's Sustainable Development Goals'*. AJB-6495–1/2017. 34–35. https://www.ajbh.hu/documents/10180/279 1084/SDG_elvi+%C3%A1ll%C3%A1sfoglal%C3%A1s_2017_12_19_kiadott.pdf/33b3f4e6-ae40-e743-8d32-2c98b8baea4c. Accessed 6 Jan 2020

European Roma Rights Centre (2017) Thirsting for justice—Europe's Roma denied access to clean water and sanitation, Budapest

Fukuda-Parr S, Messineo C (2012) Human security: a critical review of the literature. CRPD Working Paper, No. 11 January 2012

Fukuda-Parr S, Ely Yamin A, Greenstein J (2014) The Power of numbers: a critical review of millennium development goal targets for human development and human rights. J Human Dev Capab 15:2–3

Gradwohl CS , Vámosi T (2012) A fogyatékossággal élők egyenlő esélyű hozzáférés biztosítására irányuló törekvésekről és azok akadályairól. Tudásmenedzsment: A Pécsi Tudományegyetem TTK Felnőttképzési és Emberi Erőforrás Fejlesztési Intézetének periodikája 13(2)

Kaufman ER (2017) Localizing human rights in the United States through the 2030 sustainable development agenda. Columbia Human Rights Law Rev 49(1)

Langbroek P, van den Bos K, Thomas MS, Milo M, van Rossum W (2017) Methodology of legal research: challenges and opportunities. Utrecht Law Rev. 13(3)

Oomen B, Baumgärtel M (2014) Human rights cities. In: Mihr A, Gibney M (eds) The SAGE handbook of human rights, vol 2. SAGE Publications Ltd., London, pp 709–730

Tomasevski K (1992) Monitoring human rights aspects of sustainable development. Am Univ J Int Law Pol 8(1)

United Nations (2015) Transforming our world: the 2030 agenda for sustainable development. A/RES/70/1

United Nations, Department of Economic and Social Affairs, Population Division (2019) World Urbanization Prospects: The 2018 Revision (ST/ESA/SER.A/420). United Nations, New York

United Nations, Report of the World Commission on Environment and Development (1987): Our Common Future. Art. I. Para 3. Subpara 27. https://sustainabledevelopment.un.org/content/doc uments/5987our-common-future.pdf. Accessed 6 Jan 2020

United Nations, Secretary-General (2019). Report of the Secretary-General on SDG Progress 2019 Special Edition. https://sustainabledevelopment.un.org/content/documents/24978Report_ of_the_SG_on_SDG_Progress_2019.pdf. Accessed 6 Jan 2020. 01. 06.

Veres, V. (2013). Népszámlálás 2011: A népességszám, foglalkozásszerkezet és iskolázottság nemzetiség szerinti megoszlása Romániában. Census 2011: population size, occupational structure and level of education by ethnicity Romania. Erdélyi Társadalom. 11/2.

Tackling Energy Poverty

Exposure and Vulnerability Toward Summer Energy Poverty in the City of Madrid: A Gender Perspective

Miguel Núñez-Peiró, Carmen Sánchez-Guevara Sánchez,
Ana Sanz-Fernández, Marta Gayoso-Heredia, J. Antonio López-Bueno,
F. Javier Neila González, Cristina Linares, Julio Díaz,
and Gloria Gómez-Muñoz

Abstract Recent research has addressed the special relationship between energy poverty and women. Despite that not many studies are yet available, results show that there might be strong gender inequalities connected with household's energy deprivation. Furthermore, differentiated health impacts have been detected between men and women, putting women into a more vulnerable position. In this sense, the so-called feminization of energy poverty is urging a revision of the existing studies from a gender perspective to foster its inclusion within energy poverty alleviation policies. The present study explores the links between summer energy poverty and gender in the city of Madrid. Summer energy poverty is considered another variety of energy deprivation particularly relevant within mid- and low-latitude countries, in which energy consumption for cooling is heavily increasing. It also seems to be particularly relevant in cities in which the urban heat island introduces relevant variations in the microclimatic conditions that might increase the housing-cooling demand. Following the methodology developed in previous studies, the risk of suffering from summer energy poverty is, in this paper, explored considering the household's gender composition. The geospatial distribution of their vulnerability is compared with other indicators related to their exposure to high temperatures: the housing energy efficiency and the cooling degree hours. The evaluation at the sub-municipal scale is carried out among the different subgroups in which a woman is the main breadwinner: single women with children and single women over 65 years old. Their situation is also compared to those households in which a man is the main breadwinner. The analysis of the selected variables is conducted using a *hot spot* analysis, which evaluates the autocorrelation of each variable according to its spatial distribution. Results show

M. Núñez-Peiró (✉) · C. S.-G. Sánchez · A. Sanz-Fernández · M. Gayoso-Heredia ·
J. A. López-Bueno · F. J. N. González
School of Architecture, Universidad Politécnica de Madrid, Avda Juan de Herrera 4, 28040
Madrid, Spain
e-mail: miguel.nunez@upm.es

J. A. López-Bueno · C. Linares · J. Díaz
National School of Public Health, Carlos III Institute of Health, Avda. Monforte de Lemos 5,
28029 Madrid, Spain

G. Gómez-Muñoz
Fundación Arquitectura y Sociedad, Calle Bahína Valverde 17, 28002 Madrid, Spain

that women living alone and above 65 years old seem to be under the highest risk. They concentrate in areas with low energy-efficient housing stock and strong urban heat island intensities. On a general basis, the income gap between women and men makes it advisable to address energy poverty with a gender perspective.

Keywords Summer energy poverty · Gender perspective · Intra-urban variations · Feminization · Urban heat island

1 Introduction

Climate change projections for Europe suggest that the current temperature rise will continue throughout this century (IPCC 2013), which will coincide with an increase in the frequency and intensity of heat waves (Tebaldi et al. 2006). These temperatures will be higher in urban areas, where the urban heat island (UHI) phenomenon produces temperature increases that can exceed 12 °C in comparison with their immediate rural areas (Gago et al. 2013; Oke et al. 2017). In that sense, and despite having focused its development on the inability to keep households at an adequate temperature during winter, the analysis of energy poverty is beginning to extend beyond under-heated months (Moore 2012). This new dimension of fuel poverty is known as *summer energy poverty.*

1.1 Summer Energy Poverty

Despite that there is not a specific definition, summer energy poverty can be understood as the inability to keep the house at an adequate temperature during the hottest months. According to the Energy Poverty Observatory (European Commission 2020), households might experience this situation due to a combination of factors: low income, inefficient building and appliances, high energy expenditures and specific energy needs. Although the definition of energy poverty includes heating and cooling needs in both the European and Spanish context (European Commission 2020; Ministerio para la Transición Ecológica 2019), the statistical databases do not reflect these situations equally. An example of this is that the last (and only) time European citizens were asked about their inability to keep their homes at a comfortable temperature during summer was back in 2012 (European Commission 2012). In a similar way, the last time Spanish households were asked about the availability of cooling systems was in the 2001 Census (Instituto Nacional de Estadística 2004).

However, the lack of statistical information has not prevented researchers from incorporating the summer energy perspective into the study of energy poverty. Several studies from around the world, such as in Australia (Moore et al. 2017), the UK (Mavrogianni et al. 2015; Wolf et al. 2010), Chile (Rubio-Bellido et al. 2017) and Italy (Pisello et al. 2017), have evaluated the resilience of vulnerable homes to high temperatures. Others have worked on the identification of summer energy poverty in the European context (Sánchez-Guevara et al. 2019; Santamouris and Kolokotsa 2015; Thomson et al. 2019). Researchers have even begun to explore various strategies to provide fresh and safe spaces for the population (Pearsall 2017; Sanchez and Reames 2019) or to improve the local and intergenerational connections and provide community assistance to the most vulnerable people (Sampson et al. 2013). A wide range of studies is, thus, beginning to be carried out to analyze the causes, consequences and possible solutions to alleviate summer energy poverty.

1.2 The Feminization of Energy Poverty

In recent years, several studies have mainstreamed a gender perspective into energy poverty (e.g., Clancy et al. 2017; Clancy and Feenstra 2019; Gonzalez Pijuan 2017; Robinson 2019). Continuing with this work, the FEMENMAD Project (Universidad Politécnica de Madrid 2019) has evaluated the feminization of energy poverty in the city of Madrid (Sánchez-Guevara et al. 2020a; Sanz Fernández et al. 2016). The results show that, while 23% of total households of Madrid suffer from energy poverty (Sánchez-Guevara et al. 2020b), it rises to 29% in the case of those households where a woman is the main breadwinner. These values increase even more for the situation of a single woman over 65 (39%) or a single mother with children (41%) is assessed. In other words, single mothers with children are almost twice as likely to suffer from energy poverty as the average of the household of Madrid.

Risks associated with a different physiological response to high temperatures are also relevant between the sexes (Díaz et al. 2018; López-Bueno et al. 2019). Within the framework of the FEMENMAD project, the differences in terms of mortality and emergency hospital admissions were also analyzed. Although the differences found in mortality are explained by the difference in life expectancy between men and women, it was found that, with each degree exceeding 36 °C, older women corresponding to the group from 65- to 74-years old experienced an attributable risk increase of 4.6% in admissions for natural causes. In the case of mortality due to circulatory causes, this percentage increased by 11.8% with each degree, while no significant statistical association was found for men during heat waves.

Given the higher incidence of energy poverty among households in which a woman is the main breadwinner and the higher vulnerability that some of these households might experience toward high temperatures, the present study explores their risk of suffering from summer energy poverty using the following approach.

2 Methodological Approach

The risk of suffering from energy poverty can be assessed through the spatial analysis of proxy indicators, as several studies have shown in recent years (Gouveia et al. 2019;

März 2018; Tomlinson et al. 2011). These indicators are mapped and analyzed in order to find those areas where energy poor are more likely to be found. In this study, the risk of summer energy poverty is defined by the following expression:

$$Risk = Exposure \cap Vulnerability.$$

Here, the risk depends on the intersection of households' exposure with households' vulnerability to high temperatures. Both exposure and vulnerability to high temperatures were measured through several indicators that were represented and treated spatially. Their degree of autocorrelation helped determine the areas where the highest values of each one of these indicators were concentrated (ESRI 2016a). These areas of concentration, called *hot spots*, were determined with a 90% confidence interval. Then, from the overlap of these *hot spots*, the areas with a higher risk were obtained together with the elements that characterize each one of them. To facilitate the comparison between the different cases, all indicators were presented by deciles. The spatial unit of the results is the census section of 2018.

2.1 Households Exposure to High Temperatures

The indicators used in Sánchez-Guevara et al. (2019) were taken as representative of those areas where a greater expenditure of energy could be given to keep home at an adequate temperature. These are, on the one hand, the energy performance of the building during the summer and, on the other hand, the cooling degree hours (CDH), which were estimated for both day and night. Since all domestic active cooling systems rely on electricity, the price of energy was not considered.

2.1.1 Energy Performance of the Housing Stock

The energy performance of buildings during summer was indirectly derived from the year of construction, in a similar way as was done in the *Technical Study on Energy Poverty in the City of Madrid* (Sanz Fernández et al. 2016). The year of construction reflects on a general basis both the construction characteristics of buildings and the regulations in force in that time (Instituto para la Diversificación y Ahorro de la Energía 2011). It was obtained from the Spanish Land Registry (Ministerio de Hacienda 2019) for each residential building in the city of Madrid. Table 1 shows the relationship between the year of construction and the estimated energy performance of the buildings, while Fig. 1 shows the delimitation of the *hot spot* for the energy performance indicator. The housing with the worst energy performances during summer seems to concentrate around the central area of the city. These buildings, mostly constructed during the 1960s and 1970s, are characterized by the low quality of the construction, with low thermal inertia and the absence of thermal insulation.

Table 1 Energy performance of buildings during summer

	Year of construction	Percentage of total buildings (%)	Construction characteristics
High	Before 1940	11	Built of masonry, characterized by high thermal inertia and the use of shading devices to protect the openings from solar radiation
	After 2006	25	Built after the adoption of the Spanish Building Code (Código Técnico de la Edificación 2019). Energy-saving and energy-efficiency criteria were introduced
Average	1981–2006	7	Built after the approval of the basic building regulation on thermal conditions (Gobierno de España 1979), which introduced minimum insulation requirements
Low	1941–1960	17	Built of the post-Civil War period, no regulations applied. It is characterized by cheap materials with no quality standards
	1961–1980	40	Constructed during an expansive economic cycle, no regulations applied. Low-quality construction. Low thermal inertia. Absence of thermal insulation

2.1.2 CDH on the Microclimatic Scale

Cooling degree hours were estimated as the number of degrees that, for each hour of the day, the outdoor air temperature is above a certain reference temperature. Given the significant temperature differences that can be recorded in Madrid due to the UHI (Núñez Peiró et al. 2017), the data collected by a network of 20 sensors deployed during the MODIFICA Project (Universidad Politécnica de Madrid 2014) was used to determine the outdoor air temperature of each urban area. These sensors were distributed throughout the municipality following contextualization criteria that would guarantee their representativeness of the urban environment (Núñez Peiró et al. 2019) and were complemented with the records of three observatories from the Spanish Meteorological Agency (AEMET). The reference temperature, on the other hand, was defined as the comfort temperature established by the ASHRAE adaptive comfort standard (ASHRAE 2013; de Dear and Brager 1998). Since this

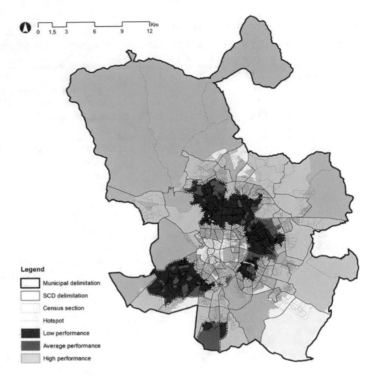

Fig. 1 Spatial distribution of the indicator associated with the energy performance of buildings during summer

standard does not contemplate sleeping hours (11–7 pm), two different reference temperatures, daytime and nighttime, were used. The daytime reference temperature was, thus, fixed at 28.3 °C for June and at 28.4 °C for July and August. On the other hand, a temperature of 27 °C was established for the nights, in accordance with the criterion of the Spanish Building Code (Código Técnico de la Edificación, 2019). Finally, the day and nighttime CDHs were spatially interpolated for the entire city using a *kriging*, a tool available in the geostatistical analysis module of *ArcGIS* (ESRI 2016b; Oliver and Webster 1990).

Figure 2 shows the corresponding *hot spots* for high temperatures on the microclimatic scale. The concentration of a higher amount of CDH in the city center during the night corresponds to the typical concentric distribution of the UHI, while during the day the highest temperatures take place in the south-central area of the city. Despite the obvious differences between the day and nighttime CDH, there are some southern areas of the city in which high temperatures might be concentrated 24 h a day.

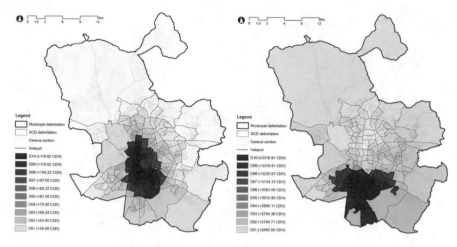

Fig. 2 Spatial distribution of CDH during the night (left) and during the day (right)

2.2 Gender-Related Vulnerability of Households

To incorporate the gender perspective in this study, the indicators of vulnerability to high temperatures were defined in relation to the concentration of households in which women are the main breadwinners. Households with a single woman over 65 and with a single woman with children were, therefore, used as indicators of vulnerability. The data used in this study was extracted from the municipal statistical database (Ayuntamiento de Madrid 2018), which provides the total number of households according to their composition and by census section. The situation of each one of these vulnerable households was compared with their male counterparts, aiming at detecting differences in their spatial distribution, degree of exposure to high temperatures and any relationship with other relevant indicators. All these indicators were treated using the same statistical analysis tools as done with the exposure indicators, thus generating *hot spots* for the identification of areas with the greatest concentration of vulnerable.

2.2.1 Households with a Single Women Over 65

The concentration of households with a single person over 65, both men and women, is shown in Fig. 3. While single men over 65 tend to inhabit the north-central part of the city, women are distributed in a ring around the city center, living in the northern, eastern and southern areas of the city. Although *hot spots* of both men and women seem to have a similar dimension, the total number of households concentrated in these areas is different. Single women over 65 represent 10% of the total households of Madrid, while this percentage reduces to 3% in the case of men.

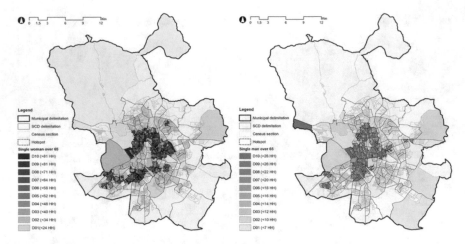

Fig. 3 On the left, the spatial distribution of the vulnerability indicator associated with single women over 65 years old. On the right, the situation of their male counterparts

2.2.2 Households of a Single Woman with Children

Figure 4 shows the concentration of single-parent households led by women and men. Both *hot spots* concentrate on the outskirts of the city. Male single-parent households are mostly found in the northern part of the city, while female single-parent households are distributed as well in the eastern and southern parts. In comparison with the total amount of households of the city, single-parent households account for 2.5%, and, again, women comprise most of them (83% of single-parent households).

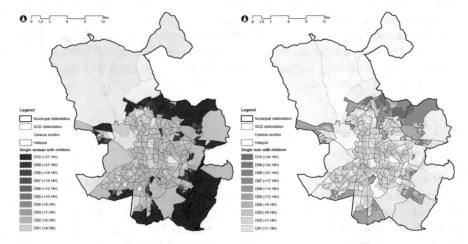

Fig. 4 On the left, the spatial distribution of the vulnerability indicator associated with single-parent households led by women. On the right, the situation of their male counterparts

3 Results

Table 2 summarizes the result of overlapping the indicators associated with a greater exposure to high temperatures with the indicators of greater vulnerability in the city of Madrid. Among single women over 65, 41.1% live in areas where there is at least one overlap with an exposure indicator. In the case of their male counterparts, this percentage drops to 33.9%. This situation is repeated when analyzing the overlaps with two (12.6% vs. 7.2%) and three (2.5% vs. 0.7%) indicators. Similarly, and even though this is not an indicator of exposure but related with the expenditure capacity, women are more concentrated in the areas of Madrid where the lowest incomes concentrate (7.7% vs. 1.0%).

Regarding their exposition to high temperatures, the situation of single women over 65 also differs from men: While both groups tend to suffer from high nighttime temperatures (21.3 vs. 23.4%), women are more prone to concentrate in areas with less efficient buildings (25.0 vs. 12.8%). Figure 5 also shows that women tend to occupy more areas to the south, where several exposure indicators overlap and, thus, where the highest risk areas are located. These, which seem to concentrate in the districts of Carabanchel, Usera and, to a lesser extent, Puente de Vallecas, Retiro and Tetuán, are coincident with other vulnerable population areas revealed in previous studies (Sánchez-Guevara Sánchez et al. 2017).

Table 2 Percentage of households living in areas with a higher risk of suffering from summer energy poverty

Exposure indicators	Single women		Single men	
	Over 65 (%)	With children (%)	Over 65 (%)	With children (%)
One overlap	**41.10**	**4.70**	**33.90**	**3.70**
Daytime CDH	9.80	3.30	5.60	1.10
Nighttime CDH	21.30	1.10	23.40	0.70
Buildings performance	25.00	1.80	12.80	2.80
Two overlaps	**12.60**	**1.30**	**7.20**	**0.70**
Daytime and nighttime CDH	7.30	1.10	5.50	0.70
Daytime CDH and buildings performance	4.80	0.40	0.70	0.30
Nighttime CDH and building performance	5.40	0.20	2.40	0.30
Three overlaps	**2.50**	**0.20**	**0.70**	**0.30**
Daytime CDH, nighttime CDH and building performance	2.50	0.20	0.70	0.30
Low income indicator	**7.70**	**1.00**	**3.60**	**0.00**

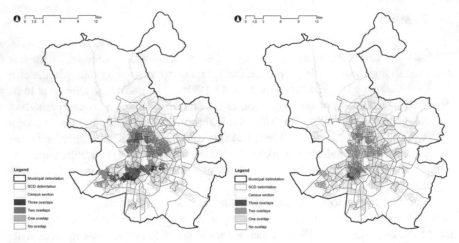

Fig. 5 On the left, the risk of suffering from summer energy poverty for single women over 65. On the right, results obtained for their male counterparts

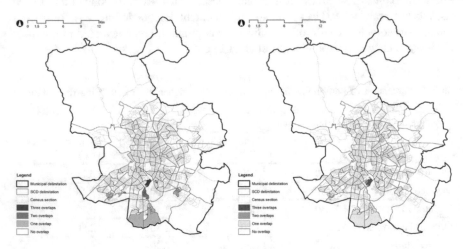

Fig. 6 On the left, the risk of suffering from summer energy poverty for single women with children. On the right, results obtained for their male counterparts

As for single-parent households, Fig. 6 shows the distribution for both headed by a woman and those headed by a man. Despite the fact that these households tend to be concentrated in the outskirts of the city, where they are less likely to suffer the highest temperatures during summer or to inhabit buildings with the poorest energy performance, the situation of women is still relatively worse than the situation of men. Single women with children concentrate on 4.7% of occasions in areas with a certain degree of exposure and opposed to 3.7% in the case of single men with children. This situation is reproduced when analyzing higher risks, i.e., the overlap of two indicators (1.3% vs. 0.7%).

4 Discussion

Despite that single women with children seem to face relatively low risks from summer energy poverty, almost half of them are under *general* energy poverty since most of them are below the monetary poverty line (Sánchez-Guevara et al. 2020a; Sanz Fernández et al. 2016). While, in terms of exposure to high temperatures, these households might be relatively better than the rest of the municipality, their vulnerability puts them at a higher risk of suffering energy poverty during heat waves episodes, which are expected to become more frequent in the next decades due to climate change. In this scenario and given the increasing penetration of cooling systems in Spain (Idealista 2019), the use of social tariffs might help to reduce the risk within this group. Cooling centers, which provide air-conditioned spaces during heat waves, have proven to be effective as well (Sanchez and Reames 2019).

Although fostering the use of air conditioning among certain groups might be protective during heat waves, it might be a maladaptive strategy in the long run. The challenge might be not to increase the cooling energy consumption but to lower indoor temperatures in the most efficient way. In that sense, since the risk of suffering summer energy poverty among single women over 65 seems to be associated with an inefficient housing stock, energy retrofitting of buildings seems to be a reasonable approach to alleviate their situation. When possible, retrofitting should be accompanied by interventions in the public space to help mitigate the UHI. Reducing the outdoor temperature would increase the effectiveness of passive strategies, such as natural ventilation or evaporative cooling, and enhance the attractiveness of ventilators, which are low-energy consumption cooling systems. Additionally, avoiding the use of air conditioners would help to not contributing to the increase in the outdoor temperature (Sampson et al. 2013).

Regarding the limitations of this study, these are mainly related with data availability and disaggregation barriers found both spatially and by gender. In that sense, socioeconomic indicators could not be included to further explore the characteristics of each vulnerable group at the various locations. Despite that these groups were derived from previous studies for the municipality of Madrid in which socioeconomic indicators were used, relevant socioeconomic variations could be expected and should be analyzed in future studies as soon as disaggregated data becomes available. Another source of uncertainty relates to the building's energy performance indicators, which is only based on the year of construction. Further research should incorporate other variables such as the orientation, the relative position within the same building or the glazing area of the thermal envelope of the building, which does certainly have an impact into the cooling loads.

Finally, and beyond the limitations of not including in the present study neither the intra-household differences (Haddad and Kanbur 1992; Ponthieux and Meurs 2015) nor the greater risks associated with a different physiological response to high temperatures (see Sect. 1.2), it should be noted that the combination of all single women over 65 and single-mother households in the city of Madrid accounts for only one third of the total households in which a woman is a breadwinner. Derived from a lack of disaggregation by gender, this situation could hide the vulnerabilities associated with other household compositions. Gender-disaggregated statistical data, together with a greater methodological interconnection between the different databases and greater consistency and frequency of the provided data would, consequently, not only facilitate mainstreaming the gender perspective into energy poverty studies but increase the robustness of the results and ensure that they are monitored over the years.

5 Conclusions

This study is a first exploration of the risks that households led by women might face regarding summer energy poverty. Two household typologies, single women over 65 and single women with children, were analyzed spatially at the sub-municipal scale. A methodology based on the intersection of vulnerability and exposure to high-temperature indicators was used, showing that households led by women, together with a greater vulnerability in comparison with those led by men, also tend to concentrate in areas in which a higher exposure to high temperatures can be expected.

Regarding the different household typologies, single women over 65 seem to accumulate the highest risk of suffering from summer energy poverty. This risk is mostly associated with the energy performance of their buildings during the summer months, but it is also related to high outdoor temperatures due to the UHI effect. Since reducing the outdoor temperature would promote the use of passive strategies, housing interventions should be coordinated with urban adaptation and mitigation strategies to improve microclimatic conditions. On the other hand, single women with children tend to concentrate on the outskirts of the city and live in relatively new housing stock. They seem to face a relatively low risk toward summer energy poverty, although they might encounter relevant risks during heat waves given their limited economic capacity to cope with unforeseen energy expenditures. In this context, providing financial assistance might be protective for those households with air conditioning systems. For those without any cooling device, setting up cooling centers might be effective as well.

Despite the limitations of not including other related variables, such as socioeconomic indicators, intra-household differences or the different physiological response toward high temperatures, the geospatial analysis based on proxy indicators has proved to be a useful tool to evaluate the relative risk of suffering from summer energy poverty at the sub-municipal scale and integrate a gender perspective. In that sense, policies aiming at mitigating summer energy poverty should consider the intra-urban variability of the phenomenon, prioritizing the most vulnerable areas and mainstreaming a gender perspective. To do so, further gender-disaggregated data should be collected in order to explore the situation of all household typologies.

Acknowledgements This research was funded by the Municipal Consumption Institute of Madrid City Council and the project FEMENMAD—*Feminización de la pobreza energética en Madrid.*

Exposición a extremos térmicos—unded by Madrid City Council under the call *Subvenciones 2018 para la realización de proyectos de investigación en materia de ciudadanía global y cooperación internacional para el desarrollo.* It was also partially funded by an FPU research grant (FPU15/05052) from the Spanish Ministry of Science, Innovation and Universities. The Spanish Meteorological Agency (AEMET) provided weather data, and Madrid Local Council helped with the installation of the urban network.

References

ASHRAE (2013) ANSI/ASHRAE Standard 55–2013. Thermal Environ Conditions Human Occupancy

Ayuntamiento de Madrid (2018) Hogares por tamaño, composición del hogar, nacionalidad y sección según distrito. Padrón Munic, Habitantes (explotación estadística)

Clancy J, Daskalova V, Feenstra M, Franceschelli N, Sanz Blomeyer M (2017) Gender perspective on access to energy in the EU. Belgium, Brussels

Clancy J, Feenstra M (2019) Women, gender equality and the energy transition in the EU

Código Técnico de la Edificación (2019) Documento Básico de Ahorro de Energía. Anejo D, Condiciones operaciones y perfiles de uso

de Dear RJ, Brager GS (1998) ASHRAE RP-884: developing an adaptive model of thermal comfort and preference

Díaz J, López IA, Carmona R, Mirón IJ, Luna MY, Linares C (2018) Short-term effect of heat waves on hospital admissions in Madrid: analysis by gender and comparison with previous findings. Environ Pollut 243:1648–1656. https://doi.org/10.1016/j.envpol.2018.09.098

ESRI (2016a) ArcGIS spatial statistics toolbox for ArcMap

ESRI (2016b) How kriging works. https://desktop.arcgis.com/en/arcmap/10.5/tools/3d-analyst-toolbox/how-kriging-works.htm. Accessed 1.22.20

European Commission (2020) What is energy poverty? EU Energy Poverty Obs

European Commission (2012) Share of population living in a dwelling not comfortably cool during summer time by income quintile and degree of urbanisation. Eurostat

Gago EJ, Roldan J, Pacheco-Torres R, Ordóñez J (2013) The city and urban heat islands: a review of strategies to mitigate adverse effects. Renew Sustain Energy Rev 25:749–758. https://doi.org/10.1016/j.rser.2013.05.057

Gobierno de España (1979) Real Decreto 2429/1979, de 6 de julio, por el que se aprueba la norma básica de edificación NBE-CT-79, sobre condiciones térmicas en los edificios

Gonzalez Pijuan I (2017) Desigualdad de género y pobreza energética

Gouveia JP, Palma P, Simoes SG (2019) Energy poverty vulnerability index: a multidimensional tool to identify hotspots for local action. Energy Rep 5:187–201. https://doi.org/10.1016/j.egyr.2018.12.004

Haddad L, Kanbur R (1992) Intrahousehold inequality and the theory of targeting. Eur Econ Rev 36:372–378. https://doi.org/10.1016/0014-2921(92)90093-C

Idealista (2019) Solo una de cada tres casas en España tiene aire acondicionado

Instituto Nacional de Estadística (2004) Censo 2001. Resultados definitivos, Censos Población y Viviendas

Instituto para la Diversificación y Ahorro de la Energía, (2011) Proyecto Sech-Spahousec. Análisis del consumo energético del sector residencial en España. Informe final, Madrid

IPCC (2013) Climate change 2013: the physical science basis. Contribution of Working Group I to the Fifth Assessment Report of the Intergovernmental Panel on Climate Change. Cambridge University Press, Cambridge, United Kingdom and New York, NY, USA

López-Bueno J, Díaz J, Linares C (2019) Differences in the impact of heat waves according to urban and peri-urban factors in Madrid. J Biometeorol Int . https://doi.org/10.1007/s00484-019-01670-9

März S (2018) Assessing the fuel poverty vulnerability of urban neighbourhoods using a spatial multi-criteria decision analysis for the German city of Oberhausen. Renew Sustain Energy Rev 82:1701–1711. https://doi.org/10.1016/j.rser.2017.07.006

Mavrogianni A, Taylor J, Davies M, Thoua C, Kolm-Murray J (2015) Urban social housing resilience to excess summer heat. Build Res Inf 43:316–333. https://doi.org/10.1080/09613218.2015.991515

Ministerio de Hacienda de Hacienda (2019) Catastro inmobiliario. Dirección General del Catastro

Ministerio para la Transición Ecológica (2019) Estrategia nacional contra la pobreza energética 2019–2024

Moore R (2012) Definitions of fuel poverty: implications for policy. Energy Pol 49:19–26. https://doi.org/10.1016/j.enpol.2012.01.057

Moore T, Ridley I, Strengers Y, Maller C, Horne R (2017) Dwelling performance and adaptive summer comfort in low-income Australian households. Build Res Inf 45:443–456. https://doi.org/10.1080/09613218.2016.1139906

Núñez Peiró M, Sánchez-Guevara Sánchez C, Neila González FJ (2019) Source area definition for local climate zones studies. A Syst Rev Build Environ 148:258–285. https://doi.org/10.1016/j.buildenv.2018.10.050

Núñez Peiró M, Sánchez-Guevara Sánchez C, Neila González FJ (2017) Update of the urban heat Island of Madrid and its influence on the building's energy simulation. Sustain Dev Renov Archit Urban Eng. https://doi.org/10.1007/978-3-319-51442-0_28

Oke TR, Mills G, Christen A, Voogt JA (2017) Urban climates. Cambridge University Press. https://doi.org/10.1017/9781139016476

Oliver MA, Webster R (1990) Kriging: a method of interpolation for geographical information systems. Geogr Inf Syst 4:313–332. https://doi.org/10.1080/02693799008941549

Pearsall H (2017) Staying cool in the compact city: vacant land and urban heating in Philadelphia. Pennsylvania Appl Geogr 79:84–92. https://doi.org/10.1016/j.apgeog.2016.12.010

Pisello AL, Rosso F, Castaldo VL, Piselli C, Fabiani C, Cotana F (2017) The role of building occupants' education in their resilience to climate-change related events. Energy Build 154:217–231. https://doi.org/10.1016/j.enbuild.2017.08.024

Ponthieux S, Meurs D (2015) Gender inequality, 1st edn. Elsevier B.V, Handbook of income distribution. https://doi.org/10.1016/B978-0-444-59428-0.00013-8

Robinson C (2019) Energy poverty and gender in England: a spatial perspective. Geoforum 104:222–233. https://doi.org/10.1016/j.geoforum.2019.05.001

Rubio-Bellido C, Fargallo A, Pulido Arcas J, Trebilcock M (2017) Application of adaptive comfort behaviors in Chilean social housing standards under the influence of climate change. Build, Simul, p 10

Sampson NR, Gronlund MA, Buxton L, Catalano J, White-Newsome JL, Conlon MS, O'Neill S, McCormick E (2013) Staying cool in a changing climate: reaching vulnerable populations during heat events. Glob Environ Chang 23:475–484. https://doi.org/10.1016/j.gloenvcha.2012.12.011

Sánchez-Guevara Sánchez C, Núñez Peiró M, Neila González FJ (2017) Urban heat island and vulnerable population. The case of madrid. In: Mercader-Moyano P (ed) Sustainable development and renovation in architecture, urbanism and engineering. springer international publishing, Seville, pp. 3–13. https://doi.org/https://doi.org/10.1007/978-3-319-51442-0_1

Sánchez-Guevara C, Núñez Peiró M, Taylor J, Mavrogianni A, Neila González J (2019) Assessing population vulnerability towards summer energy poverty: case studies of Madrid and London. Energy Build 190:132–143. https://doi.org/10.1016/j.enbuild.2019.02.024

Sánchez-Guevara CS, Sanz Fernández A, Núñez Peiró M, Gómez Muñoz G (2020a) Feminisation of energy poverty in the city of madrid. Energy Build 223, https://doi.org/10.1016/j.enbuild.2020.110157

Sánchez-Guevara CS, Sanz Fernández A., Núñez Peiró M, Gómez Muñoz,G (2020b) Energy poverty in madrid: Data exploitation at the city and district level. Energy Policy 144, https://doi.org/10.1016/j.enpol.2020.111653

Sanchez L, Reames TG (2019) Cooling Detroit: a socio-spatial analysis of equity in green roofs as an urban heat island mitigation strategy. Urban Urban Green 44:126331. https://doi.org/10.1016/j.ufug.2019.04.014

Santamouris M, Kolokotsa D (2015) On the impact of urban overheating and extreme climatic conditions on housing, energy, comfort and environmental quality of vulnerable population in Europe. Energy Build 98:125–133. https://doi.org/10.1016/j.enbuild.2014.08.050

Sanz Fernández A, Gómez Muñoz G, Sánchez-Guevara Sánchez C, Núñez Peiró M (2016) Estudio técnico sobre pobreza energética en la ciudad de Madrid. Ayuntamiento de Madrid, Madrid

Tebaldi C, Hayhoe K, Arblaster JM, Meehl GA (2006) Going to the extremes: an intercomparison of model-simulated historical and future changes in extreme events. Clim Change 79:185–211. https://doi.org/10.1007/s10584-006-9051-4

Thomson H, Simcock N, Bouzarovski S, Petrova S (2019) Energy poverty and indoor cooling: an overlooked issue in Europe. Energy Build 196:21–29. https://doi.org/10.1016/j.enbuild.2019.05.014

Tomlinson CJ, Chapman L, Thornes JE, Baker CJ (2011) Including the urban heat island in spatial heat health risk assessment strategies: a case study for Birmingham. UK Int J Health Geogr 10:42. https://doi.org/10.1186/1476-072X-10-42

Universidad Politécnica de Madrid (2019) FEMENMAD project: feminisation of energy poverty in the city of Madrid. s= Feminización de la pobreza energética en Madrid. Exposición a extremos térmicos. Madrid City Counc. https://abio-upm.org/project/proyecto-femenmad/

Universidad Politécnica de Madrid (2014) MODIFICA project: predictive model for dwellings energy performance under the urban heat island effect. Minist Econ Compet BIA2013–41732-R

Wolf J, Adger WN, Lorenzoni I (2010) Heat waves and cold spells: an analysis of policy response and perceptions of vulnerable populations in the UK. Environ Plan A 42:2721–2734. https://doi.org/10.1068/a42503

The Ecobonus Incentive Scheme and Energy Poverty: Is Energy Efficiency for All?

Chiara Martini

Abstract With 2017, National Energy Strategy, Italy introduced a definition of energy poverty, combining three elements: the presence of a high level of energy expenditure, an amount of total expenditure below the relative poverty threshold and a null value for the purchase of heating products. The Integrated National Plan for Energy and Climate adopts the same definition, provides an estimation on the evolution of energy poverty in 2030 and lists the tax deduction scheme for energy renovation of the existing building stock (Ecobonus) among the specific measures dedicated to energy poverty. Implemented as an alternative measure under Article 7 of EED (European Energy Efficiency Directive), Ecobonus enables the households in the no-tax area—which are likely to be energy poor—to transfer their tax credits to financial institutions, work suppliers or other private entities, reducing the investment cost for energy/efficiency interventions. Based on information at the regional level, namely ENEA microdata on Ecobonus, this paper examines the possible relationship between indicators such as household income and the access to Ecobonus. Additionally, the study analyzes if this relationship changes for the different categories of interventions incentivized by Ecobonus, such as the replacement of windows and shutters or of heating systems. The hypothesis is that the incentive measure, with its current approach, has a regressive distributive effect on households, and it does not effectively support energy-poverty eradication. To our knowledge, the relationship between income and interventions incentivized by Ecobonus has not been investigated before, neither at the regional level nor in an energy-poverty framework.

Keywords Energy poverty · Energy policy · Energy efficiency · Distributive effect

1 Introduction

Energy poverty affects roughly 50 million people in Europe, representing a critical issue with health, social, economic and environmental implications. In the European

C. Martini (✉)
Italian Energy Agency for New Technologies, Environment and Sustainable Economic Development (ENEA), Energy Efficiency Unit Department, Roma, Italy
e-mail: chiara.martini@enea.it

© The Author(s), under exclusive license to Springer Nature Switzerland AG 2021   497
A. Bisello et al. (eds.), *Smart and Sustainable Planning for Cities and Regions*,
Green Energy and Technology, https://doi.org/10.1007/978-3-030-57332-4_35

Member States' strategies to tackle energy poverty, energy-efficiency measures are more and more recognized as a long-term solution, to accompany and complement social security policies.

To address energy poverty, it is compelling to figure out if the energy policies in force are effective to address the phenomenon, as recommended by the European Energy Network (EnR 2019). In other words, a crucial aspect is to estimate if existing policy measures have differentiated impacts on different income groups, in terms of who is paying their cost or who has access to the financial incentives. In this vein, our study concentrates on the main energy-efficiency policy for residential sector in Italy, namely the tax deduction scheme for energy renovation of existing building stock (Ecobonus), and it investigates if it is an effective policy measure to mitigate energy poverty.

At the European level, while there is common agreement on the main components of energy poverty (among which is the poor energy performance of buildings), there is not a shared definition of the issue. In the new directives adopted after the Winter Package, energy poverty has a key role,[1] which is currently reflected in national policy strategies, such as the Integrated National Climate and Energy Plans (NECPs) and Long-Term Renovation Strategies of the Building Stock (LTRS).

An integrated approach could successfully deal with energy poverty, in particular by: choosing a comprehensive definition and to compare countries/regions; improving data availability and to integrate database; creating enabling conditions for energy-efficiency potential to be exploited; implementing measures to address all relevant dimensions (split incentives, appliances, transport, etc.); recognizing the role of non-technological actions; and measuring energy-poverty trends to identify its main drivers and to elaborate robust predictions.

The European Energy Network elaborated five recommendations for the European Commission (EnR 2019), which can be summarized as follows:

1. To introduce a unique EU energy-poverty measure that could be a low-income high-cost (LIHC) measure, and accompanying it by country-specific indicators, to be set according to country-specific characteristics;
2. To promote energy-efficiency measures as key solutions to energy poverty, allowing for multiple benefits and structural change and to act at the local level;
3. To develop an integrated approach to tackle energy poverty and to elaborate policy responses at the country level;
4. To examine energy-poverty implications in terms of cost distribution of the measures adopted to achieve long-term energy and environmental objectives;
5. To recognize that training and information campaigns are essential to achieve a behavioral change and then boost the rate of energy renovation of dwellings of household in energy poverty.

In such a framework, it appears to be important to analyze the effectiveness of energy-efficiency policies, when targeted also to energy-poor households, namely

---

[1] In particular, energy poverty is mentioned in Art. 7 of Energy Efficiency Directive (2018/2002), Art. 2 of Energy Performance of Buildings Directive (2018/844), Art. 28 and 29 of Electricity Directive (2019/944) and Art. 2 of Governance Regulation (2018/2002).

to households facing difficulties in satisfying their energy needs. The objective of this paper is to investigate the regional differentiation regarding access to the tax deduction scheme for energy efficiency, which is mentioned in the Italian NECP as a key policy to achieve the 2030 energy-savings target. First, the adopted definition of energy poverty and the investigation method will be outlined; then, results will be described, discussion will follow, and conclusions will present further research issues.

2 Method

To answer our research question on the effectiveness of an existing policy measure to tackle energy poverty in Italy, a descriptive statistical analysis was applied and regional maps of access to Ecobonus were elaborated. To do this, the following methodological steps were adopted: first, looking at the definition of energy poverty in Italy and at the incidence of this phenomenon and then choosing the policy measure and setting the framework in which to analyze it.

First, the definition and incidence of energy-poverty phenomenon in Italy should be described. The work developed by the European Energy Poverty Observatory shows that, to measure energy poverty, both objective and subjective indicators can be adopted; the share of households with an absolute energy expenditure below half the national median is an example of the first indicator type, while the share of households auto-reporting an inability to keep the home adequately warm is the second one. In Italian National Energy Strategy, and later on in NECP, an ad hoc objective indicator was adopted, based on Faiella and Lavecchia (2015); in this work, we refer to this definition. The measure is a low-income high-cost indicator, which considers three dimensions: (1) a share of energy costs more than twice the average share of energy expenditure; (2) a household budget, after energy costs are deducted, below the national (relative) poverty line set by the National Statistical Institute; and (3) null heating purchases when the total expenditure is below the median.

According to this measure, in 2017 there were 2.2 million energy-poor households (more than 5 million persons), equal to 8.7% of the total population. Energy poverty has a higher incidence in southern Italy and in larger households.[2] In the 2007–2017 decade, the share of energy expenditure of the total has increased from 4.7 to 5.1%. This share is higher (around 8%), and it increased more (almost +1%) for households in the first quintile of equivalent expenditure. Energy-poor households are very often located in inefficient buildings, use obsolete appliances and are also likely to suffer social-exclusion problems, associated with thermal discomfort or a television turned on for even eleven hours/day (Fondazione Di Vittorio 2018). A recent study has investigated the energy demand needed to reach standard comfort levels in different

---

[2]More information can be found in Energy Efficiency Annual Report (ENEA 2019) and in Rapporto sullo stato della povertà energetica in Italia (OIPE 2019).

building typologies, trying to assess in an objective way the difficulties with energy-poverty subjective indicators in keeping the home adequately heated (Faiella et al. 2017).

The main drivers for energy poverty considered in NECP to develop energy-poverty projections for 2030 are: the expected price trends for energy products; the trends in overall household expenditure; demographic changes; and the trends in residential energy consumption and the associated mix. The renovation rate of the building stock is also relevant, as well as indirect benefits on sanitary system associated with a reduction in diseases related to living in inadequately heated apartments.

According to the projections in NECP, in 2030, the incidence of energy poverty will remain in the 7–8% range. This means that energy poverty will decrease by approximately one percentage point compared to 2016, corresponding to approximately 230,000 households, caused by, among others, the number of people over the age of 65 reaching a quarter of the total in 2030 and by the fall of 15.5% in residential consumption in 2030 relative to 2016, with a growth in the electricity component against a reduction in natural gas. According to the LTRS draft, an annual renovation rate in the range 0.6%-0.8% would be needed in residential sector to reach the 2030 NECP objective; clearly, apartments inhabited by energy-poor households should be involved in such renovations.

Second, all relevant policy measures should be identified, and the one to be analyzed should be chosen. The Italian NECP mentions various measures to tackle energy poverty, which vary in nature. In particular, they are represented by:

- social measures, namely the electricity, gas and physical ailment bonuses;
- structural measures, the tax deduction scheme for energy renovation of buildings;
- fiscal measures, namely the exemption from excise duties for electricity and heating fuels, respectively, for households in the first consumption bracket and for households living in disadvantaged geographical areas.

Among these measures, this work examines Ecobonus, the tax deduction scheme for energy renovation of existing buildings. ENEA collects the applications to access the incentive mechanism and is also charged of managing the monitoring system. The Ecobonus incentive scheme has been in force since 2007: during its existence, it has indeed been extended by several Budget Laws, which introduced new features concerning, for some specific cases, rates of deduction, eligible actions and technical or performance requirements. In general, Ecobonus applies a tax deduction on income tax paid by physical persons (or by companies), and the tax deduction rate changes according to the eligible action considered.

Some examples of deduction rates are[3]:

---

[3]These rates have been modified by 2018 Budget Law (27 December 2017 no. 205) and confirmed by 2019 Budget Law (30 December 2018, no. 145). The recent 2020 Budget Law (27 December 2019, no. 160) is not taken into account here since the analysis is concentrated on 2018 data.

- 50% for the expenses incurred for replacing windows and shutters, installing solar shading, replacing heating systems with at least class-A energy-efficient condensation boilers;
- 65% for replacing heating systems with at least class-A energy-efficient condensation boilers and also installing an advanced thermoregulation system with efficiency classes V, VI or VIII, as indicated in Commission Communication 2014/C 207/02;
- 65% for expenses to install solar panels.

The 2018 Budget Law introduced a higher rate for energy-efficiency actions on the building block and also for actions combined with anti-seismic interventions, for which the deduction ranges from 70 up to 85% of the expense, depending on specific conditions.

The possibility of tax credit transfer, for all eligible energy-efficiency actions, for people in the no-tax area and social housing institutes, was introduced in the 2016 Budget Law (28 December 2015, no. 209), and it was limited exclusively to suppliers who implemented works. For people in the no-tax area, the tax-credit transfer has been extended to other private entities, banks and financial institutions by the 2017 Budget Law (December 11, 2016, no. 232). The possibility of a tax-credit transfer is intended to increase the access to the Ecobonus scheme for households in difficult economic conditions, among which energy-poor households are likely to be included.

Last, the most appropriate way to analyze available data should be defined. The database of the Ecobonus incentive scheme includes data on investments in various types of interventions, based on various technologies and on the associated energy savings. To examine the results of the Ecobonus incentive scheme at the national level, a cost-effectiveness indicator can be computed, as the ratio of EUR spent and kWh saved. This indicator shows better values for interventions on envelope insulation, windows and shutter replacement and solar panels. Envelope insulation and windows and shutters replacement are also associated with a higher share of savings of the total; significant savings are also generated by replacing heating systems, in particular, installing condensation boilers. The analysis of intervention distribution at the national level also shows a relevant share of investment on buildings built before 1980 (77% of the total), which is consistent with a higher energy-saving potential in these buildings. Finally, it is worth specifying that no information on households having transferred the tax credit is currently available in the database managed by ENEA.

The following analysis began with the recognition of the higher incidence of energy poverty in southern Italy revealed by the adopted definition, and it examines, from this basis, the regional distribution of investments activated by Ecobonus, considering different technologies. In particular, the ratio between regional investments, normalized (where relevant) to correct for climatic effects, and regional net available income will be shown in maps,[4] developed with a free online tool. This geographical representation is aimed at assessing the access to Ecobonus, showing

---

[4]Maps are a tool more and more used to describe a wide range of phenomena, also thanks the availability of georeferenced data. Among other research, it is interesting to mention the work by

evidence at the qualitative and descriptive level. Some first insights into the effectiveness of Ecobonus in addressing energy poverty can be derived by connecting this evaluation with the available information on the geographical pattern of energy-poverty incidence.

3   Results

At the national level, €3.3 billion of investments were activated by Ecobonus in 2016, of which 1.5 billion were associated to windows and shutters replacement and 950 million to envelope insulation. In 2018, the total investment level was the same, amounting to slightly more than €3.3 billion; the replacement of windows and shutters was again the main component, with more than one billion of investment, followed by envelope insulation (900 million) and the replacement of heating systems (slightly more than 870 million).

In 2018, regional total investments activated by the Ecobonus incentive scheme ranged between a maximum of €785 million and a minimum of €8 million, and they show an asymmetric distribution when normalized by regional net available income, provided by Italian National Statistical Institute.[5] In fact, in 2016, only one region in southern Italy was in the second quartile of the distribution, while all the others were the first. The geographical incidence of energy poverty follows the opposite pattern, with a higher share of energy-poor households in southern Italy. The distribution of investments activated by Ecobonus slightly changed in 2018, with three regions in southern Italy included in the second quartile (Fig. 1).

In terms of deviation from the average, regions in southern Italy are always below the national average, and this pattern remained unchanged between 2016 and 2018. For example, Campania has the higher negative deviation from the average, followed by Sicilia and Sardinia, respectively, in second and third position in 2018. By contrast, regions having a higher positive deviation from the average are, in decreasing order, Trentino Alto-Adige, Valle d'Aosta and Piemonte.

The results can be mapped also relative to specific technologies, comparing 2016 and 2018: given their high share of total investments, windows and shutters, building envelopes and heating systems will be shown. In all cases, the investment activated by Ecobonus has been normalized by regional Heating Degree Days (HDD), available from Eurostat.[6]

In 2018, regional investments in the replacement of windows and shutters, normalized by HDD, range €295–920 per billion of net available income. The geographical

---

Hills (2012) devoted to mapping energy poverty in the United Kingdom and the work by Lelo et al. (2019), aimed at mapping a wide range of social phenomena in Rome.

[5]https://dati.istat.it/Index.aspx?DataSetCode=DCCN_SEQCONTIRFT, last accessed 2/12/2019.

[6]https://ec.europa.eu/eurostat/web/products-datasets/product?code=nrg_chddr2_m, last accessed 2/12/2019. In 2018, HDD in Italian regions ranged between a maximum value of 4,184 (Valle d'Aosta) and a minimum value of 946 (Sardinia). These two regions had the highest and lowest values also in 2016.

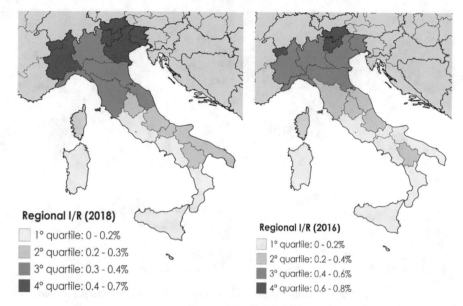

Fig. 1 Ratio between total investments activated by Ecobonus and net available income by region (I/R), 2018 and 2016

asymmetry is less pronounced than for the total investment, since, already in 2016, three southern regions are in the second quartile and another in the third one. An improvement in positioning of southern regions is observed in 2018, with two regions in the third quartile and another in the fourth one (Fig. 2).

Regional investments in building envelope insulation, normalized by HDD, range from €120 to €965 per billion of net available income in 2018. A pronounced geographical pattern and a very slight improvement are observed, with a southern region moving into the third quartile and the number of regions in the second quartile of the investment distribution remaining unchanged (Fig. 3).

Looking at the replacement of heating system with a more efficient one, investments normalized by HDD range from €195 to €635 per billion of net available income in 2018. The geographical asymmetry is again observed, as well as the improvement of southern regions positioning in the quartiles. The number of southern regions in the second quartile increases, and two regions pass in the third and fourth quartile (Fig. 4).

Finally, specific technologies that could theoretically have a larger potential in southern than in northern Italy, due to higher solar radiation, can be investigated. This is the case, for example, with solar panels and solar shading; in 2018, total investments activated by Ecobonus in Italy amounted to €36 million for solar panels and €128 million for solar shading. For solar panels, regional investments range from zero to a maximum of €179 per million of net available income; for solar shading, the range is from €10 to €223 per million of net available income. The two maps in Fig. 5, relative to 2018 data, seem to suggest that a higher technology potential is not

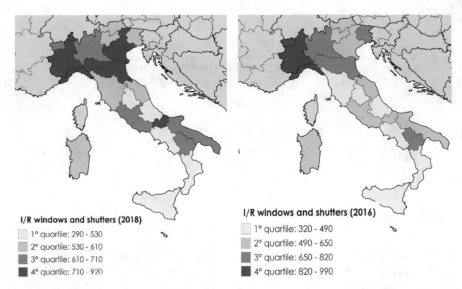

Fig. 2 Regional investments activated by Ecobonus normalized by regional HDD per billion of net available income (I/R), 2016 and 2018, windows and shutters

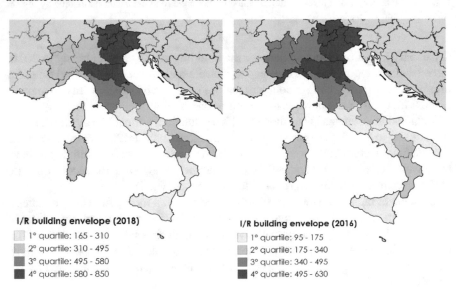

Fig. 3 Regional investments activated by Ecobonus normalized by regional HDD per billion of net available income (I/R), 2016 and 2018, building envelope

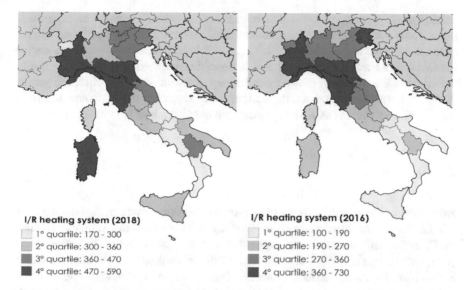

Fig. 4 Regional investments activated by Ecobonus normalized by regional HDD per billion of net available income (I/R), 2016 and 2018, heating system

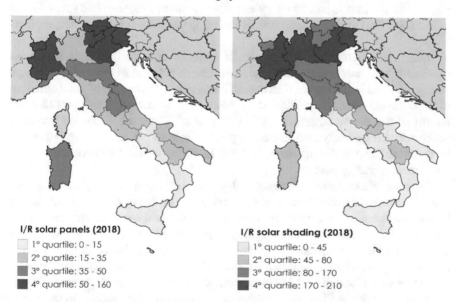

Fig. 5 Regional investments activated by Ecobonus normalized per million of net available income (I/R), 2018, solar panels and solar shading

enough to support access to the Ecobonus incentive scheme in southern regions. In other words, the higher potential in southern Italy is not followed by a higher demand of tax deductions at household level. This result is particularly relevant considering the fact that in energy-poor households, renewable energy sources could represent a structural solution, similar to energy-efficiency interventions.[7] The observed pattern could be due to, among others, the difficulties in accessing the Ecobonus incentive scheme of households having a high potential but also a low-income level.

4   Discussion

According to the results shown in the previous section, southern regions are very often in the lower distribution quintile of the ratio between investments activated by Ecobonus and household net available income. This is true at the overall level (total investments activated) but also for specific technologies, both those for which a correction for climate effects is needed (interventions on windows and shutters, envelope insulation and heating system) and those for which a higher potential exists in southern Italy (solar panels and solar shading). The comparison between the first year in which the tax-credit transfer was made available (2016) and two years after its introduction (2018) shows small improvements in the access pattern at geographical level.

Italian NECP confirms that the results obtained through Ecobonus have been significant until now and that the incentive scheme remains associated with a high saving potential in next years. The overall cumulated contribution of tax deductions to 2030 targets would be around 18.15 Mtoe of final energy, which would cover almost all the saving target for residential sector.[8] Then, the Ecobonus incentive scheme would certainly continue to play a key role to enhance energy efficiency in residential sector. The question is if the incentive would also contribute to the alleviation of energy poverty.

Several development trends at policy level are envisaged in NECP to ensure and reinforce Ecobonus effectiveness in generating energy savings. Also in this case, it is worth to assess to what extent these interventions could contribute to increase the potential of the scheme in mitigating energy poverty. First of all, the tax deductions schemes for energy renovation and for refurbishment of existing buildings would be optimized by integrating them into a single scheme. Additionally, this new scheme should provide a benefit scalable in relation to the expected saving to reward those interventions with the best cost-efficiency ratio and to increase the trend toward

---

[7] It is interesting to mention the local initiative "Reddito Energetico", financing small photovoltaic installations in buildings inhabited by energy-poor households and also introducing a revolving fund. This has been implemented by Gestore Servizi Energetici in Sardinia and more information can be found in Energy Efficiency Annual Report (ENEA 2019).

[8] Also, the tax deductions for refurbishment of existing buildings (Bonus Casa) would contribute to reach this overall figure. In 2018, interventions incentivized through Bonus Casa saved 0.225 Mtoe/year, whereas those incentivized by Ecobonus were 0.106 Mtoe/year.

deep renovation of buildings and seismic improvement. Finally, provisions aimed at promoting initial investments should be introduced, such as extending the transferability of the tax credit and implementing a guarantee fund on green financing issued by credit institutions. This last intervention could modify the pattern in accessing the Ecobonus incentive scheme for energy-poor households.

Ensuring further adaptation of the Ecobonus incentive scheme to improve the access of energy-poor households would be consistent also with the general approach proposed by EnR in its 2019 position paper. This analysis on the effectiveness of the Ecobonus incentive scheme in tackling energy poverty, and the eventual need to make it more suitable for energy-poor households is indeed aligned with several EnR recommendations.

First, it is obviously in line with energy efficiency being a structural solution to energy poverty, not only alleviating it but acting on its causes and potentially allowing people to definitively exit their energy-poverty condition. Supporting investments in building renovation would enable strategies to counter energy poverty to take into account that energy needs change in an objective way according to technical building characteristics. As widely known, energy efficiency is also associated with multiple benefits, such as social and health benefits, which are even more evident when energy-efficiency interventions are implemented in the energy-poverty context (BPIE 2014; Liddell and Guiney 2015; Ntaintasis et al. 2019). If such benefits are opportunely translated into the investments' business plan, they may shorten payback periods and also increase the credit worthiness of low-income households. Besides, poorest deciles of the population are those where retrofit actions are usually more urgent, being more likely to live in non-refurbished homes with high fuel costs (Schleich 2019). Ownership is another delicate issue to be taken into account: the energy-poverty condition could arise both in the private residential sector, relative to households owning or renting their apartment, or in public residential sector, relative to social housing. Including renters among the eligible subjects of energy-efficiency policies, as is the case with Ecobonus, could turn out to not be effective due to the split incentive dilemma. Owners have no incentive to make investments whose benefits are mainly enjoyed by tenants, and this problem could be especially acute in the energy-poverty context. Sound solutions should provide incentives to both the owner and tenant, defining how multiple benefits due to energy efficiency could be split among the two parties (Bird and Hernandez 2012). Further research could extend this analysis to consider ownership information at the regional level, to detect if relevant differences exist in accessing the Ecobonus incentive scheme. In Liguria region, the EnerSHIFT project has promoted the use of Energy Performance Contract in social housing, according to an innovative financial mechanism that links energy-efficiency incentives with the associated savings.[9]

Second, the regional figures shown in this work confirm the need of an integrated approach, where energy agencies work with regional and local institutions to promote and target the use of existing mechanism such as Ecobonus. Additional resources can

---

[9] Such a tool is perfectly in line with Art. 10 of Energy Performance of Building Directive (2018/844). More information can be found on the project website www.enershift.eu.

be provided by European structural funds, to be used both in private and public residential sector, in particular in social housing. For example, several regional call for tender have been published to finance energy renovation in social housing, and often these opportunities can also be associated with existing energy-efficiency incentives such as thermal account.[10] Energy agencies could also contribute to the identification of consumers eligible for measures against energy poverty, for example by looking at Energy Performance Certificates (EPCs) and to the associated integrated database at the national level.[11] EPCs information could usefully complement the regional analysis developed here. Although EPC certainly is a useful tool, it should be considered that referring to EPCs could provide partial information since they are computed in simulated conditions, and they not refer to the effective energy use of a building.

Third, the analysis developed here is consistent with the need to deepen existing knowledge on how existing energy-policy measures could have differentiated impacts on income groups, in particular in terms of who is paying the cost or who has access to the financial incentives. If the distributive effects of energy policies are regressive, that is to say low-income households have a higher burden compared to richer ones, compensation should be envisaged or policy reforms should be implemented. Regressive effects of policy measures may worsen energy-poverty incidence, as well as deteriorate indoor environmental conditions and, more in general, the well-being of households (Berry 2019).

Fourth and last, efforts to improve the access to the Ecobonus incentive scheme by energy-poor households could also include training, information, dissemination and awareness-raising activities. To date, little attention has been given to the dissemination and public awareness of the energy-poverty issue (Bartiaux et al. 2016), as well as the way the topic is dealt with by the media (Scarpellini et al. 2019).

5   Conclusions

Results of the study could have significant implications for policy makers to better understand how adjusting energy-efficiency measures can address energy poverty. The maps suggest that there is room for maneuver in further modifying the possibility of the tax-credit transfer, introduced in 2016, to facilitate lower-income households in accessing the Ecobonus incentive scheme. Indeed, two years after the introduction of tax credit transfer, a lower access to Ecobonus is still observed in southern Italy regions, where the incidence of energy poverty is higher. As suggested in the EnR position paper, energy-efficiency measures could represent a structural solution to energy poverty. The low access to the Ecobonus incentive scheme in southern regions confirms the need to apply a distributive analysis of the policy measures adopted

---

[10]More information on this incentive scheme for energy efficiency and renewable energy sources can be found in Ministerial Decree 16/02/2016.

[11]In Italy, an EPC integrated database exists and is managed by ENEA: it currently includes the EPCs from seven regions and two autonomous provinces.

to achieve long-term objectives. Clearly, regional policy action, in particular that associated with a targeted use of European structural funds, could help in making existing energy-efficiency measures more effective in tackling energy poverty. An integrated policy approach, as well as action at local level and information and training campaigns, could help in improving the access to the Ecobonus incentive scheme for energy-poor households. Further research should explore which compensation instruments could be adopted to reduce the distributive imbalances potentially associated to existing energy-efficiency measures, such as tax deductions for energy renovation of existing buildings.

Acknowledgements I would like to thank all ENEA colleagues who contributed to the "2017 Annual Report—65% Tax deduction scheme—Ecobonus," to the "2019 Annual Report–Tax deduction schemes" and to the "EnR Position paper on Energy Poverty in the European Union."

References

Bartiaux F, Schmidt L, Horta A, Correia A (2016) Social diffusion of energy-related practices and representations: patterns and policies in Portugal and Belgium. Energy Pol 88:413–421

Berry A (2019) The distributional effects of a carbon tax and its impact on fuel poverty: a microsimulation study in the French context. Energy Pol 124:81–94

Bird S, Hernandez D (2012) Policy options for the split incentive: Increasing energy efficiency for low-income renters. Energy Pol 48:506–514

Buildings Performance Institute Europe—BPIE (2014) Alleviating fuel poverty in the EU. Investing in home renovation. A sustainable and inclusive solution

European Energy Network—EnR (2019) Energy poverty in the European Union

European Union (2018a) Directive (EU) 2018/844 of the European Parliament and of the Council of 30 May 2018 amending Directive 2010/31/EU on the energy performance of buildings and Directive 2012/27/EU on energy efficiency (Energy performance in buildings directive 2018/844, EPBD)

European Union (2018b) Regulation (EU) 2018/1999 of the European Parliament and of the Council of 11 December 2018 on the Governance of the Energy Union and Climate Action, amending Regulations (EC) No 663/2009 and (EC) No 715/2009 of the European Parliament and of the Council, Directives 94/22/EC, 98/70/EC, 2009/31/EC, 2009/73/EC, 2010/31/EU, 2012/27/EU and 2013/30/EU of the European Parliament and of the Council, Council Directives 2009/119/EC and (EU) 2015/652 and repealing Regulation (EU) No 525/2013 of the European Parliament and of the Council (Governance Regulation)

European Union (2018c) Directive (EU) 2018/2002 of the European Parliament and of the Council of 11 December 2018 amending Directive 2012/27/EU on energy efficiency (Energy Efficiency Directive 2018/2002, EED)

European Union (2019) Directive (EU) 2019/944 of the European Parliament and of the Council of 5 June 2019 on common rules for the internal market for electricity and amending Directive 2012/27/EU (Electricity Directive 2019/944)

Faiella I, Lavecchia L (2015) Energy Poverty in Italy. Polit Econ 1:27–76

Faiella I, Lavecchia L, Borgarello M (2017) Una nuova misura della povertà energetica delle famiglie. Roma: Banca d'Italia, QEF 404

Fondazione Di Vittorio (2018) La povertà energetica e gli anziani

Hills J (2012) Getting the measure of fuel poverty: final report on the fuel poverty review. CASE report 72. Centre for Analysis of Social Exclusion The London School of Economics and Political Science

Italian Energy Agency for New Technologies, Energy and Sustainable Economic Development—ENEA (2019a). Energy Efficiency Annual Report

Italian Energy Agency for New Technologies, Energy and Sustainable Economic Development—ENEA (2019b). Annual Report 2019—Tax deduction schemes.

Italian Energy Agency for New Technologies, Energy and Sustainable Economic Development—ENEA (2017). Annual Report 2017—65% Tax Deduction Scheme—Ecobonus

Lelo K, Monni S, Tomassi F (2019) Le mappe della disuguaglianza. Donzelli Editore

Liddell C, Guiney C (2015) Living in a cold and damp home: frameworks for understanding impacts on mental well-being. Publ Health 129:191–199

Ntaintasis E, Mirasgedis S, Tourkoliasb C (2019) Comparing different methodological approaches for measuring energy poverty: Evidence from a survey in the region of Attika, Greece. Energy Pol 125:160–169

Osservatorio Italiano sulla Povertà Energetica—OIPE (2019). Rapporto sullo stato della povertà energetica in Italia

Scarpellini S, Sanz Hernández MA, Moneva JM, Portillo-Tarragona P, López Rodríguez ME (2019) Measurement of spatial socioeconomic impact of energy poverty. Energy Pol 124:320–331

Schleich J (2019) Energy efficient technology adoption in low-income households in the European Union—What is the evidence? Energy Pol 125:196–206

A Behavioral Model for In-Home Displays Usage in Social Housing Districts

Valeria Fanghella and Nives Della Valle

Abstract The SINFONIA project is one of the first attempts to combine technological and behavioral policy levers to fight energy poverty in social housing districts. Tenants of Bolzano social housing are provided with renovated dwellings. To enhance the management of the renovated capital stock, they are also supplied with an in-home display (IHD) that provides real-time feedback on energy consumption and indoor parameters. But how will tenants react to IHDs? Previous studies investigate which features and benefits of IHDs generate engagement, but they yield little useful information on their effectiveness in low socioeconomic-status settings. With this study, we examine the behavioral process underlying tenants' usage of IHDs. In contrast to the existing literature, we consider how cognitive biases, specifically, locus of control and present bias, affect the degree of interaction with IHDs. Their consideration is particularly important in this setting: Scarcity affects the cognitive process in a way that may undermine the effectiveness of projects requiring active behavioral change (such as IHDs). To integrate the various elements and account for their relative importance, we develop a theoretical model of the decision to interact with in-home displays (IHDs). On the one hand, by interacting with IHDs, tenants reduce their energy bills and CO_2 emissions, deriving economic and moral utility. On the other hand, interacting with the IHDs generates disutility, for instance, in terms of opportunity cost of time to put in place their feedback. The interaction will occur only if the expected benefits are higher than the expected costs. We argue that such cost-benefit evaluation is further affected by present bias and locus of control. First, a stronger present bias may lead to higher discounting of such benefits and make them loom weaker than the immediate effort required to use the IHD. Second, a more external locus of control may downgrade the perception of energy saving resulting from IHDs usage, thereby reducing the expected economic and environmental benefits associated with a specific level of interaction. Through a theoretical discussion,

V. Fanghella (✉)
Department of Economics and Management, University of Trento, Trento, Italy
e-mail: valeria.fanghella@unitn.it

N. Della Valle
Eurac Research, Institute for Renewable Energy, Bolzano, Italy
e-mail: nives.dellavalle@eurac.edu

our work contributes to informing the design of policies aimed at tackling energy poverty.

Keywords Energy poverty · Energy saving · Real-time feedback · Cognitive biases

1 Introduction

In the European Union, more than 50 million households suffer from energy poverty, making it one of the most challenging priorities of the European Commission.[1] In particular, energy poverty is a condition in which individuals cannot access the essential services of lighting, power for standard appliances and adequate warming and cooling (Thomson et al. 2017). Among other factors, the inability to keep one's house warm and cold is caused by the inadequate energy performance of the dwellings (Ugarte et al. 2016). This is especially true for low-income households, who are less likely to have the resources to invest in efficiency renovations (Schleich 2019). The European Commission has therefore identified energy-efficiency improvements as a key policy lever to tackle energy poverty.[2] Improving the energy performance of buildings not only enhances the quality of life of vulnerable households (Jenkins 2010) but also contributes to the EU climate-related goals (Charlier, Risch and Salmon 2018).

However, the built environment constitutes only part of household energy consumption, and how tenants use it also represents a significant share (Guerra-Santin and Itard 2010). As an example, behavioral factors explain between 30 and 50% of the variance of overall heating and cooling consumption (Mansouri et al. 1996; Sonderegger 1978; Steemers and Yun 2009). Hence, the effectiveness of technological interventions depends on the behavior of individuals who daily interface with them (Gillingham and Palmery 2014). Ignoring the human dimension reduces the likelihood of achieving significant energy savings (Allcott et al. 2014), which is usually referred to as the energy-efficiency gap (Dietz 2010; Jaffe and Stavins 1994) and rebound effects (Belaïd et al. 2018; Milne and Boardman 2000; Nässén and Holmberg 2009). These behavioral factors weigh even more among vulnerable households, who usually display low socioeconomic status and have limited access to essential resources (DellaValle 2019). Moreover, such a resource scarcity affects the cognitive process (Shah et al. 2015) in a way that makes occupant behavior the "fourth driver of fuel poverty" (Kearns et al. 2019).

Acknowledging the interconnections between the built environment and tenants' behavior, SINFONIA[3] combines technological and behavioral policy levers to

---

[1] https://ec.europa.eu/energy/en/topics/energy-strategy-and-energy-union/clean-energy-all-europe ans%20.

[2] https://ec.europa.eu/energy/en/topics/energy-efficiency/targets-directive-and-rules/energy-effici ency-directive.

[3] SINFONIA is one of the pioneer projects promoting the retrofitting of social housing in Italy. Thanks to funding from the Seventh Framework Program of the European Union for Research,

address energy poverty. In the pilot case of Bolzano, coordinated by the Institute of Renewable Energy-Eurac Research, retrofitting interventions were implemented to enhance the energy performance of social housing buildings (see DellaValle et al. (2018) for further details). In addition to other activities that engage tenants in learning how to manage the renovated infrastructure, in-home displays (IHDs) are installed in the dwellings of tenants who gave their consent.[4] The IHDs provide real-time feedback on indoor parameters and energy consumption. Notably, every time a parameter overcomes an optimal threshold, the IHD sends out a pop-up specifying the problem and providing tips on how to solve it. The tenant can solve the pop-up by clicking on the relevant button.

This study focuses on the behavioral process underlying tenants' usage of IHDs. Our study contributes to energy poverty and IHDs research in two ways. First, we capitalize on the SINFONIA project, which is one of the first attempts to use real-time feedback in a setting characterized by low socioeconomic status (Warnierand Nurminen 2017), to advance the understanding on whether interventions that require active users' engagement can tackle energy poverty. Considering how the context shapes behavior not only makes policy making more informed, but also increases the likelihood to achieve projects' goals (Westskog et al. 2015). Second, we model the decision to use the IHDs by accounting for the role of cognitive biases. Previous studies on IHDs mainly focused on their environmental and economic benefits and their technological features as sources of engagement (e.g., Pyrko 2011; Van Dam et al. 2010; Westskog et al. 2015). However, cognitive limitations affect the cost-benefit considerations of using energy-efficient technology (Spandagos and Ng 2018), especially among vulnerable households (Mills and Schleich 2012). While it is true that IHDs embed a huge potential to bring economic and environmental benefits (Fischer 2008), such a potential can be undermined by, among others, behavioral factors.

The remainder of the paper proceeds as follows. In Sect. 2, we review the relevant literature. Section 3 sets up the theoretical model and hypotheses. Conclusions and implications of our work are drawn in Sect. 4.

---

researchers from the Institute for Renewable Energy of the European Academy of Bolzano (Eurac Research), in collaboration with the Municipality of Bolzano/Bozen, supported by experts from IDM Südtirol—South Tyrol, the Institute for Social Building IPES, Alperia and Agenzia CasaClima are transforming some areas of Bolzano/Bozen. At the core of this transformation is the renovation of a number of social housing districts, through the introduction of several technologies. These interventions aim to improve the quality of life at home, increasing energy saving and comfort for the residents. For additional details see: http://www.sinfonia-smartcities.eu/en/project.

[4] http://www.sinfonia-smartcities.eu/en/blog/post/behaviour-change-how-to-increase-the-impact-of-energy-efficient-renovation projects.

2 Cognitive Biases and In-Home Displays

For decades, policy makers have relied on the assumption that individual behavior complies with rational choice theory (Von Neumann and Morgenstern 1947). However, extensive experimental and empirical evidence has shown that individuals sistematical deviate from the assumptions of this theory (Camerer et al. 2004). In particular, to perform rational calculations under limited cognitive capacity (Simon 1955), individuals use the tools of heuristics (Tversky and Kahneman 1974). However, these may often lead to systematic errors (e.g., cognitive biases).

Among the most prominent deviations from rational choice theory, other-regarding preferences certainly have received the greatest attention in the experimental literature. In particular, studies on the topic have confronted the classical assumption that individuals display strictly selfish preferences (i.e., individuals only care about their own well-being) with the evidence that they also care about the well-being of others and the society as a whole. These other-regarding concerns are especially relevant in the context of energy consumption, and this is well-acknowledged in the related literature (Frederiks et al. 2015a). Indeed, being the climate the most prominent public good (Brekke and Johansson-Stenman 2008), the decision to conserve energy is also explained by a motivation to contribute to others' well-being (Bénabou and Tirole 2006; Farrow et al. 2017). In our study, we account for these preferences.

Another deviation from rational decision-making is the tendency to place attention on objectives that are close in time and disregard those that are distant (also called *present bias*). Such a tendency results in decisions that are guided by a preference for options that provide immediate benefits, even when other alternatives would provide higher benefits in the future (Loewenstein and Prelec 1992). In the context of energy consumption, the economic (Harding and Hsiaw 2014) and environmental (Brekke and Johansson-Stenman 2008) consequences associated with the effort required to consume less energy are delayed. Therefore, the decision to use energy efficiently can be seen as an intertemporal decision that may be affected by present bias (Harding and Hsiaw 2014).

While present bias has been extensively studied in the context of energy efficiency (Newell and Siikamki 2015), to the best of our knowledge, this is the first paper that investigates how it explains the engagement with IHDs. As for the decision to consume less energy, also using IHDs can be seen as an intertemporal choice associated with immediate cost and delayed benefit. Regarding the cost, it has two main components. The former refers to the opportunity cost of time to interact with the monitor, understand its feedback and apply it to daily behavior (Sintov and Schultz 2017). The latter is caused by the reduction of energy consumption, which generates lower comfort and social status (Pierce et al. 2010). In contrast, the benefits of IHDs are delayed and made salient only at the end of the billing period. Hence, those who assign higher values (namely, are more present biased) to the cost of using IHDs than to its expected benefits might be less willing to use them.

At the same time, IHDs are likely to reduce the negative consequences associated with present bias by making salient the energy consumed and the associated outcomes (Hargreaves et al. 2010). For this and other reasons, it is thus crucial to better understand the interplay between present bias and IHDs usage. And this is particularly true for social housing contexts. As the surrounding social and economic backgrounds generally influence intertemporal decisions (Watts et al. 2018), energy-related behaviors also may be affected by living in conditions of resource scarcity (DellaValle 2019).

Finally, in this study, we account for the role of *locus of control*, which is "a generalized attitude, belief, or expectancy regarding the nature of the causal relationship between one's own behavior and its consequences" (Rotter 1966: 2). Locus of control constitutes the extent to which individuals perceive outcomes in their life as caused by their actions (Rotter 1966). Specifically, locus of control is internal if one feels control over outcomes and external if one perceives that outcomes are caused by factors beyond her control. Having an internal locus of control relates to the belief that there is a higher return in doing an activity (Caliendo et al. 2015) and to higher persistence in trying to achieve a goal (Ng et al. 2006). In the context of IHDs, behavioral change may be thwarted by perceived action potential (Frederiks et al. 2015b).

This psychological concept has only recently been introduced in the economic literature to explain human behavior. However, attitudes affect the link between preferences and behavior and cannot be disregarded while investigating the heterogeneity behind IHDs engagement (Buchanan et al. 2015). With this study, we introduce locus of control as a driver of engagement with IHDs. This question is particularly interesting for vulnerability contexts in which individuals are more likely to display a more external locus of control (Sheehy-skeffington and Rea 2017). In particular, if occupants with a more external locus of control do not feel in control over their energy consumption, they may perceive that using IHDs will not generate significant energy saving. This, in turn, may undermine the perceived economic and environmental benefits of using IHDs and the motivation to use them. Moreover, it is likely that an external locus of control also predicts low persistence in interacting with the IHD and reacting to its feedback. On the other hand, however, IHDs embed the potential to empower consumers to take action with respect to lowering their energy consumption (Grønhøj and Thøgersen 2011). Hence, it may also be the case that IHDs help overcome the behavioral consequences of an external locus of control.

3 Theoretical Framework

3.1 Tenants' Utility

The theoretical framework focuses on tenants' decision to use the IHDs. The engagement with IHDs is essential because, alone, this technology does not conserve energy.

Rather, it provides information that tenants need to process and implement in order to decrease their energy consumption (Buchanan, Russo and Anderson 2014). Hence, without engagement and interaction with IHDs, energy saving is not achievable (Matsukawa 2004; Nunes et al. 2011).

We assume that interaction with IHDs generates energy saving compared to the baseline of no intervention. Without IHDs, individuals consume energy e_0. We assume that $e_0 * p_e \leq y$, where p_e is the price of energy and y is income. Namely, individuals can afford the energy costs they faced under no intervention.[5] Individuals can choose to interact with IHDs to reduce their consumption. Their decision variable is the level of interaction with IHDs, x. x is a continuous variable; $x \in [0; 1]$, so that $x = 0$ means no interaction, and $x = 1$ full interaction. In practical terms, this variable represents whether a tenant interacts with the IHD and solves the pop-ups. One means that the tenant solves a problem every time a new one arises, and zero that the tenant never touches the screen.

An interaction with IHDs equal to x leads to an energy consumption of $e(x)$; $e' < 0$ and $e'' > 0$, so that more interaction leads to lower consumption, with diminishing return. Diminishing return is consistent with previous literature showing that, after repeated interactions with the IHD, this technology does not offer new information any longer (Buchanan et al. 2015). If $x = 0$, $e(x) = e_0$; if $x = 1$, $e(x) = e_{min} < e_0$, which represents the technically minimum possible level of final energy consumption achievable through behavioral change;[6] $e(x) \leq e_0$ because we assume that interacting with IHDs generates energy savings equal to: $\Delta e = e_0 - e(x)$.

Energy savings have two main effects on individuals' utility. On the one hand, savings generate economic and moral utility because they reduce the energy bills and CO_2 emissions associated with one's consumption. On the other hand, they generate disutility, for instance, in terms of opportunity cost of time to use the IHD and of reduced comfort associated with lower energy consumption.

Tenants' utility is made of two components, an economic and a moral one. We model economic utility as $g(\Delta e * p_e; y)$, where: p_e is the price of energy and y is individuals' current income. Since individuals who can afford their current expenses are less interested in interacting with IHDs to reduce their bills (Pyrko 2011; Westskog et al. 2015), the utility derived from economic savings is assumed to depend on current income. $\frac{\partial g}{\partial e} > 0$; $\frac{\partial^2 g}{\partial e^2} < 0$, so that utility is quasi linear in energy saving; $\frac{\partial g}{\partial y} > 0$; $\frac{\partial^2 g}{\partial y^2} < 0$, to capture the fact that sensitivity reduces with higher income.

We model moral utility as a moral subsidy, in which consumers receive moral utility μ for every unit of e not consumed (Allcott and Kessler 2019): $M = \mu * (\Delta e)$.

---

[5] We consider the situation where tenants of social housing have full responsibility of their energy bills, as this is the case in Bolzano. However, in other social housing districts, tenants might not be charged with any energy bill, or only up to a certain amount. It would be possible to adapt the model considering these variations, for instance, by eliminating the economic incentive as in Myers and Souza (2020) or by modeling the economic utility as a step function.

[6] We consider a setting where energy-efficiency investments are exogenous. This generally applies to vulnerability contexts, where households cannot afford this kind of investment, and can achieve energy saving mostly by changing their behavior (Dillahunt et al. 2009).

Individuals with higher μ perceive greater moral utility from saving energy because energy conservation can be intended as a form of contribution to the common good, notably the environment. Therefore, they have a strong incentive to use IHDs to reduce their consumption. Finally, we model disutility as $f(\Delta e, \alpha)$, where α is a taste parameter. It considers both the negative utility generated by using IHDs and by reducing own energy consumption. To represent that there is a diminishing return on effort in the process of domestic energy conservation (Oikonomou et al. 2009), we assume that the marginal disutility is positively and weakly increasing in Δe (and therefore in x): $f' > 0$, $f'' \geq 0$.

3.2 Equilibrium Choice Without Cognitive Biases

Let $\theta = \{y, p_e, \alpha, \mu\}$ be the vector of factors that affect utility. The consumer maximizes

$$\max_x U(\theta) = g(\Delta e * p_e; y) + \mu * (\Delta e) - f(\Delta e, \alpha) \tag{1}$$

Consumers' equilibrium choice of x, denoted $x^*(\theta)$, is determined by the following first-order condition:

$$g'(\Delta e * p_e; y) * e'(x) + \mu * e'(x) = f'(\Delta e, \alpha) * e'(x) \tag{2}$$

Simplifying, the equilibrium choice becomes

$$g'(\Delta e * p_e; y) + \mu = f'(\Delta e, \alpha) \tag{3}$$

Such that the economic plus the moral utility should equal the marginal disutility.

3.3 The Role of Locus of Control and Present Bias

We extend the framework to consider how present bias and locus of control affect the usage of IHDs. Cognitive biases affect choice but not experienced utility (Allcott and Kessler 2019).

Present bias. To consider the effect of present bias, the terms that affect utility with delay are discounted at rate $\beta * \delta$, where δ denotes the discount factor for exponential time-consistent impatience and β represents a preference for immediate gratification (Rabin and O'Donoghue 2000). Energy saving yields immediate disutility and delayed economic and environmental benefits (Harding and Hsiaw, 2014). Hence, in the present, individuals face discounted economic and moral utility, respectively, of $\beta * \delta * g(\Delta e * p; y)$ and $\beta * \delta * \mu * (\Delta e)$. We make two simplifying assumptions: First, all aspects in any future period, viewed from today, are discounted at rate $\beta \leq 1$. Second, subjects discount economic and moral outcomes at the same rate (Hardisty and Weber 2009). Disutility $f(\Delta e, \alpha)$ is not discounted because it is simultaneous with the decision to interact with IHDs.

Locus of control. We include locus of control by considering the relation between actual and perceived energy saving generated by IHDs usage. We assume that individuals do not know how their interaction with IHDs affects their energy consumption—namely, $e(x)$ is unknown. This assumption is realistic because individuals are often unconscious of how to reduce their energy consumption or the effort required to do it (Mizobuchi and Takeuchi 2013). Each individual has a subjective belief concerning this relationship $\tilde{e}(x, loc)$, which depends on the degree of internalization of the locus of control loc (low loc corresponds to an internal locus of control, high loc to an external locus of control).

We assume that locus of control remains constant over time and, therefore, is not affected by IHDs. One may argue that by improving the knowledge on the functional relation between x and e, IHDs empower occupants and reduce the influence of locus of control. However, previous evidence shows that, despite affecting other outcomes, such as awareness about own and appliances' consumption, IHDs do not increase tenants' sense of personal control (Buchanan et al. 2014). Locus of control is indeed a dispositional trait, which is hardly affected by short-term interventions and is independent from realistic behavior-outcome relation (Ajzen 2002).

Similarly to Caliendo et al. (2015), we assume a multiplicative effect of locus of control: $\tilde{e}(x, loc) = e\left(\frac{x}{h(loc)}\right)$, with $h'(loc) > 0$. In this way, for the same level of x, an individual with an external locus of control perceives a lower effect on energy consumption compared to a peer with a more internal one. In other words, the more external the locus of control, the higher the perceived level of interaction required to achieve a specific level of energy saving. Locus of control has an effect on the elements described above: $\tilde{e}(x, loc)$ substitutes $e(x)$ in the expression of Δe, that becomes: $\Delta \tilde{e} = e(0) - \tilde{e}(x, loc)$. Economic and moral utilities therefore become $g(\Delta \tilde{e} * p; y)$ and $M = \mu * (\Delta \tilde{e})$, so that the more external the locus of control, the lower the perception of energy saving from a specific level of interaction x and the lower the resulting economic and moral utility. In the same way, a more external locus of control increases the perceived amount of interaction required to achieve a saving equal to x, affecting the disutility as follows: $f(\Delta \tilde{e}, \alpha)$. Hence, a more external locus of control is likely to increase the perceived opportunity cost of time and effort of using IHDs. Taken together, these effects may result in lower HIDs usage.

3.4   Equilibrium Choice with Cognitive Biases

Let $\theta = \{y, p_e, \alpha, \mu, \delta, \beta, loc\}$ be the new vector of factors that affect utility. The consumer maximizes

$$\max_{x} U(\theta) = \beta * \delta * (g(\Delta\tilde{e} * p_e; y) + \mu * (\Delta\tilde{e})) - f(\Delta\tilde{e}, \alpha) \tag{4}$$

where $\Delta\tilde{e} = e_0 - e\left(\frac{x}{h(loc)}\right)$.

Consumers' equilibrium choice of x, denoted $x^*(\theta)$, is determined by the following first-order condition:

$$\beta * \delta * e'\left(\frac{x}{h(loc)}\right) * \left(\mu_e + g'\left(e_0 - e\left(\frac{x}{h(loc)}\right) * p_e; y\right)\right)$$
$$= f'\left(e_0 - e\left(\frac{x}{h(loc)}\right), \alpha\right) * e'\left(\frac{x}{h(loc)}\right) \tag{5}$$

Simplifying, the equilibrium choice becomes

$$\beta * \delta * \left(\mu_e + g'\left(e_0 - e\left(\frac{x}{h(loc)}\right) * p_e; y\right)\right) = f'\left(e_0 - e\left(\frac{x}{h(loc)}\right), \alpha\right) \tag{6}$$

Such that the economic and moral utility, discounted at rate $\beta * \delta$ and affected by the locus of control $h(loc)$, should equal the marginal disutility from x^*, in turn biased by $h(loc)$.

3.5 Hypotheses

From the equilibrium analyses reported in Sects. 3.2 and 3.4, we draw the hypotheses on the role of economic and moral utility. In particular, equilibrium choices show that the sum of economic and moral components has to be higher than the disutility for interaction to happen. Hence, we hypothesize the following:

H1. *The more stringent the budget constraint, the higher the usage of IHDs.*

H2. *The higher the moral tax associated with energy consumption, the higher the usage of IHDs.*

By comparing the two equilibrium choices, we draw the following hypotheses on the effect of cognitive biases:

H3. *The higher the focus on the present, the lower the usage of IHDs.*

Namely, myopia discounts the utility derived from events occurring in the future. Economic and moral utilities are discounted in Eq. (6) but not in Eq. (3). Consequently, a present biased person draws less utility from a level of interaction x compared to another who is time consistent.

H4. *The more external the locus of control, the lower the usage of IHDs.*

Notably, an external locus of control reduces the perceived amount of energy saved from a specific level of interaction x. Hence, for the same x, when the subject has an external locus of control, the utility in Eq. (3) is higher, and the disutility is lower, than in Eq. (6).

H5. *Tenants will use the IHDs if the perceived discounted economic and environmental utility achieved with a specific level of interaction, biased by the locus of control, is higher or equal than the perceived disutility associated with that amount of interaction.*

When we account for cognitive biases, tenants' motivation to use the IHDs needs to be higher than in the no-bias scenario. Indeed, the utility not only needs to overcome the effort required to interact with the IHDs, but also the distortion added by cognitive biases. The difference between the choices without and with biases augments in the severity of biasedness.

4 Discussion and Conclusion

In this study, we propose a theoretical framework that constitutes a flexible tool to assess the effect of cognitive biases on IHDs usage in social housing districts and beyond. The next step of this study will consist in testing the model with field data from Bolzano social housing districts, where IHDs are currently under installation. Interaction data will be complemented with survey information to collect the parameters of the model. We expect the validated model to provide insights on the role played by IHDs in the tackling of energy poverty, especially when tenants are required to be actively engaged. Meantime, this model provides a starting point to better inform the design of future projects aimed at tackling energy poverty, since accounting for the effect of cognitive biases generally increases the chance to achieve policies' goals (Bertrand et al. 2004).

Furthermore, our model can also be tested among high-income households. Evidence on IHDs' effectiveness is still mixed in this context (Buchanan et al. 2015), and cognitive biases may unveil an unobserved driver of (lack of) engagement. Moreover, the sources of interest in the IHDs and cognitive biases may have a different effect in high-income contexts (Dillahunt et al. 2009; Westskog et al. 2015). As an example, the economic incentive is likely to weigh more for vulnerable consumers because, by reducing the energy costs, they will have more financial resources to access other necessary goods that they might not afford otherwise (Schaffrin and Reibling 2015). On the other hand, high-income households may be more interested in the environmental (or social) return of IHDs (Kollmuss and Agyeman 2002).

This study is by no means without limitations. At this stage, this work could be perceived as a theoretical exercise, given that the hypotheses, even if grounded on empirical evidence, need to be tested in the specific social housing context to draw policy implications. Second, we focused only on a specific cognitive bias and a psychological concept, whereas other influences, such as status quo bias, social influence and loss aversion, may also play an important role in this setting. These elements could be conveniently added to the model presented here. As an example, the effect of social influence on moral utility can be modeled as in Myers and Souza (2020). Finally, we model tenants' choices as time-invariant. The model would benefit from considering the time dimension and endogenous cognitive biases. In particular, future research may be worthwhile to investigate whether real-time feedback affects the perception of control over energy consumption. If IHDs empower consumers, their benefits exceed their short-term implementation and have broader implications for the fight against energy poverty.

References

Ajzen I (2002) Perceived behavioral control, self-efficacy, locus of control, and the theory of planned behavior. J Appl Soc Psychol 32(4):665–683. https://doi.org/10.1111/j.1559-1816.2002. tb00236.x

Allcott H, Kessler JB (2019) The welfare effects of nudges: a case study of energy use social comparisons. Am Econ J Appl Econ 11(1):236–276. https://doi.org/10.1257/app.20170328

Allcott H, Mullainathan S, Taubinsky D (2014) Energy policy with externalities and internalities. J Publ Econ 112:72–88. https://doi.org/10.1016/j.jpubeco.2014.01.004

Belaïd F, Bakaloglou S, Roubaud D (2018) Direct rebound effect of residential gas demand: empirical evidence from France. Energy Pol 115:23–31. https://doi.org/10.1016/j.enpol.2017. 12.040

Bénabou R, Tirole J (2006) Incentives and prosocial behavior. Am Econ Rev. https://doi.org/10. 1257/aer.96.5.1652

Bertrand M, Mullainathan S, Shafir E (2004) A behavioral-economics view of poverty. Am Econ Rev 94(2):419–423. https://doi.org/10.1257/0002828041302019

Brekke KA, Johansson-Stenman O (2008) The behavioural economics of climate change. Oxford Rev Econ Pol 24(2):280–297. https://doi.org/10.1093/oxrep/grn012

Buchanan K, Russo R, Anderson B (2014) Feeding back about eco-feedback: How do consumers use and respond to energy monitors? Energy Pol 73:138–146. https://doi.org/10.1016/j.enpol. 2014.05.008

Buchanan K, Russo R, Anderson B (2015) The question of energy reduction: The problem(s) with feedback. Energy Pol 77:89–96. https://doi.org/10.1016/j.enpol.2014.12.008

Caliendo M, Cobb-Clark DA, Uhlendorff A (2015) Locus of control and job search strategies. Rev Econ Stat 97(1):88–103. https://doi.org/10.1162/REST_a_00459

Camerer, C. F., Loewenstein, G., and Rabin, M. (2004). *Advances in behavioral economics*. Princeton University Press

Charlier D, Risch A, Salmon C (2018) Energy burden alleviation and greenhouse gas emissions reduction: can we reach two objectives with one policy? Ecol Econ 143:294–313. https://doi.org/ 10.1016/j.ecolecon.2017.07.002

DellaValle N (2019) People's decisions matter: understanding and addressing energy poverty with behavioral economics. Energy Build 204:109515

DellaValle N, Bisello A, Balest J (2018) In search of behavioural and social levers for effective social housing retrofit programs. Energy Build 172:517–524. https://doi.org/10.1016/j.enbuild. 2018.05.002

Dietz T (2010) Narrowing the US energy efficiency gap. Proc Natl Acad Sci 107(37):16007–16008. https://doi.org/10.1073/pnas.1010651107

Dillahunt T, Mankoff J, Paulos E, Fussell S (2009) It's not all about "Green": energy use in low-income communities. In: ACM International Conference Proceeding Series, pp 255–264. https://doi.org/10.1145/1620545.1620583

Farrow K, Grolleau G, Ibanez L (2017) Social norms and pro-environmental behavior: a review of the evidence. Ecol Econ 140:1–13. https://doi.org/10.1016/j.ecolecon.2017.04.017

Fischer C (2008) Feedback on household electricity consumption: a tool for saving energy? Energ Effi 1(1):79–104. https://doi.org/10.1007/s12053-008-9009-7

Frederiks ER, Stenner K, Hobman EV (2015a) Household energy use: applying behavioural economics to understand consumer decision-making and behaviour. Renew Sustain Energy Rev 41:1385–1394. https://doi.org/10.1016/j.rser.2014.09.026

Frederiks ER, Stenner K, Hobman EV (2015b) The socio-demographic and psychological predictors of residential energy consumption: a comprehensive review. Energies 8(1):573–609. https://doi.org/10.3390/en8010573

Gillingham K, Palmery K (2014) Bridging the energy efficiency gap: policy insights from economic theory and empirical evidence. Rev Environ Econ Pol 8(1):18–38. https://doi.org/10.1093/reep/ret021

Grønhøj A, Thøgersen J (2011) Feedback on household electricity consumption: learning and social influence processes. Int J Consum Stud 35(2):138–145. https://doi.org/10.1111/j.1470-6431.2010.00967.x

Guerra-Santin O, Itard L (2010) Occupants' behavior: determinants and effects on residential heating consumption. Build Res Inf 38(3):318–338. https://doi.org/10.1080/09613211003661074

Harding M, Hsiaw A (2014) Goal setting and energy conservation. J Econ Behav Organ 107:209–227. https://doi.org/10.1016/j.jebo.2014.04.012

Hardisty DJ, Weber EU (2009) Discounting future green: money versus the environment. J Exp Psychol Gen 138(3):329–340. https://doi.org/10.1037/a0016433

Hargreaves T, Nye M, Burgess J (2010) Making energy visible: a qualitative field study of how householders interact with feedback from smart energy monitors. Energy Pol 38(10):6111–6119. https://doi.org/10.1016/j.enpol.2010.05.068

Jaffe AB, Stavins RN (1994) The energy-efficiency gap what does it mean? Energy Pol 22(10):804–810. https://doi.org/10.1016/0301-4215(94)90138-4

Jenkins DP (2010) The value of retrofitting carbon-saving measures into fuel poor social housing. Energy Pol 38(2):832–839. https://doi.org/10.1016/j.enpol.2009.10.030

Kearns A, Whitley E, Curl A (2019) Occupant behaviour as a fourth driver of fuel poverty (aka warmth and energy deprivation). Energy Pol 129(March):1143–1155. https://doi.org/10.1016/j.enpol.2019.03.023

Kollmuss A, Agyeman J (2002) Mind the Gap: why do people act environmentally and what are the barriers to pro-environmental behavior? Environ Educ Res 8(3):239–260. https://doi.org/10.1080/1350462022014540

Loewenstein G, Prelec D (1992) Anomalies in intertemporal choice: evidence and an interpretation. Q J Econ 107(2):573–597. https://doi.org/10.2307/2118482

Mansouri I, Newborough M, Probert D (1996) Energy consumption in uk households: impact of domestic electrical appliances. Appl Energy 54(3):211–285. https://doi.org/10.1016/0306-2619(96)00001-3

Matsukawa I (2004) The effects of information on residential demand for electricity. Energy J 25(1):1–17. https://doi.org/10.5547/ISSN0195-6574-EJ-Vol25-No1-1

Mills B, Schleich J (2012) Residential energy-efficient technology adoption, energy conservation, knowledge, and attitudes: an analysis of European countries. Energy Pol 49:616–628. https://doi.org/10.1016/j.enpol.2012.07.008

Milne G, Boardman B (2000) Making cold homes warmer: the effect of energy efficiency improvements in low-income homes A report to the energy action grants agency charitable trust. Energy Pol 6–7:411–424. https://doi.org/10.1016/s0301-4215(00)00019-7

Mizobuchi K, Takeuchi K (2013) The influences of financial and non-financial factors on energy-saving behaviour: a field experiment in Japan. Energy Pol 63:775–787. https://doi.org/10.1016/j.enpol.2013.08.064

Myers E, Souza M (2020) Social comparison nudges without monetary incentives: evidence from home energy Reports. J Environ Econ Manag 101:102315

Nässén J, Holmberg J (2009) Quantifying the rebound effects of energy efficiency improvements and energy conserving behaviour in Sweden. Energ Eff 2(3):221–231. https://doi.org/10.1007/s12053-009-9046-x

Newell RG, Siikamki J (2015) Individual time preferences and energy efficiency. Am Econ Rev 105(5):196–200. https://doi.org/10.1257/aer.p20151010

Ng TWH, Sorensen KL, Eby LT (2006) Locus of control at work: a meta-analysis. J Organ Behav 27(8):1057–1087. https://doi.org/10.1002/job.416

Nunes NJ, Pereira L, Quintal F, Bergés M (2011) Deploying and evaluating the effectiveness of energy eco-feedback through a low-cost NILM solution. In: Proceedings of the 6th international conference on persuasive technology, pp 2–5

Oikonomou V, Becchis F, Steg L, Russolillo D (2009) Energy saving and energy efficiency concepts for policy making. Energy Pol 37(11):4787–4796. https://doi.org/10.1016/j.enpol.2009.06.035

Pierce J, Schiano DJ, Paulos E (2010) Home, habits, and energy: examining domestic interactions and energy consumption. In: Proceedings of the SIGCHI conference on human factors in computing systems, pp 1985–194. https://doi.org/10.1145/1753326.1753627

Pyrko J (2011) Am I as smart as my smart meter is?—Swedish experience of statistics feedback to households. In: Proceedings of the ECEEE, pp 1837–1841

Rabin M, O'Donoghue T (2000) The economics of immediate gratification. J Behav Decis Mak 13(2):233–250. https://doi.org/10.1002/(SICI)1099-0771(200004/06)13:23

Rotter JB (1966) Generalized expectancies for internal versus external control of reinforcement. Psychol Monogr 80(1):1–28. https://doi.org/10.1037/h0092976

Šć S, Warnier M, Nurminen JK (2017) The role of context in residential energy interventions: a meta review. Renew Sustain Energy Rev 77:1146–1168. https://doi.org/10.1016/j.rser.2016.11.044

Schaffrin A, Reibling N (2015) Household energy and climate mitigation policies: Investigating energy practices in the housing sector. Energy Policy 77:1–10. https://doi.org/10.1016/j.enpol.2014.12.002

Schleich J (2019) Energy efficient technology adoption in low-income households in the European Union—what is the evidence? Energy Pol, 125(October 2018), 196–206. https://doi.org/10.1016/j.enpol.2018.10.061

Shah AK, Shafir E, Mullainathan S (2015) Scarcity frames value. Psychol Sci 26(4):402–412. https://doi.org/10.1177/0956797614563958

Sheehy-skeffington J, Rea J (2017) How poverty affects people' s decision-making processes decision-making processes. Joseph Rowntree Foundation

Simon HA (1955) A Behavioral Model of Rational Choice. Q J Econ 69(1):99–118. https://doi.org/10.2307/1884852

Sintov ND, Schultz PW (2017) Adjustable green defaults can help make smart homes more sustainable. Sustainability 9(4):622. https://doi.org/10.3390/su9040622

Sonderegger RC (1978) Movers and stayers: the resident's contribution to variation across houses in energy consumption for space heating. Energy Build 1(3):313–324. https://doi.org/10.1016/0378-7788(78)90011-7

Spandagos C, Ng T (2018) Fuzzy model of residential energy decision-making considering behavioral economic concepts. Appl Energy 213:611–625. https://doi.org/10.1016/j.apenergy.2017.10.112

Steemers K, Yun GY (2009) Household energy consumption: a study of the role of occupants. Build Res Inf 37(5–6):625–637. https://doi.org/10.1080/09613210903186661

Thomson H, Bouzarovski S, Snell C (2017) Rethinking the measurement of energy poverty in Europe: a critical analysis of indicators and data. Indoor Built Environ 26(7):879–901. https://doi.org/10.1177/1420326X17699260

Tversky A, Kahneman D (1974) Judgment under uncertainty: heuristics and biases. Science 185(4157):1124–1131. https://doi.org/10.1126/science.185.4157.1124

Ugarte S, Ree Van Der B, Voogt M, Eichhammer W, Ordoñez JA, Reuter M, Villafáfila R (2016) Energy efficiency for low-income households. Brussels

Van Dam SS, Bakker CA, Van Hal JDM (2010) Home energy monitors: impact over the medium-term. Build Res Inf 38(5):458–469. https://doi.org/10.1080/09613218.2010.494832

Von Neumann J, Morgenstern O (1947) Theory of games and economic behavior, 2nd rev. ed

Watts TW, Duncan GJ, Quan H (2018) Revisiting the marshmallow test: a conceptual replication investigating links between early delay of gratification and later outcomes. Psychol Sci 29(7):1159–1177. https://doi.org/10.1177/0956797618761661

Westskog H, Winther T, Sæle H (2015) The effects of in-home displays-revisiting the context. Sustainab (Switzerland) 7(5):5431–5451. https://doi.org/10.3390/su7055431

Investigating the Role of Occupant Behavior in Design Energy Poverty Strategies. Insights from Energy Simulation Results

Angela Santangelo, Simona Tondelli, and Da Yan

Abstract Energy poverty is very much interlinked with housing stock characteristics in terms of energy performance. Energy-led building renovation within the social housing stock combines the energy-saving goal with social and economic co-benefits (e.g., poverty alleviation and health improvements), thus contributing to facing stigmatization, social segregation and energy poverty, particularly prevalent in the social housing sector. However, to design renovation strategies aimed at achieving multi-benefits rather than just improving the buildings energy performance still remains a challenge, and considerations concerning energy poverty alleviation are far from being embedded. To the aim of this investigation, building energy renovation is believed to be a key opportunity to roll-out a comprehensive urban regeneration strategy with the goal of tackling energy poverty. However, at the same time, it is important to acknowledge that the gap between expected and actual energy consumption in buildings is highly dependent upon the human factor. Indeed, energy saving is not only a matter of technology, but it is influenced by the use by and the behavior of occupants. The overall aim of the paper is to provide an insight into building energy simulation and occupant behavior modeling as tools to support policymakers in making decisions on which strategies to apply to energy poverty through improvement of energy efficiency of public housing stock. To do so, this contribution investigates the impact of occupant behavior to reduce energy consumption at the household level. The Italian public housing sector is taken as a reference. A multi-family public housing building is assumed as a case study. Three dwellings with different sizes and exposures are considered, having three different occupancy patterns in turn. The results show to what extent the heating loads are influenced by occupant behavior and dwelling characteristics. The results are then discussed to form a basis for exploring how and to what extent housing policies and energy-led regeneration strategies can contribute to addressing energy poverty.

A. Santangelo (✉) · S. Tondelli
Alma Mater Studiorum—University of Bologna, CIRI Building and Construction and Department of Architecture, Viale Risorgimento 2, Bologna, Italy
e-mail: angela.santangelo@unibo.it

D. Yan
School of Architecture, Tsinghua University, Beijing, China

Keywords Occupant behavior · Energy behavior · Energy poverty · Public housing · Building renovation

1 Introduction

According to the Italian Statistic Bureau (ISTAT 2014), in Italy there are 14.5 million buildings, and more than 84% of them are residential buildings. Approximately half of the housing stock consists of apartments in multi-family buildings, a figure that is higher particularly in metropolitan areas, where this share reaches 85.5% of the total housing stock (ISTAT 2014). However, Italian multi-family buildings are rather small and low-rise, with high surface area-to-volume ratios, resulting in very likely high thermal dispersions. Commenting on the quality of the Italian building stock, it is quite old and not adequately refurbished. More than 75% of households live in buildings built before 1990[1] (ISTAT 2010), with low efficiency rates, high maintenance costs for the owners and high energy costs for the households.

Propriety fragmentation, low awareness of homeowners of energy efficiency improvements and benefits and the low skills of multi-family building managers can be considered among the main barriers to buildings retrofitting and, thus, to energy-led urban regeneration. Energy efficiency renovation rates have been positive but negligible, namely an annual average of 0.5%. Nevertheless, the potential for improving building energy performance is still substantial, particularly in the field of public housing (Copiello 2015), where the housing stock is generally characterized by increasing age and poor energy efficiency. In this situation, more than a half of the public housing stock, about 500,000 public dwellings, belong to the three lowest EU energy labels, where households spend more than 10% of their income on energy costs (Federcasa 2015).

Conversely, the renovation of the social housing stock has not only an energy-saving purpose, but it also enhances the role of energy efficiency in combination with the social and economic co-benefits (e.g., poverty alleviation and health improvements), thus contributing to avoiding stigmatization and social segregation and to reducing energy poverty.

However, as pointed out by an increasing number of scholars from various disciplines (Janda 2009; Feng et al. 2016; Della Valle et al. 2018), an improvement in energy efficiency does not automatically lead to the expected energy savings; this is due to the fact that the occupants are the ones who use energy in the buildings. In fact, energy savings through behavioral factors can be as high as those from technological ones (Lopes et al. 2012), thus giving the occupants the possibility either to reinforce the savings from energy efficiency measures or to waste them. Behavior is more likely to be deliberately considered and changed when a discontinuity occurs in the household context (Huebner et al. 2013). Therefore, building renovation programs

---

[1]In 1990 the first Italian energy efficiency regulation was adopted, although it was limited only to providing prescriptions for the heating systems in new residential buildings.

offer a key opportunity to involve households to convince them to reconsider their consumption practices (Santangelo et al. 2018).

Better energy performance of the housing stock can have a significant reduction in the amount of energy bills and can contribute to alleviating energy poverty. Nevertheless, as just framed, building characteristics alone do not explain the energy consumption, and energy behavior and practices should be taken into account, too. Energy-efficient technologies, together with awareness of energy behavior and information on how to reduce the energy consumption, have the potential to alleviate energy poverty. Nevertheless, as noted by Ürge-Vorsatz and Tirado Herrero (2012), due to a number of barriers (i.e., relatively long payback times, restricted access to credit, a lack of appropriate financing schemes, low awareness of decision-makers about the alternatives and split incentives between tenants and owners), greater efficiency is not often usually achieved in spite of the corresponding larger societal benefits. Smart metering, a growing priority of EU energy policy, holds the potential to combat energy poverty to some extent, although it also requires careful considerations of the role of user behavior (Darby 2012).

Low-carbon urban and regional development policies also proffer significant opportunities for energy poverty reduction. In Europe, the Covenant of Mayors currently requires the signatory municipalities to commit not only to take action on mitigating climate change and adapting, but also on alleviating energy poverty. Therefore, due to the great number of Italian municipalities being part of the Covenant of Mayors initiative, how to tackle energy poverty can be considered a predominant challenge of Italian cities, despite the fact that competences and awareness concerning how to deal with this issue are still limited. In addition to this, it is worth noting that public housing, social housing and housing policies have always been studied as issues contributing to achieve sustainability in cities. Therefore, demonstrating—first in the public housing stock even before the rest of the housing stock—the relevance of household energy behavior to effectively reduce energy consumption in buildings and, in turn, to contribute to the reduction of energy poverty, is the focus of this research.

The aim of this paper is to provide an insight on building energy simulation and occupant behavior modeling as powerful tools to support policymakers and planners in setting strategies to address energy poverty while improving energy efficiency. The Italian public housing sector is taken as a reference to investigate to what extent understanding the human factor can contribute to alleviating energy poverty. To do so, this research investigates the impact of occupant behavior to reduce energy consumption at the household level.

1.1 The Italian Public Housing Stock

In the European context, Italy emerges as one of the countries where social housing is less developed, and, more generally, public expenditure for housing is lower: Only 0.1% of total social expenditure is devoted to housing, less than 1% of the total expenditure of EU-28 (Eurostat 2018). Less than 4% of all households live in social housing dwellings, with the public sector having almost the exclusive role in providing social housing. Therefore, Italian social housing can be almost totally identified with public housing.

Today, Italian public housing is managed by public housing providers, either at the regional or municipality level, depending on the regional legislation. Renting is by far the most prevalent tenure (i.e., almost 95% of the total dwellings) (Nomisma 2016). However, assisted home-ownership has always been part of public housing schemes through various forms of leasing contracts (Baldini and Poggio 2014). At the end of 2013, the total available stock accounted for about 805,000 housing units; however, only 86% was properly allocated, 5% less than in 2004, resulting in an increasing share both in the number of squatters and in the number of vacant dwellings due to insufficient maintenance conditions. Fifty-five percent of public housing tenants pay less than 100 euros per month, while the rate of households paying a rent higher than 300 euros is only about 7% (Nomisma 2016). 45% of the total public housing stock is located in the 12 largest urban areas, where higher housing demand is concentrated.

Access to the Italian public housing is related to one's income level, and the selection criteria for economic and social needs are tight. The stock is allocated on the basis of availability and size in relation to the household composition. However, low-income households may experience significant differences in energy expenditures depending on the energy efficiency of their dwellings, having in turn consequences on energy poverty and causing both horizontal and vertical inequality (Santangelo and Tondelli 2017). The turnover rate of social tenants is extremely low, and most of the occupants tend to remain in public housing for their entire lives, regardless of improvements in their economic and social conditions and changes in the household size. This propensity for stability, together with the reduction in the construction of new dwellings, explains the progressive aging of occupants. Elderly people, in fact, are the main beneficiaries, with more than 44% of household heads being 65-years old or older, while the share reaches 28% for those 75-years old or older (Nomisma 2016). These figures are significantly higher in the public housing stock than in the residential sector as a whole, where the share of household heads 65-years old or older and 75-years old or older decreasing, respectively, to 35 and 19% (ISTAT 2012).

Although housing needs have constantly increased, both in terms of affordability and quality of the housing stock, and despite the fact that about 650,000 households across the entire country are subscribed to waiting lists to access public housing (Federcasa 2015), the public housing sector has continued to shrink since the 1990s. The lack of constant resources to be used to manage, maintain, and refurbish the existing stock and for the creation of new housing stock, together with cuts in public investments and the privatization plans for public housing stock, are only some of the structural issues that affect the public housing sector (Almadori and Fregolent 2020). The traditional rent-setting model raises, as well, financial sustainability issues. It operates by setting a rent to cover costs and subsequently to discount it according to the tenants' income. All the aforementioned issues result in a public housing model that is far from being sustainable.

2 Methodology and Assumptions

2.1 Methodology

A multi-family public housing building in Bologna, the capital city of Emilia-Romagna Region, is assumed as a case study to estimate the influence of three dimensions linked to occupant behavior—management of the thermostat, management of the heating system and variation of building characteristics—on energy heating consumption. The reference building is a ten-story building[2] (ground floor on *pilotis*) built in 1976, with five staircases and 92 housing units. Each story hosts from 6 to 11 housing units, while their sizes vary from 65 m² for the smallest housing unit—kitchen, living room, bedroom, bathroom, and corridor—to 116 m² for the largest ones—kitchen, living room, four bedrooms, two bathrooms, and corridor. The total heating consumption for the year 2011 was measured as 552,864 kWh. The high level of heating consumption was mainly due to the poorly insulated building materials and the lack of thermostats that would enable the occupants to regulate the temperature according to their presence and needs.

A building performance simulation tool was applied to investigate the impact of behavior and to build scenarios to inform decision-makers. The case study is modeled using the dynamic building simulation program DeST (Yan et al. 2008; Zhang et al. 2008). The model[3] was previously applied by Santangelo et al. (2018) to investigate the impact of the human factor in energy consumption at building scale and to build scenarios to support decision-makers regarding design renovation strategies to apply to the public housing sector.

2.2 Basic Assumptions

The reference building has a centralized heating system which operates for 14 h per day during the reference period (i.e., from the 15th of October to the 15th of April, according to E climatic zone[4]), from 6 to 9 am, and from 11 am to 10 pm. Since the measured consumption refers to 2011, the Bologna meteorological data of the

---

[2]Although this type of building has a higher density than the average Italian housing stock—characterized by small and low-rise multi-family buildings—it represents well the public housing stock, which from the 1960s has been the result of ad hoc urban planning instruments named *"Piani per l'Edilizia Economica e Popolare (PEEP)"*.

[3]Further details on the model and its validation are described in Santangelo et al. (2018).

[4]According to the legislative decree 412/93, the Italian municipalities are divided into six climatic zones (i.e., from A to F) based on the energy consumption needed to maintain a comfortable temperature inside the building equalling to 20 °C. The degree-day is the unit used to assign a climatic zone to each municipality. It is the sum, extended to all days in a conventional annual heating period, of positive differences between interior temperature (conventionally fixed at 20 °C) and the mean daily external temperature. Degree-day for E climatic zone varies from over 2100 3000.

entire 2011 are used for the analysis of the heating consumption. When it comes to internal gains, since the Italian average household size is 2.4 persons (ISTAT 2015), the number of people assumed for each housing unit in the model is the statistical number rounded up to three people. The thermal characteristic values of main building elements have been considered as representative of the residential buildings built between 1960s and 1980s in Italy, particularly in the public housing sector where the majority of the stock was built of a combination of precast and construction materials with poor thermal properties. Window surfaces are larger than the average for similar buildings, and windows have aluminum frames with single pane of glass, resulting in a high heat-dissipating factor.

Due to the unavailability of actual occupancy data, three different occupancy preferences (OPs) have been considered in order to take into the account variety in occupancy patterns. Each OP is representative of the number of hours that households decide to spend at home with the heating system on and the thermostat at the highest preferred temperature. OP1 considers tenants setting the heating system at the preferred temperature for 14 h per day (from 6 to 9 am and from 11 am to 10 pm), which decreases to 11 h per day for OP2 and eight hours per day for OP3.

Two extreme cases have been taking as a reference. In the first case called "no control," the central heating system of the reference building is turned on for 14 h per day and occupants are not allowed to switch on/off the heating system, while they may choose the temperature set-point and the time to spend with their preferred set-point, among the 14 h that the system is on. They choose to adopt an "energy intensive behavior," with the temperature set at 22 °C in all the rooms for the maximum number of hours defined by the occupancy preferences. In the second case called "full control, dual set-point," households decide to switch off the heating system when away from home and not to use some of the rooms (e.g., bedrooms during the day), in accordance with the assigned occupancy reference. In addition, they adopt "energy-saving behavior," setting the temperature to 20 °C when the heating system is operating, and the rooms are occupied.

These two cases were simulated according to three different building characteristics. "No retrofit" applies when no physical intervention to the building elements is foreseen. "Limited retrofit" foresees the replacement of the single-glass windows with double-glass windows and the insulation of the roof. "Total retrofit" is the deepest level of retrofit considered, when all the external elements are renovated, and the building is completely retrofitted. The heat transmission values (U) considered are the minimum requirements for the renovation of the existing buildings in the municipality of Bologna, according to the regional regulation (Regione Emilia-Romagna 2015).

The building energy simulations and occupant behavior model previously presented and already applied by Santangelo et al. (2018) to get results on energy behavior at building scale, provide also significant results at dwelling scale. These results involving dwellings with different characteristics have not been studied before; therefore, they are taken into consideration in this research to give a novel contribution about the role of occupant behavior in tackling energy poverty, an issue

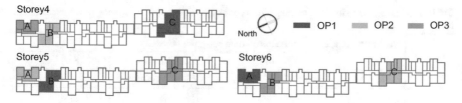

Fig. 1 Housing units and occupancy preferences distribution. *Source* author's elaboration

that has recently received attention also from other scholars (Kearns et al. 2019; Della Valle 2019).

Three levels of renovation have been considered to check whether the variations among the dwelling occupancy and types of dwelling remain stable or not as the energy efficiency of the dwelling changes.

Three different dwelling sizes have been considered. They are located in mid-stories (i.e., fourth, fifth, and sixth stories) to limit the influences of the heating loads due to the poor thermal properties of floors and roofs. Housing unit A is the smallest among the ones considered: The total net size is 65 m^2 with one bedroom, kitchen, living room, bathroom, and a corridor. It faces north and east at the building corner, with two external walls, and one of the internal walls next to the building staircase. The occupancy preference is foreseen to vary among the different stories, and therefore, OP1 is assigned to housing unit A at story 6 and OP2 at story 5, while it is OP3 at story 4 (Fig. 1). Housing unit B is located on the opposite side of the staircase: It is 79 m^2 in area, and it has two bedrooms, kitchen, living room, bathroom, and a corridor. The kitchen and one of the bedrooms face east, while the living room and the other bedroom face west. OP1 is allocated to housing unit B at story 5, OP2 at story 4 and OP2 at story 6. The final dwelling considered, housing unit C, has an area of 99 m^2, with one more bathroom and bedroom than housing unit B. It is an internal unit with only two external walls facing east and west, while internal walls border on the one side the staircase and on the other side a specular housing unit.

3 Results

Table 1 shows the heating load simulation results according to the different housing units and different occupancy preferences, also taking into consideration the building retrofit level. Similarly to what resulted from the investigation at building scale (Santangelo et al. 2018), the analysis at dwelling scale confirms that the more energy efficient the dwelling is, the greater is the share of the energy consumption that occupant behavior can affect: up to 52% of energy reduction for housing unit B with occupancy preference OP3 (i.e., "back home in the afternoon" with the heating system at the maximum preferred temperature for 8 h per day) in the case of total

Table 1 Heating load simulations related to types of dwelling and occupancy preferences (OPs) for the three levels of renovations

	No control (Case 1)		Full control, dual set-point (Case 2)					
			OP1 (14 h)		OP2 (11 h)		OP3 (8 h)	
	kWh/y	kWh/m²y	kWh/y	kWh/m²y	kWh/y	kWh/m²y	kWh/y	kWh/m²y
No Retrofit			−11–14%		−12–15%		−17–21%	
Housing unit A	4244	65.3	3779	58.1	3748	57.7	3426	52.7
Housing unit B	3667	46.4	3167	40.1	3120	39.5	2891	36.6
Housing unit C	4709	47.6	4184	42.3	4157	42.0	3896	39.4
Limited Retrofit			−15–23%		−16–24%		−23–29%	
Housing unit A	2241	34.5	1912	29.4	1886	29.0	1715	26.4
Housing unit B	1036	13.1	794	10.1	786	9.9	739	9.4
Housing unit C	1558	15.7	1252	12.6	1213	12.3	1135	11.5
Total Retrofit			−31–51%		−35–51%		−39–52%	
Housing unit A	356	5.5	247	3.8	232	3.6	217	3.3
Housing unit B	305	3.9	148	1.9	148	1.9	146	1.8
Housing unit C	311	3.1	208	2.1	200	2.0	178	1.8

retrofit. However, the no-retrofit scenario shows the highest potential for occupants to decrease the dwelling heating load in absolute terms: up to 818 kWh/year of reduction for housing unit A with OP3 by adopting an energy saver behavior, with full control and a dual set-point.

As expected, results also show that the higher the exposure to external conditions due to the location at the building corner, the higher is the heating load per square meter yearly. In fact, housing unit A can require more energy than larger dwellings such as housing units B and C, with the same building- component thermal proprieties, but located in a more favorable position within the building. In case of a limited retrofit, the simulated heating load for housing unit A is 2.2 times higher than the one for housing unit B, and 2.6 times more than the one for housing unit C.

The occupancy preferences play also a role in the definition of the dwelling energy demand. Regardless of the retrofit level of the dwelling, there is a common tendency for an energy demand decrease as the number of hours the heating system is operating decreases. Taking into account Case 2, where households adopt an energy saver behavior compared to Case 1, results show that, in the case where no retrofit occurs, the energy reduction is in the range of 11–14% for OP1, 12–15% for OP2, and 17–21% for OP3, depending on the housing unit considered. Energy savings increase as the energy efficiency of dwellings increases. Therefore, they are in the range of 15–23% for OP1, 16–24% for OP2, and 23–29% for OP3 when the limited retrofit occurs; and 31–51% for OP1, 35–51% for OP2, and 39–52% for OP3 when the dwellings are totally retrofitted. However, for all the three retrofit levels and all the dwellings considered, less time spent at home with the heating system on generates higher energy savings.

When it comes to the rebound effect identification, the investigation at dwelling level shows that it can be even higher than the one identified with heating load simulations at building level as presented in Santangelo et al. (2018). In fact, if occupants just decide to change the thermostat from 20 °C in the day-zone and 18 °C in bedrooms when occupied to 22 °C in all rooms for the maximum working hours of the heating system (i.e., from Case 2 to Case 1), then the consumption increases, and the rebound effect reaches up to 23%, 29% and 27%, respectively, for limited retrofits of housing unit A, B, and C. When the total retrofit occurs, the rebound effect can increase further up to 39%, 52%, and 43%, respectively, for housing unit A, B, and C, in comparison with the expected consumption after the renovation.

4 Discussion

Occupant behavior modeling is an effective tool to make explicit the impact of the human factor on the energy saving potential. It also enables quantifying the impact of the rebound effect in case of a building retrofit, as well as the impact of "green behavior" (Ben and Steemers 2014) when positive behavioral changes occurs. The findings at dwelling level show that impact of occupant behavior can reach up to 52% of energy savings for the case study considered. As for the analysis at building scale (Santangelo et al. 2018), this impact increases as the retrofit level increases in terms of percentage on the total heating consumption, while the energy savings due to behavior change decreases as the retrofit level increases. Despite that the results are not surprising, they deserve to be outlined since they show variance in the energy saving potential of different households living in different dwellings. Therefore, it is of key importance to understand a household's saving potential in relation to the dwelling characteristics they lived in and to avoid "blaming the victim" (Stevenson and Leaman 2010), especially in the public housing sector where the random allocation of dwellings generates inequalities in energy costs and energy poverty conditions. Indeed, considering three different dwellings that have in turn different occupancy preferences (OPs) shows that, income and energy poverty conditions being equal, people living in public housing units experience differences in energy demands and, therefore, in energy bills. This is evident not only for households living in dwellings located in buildings with different characteristics, but also for those living in the same multi-family building but on different floors and in different apartments. Due to the fact that public housing is allocated according to the income level and the household size, but not the housing unit characteristics, and the social rent is calculated mainly according to the income rather than the energy efficiency of the dwellings, inequalities among low-income families might be exacerbated. Therefore, strategies addressing energy poverty should take into account the challenges linked to the allocation system and the rent calculation system.

When limited resources are available, investing in tackling behavior change through information provision and education, rather than building retrofit, has resulted in a convenient strategy to implement energy efficiency while addressing energy poverty. However, as soon as other funds become available, it is important to combine informative and feedback strategies with the renovation of the building. While the former affects the energy consumption and, in turn, energy poverty, the latter has the potential to lead to higher energy savings. Especially for those dwellings located on the first or top floor, any retrofit could improve the household energy poverty situation considerably more than in the other cases, while the influence of the household behavior might be more limited. The combination of informative strategies and retrofit solutions should be considered the most favorable option, one the one hand, to maximize energy saving and on the other hand to avoid overestimations of the saving potential of retrofitting solutions due to underestimations of the occupant behavior impact (Santangelo and Tondelli 2018). Moreover, the combination of such measures is expected to reduce both the gap between expected and actual heating consumption and the rebound effect. However, as pointed out by Della Valle (2019), these measures implicitly rely on the rational choice model of decision making, while extensive evidence has shown that the capacity to make rational decisions depends on situational factors that should be studied by applying behavioral economics. In addition to this, it should be noted that, although the rebound effect—as the offset of the beneficial effects of energy retrofitting—increases when the energy efficiency of buildings increase, the building retrofit may alone reduce the energy savings gained thanks to a greener behavior due to the higher impact of retrofit solutions on energy saving compared to the ones addressing behavior only.

Even more than the Jevons' paradox (Alcott 2005) explaining the rise of quantity demanded although the promotion of different types of measures should result in higher energy efficiency and saving, it is actually the prebound effect that is the one more closely linked to energy poverty. It can be framed as households using less energy than foreseen due to an energy poverty condition and low awareness of the use of energy efficient technology. The prebound effect can to some extent explain the reason why public housing providers are reluctant to implement behavior change strategies, or to involve Energy Service Companies (ESCo) in the renovation of the public housing stock, since there is a limited knowledge on how to deal with people experiencing energy poverty conditions. To this aim, more effort has to be invested to show that the prebound effect and energy poverty condition are interlinked with housing policies and housing-welfare strategies. In fact, in order to both ensure equality among public housing tenants, and to tackle energy poverty, housing policies cannot be seen as apart from the energy performance of the housing stock. Other measures related to the housing sphere (e.g., rent re-calculation according to the energy demand of the housing unit), as well as involving health, education or transport services should be considered. Any single solution cannot address a multi-faceted issue such as energy poverty.

5 Conclusions and Policy Implications

This research has contributed to show that occupant behavior is worth being investigated when it comes to energy poverty. In the case of the Italian public housing sector—which is allocated to low-income households matching certain characteristics—when the income and family size are equal, households living in dwellings with different physical and thermal characteristics experience differences in energy bills. This situation creates inequalities that can lead to the exacerbation of energy poverty conditions, and, therefore, they need to be taken into account by municipalities.

Energy poverty policies should embed household behavior in their considerations as a driver to alleviate the energy poverty conditions, in particular, in the absence or scarcity of structural policies. Among the types of policy instruments widely used in Europe to tackle energy poverty (Kyprianou et al. 2019), energy savings measures—including energy efficiency and renewable energy sources—and information provision should both include references and specific measures to address behavior and mitigate energy poverty.

The extreme fragmentation of the housing demand, the increasing energy poverty of tenants living in public buildings, the overall decline of public spending on the housing sector and the aging of the building stock are only some of the elements describing the Italian housing issue that need to be tackled when structuring strategies to alleviate energy poverty. New conceptual and methodological advancements are needed in order to tackle the housing issue, where the provision of information, awareness raising and new skills can be the drivers not only of the renovation of the public housing stock, but also for a new housing policy based on providing housing as a service, rather than just an asset.

Embedding the energy poverty issue into the ones driving the decision to renovate the housing stock can enhance social sustainability and livability in cities. Urban planning and energy-led regeneration strategies are increasingly seeking criteria to renovate the existing housing stock, where the energy saving is only one of the benefits that cities aim at achieving. The Covenant of Mayor is certainly pushing the transition of cities toward agents for the reduction of energy poverty. However, more research in the field of urban policies and urban planning is needed in order to understand how to integrate energy poverty measures into the city planning systems.

Acknowledgements The authors of this paper received funding from the EU–China "IRES-8" Mobility Project under EU Contract Number No. ICI+ /2014/347-910. The contents reflect only the authors' view, and the European Union is not liable for any use that may be made of the information contained therein.

References

Alcott B (2005) Jevons' paradox. Ecol Econ 54:9–21

Almadori A, Fregolent L (2020) Condizioni, pratiche e prospettive degli enti gestori dell'edilizia residenziale pubblica. In: Urban@it, Quinto Rapporto sulle città. Politiche urbane per le periferie. Il Mulino, Bologna

Baldini M, Poggio T (2014) The Italian housing system and the global financial crisis. J Housing Built Environ 29:317–334. https://doi.org/10.1007/s10901-013-9389-7

Ben H, Steemers K (2014) Energy retrofit and occupant behaviour in protected housing: a case study of the Brunswick centre in London. Energy Build 80:120–130. https://doi.org/10.1016/j.enbuild.2014.05.019

Copiello S (2015) Achieving affordable housing through energy efficiency strategy. Energy Policy 85:288–298. https://doi.org/10.1016/j.enpol.2015.06.017

Darby SJ (2012) Metering: EU policy and implications for fuel poor households. Energy Policy 49:98–106. https://doi.org/10.1016/j.enpol.2011.11.065

Della Valle N (2019) People's decisions matter: understanding and addressing energy poverty with behavioral economics. Energy Build 204:109515. https://doi.org/10.1016/j.enbuild.2019.109515

Della Valle N, Bisello A, Balest J (2018) In search of behavioural and social levers for effective social housing retrofit programs. Energy Build 177:91–96. https://doi.org/10.1016/j.enbuild.2018.05.002

Eurostat (2018). Government expenditure on social protection. Data refer to 2016. https://ec.eur opa.eu/eurostat/statistics-explained/index.php?title=Government_expenditure_on_social_pro tection#Expenditure_on_.27social_protection.27. Accessed 13th Feb 2020

Federcasa (2015) L'Edilizia residenziale pubblica: Elemento centrale della risposta al disagio abitativo e all'abitazione sociale. Federcasa, Rome

Feng X, Yan D, Wang C, Sun H (2016) A preliminary research on the derivation of typical occupant behavior based on large-scale questionnaire surveys. Energy Build 117:332–340. https://doi.org/10.1016/j.enbuild.2015.09.055

ISTAT (2010) L'abitazione delle famiglie residenti in Italia. Anno 2008

ISTAT (2012) 15° Censimento della popolazione e delle abitazioni 2011

ISTAT (2014) I consumi energetici delle famiglie italiane. Anno 2013

ISTAT (2015) Annuario Statistico Italiano 2015

Huebner GM, Cooper J, Jones K (2013) Domestic energy consumption—what role do comfort, habit, and knowledge about the heating system play? Energy Build 66:626–636. https://doi.org/10.1016/j.enbuild.2013.07.043

Janda KB (2009) Buildings don't use energy: people do. In: PLEA 2009—26th conference on passive and low energy architecture. Quebec City Canada, 22–24 June 2009

Kearns A, Whitley E, Curl A (2019) Occupant behaviour as a fourth driver of fuel poverty (aka warmth and energy deprivation). Energy Policy 129:1143–1155

Kyprianou I, Serghides DK, Varo A, Gouveia JP, Kopeva D, Murauskaite L (2019) Energy poverty policies and measures in 5 EU countries: a comparative study. Energy Build 196:46–60. https://doi.org/10.1016/j.enbuild.2019.05.003

Lopes MAR, Antunes CH, Martins N (2012) Energy behaviours as promoters of energy efficiency: a 21st century review. Renew Sustain Energy Rev 16:4095–4104. https://doi.org/10.1016/j.rser.2012.03.034

Nomisma (2016) Dimensione e caratteristiche del disagio abitativo in Italia e ruolo delle Aziende per la casa. Documento di sintesi. https://www.federcasa.it/allegati/KZvkdtbvyC8. Accessed 12th Feb 2020

Regione Emilia-Romagna (2015) Deliberazione della giunta regionale 20 luglio 2015, n. 967. Approvazione dell'atto di coordinamento tecnico regionale per la definizione dei requisitiminimi di prestazione energetica degli edifici. Available at: https://energia.regione.emilia-romagna.it/leggi-atti-bandi-1/norme-e-atti-amministrativi/certificazione-energetica/certificazione-energe tica/dgr-967_2015-_allegati. Last access: 30th April 2020

Santangelo A, Tondelli S (2017) Equità e qualità degli interventi di rigenerazione del patrimonio ERP: dallo studio del caso olandese, verso la definizione di un modus operandi. In: XX Conferenza Nazionale SIU Urbanistica E/È Azione Pubblica. La Responsabilità Della Proposta. Planum Publisher, Rome 12–14 June. 414–419

Santangelo A, Tondelli S (2018) Embedding energy user's behaviour into multi-criteria analysis: providing scenarios to policy-makers to design effective renovation strategies of the housing stock. In: Proceedings of 54th ISOCARP Congress Bodø, Norway, October 1–5, 2018, Cool planning: changing climate and our urban future. The Hague, 1414–1424

Santangelo A, Yan D, Feng X, Tondelli S (2018) Renovation strategies for the Italian public housing stock: applying building energy simulation and occupant behaviour modelling to support decision-making process. Energy Build 167:269–280. https://doi.org/10.1016/j.enbuild.2018.02.028

Stevenson F, Leaman A (2010) Evaluating housing performance in relation to human behaviour: new challenges. Build Res Info 38:437–441. https://doi.org/10.1080/09613218.2010.497282

Ürge-Vorsatz D, Tirado Herrero S (2012) Building synergies between climate change mitigation and energy poverty alleviation. Energy Policy 49:83–90. https://doi.org/10.1016/j.enpol.2011.11.093

Yan D, Xia J, Tang W, Song F, Zhang X, Jiang Y (2008) DeST—An integrated building simulation toolkit Part I: fundamentals. Build Simul 1:95–110. https://doi.org/10.1007/s12273-008-8118-8

Zhang X, Xia J, Jiang Z, Huang J, Qin R, Zhang Y, Liu Y, Jiang Y (2008) DeST—An integrated building simulation toolkit Part II: applications. Build Simul 1:193–209. https://doi.org/10.1007/s12273-008-8124-x

Energy Retrofitting in Public Housing and Fuel Poverty Reduction: Cost–Benefit Trade-Offs

Chiara D'Alpaos and Paolo Bragolusi

Abstract The Italian housing stock is one of the least energy-efficient in Europe. The residential sector accounts for 36% of primary energy consumption, and nearly 76% of Italian dwellings were built before 1981 (49% are more than 50 years old). According to the Italian Ministry of Economic Development, almost 90% of the Italian building stock exhibits an excessive energy demand. This condition widely affects public properties and specifically public housing. Due to the lack of financial resources, public-housing energy retrofitting is nowadays a critical issue in Italy. Nonetheless, energy retrofitting may play a key role in reducing fuel poverty, especially in public-housing contexts where there is a convergence of factors aggravating it: low-income households, which cannot afford high energy prices, and poor energy efficiency of homes due to the lack of insulation and/or inefficient heating systems. Although the urgent need for investments in the improvement of public-housing energy performance to comply with the Directive 2010/31/EU (recast in 2018—Directive 2018/844/EU) is widely recognized, the spread of good practices is strongly hampered by their cost-effectiveness and the split incentive issue. The aim of the paper is to provide a methodological framework to identify cost-effective and cost-optimal strategies of intervention that match technological advancements and knowledge in energy retrofitting, with both social and environmental needs and end-users behavior. We take into consideration the adoption of basic energy efficiency measures involving the building envelope, HVAC and domestic hot-water systems. We compare costs (e.g., investment and operating costs) and benefits (e.g., energy cost savings) and determine how far and how much it is optimal to push on retrofitting of public-housing assets. To solve this issue, we proposed to split monetary benefits due to energy savings between landlords and tenants to find a compromise solution that accounts for both landlords' and tenants' interests. Our results show that installation of building envelope thermal insulation guarantees paying back investment costs as quickly as possible. Nonetheless, this investment is not the most profitable

C. D'Alpaos (✉) · P. Bragolusi
ICEA Department, Via Venezia 1, 35131 Padua, Italy
e-mail: chiara.dalpaos@unipd.it

P. Bragolusi
e-mail: paolo.bragolusi@unipd.it

from tenants' perspective, due to relatively small monetary benefits obtained after a discounted payback period. From the tenants' perspective, the replacement of existing boilers and the installation of building envelope thermal insulation proved to be the most profitable with respect to our split incentive hypothesis.

Keywords Public housing · Fuel poverty · Buildings energy retrofitting · Global cost · Discounted payback period

1 Introduction

The Italian Action Plan for Energy Efficiency (ENEA 2017) and the Italian Integrated National Plan for Energy and Climate 2020 (MISE 2020) promote energy efficiency of buildings as a key driver of environmental sustainability and energy cost reduction, especially for low-income households. According to recent estimates (CRESME 2018), the residential sector in Italy, which accounts for more than 12.2 million buildings corresponding to 31.2 million dwellings, is responsible for 36% of primary energy consumption as most of the assets (i.e., 70%) were built before the entry into force of first regulation on buildings energy performance (Act n. 373/1976).

Almost 90% of the Italian building stock exhibits excessive energy demand, and this condition largely affects public properties and specifically public housing (D'Alpaos and Bragolusi 2018b, 2019; Bottero et al. 2019). In this respect, improvement in energy efficiency of buildings through energy-efficient renovations is a crucial issue in Italy, where the existing stock has a great potential for energy savings, sustainable development and reduction of greenhouse gas (GHG) emissions (Ferrari and Beccali 2017; Mangialardo et al. 2018; Mauri et al. 2019; Mutani et al. 2019; Jafari et al. 2019).

In addition to changes in end-user behavior, energy saving and buildings efficiency may be significantly increased by implementing energy-efficient technologies (EETs), the selection of which involves the solution of an optimization problem (Knobloch et al. 2019). Starting from a set of implementable EETs, which define a set of feasible buildings energy retrofit projects (BERPs), the objective is to determine the cost-optimal EET in a cost-effective perspective. This is a complex issue which requires to consider all related costs through the life cycle of the building, complies with energy efficiency performance standards set by laws and regulations and ensure acceptable thermal comfort levels (Ma et al. 2012; D'Alpaos 2021).

Building energy retrofitting (BER) is a critical issue in Italy due to stringent public budget constraints and lack of financial resources. Nonetheless, energy-retrofit may play a key role in reducing fuel poverty, especially in public-housing contexts, where there is a convergence of factors accruing it: low-income households, which cannot afford high energy prices, poor energy efficiency of homes due to lack of insulation and/or inefficient heating systems and social exclusion (Vilches et al. 2017; Littlewood et al. 2017; D'Alpaos and Bragolusi 2019). In the public-housing sector, in fact, specific problems sum up to common problems such as the landlord–tenant dilemma (Ástmarsson et al. 2013; Monteiro et al. 2017). The landlord–tenant dilemma represents one of the larger barriers in BER implementation. It is a split-incentive issue, which arises when the interests of homeowners (i.e., landlords) and tenants misalign and agents involved do not reach consensus on renovation strategies (IEA 2008; Ástmarsson et al. 2013; Brown 2015; Monteiro et al. 2017). In residential rentals, tenants usually pay for energy consumption (i.e., energy bills), and they might be strongly interested in property energy renovations, whereas landlords, who do not pay for energy consumption and are not affected by cost-saving concerns, might not be willing to invest and pay high upfront investment costs. In other words, tenants may gain immediate benefits (i.e., reduction in energy costs) and do not pay renovation costs, whereas landlords pay renovation costs and may obtain benefits in the long-run (e.g., a property market value increase). In addition, tenants are not the property owners, and usually, they not conceive the idea to fully or partially finance investments, which they do not capitalize in terms of property market-price premiums (Canesi et al. 2016; D'Alpaos and Bragolusi 2018a, 2018b, 2019; Mangialardo et al. 2018, Dell'Anna et al. 2019; Bottero et al. 2020). The split-incentive issue, which underpins the landlord-tenant dilemma, plays therefore a major role in preventing investments in energy renovations, and it becomes more crucial when financial resources are scarce and budget constraints are stringent, as in public-housing contexts.

Although to comply with the Directive 2010/31/EU (recast in 2018—Directive 2018/844/EU), the urgent need for investments in improvements in public-housing energy performance is widely recognized, the spread of good practices is strongly hampered by their cost-effectiveness.

The Italian Government Decree n. 192/2005, which transposed Directive n. 2002/91/EC, and the Commission Delegated Regulation (EU) n. 244/2012, was established to implement the Global Cost (GC) method, also known in the literature as Life Cycle Cost (LCC), for the valuation of BER investments. For each implementable BERP, it is necessary to estimate its GC, which represents the present value of investment costs paid to install EETs, running costs (energy, operating and maintenance costs), and replacement costs, as well as disposal costs (if applicable). BERPs, which minimize GC, are "cost-optimal" solutions to be selected and implemented.

Monetary valuation of co-benefits generated by BER, such as improvements in indoor air quality and thermal comfort, improvements in aesthetic appearance of the building and increases in social inclusion and reputation, challenges BERPs economic valuation (Banfi et al. 2008; Cappelletti et al. 2015; Ferreira and Almeida 2015; D'Oca et al. 2018). These benefits and co-benefits may be key drivers in BER investment decisions.

In the light of these considerations, the adoption of the GC method to select cost-optimal BERPs should therefore be integrated with other economic evaluation methodologies, such as the Discounted Payback Period (DPB) and the Net Present Value (NPV) rules and Stated Preference Approaches (Guardigli et al. 2018; Alberini et al. 2018). With respect to public housing, the valuation of these co-benefits and of the trade-offs between costs and benefits of renovations may contribute to solve the split incentive and the fuel poverty issues and accelerate investments.

In this respect, a recent strand of literature has focused on the economic analysis of BER in public-housing contexts. As an example, Preciado-Pérez and Fotios (2017) implemented the NPV rule to evaluate the cost-effectiveness of specific EETs in a social housing case study in Mexico. Gillich et al. (2018) analyzed the US Better Buildings Neighborhood Program (BBNP) and identified best-practice principles associated with each program step, which provide a template for an optimal program design for retrofit intervention plans with stated objectives similar to the US BBNP. Guardigli et al. (2018) proposed a decision support system (DSS) to select optimal EETs, by taking into account their economic-sustainability and energy efficiency targets. In detail, they integrated the GC method with the Net Present Value (NPV) and the Pay Back Period (PBP) rules to support ownership decisions on EETs selection. D'Alpaos and Bragolusi (2019) provided a multiple criteria model, based on the Analytic Hierarchy Process (AHP), to identify relevant key factors in BER and rank (from best to worse) alternative EETs. According to their findings, the thermal insulation of the building envelope was determined to be the most important EET because it contributes to minimizing operating and maintenance costs and occupants' decanting costs, while ensuring good indoor thermal comfort and external noise reduction (D'Alpaos and Bragolusi 2019). de Feijter et al. (2019) investigated the characteristics of retrofit governance frameworks aimed at householders' inclusion and strategic improvements for intermediation. More recently, Rolim and Gomes (2020) presented a methodology to evaluate the impact of retrofit measures, which considers citizens as crucial retrofit stakeholders; whereas Sangalli et al. (2020) argued that the techno-centric approach, usually adopted during works, tends to exclude occupant involvement and information and leads to occupants' behavior not consistent with optimal uses of building services and components[1].

This paper contributes to this strand of the literature. We analyze different EETs to be implemented in public housing, evaluate their impact on buildings' energy performance and determine the relative cost–benefit trade-offs. The aim of the paper is to provide a methodological framework to identify cost-effective and cost-optimal strategies of intervention that match technological advancements and knowledge in energy retrofitting with both social and environmental needs and end-users behavior. We take into consideration the adoption of basic energy efficiency measures involving the building envelope, heating, ventilation, and air-conditioning (HVAC) systems and domestic hot-water systems. We compare costs (e.g., investment and operating costs) and benefits (e.g., energy cost savings) and determine how far and how much it is optimal to push the retrofitting of public-housing assets.

The remainder of the paper is organized as follows. Section 2 provides the methodological framework; Sect. 3 illustrates the case study; in Sect. 4, results are presented and discussed, and Sect. 5 summarizes and concludes.

---

[1]Occupants' behavior may in fact significantly affect building energy consumption and generate uncertainty in energy-use predictions (Haas et al. 1998; Sunikka-Blank and Galvin 2012; Hopfe et al. 2013; Hong et al. 2016; Gucyeter 2018; Laaroussi et al. 2019; Ali et al. 2020).

2   Methodological Framework

One of the main deterrents of BERPs implementation are the high upfront invest-ment costs to install EETs and their long payback period (Monteiro et al. 2017; Penna et al. 2019). To tackle the issue of financial barriers and promote investments in BER, over the last decade governments introduced incentive schemes that targeted specific home renovations and equipment. Starting in 2006, the Italian government has introduced fiscal incentive programs to favor investments in improvement of the energy efficiency of existing buildings (D'Alpaos and Bragolusi 2018b; Bottero et al. 2019; D'Alpaos 2021). According to these policy incentives, homeowners (i.e., land-lords) can deduct from their income taxes a fixed percentage of the expenses incurred to implement specific EETs (Agenzia delle Entrate 2019) and consequently reduce investment costs. By contrast, as proved in the literature, along with a wide range of co-benefits, energy cost savings represent the largest monetary benefit arising from BER. Recent forecasts show that cost-effective energy improvements may reduce up to 25% the households' energy demand (Rosenow et al. 2018; Gillich et al. 2019).

To make explicit the cost–benefit trade-offs of BERPs in public housing and provide a decision support in implementation and acceleration of BER, as well as in the solution of the landlord–tenant dilemma, we propose a methodological framework in which we focus on two groups of stakeholders involved in BER of public housing: tenants and landlords, who may be a company or an individual. As already mentioned, the GC method used to determine the cost-optimal solution may be integrated with other economic valuation approaches (e.g., NPV and DPB) to account for all costs and benefits and co-benefits generated by undertaking building energy renovations. It is widely recognized that the higher the buildings energy efficiency, the higher the costs savings on energy consumption (Rathore and Shukla 2019; D'Amico et al. 2019; Zheng et al. 2019; Berseneva et al. 2020; Shao and Liu 2020). It is also a well-established evidence from the literature that BER can increase property market value in the form of a market-price premium (Banfi et al. 2008; Achtnicht 2011; Popescu et al. 2012; Zalejska-Jonsson 2014; Bottero et al. 2018; Dell'Anna et al. 2019) or of a potential higher rent (Ástmarsson et al. 2013; Monteiro et al. 2017)[2]. Furthermore, as BER generates improvements in building thermal comfort and indoor air quality and provides better protection against external noise, this may widen the set of monetary (and non-monetary) benefits that can offset investment cost, in addition to energy cost savings (Nicol and Humphreys 2002; De Dear and Brager 2002; Balaras et al. 2016; Monteiro et al. 2017; Escandón et al. 2019; Papadopoulos et al. 2019; Che et al. 2019). There is also evidence in the literature that BER may play a key role in reducing fuel poverty and social exclusion (Gillard et al. 2017; Littlewood et al. 2017; Monteiro et al. 2017; D'Alpaos and Bragolusi 2019) that is due to reduction

---

[2]In Italy, public-housing assets, subject to the regional authorities' permission, can be sold via auctions and the listing price is determined according to the asset market price (see, e.g., Legge Regionale della Regione Veneto 39/2017). Public regional authorities can also revise public-housing rents depending on building quality and finishing, including energy performances (see e.g., Legge Regionale Regione Friuli Venezia Giulia 1/2016 and http://trionto.atorfvg.it/index.php?id=32322).

in payments for energy bills, which in turn provides an increase in available income for the households' current expenses.

In our valuation framework with specific reference to the Italian context, tenants benefit from immediate cost savings due to the implementation of specific EETs, whereas landlords pay investment costs, which are reduced thanks to fiscal incentives. We investigate to which extent landlords and tenants may reach a consensus balancing their interests, and we try to figure out how much value for money these two groups of stakeholders really obtain after completion of energy renovations. Due to the lack of public financial resources, in our decisional framework, we favor the selection of cost-effective EETs that involve low installation costs and high energy savings (D'Alpaos and Bragolusi 2019) and may, nonetheless, significantly reduce fuel poverty and social exclusion (Gillard et al. 2017; Littlewood et al. 2017; Monteiro et al. 2017). The theoretical framework grounds indeed on the principle-agent theory and related concepts (Milgrom and Roberts 1992).

To investigate this issue, we implement the GC method (1) and we analyze the investment profitability according to the NPV and the DPB approaches:

$$GC(T) = IC + DC - \sum_{j=1}^{k} \sum_{n=1}^{n_k} \frac{FI_j}{(1+r)^n} + \sum_{i=1}^{T} \frac{EC_i + OC_i + MC_i + RC_i}{(1+r)^i} \quad (1)$$

where

T is the analysis time horizon (30 years according to European Directive 2012/27/EU)

$GC(T)$ is BERP Global Cost

FI_j is annual installment (of equal value each) of fiscal incentive (i.e., tax deduction) relative to the j-th EET

k is the number of alternative EETs to be implemented in BERP under investigation

n_k is the number of annual installment relative to the j-th EET

IC   are BERP investment costs

DC   are BERP costs to decant occupants

EC_i are BERP energy costs at i-th year

OC_i are BERP operating costs at i-th year

MC_i are BERP maintenance costs at i-th year

RC_i are BERP replacement costs at i-th year

r is the discount rate.

We thus complement the GC method with the DPB rule to determine the cost-optimal solution and suggest a compromise solution to the split-incentive problem (Fig. 1). Landlords pay investments costs, but benefit in the short–mid run from tax deductions (Agenzia delle Entrate 2019; MISE 2020), which reduces de facto investment costs. Tenants benefit from energy cost savings thanks to EETs implementation. In addition, as just mentioned, landlords may gain from potential increases in property market value in the case of a future sale of renovated assets or from potential

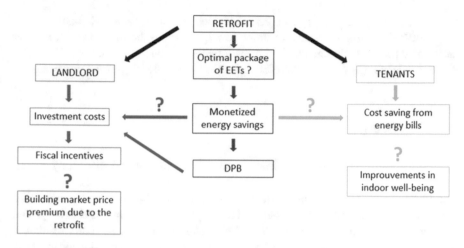

Fig. 1 Methodological framework

increases from rental fees (Ástmarsson et al. 2013; Monteiro et al. 2017). On the other hand, tenants can gain from improvements in well-being due to increase in thermal comfort, indoor air quality, and the social inclusion related to improvements in the aesthetic appearance of the building and additional available income. In detail, we propose that tenants continue to pay the same for energy consumption as prior to retrofitting until the investment (net of fiscal incentives) discounted payback time (DPB) is reached. Afterward, tenants will benefit from net gains due to energy cost savings throughout the EETs lifetime. In turn, this will contribute to reduce BERPs payback time and reduce landlords' financial exposure.

The discounted payback period (DPB) of energy renovations can be calculated as follows:

$$
\text{NPV} = \left(-\text{IC} + \sum_{j=1}^{k} \sum_{n=1}^{n_k} \frac{\text{FI}_j}{(1+r)^n} - \text{DC} \right) + \sum_{i=1}^{\text{DPB}} \frac{\text{ES}_i - \text{OC}_i - \text{MC}_i - \text{RC}_i}{(1+r)^i} = 0
$$

(2)

where

NPV   is BERP Net Present Value
ES_i is monetary benefits from energy savings at i-th year
DPB   is BERP discounted payback period

In our framework, the optimal BERP is the combination of EETs that minimize investment DPB. This solution provides potential incentives to both landlords and tenants in reaching a consensus as, on the one hand, landlords are willing to payback investment costs as soon as possible and, on the other hand, tenants are willing to reduce payments on energy bills as soon as possible and for the longest time period.

It is worth noting that in (2) property market price premiums[3] and co-benefits arising from BER (e.g., improvements in indoor air quality and thermal comfort) are not accounted for. By including these additional benefits in our analysis, the DPB can be further reduced; (2) provides therefore a cautious estimate of split-incentive shares.

3   Case Study

To test the above framework, we implemented it in a real-world case study, which consists of a four-story condominium located in a suburban area in the city of Padua (northeast Italy). The building, built in 1984, is a public-housing asset comprising 15 apartments and is currently owned and managed by a public regional agency. The building's internal and external walls, ceilings, and roof are not thermally insulated, HVAC systems are obsolete and are low-efficiency systems, and cooling systems are not installed. The 15 apartments have an autonomous heating and domestic hot-water system. The building requires energy efficiency renovations, but the agency is facing stringent budget constraints. The selection of EETs must be therefore cost-effective and preferably avoid additional costs for relocating tenants in other locations during retrofit works. Consequently, we selected three basic EETs and their combinations and obtained seven implementable BERPs (i.e., valuation scenarios).

The first BERP consists in the replacement of existing boilers and the installation of condensing high-efficiency boilers (CB). The second and third BERPs involve the replacement of existing doors and windows with double-glazed low emittance ones (LEDW) and the installation of a thermal insulation on the building envelope (ETI), respectively. These EETs generate lower operating and maintenance costs (nearly zero) throughout the building's life cycle, ensure technical performances over the years and provide high potential energy savings (Zangheri et al. 2018; D'Alpaos and Bragolusi 2018b, 2019). These three basic EETs meet technical requirements set by current regulation (Government Decree n. 192/2015). EETs investment costs for the selected EETs were estimated according to the Veneto region pricelist of public works. ETI operating, maintenance and replacement costs and LEDW operating and maintenance costs were set equal to zero over the analysis time horizon (i.e., 30 years as set by EU Directives 2010/31/EU and 2018/844/EU). Maintenance costs related to condensing boilers (boilers check-up and smoke control) were estimated by consulting technicians and service companies. Table 1 summarizes EETs costs.

---

[3] According to Bonifaci and Copiello (2015), property market-price premiums relative to different energy-performance certificates range from 2.3% for F-EL to 21.9% for A-EL class in the city of Padua (northeast Italy).

Table 1 EETs costs

EET	Installation cost	Maintenance costs	Replacement costs
Condensing boiler (CB)	€3100 per condensing boiler, i.e., €46,500	€75 annually for single boiler check-up; 120 € every 4 years for single boiler smoke control	€46,500 every 15 years for condensing boilers replacement
Double-glazed low emittance doors and windows (LEDW)	€67,847	–	–
Building envelope thermal insulation (ETI)	€56,279	–	–

Table 2 Estimated annual energy consumption in each valuation scenario (BERP)

Scenario (BERP)	EET	Annual energy consumption (kWh/m^2y)	Annual gas consumption (Smc)
SQ	Status quo	168.9	16,578.68
S1	CB (Condensing boiler)	140.3	13,121.58
S2	LEDW (Double-glazed low emittance doors and windows)	152.1	14,768.42
S3	ETI (Building envelope thermal insulation)	123.2	11,579.69
S4	CB + LEDW	127.7	11,785.42
S5	CB + ETI	104.1	9259.03
S6	LEDW + ETI	108.0	9922.23
S7	CB + LEDW + ETI	91.2	7913.28

For each valuation scenario, which identified a feasible BERP, we estimated the building annual energy consumption for heating and domestic hot-water production by implementing the *TerMus* software[4]. This software respects energy performance requirements set by current regulation (Government Decree n. 192/2015) and provides energy performance certifications and relative Energy Labels (ELs) for each dwelling. We assumed that annual energy consumption and, in turn annual energy savings ΔE, are constant over time. Table 2 shows thermal analysis results for the entire building in each valuation scenario:

To estimate annual energy costs, we analyzed natural gas market prices in the time period 2007–2019, provided by the Italian Regulatory Authority for Energy, Networks and Environment (ARERA). ES_i were estimated as follows:

---

[4]Termus software is an energy performance calculation software (https://www.acca.it/software-certificazione-energetica)

$$ES_i = p_{gi} \Delta E \tag{3}$$

where

p_{gi} is gas price at i-th year
ΔE is annual energy savings.

We performed our analysis by considering real cash flows and real discount rates to provide a cautious estimate of both benefits and DPB as gas prices are expected to increase in the future. Therefore, (3) changes into:

$$ES = ES_i = p_{g0} \Delta E \tag{4}$$

where

p_{g0} is current gas price

The present value of energy savings is:

$$ES_{PV} = \sum_{i=1}^{T} \frac{ES_i}{(1+r)^i} \tag{5}$$

whereas the present value of energy savings after the DPB period is determined as follows:

$$ES_{PV,DPB} = \frac{1}{(1+r)^{DPB}} \cdot \sum_{i=1}^{(T-DPB)} \frac{ES_i}{(1+r)^i} \tag{6}$$

and, analogously, the present value of fiscal incentives FI_{PV} is:

$$FI_{PV} = \sum_{j=1}^{k} \sum_{n=1}^{n_k} \frac{FI_j}{(1+r)^n} \tag{7}$$

To estimate the current natural gas price, we considered the average gas price provided by ARERA over the period January 2018–January 2020, and we set $p_{g0}= 0.784$ €/Smc (ARERA 2020).

The present value of fiscal incentives was estimated according to current regulation (Agenzia delle Entrate 2019; MISE 2020). In detail, taxpayers can pay back 65% of investment costs relative to CB installation by ten annual equal value installments, 75% of investment costs relative to ETI installation by ten annual equal value installments and the 50% of investment costs relative to LEDW installation by ten annual equal value installments.

4 Results and Discussion

According to EU Directive 2012/27/EU, we considered a time horizon T equal to 30 years, a discount rate to $r = 3\%$ and solved Eqs. (1), (2), and (4) for our parameters estimates. In our simulations, we did not account for property price premiums nor for tenants' co-benefits in terms of, e.g., improvements in indoor air quality or thermal comfort. In addition, we assumed that costs for relocating tenants are null, as selected EETs can be implemented without imposing occupants to leave their homes. Tables 3 and 4 summarize the results. In detail, Table 3 reports, for each scenario, annual energy savings per household (ΔE_{HOUS}), total annual energy savings (ΔE) and percentage of energy savings with respect the status-quo building and global cost (GC). Table 4 reports, for each scenario, the percentage of energy savings with respect the status-quo building, EETs installation costs, present value of fiscal incentives FI_{PV} calculated according to Eq. (7), present value of energy savings ES_{PV} calculated according to (5), DPB calculated according to Eq. (2), present value of

Table 3 Annual ΔE, annual ΔE per household and GC in each valuation scenario (BERP)

Scenario (BERP)	Households (number)	ΔE_{HOUS} per household per year (Smc/year)	ΔE (Smc/year)	ΔE (%)	GC (€)
Status quo	15	–	–	–	309,757
S1	15	180.74	3457.10	16.98	277,335
S2	15	94.64	1810.26	9.97	320,841
S3	15	261.35	4998.99	27.04	253,191
S4	15	250.60	4793.26	24.43	295,706
S5	15	382.68	7319.65	38.39	238,237
S6	15	348.01	6656.45	36.05	266,623
S7	15	453.04	8665.4	45.99	256,461

Table 4 ΔE, IC, FI_{PV}, ES_{PV}, DPB, $ES_{PV,DPB}$ in each valuation scenario (BERP)

Scenario (BERP)	ΔE (%)	IC (€)	FI_{PV} (€)	ES_{PV} (€)	DPB (year)	$ES_{PV,DPB}$ (€)	$ES_{PV,DPB}$ (€/household)
Status quo	0	0	0	0	–	0	0
S1	16.98	46,500	25,783	53,139	10	30,015	2001
S2	9.97	67,847	28,937	27,826	> 30	–	–
S3	27.04	56,279	36,005	76,840	6	55,605	3707
S4	24.43	114,347	54,720	73,678	>30	–	–
S5	38.39	102,779	61,788	112,511	10	63,540	4236
S6	36.05	124,126	64,943	102,317	22	19,125	1275
S7	45.99	170,626	90,725	133,197	24	18,105	1207

energy savings after the DPB $ES_{PV,DPB}$ calculated according to Eq. (6) and present value of energy savings after the DPB $ES_{PV,DPB}$ per household.

Our results reveal that scenario S5 incurs the lowest cost because its GC is equal to €238,237; nonetheless scenario S3 minimizes DPB and allows the landlord to pay back investment costs as quickly as possible, although it does not maximize the present value of monetary benefits from energy savings. In fact, the installation of building envelope thermal insulation provides large energy savings (about 27%) and it does not involve operating, maintenance, and replacement costs. Thus, installation of building envelope thermal insulation pays back investment costs faster than other EETs selected. According to current regulation, as scenario S5 is the least GC, it represents the optimal BERP to be implemented. Its DPB is equal to ten years, and it provides the highest present value per household of future monetary energy savings after DPB (i.e., €4236). From the tenants' perspective, this scenario is the most profitable.

Scenarios S4 and S2 are characterized by DBPs higher than 30 years, and consequently, they are not considered as profitable solutions according to the GC approach nor to the landlord's perspective. Scenario S7 is the solution that provides the highest rate of energy savings (about 46%). Nevertheless, its DPB is equal to 24 years, and it is not profitable for the landlord with respect to other solutions. By comparing the GC of scenarios S7 and S3 (which is the most profitable if the DPB rule is considered), GC values are similar (€256,461 and €253,191, respectively), but their DPB values differ significantly (24 years and six years, respectively). This result reveals that solutions whose GC is comparable may generate opposite effects in terms of investment payback time, which strongly affects landlords decisions to invest. It proves also that the sole GC is not a proper metric to select optimal retrofit strategies because it counterbalances the effects of a wide range of various costs, but does not account for cost-effective solutions due to the lack of minimum energy-performance target requirements set in current Italian regulation.

5 Conclusions

In this paper, we provided a methodological framework to identify cost-effective and cost-optimal strategies of BER in public housing. To develop our framework, we firstly identified benefits and co-benefits that may boost investment in BERPs and investigated key issues in fostering BER in public housing, thus reducing fuel poverty. We found that the landlord–tenant dilemma plays a major role in hampering investments, which in turn might significantly contribute to address fuel poverty and social inclusion concerns. To solve this issue, we proposed to split monetary benefits due to energy savings between landlords and tenants to reach a compromise solution that accounts for both landlords' and tenants' interests. We integrated usual economic valuation of BERPs by coupling the GC method with the DPB rule. Minimization of investment payback time represents in this context the optimal investment strategy, according to which landlords may decide to undertake and/or accelerate investments in BER and thus contribute to the mitigation of the fuel poverty issue, which severely affects low-income households in Italy and worldwide. To test our decisional framework, we implemented it to a real world case study, which involves a four-story condominium located in a suburban area in the city of Padua, owned by a public-housing regional agency. We selected three basic implementable EETs and their combinations for a total set of seven implementable BERPs and related valuation scenarios. We found that installation of building envelope thermal insulation enables minimizing DPB and paying back investment costs (net of fiscal incentives) as quickly as possible. This valuation scenario is not the most profitable from tenants' perspective, due to relatively small monetary benefits obtained after DPB time. From the tenants' perspective, the BERP that involves the replacement of existing boilers and the installation of building envelope thermal insulation proved the most profitable with respect to our split-incentive hypothesis. Nonetheless, its DPB is higher with respect to the BERP that involves the installation of building envelope thermal insulation. Our results reveal that the GC method needs to be integrated with different approaches to identify key drivers in BER investments and energy-retrofit renovations. Another challenge in the promotion of investments aimed at reducing fuel poverty is the quantification of non-monetary benefits that may be obtained due to BER. As future developments of this contribution, co-benefits of BER should be considered and their monetary values estimated. In addition, as occupants' behavior affects predictions on energy consumption and can contribute to reduce it, in addition to the adoption of EETs, future technical research should also be devoted to develop multi-criteria decision making models to integrate our decisional framework with the occupants' behavioral patterns, which can be considered as non-technological energy efficiency measures.

References

Achtnicht M (2011) Do environmental benefits matter? Evidence from a choice experiment among house owners in Germany. Ecol Econ 70(11):2191–2200

Agenzia Delle Entrate (2019). Le agevolazioni Fiscali per il Risparmio energetico. Ministero dell'Economia e delle Finanze. https://www.agenziaentrate.gov.it/portale/schede/agevolazioni/detrazione-riqualificazione-energetica-55-2016/cosa-riqualificazione-55-2016. Last Accessed on 12 April 2019

Alberini A, Bigano A, Ščasný M, Zvěřinová I (2018) Preferences for energy efficiency vs. renewables: what is the willingness to pay to reduce CO_2 emissions? Ecol Econ 144:171–185

Ali Q, Thaheem MJ, Ullah F, Sepasgozar SME (2020) The performance gap in energy-efficient office buildings: how the occupants can help? Energies 13(6):1–27

ARERA- Autorità di Regolazione per Energia Reti e Ambiente (2020). https://www.arera.it/it/dati/gs3.htm. Accessed 19 April 2020

Ástmarsson B, Jensen PA, Maslesa E (2013) Sustainable renovation of residential buildings and the landlord/tenant dilemma. Energy Policy 63:355–362

Balaras CA, Dascalaki EG, Droutsa KG, Kontoyiannidis S (2016) Empirical assessment of calculated and actual heating energy use in Hellenic residential buildings. Appl Energy 164:115–132

Banfi S, Farsi M, Filippini M, Jakob M (2008) Willingness to pay for energy-saving measures in residential buildings. Energy Econ 30(2):156–503

Berseneva M, Vasilovskaya G, Danchenko T, Inzhutov I, Amelchugov S, Yakshina A, Danilovich H (2020) Energy-saving technologies in design and construction of residential buildings and industrial facilities in the far North. Adv Intell Syst Comput 982:59–68

Bonifaci P, Copiello S (2015) Price premium for buildings energy efficiency: empirical findings from a hedonic model. Valori e Valutazioni 14:5–15

Bottero M, Bravi M, Dell'Anna F, Mondini G (2018) Valuing buildings energy efficiency through Hedonic Prices Method: are spatial effects relevant? Valori e Valutazioni 21:27–39

Bottero M, D'Alpaos C, Dell'Anna F (2019) Boosting investments in buildings energy retrofit: the role of incentives. Smart Innov Syst Technol 101:593–600. https://doi.org/10.1007/978-3-319-92102-0_63

Bottero M, Bravi M, Dell'Anna F, Marmolejo-Duarte C (2020) Energy efficiency choices and residential sector: observable behaviors and valuation models. In: Mondini G, Oppio A, Stanghellini S, Bottero M, Abastante F (eds) Values and functions for future cities. Green Energy and Technology, Springer, Cham, pp 167–179

Brown M (2015) Innovative energy efficiency policies: an international review. Wiley Interdisc Rev Energy Environ 4:1–25

Canesi R, D'Alpaos C, Marella M (2016) Foreclosed homes market in Italy: bases of value. Int J Hous Sci Appl 40(3):201–209

Cappelletti F, Dalla Mora T, Peron F, Romagnoni P, Ruggeri P (2015) Building renovation: which kind of guidelines could be proposed for policy makers and professional owners? Energy Procedia 78:2366–2371

Che WW, Tso CY, Sun L, Ip DYK, Lee H, Chao CYH, Lau AKH (2019) Energy consumption, indoor thermal comfort and air quality in a commercial office with retrofitted heat, ventilation and air conditioning (HVAC) system. Energy Build 201:202–215

Centro Ricerche Economiche, Sociologiche e di Mercato nell'Edilizia—CRESME (2018). Incentivi e riduzione del rischio sismico in Italia, cosa fare, come fare. http://www.cresmeantisismica.it/doc/abstract.pdf. Last Accessed on 18 Dec 2019

D'Alpaos C, Bragolusi P (2018a) Buildings energy retrofit valuation approaches: state of the art and future perspectives. Valori e Valutazioni 20:79–94

D'Alpaos C, Bragolusi P (2018b) Multicriteria prioritization of policy instruments in buildings energy retrofit. Valori e Valutazioni 21:15–25

D'Alpaos C, Bragolusi P (2019) Prioritization of energy retrofit strategies in public housing: an AHP model. Smart Innov Syst Technol 101:534–541. https://doi.org/10.1007/978-3-319-92102-0_56

D'Alpaos C (2021) Do policy incentives to buildings energy retrofit encourage homeowners' free-rider behavior? In: Morano P, Oppio A, Rosato P, Sdino L, Tajani F (eds) Appraisal and valuation: contemporary issues and new frontiers. Green energy and technology, Springer, Cham, pp 105–116. https://doi.org/10.1007/978-3-030-49579-4_8

D'Amico A, Ciulla G, Traverso M, Lo Brano V, Palumbo E (2019) Artificial neural networks to assess energy and environmental performance of buildings: an Italian case study. J Clean Prod 239:117993

De Dear RJ, Brager GS (2002) Thermal comfort in naturally ventilated buildings: revisions to ASHRAE standard 55". Energy Build 34(6):549–561

de Feijter FJ, van Vliet BJM, Chen Y (2019) Household inclusion in the governance of housing retrofitting: analysing Chinese and Dutch systems of energy retrofit provision. Energy Res Soc Sci 53:10–22

Dell'Anna F, Bravi M, Marmolejo-Duarte C, Bottero MC, Chen A (2019) EPC green premium in two different European climate zones: a comparative study between Barcelona and Turin. Sustainability 11(20):5605

D'Oca S, Hong T, Langevin J (2018) The human dimensions of energy use in buildings: a review. Renew Sustain Energy Rev 81:731–742

ENEA- Agenzia nazionale per le nuove tecnologie, l'energia e lo sviluppo economico sostenibile (2017) Piano d'Azione Italiano per l'Efficienza Energetica. https://www.mise.gov.it/images/sto ries/documenti/PAEE-2017-completo-rs.pdf. Last Accessed on 18 Dec 2019

Escandón R, Ascione F, Bianco N, Mauro GM, Suárez R, Sendra JJ (2019) Thermal comfort prediction in a building category: artificial neural network generation from calibrated models for a social housing stock in southern Europe. Appl Therm Eng 150:492–505

Ferrari S, Beccali M (2017) Energy-environmental and cost assessment of a set of strategies for retrofitting a public building toward nearly zero-energy building target. Sustain Cities Soc 32:226–234

Ferreira M, Almeida M (2015) Benefits from energy related building renovation beyond costs, energy and emissions. Energy Procedia 78:2397–2402

Gillard R, Snell C, Bevan M (2017) Advancing an energy justice perspective of fuel poverty: Household vulnerability and domestic retrofit policy in the United Kingdom. Energy Res Soc Sci 29:53–61

Gillich A, Sunikka-Blank M, Ford A (2018) Designing an 'optimal' domestic retrofit programme. Build Res Info 46(7):1–12

Gillich A, Saber EM, Mohareb E (2019) Limits and uncertainty for energy efficiency in the UK housing stock. Energy Policy 133:110889

Guardigli L, Bragadin MA, Della Fornace F, Mazzoli C, Prati D (2018) Energy retrofit alternatives and cost-optimal analysis for large public housing stocks. Energy Build 166:48–59

Gucyeter B (2018) Evaluating diverse patterns of occupant behavior regarding control-based activities in energy performance simulation. Front Arch Res 7(2):167–179

Haas R, Auer H, Biermayr P (1998) The impact of consumer behavior on residential energy demand for space heating. Energy Build 27(2):195–205

Hong T, Taylor-Lange SC, D'Oca S, Yan D, Corgnati SP (2016) Advances in research and applications of energy-related occupant behavior in buildings. Energy Build 116:694–702

Hopfe CJ, Augenbroe GLM, Hensen JLM (2013) Multi-criteria decision making under uncertainty in building performance assessment. Build Environ 69:81–90

International Energy Agency–IEA (2008) Promoting energy efficiency investments: case studies in the residential sector. France. https://webstore.iea.org/promoting-energy-efficiency-investments, last accessed on 18 Dec 2019

Italian Legislative Decree of 19 August 2005, n. 192. Attuazione della direttiva 2002/91/CE relativa al rendimento energetico nell'edilizia

Italian Ministerial Decree DM 25/06/2015. Adeguamento linee guida nazionali per la certificazione energetica degli edifici, 25 June 2015

Jafari A, Valentin V, Bogus SM (2019) Identification of social sustainability criteria in building energy retrofit projects. J Constr Eng Manage 2(145):04018136

Knobloch F, Pollitt H, Chewpreecha U, Daioglou V, Mercure J (2019) Simulating the deep decarbonisation of residential heating for limiting global warming to 1.5 °C. Energy Efficiency 2(12):521–550

Laaroussi Y, Bahrar M, Elmankibi M, Draoui A, Si-Larbi A (2019) Occupant behaviour: a major issue for building energy performance. IOP Conf Ser Mater Sci Eng 609(7):072050

Littlewood JR, Karani G, Atkinson J, Bolton D, Geens AJ, Jahic D (2017) Introduction to a Wales project for evaluating residential retrofit measures and impacts on energy performance, occupant fuel poverty, health and thermal comfort. Energy Procedia 134:835–844

Ma Z, Cooper P, Daly D, Ledo L (2012) Existing building retrofits: methodology and state-of-the-art. Energy Build 55:889–902

Mangialardo A, Micelli E, Saccani F (2018) Does sustainability affect real estate market values? Empirical evidence from the office buildings market in Milan (Italy). Sustainability (Switzerland) 11(1):1–14

Mauri L, Vallati A, Ocłoń P (2019) Low impact energy saving strategies for individual heating systems in a modern residential building: a case study in Rome. J Clean Prod 214:791–802

Milgrom P, Roberts J (1992) Economics, organization and management. Prentice Hall Inc, New Jersey

MISE–Ministero Dello Sviluppo Economico (2020). Piano Nazionale Integrato per l'Energia e il Clima (PNIEC) https://www.mise.gov.it/images/stories/documenti/PNIEC_finale_17012020.pdf. Last Accessed on 18 Dec 2019

Monteiro CS, Causone F, Cunha S, Pina A, Erba S (2017) Addressing the challenges of public housing retrofits. Energy Procedia 134:442–451

Mutani G, Todeschi V, Kampf J, Coors V, Fitzky M (2019) Building energy consumption modeling at urban scale: yhree case studies in Europe for residential buildings. In: International telecommunications energy conference—INTELEC, 7–11 October 2018. Italy

Nicol JF, Humphreys MA (2002) Adaptive thermal comfort and sustainable thermal standards for buildings. Energy Build 34(6):563–572

Papadopoulos S, Kontokosta CE, Vlachokostas A, Azar E (2019) Rethinking HVAC temperature setpoints in commercial buildings: the potential for zero-cost energy savings and comfort improvement in different climates. Build Environ 155:350–359

Penna P, Schweigkofler A, Brozzi R, Marcher C, Matt DT (2019) Social housing energy retrofitting: business model and supporting tools for public administration. IOP Conf Ser Earth Environ Sci 323(1):012167

Popescu D, Bienert S, Schutzenhofer C, Boazu R (2012) Impact of energy efficiency measures on the economic value of buildings. Appl Energy 89(1):454–463

Preciado-Pérez OA, Fotios S (2017) Comprehensive cost-benefit analysis of energy efficiency in social housing. Case study: Northwest Mexico. Energy Build 152:279–289

Rathore PKS, Shukla SK (2019) Potential of macroencapsulated pcm for thermal energy storage in buildings: A comprehensive review. Constr Build Mater 225:723–744

Rolim C, Gomes R (2020) Citizen engagement in energy efficiency retrofit of public housing buildings: a Lisbon case study. In: Littlewood J, Howlett R, Capozzoli A, Jain L (eds) Sustainability in energy and buildings. smart innovation, systems and technologies, vol 163. Springer, Singapore, 421–431

Rosenow J, Guertler P, Sorrell S, Eyre N (2018) The remaining potential for energy savings in UK households. Energy Policy 121:542–552

Sangalli A, Pagliano L, Causone F, Salvia G, Morello E, Erba S (2020) Behavioural change effects on energy use in public housing: a case study. In: Littlewood J, Howlett R, Capozzoli A, Jain L (eds) Sustainability in energy and buildings. smart innovation, systems and technologies, vol 163. Springer, Singapore, 759–768

Shao B, Liu X (2020) Calculation of energy saving based on building engineering. Adv Intell Syst Comput 921:837–844

Sunikka-Blank M, Galvin R (2012) Introducing the prebound effect: the gap between performance and actual energy consumption. Build Res Info 40(3):260–273

Vilches A, Barrios Padura Á, Molina Huelva M (2017) Retrofitting of homes for people in fuel poverty: Approach based on household thermal comfort. Energy Policy 100:283–291

Zalejska-Jonsson A (2014) Stated WTP and rational WTP: willingness to pay for green apartments in Sweden. Sustain Cities Soc 13:46–56

Zangheri P, Armani R, Pietrobon M, Pagliano L (2018) Identification of cost-optimal and NZEB refurbishment levels for representative climates and building typologies across Europe. Energ Effi 11(2):337–369

Zheng D, Yu L, Wang L (2019) Research on large-scale building energy efficiency retrofit based on energy consumption investigation and energy-saving potential analysis. J Energy Eng 145(6):04019024

Rural-Urban Relationships for a Better Territorial Development

Rural–Urban Relationships for Better Territorial Development

Elisa Ravazzoli, Christian Hoffman, Francesco Calabrò,
and Giuseppina Cassalia

Abstract Well-established relationships between urban and rural areas are key drivers for a balanced territorial development. By promoting spatial and functional interdependencies between cities and countryside, they provide opportunities for inclusive, smart, and sustainable development. The chapter presents different contributions related to various types of urban–rural relationships among different types of spaces, across different sectors, using different instruments or initiatives, which overall contribute to targets of territorial cohesion. It collects reflections on the following inquiry: theoretical, methodological, or empirical reflection on smart rural–urban relationships; digital technological developments, to incorporate digital tools as opportunities for smart solutions to live, work, and move; entrepreneurship and new business models to attract labor markets and ways of cooperation between urban and rural actors as an opportunity for new value-added chains; social innovation as initiatives and creative ways to improve social services across rural–urban areas, to reduce, adapt, or address natural risks. Finally, the chapter shed light on some of the relevant future trends in the urban–rural relationship.

Keywords Rural–urban relationships and links · Territorial cohesion · Spatial planning · Social innovation · New business models

E. Ravazzoli · C. Hoffman
Institute for Regional Development, Eurac Research, Viale Druso 1, Bolzano/Bozen, Italy
e-mail: elisa.ravazzoli@eurac.edu

C. Hoffman
e-mail: Christian.Hoffmann@eurac.edu

F. Calabrò (✉) · G. Cassalia
Department of Architectural and Urban Heritage (PAU), Reggio Calabria Via Melissari,, Italy
e-mail: fracesco.calabro@unirc.it

G. Cassalia
e-mail: giuseppina.cassalia@unirc.it

1   Introduction

When you are inside the processes, it is difficult to read and interpret them with the same gaze that, for example, the historian turns to the events of the past, from which he derives the causal relationships of the phenomena contemporary to him.

But those who deal with the issues that concern the government of territorial transformations in a broad sense have the obligation to try to identify, within the processes in progress, the dynamics that will determine future processes and, through them, try to understand the territorial structures of tomorrow.

It is an indispensable exercise to identify any corrections necessary to correct the hypothetical trajectories, where they deviate from the desirable ones.

This exercise, however, due to the technological revolution in progress, appears much more difficult than in the past because of what could be called the "temporal misalignment", that is, the profoundly different speeds with which transformations of the physical space occur, measurable in years, and those of the virtual space, measurable in days. This data, considering the profound effects of new technologies on human behaviour, makes the exercise of physical space planning even more difficult, which should have time horizons of the order of decades.

Despite all the limitations of an anthropocentric vision, there is no doubt that the physical space interests the planner to the extent that it must be organized according to the needs of mankind, both as an individual and, in an Aristotelian sense, as a social animal.

The physical space, i.e. concerns the planner as a place for carrying out human activity, first of all in its two main components, work and social relations, leaving out in this context, for obvious reasons, the most intimate dimensions of the individual.

For the first time in the history of humanity, however, the Technological Revolution challenges, in perspective, the very reason for the existence of the human species.

Thanks to (or because of?) the progressive spread of robotics and artificial intelligence, in the working field, man is no longer indispensable for almost any activity, now being replaceable in almost all fields by machines.

On a different level, then, social relations are progressively moving from physical to virtual places; just think of how much the place par excellence for relations, i.e. the square, has dematerialized: the spread of the use of social media has now largely transferred the exchanges that took place in this public space to the virtual world.

There is a term to name the new world we live in: the infosphere (Toffler 1987). It is a large grey area that presupposes a continuity between the material and technological world: it represents the contemporary version of alienation, as the distinction between real and virtual is broken down. Both biological organisms and engineered, analogue and digital artefacts operate in the infosphere, the ultimate purpose of which is the supply of information; another characteristic element is that it becomes progressively normal for technological tools to dialogue directly with each other.

"At a minimum level, the infosphere indicates the entire information environment made up of all information bodies, their properties, interactions, processes and mutual relations. It is an environment comparable to, at the same time different from, cyberspace, which is only one of its regions, since the infosphere also includes offline and analogue information spaces.

At a maximum level, the infosphere is a concept that can also be used as a synonym of reality, where we interpret the latter in informational terms. In this case, the idea is that what is real is informational and what is informational is real" (Floridi 2017). According to Floridi, we are changing the idea we have of the ultimate nature of reality, moving from a materialistic metaphysics to a metaphysics of information: we live suspended in an informational cloud, not delimited by space or borders.

With respect to all these issues, the session "Rural–Urban Relationships for a better territorial development", organized by the authors of this paper as part of the 3rd International Conference SSPCR—"Smart and Sustainable Planning for Cities and Regions 2019", enabled gathering elements of great interest: the element that, albeit implicitly, unites all the contributions presented is the profound change that is taking place in the dialectical relationship between man and physical space.

Wanting to bring to the possible extreme consequences the reading of the dynamics underway provided by the presentations, it would be possible to derive, as the main summary, the announcement of the imminent death of the city, at least in the form that we inherited from the Industrial Revolution, i.e. a place of production and exchange.

The demographic trends underway, at a planetary level, are of the diametrically opposite sign, with the progressive concentration of the population in increasingly dense and congested metropolitan areas; the contributions presented, however, implicitly indicate the existence, albeit still in an embryonic stage, of processes that, in the long run, will most likely reverse these dynamics.

2 Session's Main Addresses and Topics for Discussion

The contributions presented in the session[1] addressed a wide variety of different aspects that pertain to the urban–rural relationship, reflecting the extreme complexity

---

[1]The contributions were presented During the "Smart and Sustainable Planning for Cities and Regions 2019" Conference in the session "Rural-Urban Relationships for a better territorial development" were: Laner, Peter; Hoffmann, Christian—*Integrated solutions for the provision of Services of General Interest in peripheral mountain regions;* Moreira, Maria da Graça–*Tourism in the time of the smart regions;* Lengyel, Janka; Friedrich, Jan— *Multiscale Urban Analysis and Modelling for Local and Regional Decision Makers;* Calcatinge, Alexandru—*Relevance of cultural landscape policies and the need for smart and open source frameworks in cultural urban development processes;* Dalvit, Giorgio—*Alperia Smart Region: the region-wide approach for South Tyrol;* Mertens, Giel; Bastiaanssen—*Making cities informed implementers of autonomous mobility;* Pagliuso, Ana; Cavaco, Cristina; Mourato, João; Pereira, André—*Integrated territorial approaches in Portugal: between EU-led policy initiatives and national statutory contexts;* Membretti, Andrea—*Innov–Aree: accompanying mountain imaginary to become sustainable reality;* Pisman, Ann; Vanacker, Stijn—*Reporting about the development of new maps and indicators for Flanders and the challenge to get these new insights implemented in policy documents;* Grilli, Gianluca; Capecchi, Irene; Barbierato, Elena; Sacchelli, Sandro—*Preference-Based Planning Of Urban Green Spaces: A Latent Class Clustering Approach;* Cattivelli, Valentina—*Defining urban and rural areas: Characteristics and problems related to the methods currently adopted.*

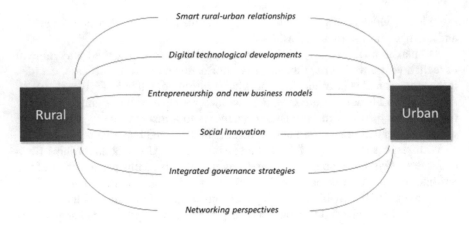

Fig. 1 Graph illustrating the topics discussed in the session "rural–urban relationships for a better territorial development." *Source* authors' own elaboration

inherent in the topic. The topic discussed were: *Theoretical, methodological or empirical reflections* on smart rural–urban relationships; *innovative tools for integrated governance strategies or policies* able to promote rural–urban connections at regional, national and cross-border levels; *Digital technological developments*, to incorporate digital tools as opportunities for smart solutions to live, work and move; *Entrepreneurship and new business models* to attract labour markets and ways of cooperation between urban and rural actors as an opportunity for new value-added chains; *Social innovation* as initiatives and creative ways to improve social services across rural–urban areas, to reduce, adapt or address natural risks and socio-demographic changes (e.g. internal and international migration) and achieve social justice; *Networking Perspectives* towards sustainable territorial cohesion, in terms of multidimensional approaches to enhance local identities, foster collective values and empower community participation (Fig. 1).

The element that unites the contribution presented, albeit to an extent and in different ways and with few exceptions, is the **role of innovation and the spread of new technologies**. ICT and artificial intelligence are the tools that are profoundly affecting both people's lifestyles and the solutions available to the planner to meet the needs of contemporary society. Some of the presentations focus more on the first aspect: this is the case, for example, of Laner, who underlines the importance of these tools in the provision of services of general interest in marginal areas; Moreira also addresses this aspect, referring in particular to the use of ICT in tourism, while Dalvit deals with the use of ICT and new technologies in general to address some environmental issues, such as, for example, the efficiency in the use of scarce resources such as water or the reduction of polluting emissions; Mertens illustrates some of the potential of self-driving vehicles. Membretti looks at a dimension of innovation, i.e. of a social nature, seen as a tool capable of counteracting the anthropic desertification of the Inner Areas. On the other hand, Lengyel looks at the use of new technologies in the field of planning, in particular, in the structuring of Decision Support

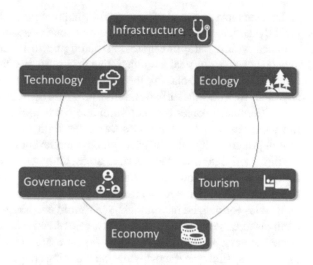

Systems (Lengyel and Friedrich, 2021); Grilli also works on the construction of
DSS, in particular, for the management and enhancement of natural resources (Grilli
and Curtis 2021); Calcatinge emphasizes the importance of the open dimension of
innovation, referring in the specific case to the application of big data to the theme of
cultural landscapes (Calcatinge 2021); Pisman focuses on the construction of maps
and indicators to address a very delicate environmental issue, that of sprawl, from
the perspective of zero soil consumption (Pisman and Vanacker 2021). The presentations of Pagliuso and Cattivelli, on the other hand, differ from the others, addressing
methodological issues: they highlight, first of all, the overcoming of the traditional
classifications of urban and rural space, also highlighting the growing diffusion of
planning tools, such as, for example, the Strategic Plans or the programs promoted
by the EU, which go beyond traditional administrative boundaries and the rigidity
of planning approaches aimed exclusively at regulating land use (Cattivelli 2021)
(Fig. 2).

3 Discussion and Concluding Remarks

Well-established relationships between urban and rural areas are key drivers for
balanced territorial development. They promote spatial and functional interdependencies between cities and countryside counteracting diverging regional disparities and the feeling of inequality and (Copus 2013; Habitat III 2015). Dissolving
borders—that are physical, mental, social—and promoting strong mutual rural–
urban linkages are today essential for the creation of inclusive, smart and sustainable
growth in both rural and urban Europe and to achieve territorial cohesion (Hjalager
2016). The distinction between rural and urban areas is no longer actual and useful to

understand current complexities nor for planning purposes in smart regions. Settlement dynamics, demographic changes and socioeconomic developments cannot be bounded within static thresholds, as belonging to urban or rural spaces, because they do not have physical administrative limits. So, the distinction between rural and urban and the need to classify the territory within these two categories are not useful anymore. We know that urban and rural areas are complementary, and they need each other. The interlinkages between rural and urban areas enable an integrated territorial approach, one that addresses the accessibility to services of general interest, spatial justice or digital solutions, which are relevant for the living standard and for leading a self-determined life in rural areas. Only when we handle these phenomena adequately can we counteract: the rural–urban divide; regional disparities; feelings of spatial and social inequality; and the perception of being a second-class citizen.

Besides formalized instruments, we should encourage place-based soft-planning methods to enhance self-responsibility, to overcome sectorial thinking and to overcome statutory planning procedures. It is no useful anymore to produce in 20 years the perfect spatial plan if meanwhile the planning conditions have changed. Although those thoughts and phenomena are not new, we are still at the beginning of transferring them into practical planning tools and governance instruments. Digitalization in the access

to services of general interest, technical infrastructures covering the local, regional, national and cross-border dimensions that enable spatial justice offer equipollent living conditions and promote a high quality of life in marginalized, rural mountain areas (Küpper and Steinführer 2017; ARL 2016, 2017; Magel 2016; Malý 2016; Fassmann et al. 2015; Miosga 2016; Littke et al. 2013).

The 12 contributions given in the session discussed the relationship between urban and rural areas from different perspectives and shed light on some of the relevant future trends in the urban–rural relationship (Fig. 3), which are summarized below.

Innovative tools and instruments:

Fig. 3 Future of urban and rural relationships. *Source* authors' own elaboration

ICT, digitalization, open-data/open-sources (software) are crucial for enhancing mutual interdependencies between rural and urban areas and to open new opportunity especially for rural dwellers.

Innovative Methods and Modelling:

GIS modelling and scenario building offer planners the possibility to integrate future trends within planning instruments and to envision and think about spaces in a new way. These methods make possible seeing future tangible changes and their impacts (such as, e.g., automotive driving, digital connectivity, functional small areas, decentralized smart energy supply; or Stop-Sealing in 2040, or green-spaces for leisure or health purpose) but also enable predicting intangible changes (e.g. social behaviour, loss of local knowledge and traditions, working conditions or jobs—Profiles need to be therein considered).

Crosscutting Items:

Cross-border cooperation: for integrated delivery of service of general interest, inter-communal services, cross-regional collaboration are already applied—but at the transnational level—that is still problematic.

Cross-sectorial approaches: at the professional (agriculture, social affairs and health, etc.) level but also at the governance level (between ministries).

Cross-scale: from container space to relational space perspective; from spatial proximity to organizational proximity; interlinking and connecting entities of different scales and not neighbouring anymore.

Empowering local community to provide vision:

Mediation tools: To create a fruitful environment between rural residents and urban newcomers to establish new opportunities, but also to avoid conflicts among them.

Appropriate governance instruments:

Instruments to implement local and regional visions: CLLD, ITI or EGTC (European Grouping of Territorial Cooperation);

These future trends refer to interrelationships among different types of spaces, sectors and methods and would be key to promote opportunities that increase societal well-being and the living standard across urban and rural spaces.

Further research may address the relations between socio-spatial justice and spatial planning

- What is the relationship between planning theories and practices in relation to spatial justice—for example, in relation to access to services (social, ICT) and equivalence of living conditions?
- To what extent are spatial functions, opportunities and burdens distributed in a socially just manner? How can distributive justice be achieved in spatial development (e.g. with regard to services of general interest)? To what extent are problems of common good relevant in the use of infrastructures, resources or landscapes

and how are these linked to questions of justice? Who can use spaces in what way and who is excluded?

- Which instruments can promote participatory planning procedures? How can we planning be more inclusive and socio-spatially justly? A fair distribution of spatial resources and functions and access to them is considered a basic human right. Idea of "right to the city", "equal living conditions" handling of spatial and planning conflicts. Which space may/should be used how and by whom? Who negotiates and decides on the design and use of spaces, who benefits from it—and who does not? It is therefore also about procedural and distributive justice.
- To what extent are environmental resources and burdens distributed spatially/socially fairly? Which planning instruments could contribute to environmentally sound spatial development? To what extent are global ecological challenges in the field of sustainability and climate justice met by local action?

References

Akademie für Raumforschung und Landesplanung (eds) (2016) Daseinsvorsorge und gleichwertige Lebensverhältnisse neu denken. Hanover: position paper from the ARL 108. Online: https://shop.arl-net.de/media/direct/pdf/pospaper_108.pdf. Accessed 07 June 2019

Calcatinge A (2021) A smart and open-source framework for cultural landscape policies. Green Energy Techol. https://doi.org/10.1007/978-3-030-57332-4_42

Cattivelli V (2021). Institutional methods for the identification of urban and rural areas—a review for italy. Green Energy Techol. https://doi.org/10.1007/978-3-030-57764-3_13

Copus A (2013) Urban-rural relationships in the new century: clarifying and updating the intervention logic. New Paradigm Action 7

Copus AK, de Lima P (eds) (2015) Territorial CohersionCohesion in Rural Europe: the relation turn in rural development. Routledge

Fassmann H, da Rauhut D, Costa EM, Humer A (eds) (2015) Services of general interest and territorial cohesion: European perspectives and national insights. Vandenhoeck & Ruprecht, Göttingen

Floridi L (2017) La quarta rivoluzione. Come l'infosfera sta trasformando il mondo. Raffaello Cortina Editore, Milano

Grilli G, Curtis J (2021) Preference-Based Planning of Urban Green Spaces: A Latent-Class Clustering Approach. Green Energy Technol. https://doi.org/10.1007/978-3-030-57332-4_41

Habitat UN (2015) Habitat III Issue Paper 22—Informal Settlements. UN Habitat, New York. Online: http://habitat3.org/wp-content/uploads/Habitat-III-Issue-Papers-report.pdf. Accessed 14 June 2019

Hjalager AM (2016) Rural-urban business partnerships—towards a new trans-territorial logic. In: Local economy. J Local Econ Policy Unit 32(1):4–54. Online: https://journals.sagepub.com/doi/full/10.1177/0269094216686528. Accessed 07 June 2019

Küpper P, Steinführer A (2017) Daseinsvorsorge in ländlichen Räumen zwischen Ausdünnung und Erweiterung: Ein Beitrag zur Peripherisierungsdebatte. Europa Regional 23(4):44–60. Online: https://nbn-resolving.org/urn:nbn:de:0168-ssoar-53589-9. Accessed 14 June 2019

Littke H, Rauhut D, Foss O (2013) Services of general interest and regional development in the European Union. Romanian J Reg Sci 7:88–107

Lengyel J, Friedrich J (2021) Multiscale urban analysis and modelling for local and regional decision-makers. Green Energy Technol. https://doi.org/10.1007/978-3-030-57332-4_40

Magel H (2016) Räumliche Gerechtigkeit—Ein Thema für Landentwickler und sonstige Geodäten. Zeitschrift für Vermessungswesen 141(6):377–383

Malý J (2016) Impact of polycentric urban systems on intra-regional disparities: a micro-regional approach. European Plann Stud 24(1):116–138. Online: https://www.tandfonline.com/doi/abs/10.1080/09654313.2015.1054792. Accessed 14 June 2019

Miosga M (2016) Gleichwertige Lebensverhältnisse und Arbeitsbedingungen in allen Landesteilen: Die Suche nach neuen Wegen zur Umsetzung des Verfassungsauftrags in Bayern. Vorbereitender Bericht zur gemeinsamen Jahrestagung der DASL und ARL 2016: Daseinsvorsorge und Zusammenhalt, 91–94

Pisman A, Vanacker S (2021) Diagnosis of the state of the territory in flanders. Reporting About New Maps and Indicators Differentiating Between Urban and Rural Areas Within Flanders. Green Energy Technol. https://doi.org/10.1007/978-3-030-57764-3_14

Toffler A (1987) La terza ondata. Il tramonto dell'era industriale e la nascita di una nuova civiltà. Sperling & Kupfer, Milano

Multiscale Urban Analysis and Modelling for Local and Regional Decision-Makers

Janka Lengyel and Jan Friedrich

Abstract In today's increasingly complex urban environments, decision-makers are facing far-reaching uncertainties regarding the possible impact of their implemented measures. Moreover, dynamic transformation processes involve a large number and variety of stakeholders representing intertwined and oftentimes conflicting interests and constraints. It is therefore extremely urgent that interventions in the urban realm be supported by innovative instruments. Amongst others, they must provide assistance in setting priorities for urban design and planning actions and in optimizing and foreseeing their possible effects on an extensive timescale. Against this background, we propose a comprehensive seven-step methodology for urban analysis and modelling, with a special focus on their effective integration. We believe that for accuracy, credibility and real-world-applicability purposes, spatial models must rest on a thorough investigation of existing space–time patterns and processes. Firstly, the analysis part identifies the most important regional trends and their shaping factors for our case study, the Ruhr area in Germany. Secondly, it investigates prevailing demographic, socioeconomic and economic profiles and derives their characteristic scales. Moving forward, we discuss how empirical findings may be infiltrated into the multiscale urban model (MURMO) and proceed to highlight the model's main features. Finally, we run simulations under a "Smart City" scenario and review how results can form the recommendation-basis for both spatial and non-spatial measures. We will explicitly discuss the relevance of each of the steps for urban stakeholders in their decision-making processes.

Keywords Urban planning and design · Multiscale modelling · Empirical analysis · Integrated approach

J. Lengyel (✉)
Universität Duisburg-Essen, Institut Für Mobilitäts- Und Stadtplanung, Universitätsstr. 5, 45141 Essen, Germany
e-mail: janka.lengyel@uni-due.de

J. Friedrich
Univ. Lyon, ENS de Lyon, Univ. Claude Bernard, CNRS, Laboratoire de Physique, 69342 Lyon, France
e-mail: jan.friedrich1@ens-lyon.fr

1   Introduction

The Ruhr area is one of the largest metropolitan regions in Europe, hosting more than five million inhabitants with a population density of 1152 people per km^2 (Regionalverband Ruhr 2020). In terms of spatial scales, its unique polycentric structure is of great interest as it lacks a clear political or economic centre. Moreover, and similar to the large majority of today's urban settings, due to its extremely dense interactions and interdependencies, traditional organizational boundaries are becoming fuzzier and fuzzier (Brenner and Schmid 2014; Cuthbert 2011). The question is therefore how we may better assist potential interventions in such increasingly complex urban systems. We believe that recent data availability (Batty 2013), used hand in hand with the massive rise in computational power (Roser and Ritchie 2020), now enables us designing new and innovative tools that are able to specifically target such questions in great detail and accuracy. Against this background, we propose an integrated methodology of spatiotemporal urban analysis and modelling. We contend that intrascale processes play a critical role in urban dynamics; thus, it is quintessential that they would infiltrate into endeavours concerned with this domain. Therefore, the here proposed methodology involves the development of a multiscale urban model (MURMO) that explicitly incorporates multiscale properties of urban processes and is, for instance, capable of reproducing the occurrence of strong small-scale fluctuations of land prices.

To demonstrate real-world potentials and challenges, we discuss the relevance of this study for the local and regional stakeholders involved in the interdisciplinary research project "New Emscher Mobility" (NEMO 2020). The main focus of the project rests on the envisioning of possible long-term urban development scenarios for the Ruhr area, with mobility as the core trigger of change. It sees the conversion and renaturalization of the river Emscher and its tributaries, which have for decades served as an open sewage system, as a unique opportunity to provide sustainable intermodal mobility for the Ruhr region and thereby instigate some powerful urban transformation processes.

2   Methodology

The following three subsections are dedicated to a compressed review of multiscale urban analysis and modelling. Because a highly detailed description of their corresponding results is beyond the scope of this paper, the main emphasis consists of describing the step-by-step comprehensive methodology and illustrating it with short examples. One must stress here that the empirical analysis has been carried out with respect to long-term transformation measures. Therefore, its strengths lie in trends and pattern identification and in the delineation of their main influencing factors.

2.1 Empirical Analysis

2.1.1 [Step 1]: Identify the Most Important Regional Trends and Their Shaping Factors

The first step is to try to decipher the most influential past and prevailing drivers of regional transformation processes. These may include demographic (age or ethnic structure), economic (sectoral shifts), societal (household, gender), environmental (congestion, pollution, climate), or political changes. Step one uses a combined methodology of a literature review (e.g., Bland and Hermann 2019; Wehling 2014; Reicher et al. 2011) and spatiotemporal data analysis. As small-scale divergences do not yet play a crucial role, data at this stage is collected mostly for regional and city levels and for the last 25–50 years (depending on data availability). Here, next to regional trend identification, a great emphasis must be put on untangling spatial projections of the deduced crucial development trajectories. The most important demographic trend in the Ruhr region, similar to other former industrial agglomerations of Europe (Agueda 2014), is shrinking populations which are at the same time rapidly ageing and becoming increasingly heterogeneous. Even though well-being is steadily rising (suggested for instance by the 10.6% increase in GDP per inhabitant between 2007 and 2015), so does the share of people who are living on social subsidies. In terms of social dynamics, there is the clear deducible trend of rising single households, as well as of single parents and of non-marital cohabitation. Last but not least, economic structural change, which has now been going on for more than half of a century, is still a great challenge but also a major opportunity: as an example, for the spatial impact of this trend, we may observe that former large-scale production and excavation areas are now incrementally becoming available for novel investors, which is manifested in the massive decrease of industrial land-use on the regional level. These brownfields are situated amongst others in attractive surroundings in the vicinity of newly (re-)constructed and renaturalized water bodies, see, for instance, the Emscher restoration (Der Emscher Umbau 2020) or the urban development project Phoenix See (Stadt Dortmund 2020). In sum, the relevance of step one is to clarify the main problems and potentials of the region in a larger historic context. This may provide the *overarching framework* for future urban measures and strategies.

2.1.2 [Step 2]: Analyse Current Demographic, Socioeconomic and Economic Profiles and Derive Their Characteristic Scales

The aim of [Step 2] is to identify correlations and regularities on multiple urban scales and derive characteristic spatial patterns. We must now identify how the—during [Step 1]—deduced macro-scale trends manifest themselves on different urban layers and if their anchoring may lead to distinct sub-regional demographic, socioeconomic or economic patterns. Regarding the Ruhr region, we observed that, in comparison

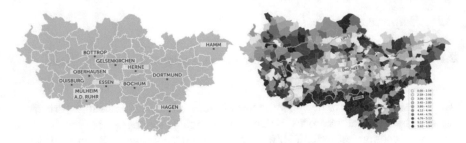

Fig. 1 Left: Administratively independent cities of the Ruhr region. **Right**: Spatial deciles of purchasing power on the district level in 2016 (Indices). Data Source: see Sect. 3.1.4

to more habitual urban dimensions (such as that of the 53 municipalities in Fig. 1, Left), the different fluvial areas (Fig. 1, Right) are of much greater importance. The historical development of these zones shows considerable divergence, and they can still be characterized by distinct social and spatial structures (Wehling 2014). It is particularly the central (Emscher and Hellweg) Ruhr region that is facing a concentration of socioeconomic challenges today (see Fig. 1, Right, for the characteristic pattern of purchasing power on a district level). Step two is of great relevance because these findings may already provide some very important clues for new, and perhaps in some cases more *appropriate, scales for future action.* This is especially true for the Ruhr region where effective collaboration is immensely hard on the large, regional level, whilst oftentimes unfruitful and insufficient on the communal one. Therefore, we need to find meaningful synergies of interest in-between the scale of more traditional urban boundaries.

2.2 [Step 3]: Compress and Infiltrate Results of the Empirical Analysis into the Modelling Process

Here, results of the previous [Step 1] and [Step 2] are brought together to unravel meaningful relationships between the various (previously identified) parameters, as well as for dimensionality-reduction and modelling purposes. To this end, we use self-organizing maps (Moosavi et al. 2014), a machine learning mechanism, which "is able to convert complex nonlinear statistical relationships between high-dimensional data items into simple geometric relationships on a low-dimensional display" (Kohonen 1990). Henceforward, the component maps of the neural network on Fig. 2 may be interpreted in the following way: neighbourhoods located in the upper-left corner are characterized by very high purchasing power, net rent price, share of German population and percentage of elderly-only households (those being in a strong positive correlation). These belong to Cluster 24 (C24) according to the Hit Map on Fig. 3, Right. A large portion of affluent C24-districts are situated in the southern regions of the city of Essen, along the Ruhr River (see Fig. 3, Left).

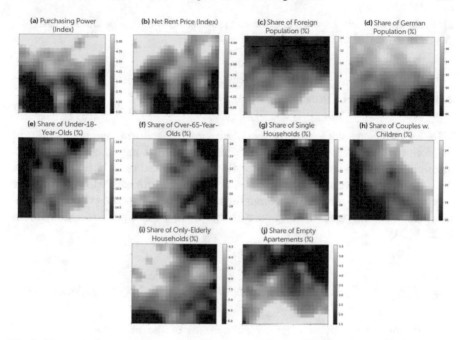

Fig. 2 The ten 2D component maps of the SOM model-space. Data Source: see Sect. 3.1.4

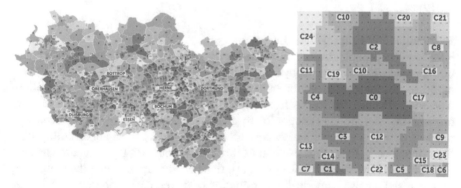

Fig. 3 Left: distribution of the 25 socioeconomic clusters throughout the Ruhr region in 2016 (left). **Right**: SOM clusters on 2D model-space (see component maps on Fig. 2). Data Source: see Sect. 3.1.4

Districts in the lower right corner are the exact opposite: low rents and purchasing power, high foreign-population and households with children rates dominate. Again, these correlate strongly with each other and are in a negative statistical relationship with features of the previously described upper-left corner. Clusters of the lower half of the output space (Fig. 3, Left) concentrate north and south of the Emscher River, justifying as well bringing together in a powerful way our observations from [Step

2]. SOM clusters will have a vital role in the model-calibration process of [Step 4]. We refer the reader to Sect. 3.1.1 for follow-up.

2.3   Multiscale Urban Modelling

MURMO simulates the long-term spatiotemporal development and interactions of the following six subsystems: demographics, residential migration, employment location, land price, land-use and accessibility. Calibrated on a 10,000-m^2 spatial grid, the model is able to delineate small-scale trends of variables, whilst at the same time to intrinsically recognize meaningful trajectories in regional economic, land-use, land price and population dynamics. In this paper, the time period for simulations is 2011–2050.

2.4   [Step 4]: Build and Calibrate the Multiscale Urban Model

This section briefly reviews the six subsystems of the multiscale urban model (MURMO) and highlights its main features and strengths for urban decision-makers. For more detail and precise numerical description, please see Lengyel and Friedrich (2019).

Residential and Employment Location
Residential and employment dynamics are captured by the well-known master equation approach from statistical physics (Haag and Weidlich 1984). The master equation describes the temporal evolution of the probability of a certain socio-configuration, whereby individual transitions from one cell to another are specified by transition probabilities that are the function of so-called utilities. If, for instance—from the point of view of a young couple—a given cell evolves to have a higher utility than their "home-cell", they might eventually, through the course of the simulations, make a decision to move there. The latter description already suggests that the method of selection and precise numerical determination of preference, cooperation and saturation parameters (Haag and Weidlich 1984)—which together define utility functions—are of outmost importance. Therefore, in the residential location sub-model, spatial preferences of inhabitants are computed separately for the 25 SOM Clusters, described in [Step 3]. This is a very crucial step because we are able to make findings of the empirical analysis to bear an immense influence on internal model-dynamics. Furthermore, the model's external driving mechanisms are based largely on trends identified in [Step 1] and on scenario presuppositions laid down in [Step 5]. To sum up, through several entry points we successfully achieved an extensive infiltration of empirical findings into the modelling process. The latter enhances the credibility of modelling results and may be extremely important for urban decision-makers to successfully test the scale, location or time period of their various interventions.

Land Price and Land-Use

In contrast to the well-known approach described in the previous section, land price is modelled by a novel method that emphasizes the analogy between urban and turbulent systems. We believe that this analogy is extremely fruitful for two reasons: First, turbulent flows, similar to urban systems, are open systems far from equilibrium, and they inherently involve transport processes in scale, thus ensuring scale-to-scale interactions. In the case of three-dimensional turbulence, this entails transfer of energy; however, in the case of cities, one could think of money, policies, large-scale master plans and so forth. Secondly, it includes and puts a strong emphasis on the so-called non-self-similar nature of (urban and turbulent) environments. The latter property implies an increased probability for the occurrence of very strong fluctuations (for instance that of land-use or land price) at small scales, whereas large-scale statistics are governed by close-to Gaussian statistics (Friedrich and Peinke 1997). This is due to strong correlations between their spatiotemporal dimensions. Here we can only highlight and emphasize some main features; however, we refer the reader for proof of concept and for more information to Lengyel and Friedrich (2019). Let us now demonstrate the use of cited analogy on the land-price subsystem: With the help of applying the phenomenological description of turbulence to urban systems, we are able to completely reconstruct the original rent-price field consisting of almost half a million points with the help of only 40 initial and 76 final values. This land-price field is then reconstructed each and every year depending on where firms and inhabitants are moving. Lastly, land-use change occurs, if the interplay between preference, cooperation and saturation parameters directs either residents or employers towards yet unsettled grid cells. Moreover, land-price values which are not available, in yet inhabited but potentially available areas, are evaluated via the synthetic land-price field discussed above.

Accessibilities

As already mentioned in the introduction section of this paper, accessibility values are provided externally by the MatSIM model (MatSIM 2020). MatSIM is an open-source platform for implementing agent-based transport simulations, where every agent is assigned an individual daily travel behaviour and, over the course of an average workday, each of them strives towards the maximization of their individual benefits. Please see Kaddoura et al. (2019) for further details on the role of MatSIM in the NEMO project and for specific simulation methods.

Data Sources

In this study, all city level information stems from the official database of North Rhine–Westphalia (Landesdatenbank NRW 2017). For more detailed analysis, we subdivided the 53 cities into their 736 municipal districts. Values on this scale were either provided by local municipal authorities or aggregated using higher-resolution data from the following sources. Socioeconomic and employment data were acquired from the German 2011 census database (Zensus 2011), the micro-dialog data of the German post and from the Hoppenstedt firm database (Bisnode Firmendatenbank 2017). We collected land-use material from the urban atlas of the Copernicus

Land Monitoring services (Copernicus 2018) and from the database of Regionalverband Ruhr (Flächennutzungskartierung RVR 2020). Finally, in accordance with the 10,000-m² grid of the German Population Census in 2011 (Zensus 2011), we further subdivided city districts and obtained almost half a million grid cells for the whole Ruhr area. Data sources of this level are either directly from the German Census Database or accumulated using same data sources as described above.

3 [Step 5]: Delineate Meaningful Future Development Trajectories

The goal of step five is the envisioning of possible long-term urban development scenarios for the study area and the testing of solutions and impulses for prevailing challenges and potentials. Here, the very important suggestion with regard to decision-maker involvement is the following: One might try to integrate issues of great societal relevance and interest at the time of scenario conception. This has the immense potential to increase public and political interest and thus to enhance real-world applicability. Regarding the NEMO project, the main focus rests on the envisioning of possible long-term urban development scenarios for the Ruhr Area, with mobility as its core trigger of change. This already picks up on the currently highly relevant topic of the "Verkehrswende", or the transition towards more sustainable mobility patterns. Secondly, the starting point of the "Smart City" scenario—for which this paper discusses results—is the so-called digital transformation, which is also on the very top of the German political agenda at the time of the writing this paper. The scenario assumes (please see Schmidt et al. 2019 for more on scenarios in the NEMO project) that by 2035 leading companies will have developed new and cutting-edge technologies for the Smart City. In the beginning, these are being implemented solely for affluent districts, cities and regions, creating and further amplifying segregation patterns and social tensions. Smart and affordable city for all may only be reachable in the long run. By then, digitization and high-tech will be used in many areas of life, supported by the belief that high data protection requirements will be met by large corporations. At the same time, new business models are pervading the mobility sector, offering all-encompassing integrated services, such as car, bike and taxi sharing or on-demand mobility, whilst these are being effectively combined with highly efficient public-transport services.

4 [Step 6]: Translate Scenarios into Modelling and Simulate Them with the Careful Integration of Existing Design Concepts in the Region

Let us now discuss the implementation of the upper scenario–narrative into the multi-scale urban model (MURMO). Firstly, mobility-presuppositions laid down in Step 5 are implemented in the MatSIM transport model, which then externally provides accessibility values for MURMO. Secondly, we make estimates and assumptions for the remaining five subsystems based on assumptions of Step 5 and insights and results of the empirical analysis ([Steps 1–2]). We do this in pursuance of customizing the scenario as far as possible for the prevailing challenges and potentials of the Ruhr area itself. We strongly contend that, if we are to use a combined method of urban modelling and scenario approach to inform urban design and planning decisions, then this may very well be the most adequate way to come up with meaningful recommendations. Moreover, to further increase interest and applicability, we take already existing design concepts of the region into consideration to showcase their potential strength and weaknesses from a variety of points of view (environmental, economic, societal, governance, etc.). In this way, we may delineate and emphasize new common grounds for the multiple stakeholders which might then enable new powerful alliances to emerge. For the example used in this paper, land is made available for the International Garden Expo Ruhr, which is scheduled for 2027 (IGA 2020), and for the urban planning project Smart Rhino Dortmund (Smart Rhino 2020). Attractivities (see Sect. 3.1.1) are adjusted for the corresponding cells accordingly. In terms of demographics, we apply the official forecasts of the National Statistical Office. In the case of the residential migration sub-model, we assume densification tendencies in already populous sub-regions and thus a five per cent increase in the attractivity of cells in and around dense and medium dense urban areas (according to the Corine Land Use database (Corine 2017)). Additionally, we assume an overall 10% increase in the attractiveness of disadvantaged neighbourhoods (identified via their SOM Clusters in Sect. 2.3) until 2050. The number of jobs is to increase 0.2–0.7% annually. Finally, we identify cells for future expansion: areas for residential, commercial, industrial and mixed use separately, according to the official regional land-use plan (Flächennutzungskartierung RVR 2020).

Modelling Results under a Smart City Scenario

The following part of this paper seeks to provide a short account of the results under the selected Smart City scenario. What can be clearly seen is the—not so surprising—overall trend of decreasing populations and increasing employment numbers, according to priory assumptions. The somewhat more interesting observation is that central zones (Ruhr, Hellweg, Emscher) suffered 4.76% lower loss of inhabitants than peripheral ones (Lippe, Kreis Moers and Südosten). For a detailed discussion of the historical zones of the Ruhr area, we refer the reader to Wehling (2014, 2016). At the same time, the number of central-zone employees increased by 11.82% which is slightly higher than the Ruhr average of 10.51%. The latter

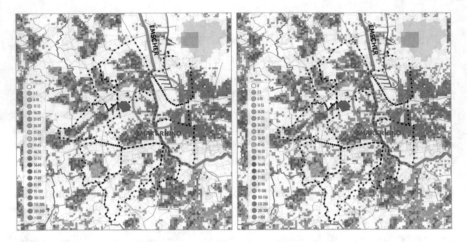

Fig. 4 Number of residents per grid cell in 2011 (left) and 2050 (right) around Dortmund Dorstfeld, Germany

trend was most pronounced for neighbourhoods along the Emscher with an 14.03% increase. Taken together, these results suggest a rather strong densification effort by both employers and residents. These may be a response to a combination of improved accessibility values and the rising attractivities of cells with above average medium- and high-density urban land-use, as well as of low real-estate prices at initial central locations. In terms of the micro-scale case study highlighted in Fig. 4, there is a noticeable amount of newly inhabited grid cells, concentrating especially around the Emscher River, including the project site of Smart Rhino (see Fig. 4 and the intro-ductory part of this section). Similarly as above, these cells most probably profited from increased attractiveness and accessibility values, as well as from dissolving large-scale industrial areas on the banks of the Emscher. We can use these initial findings to assess the specific land-use demand (residential, commercial or indus-trial) and land-use availability for each of the grid cells of our study area, which may lay fruitful grounds for future recommendations.

5 Interpreting Results and Formulating Recommendations [Step 7]

The overarching goal and the final task are to synthesize modelling results to form a basis of recommendation for both spatial (e.g., urban design and planning strategies) and non-spatial (e.g., policies) measures under the selected scenario. Let us now give some short examples.

In terms of land-use development, there are two tendencies of utmost interest for regional decision-makers. First, voids between the very characteristic "innere Ränder" (Reicher et al. 2011) or inner borders of the Ruhr region are starting to fill up–whilst at the same time these inner edges incrementally dissolve–due to strong densification tendencies in already populous cities. Second, previously large-scale mono-functional areas are being slowly replaced by mixed land-use, a tendency especially strong on the shores of the Emscher River. The last two observations may demand entirely new spatial strategies (including both push and pull actions), whilst providing invaluable clues for the extent and location of future infrastructure measures. Finally, we briefly demonstrate how modelling results can be applied to identify small-scale anchoring of major urban development processes under the Smart City scenario. Due to its very high spatial resolution, the model is able to delineate hot and cold spots for rent-price changes on the district level. In our case, hotspots concentrate in central Oberhausen, in the north of Essen, south of Bottrop, in Gelsenkirchen and Herne, around the western neighbourhoods of Dortmund and in the urban cores of Hamm and Hagen. In contrast, cold spots mostly accumulate in the outskirts of the Ruhr area. These may be important indicators for the amplitude and location of potential gentrification processes and thus may allow decision-makers to apply timely alleviation measures.

6 Conclusions and Outlook

This present study set out to give an overview of the seven-step integrated methodology for unravelling and modelling urban processes between spatiotemporal scales and illustrated it with practical examples from the Ruhr area in Germany. Secondly, it made a case for effectively connecting analysis findings with the internal dynamics of the multiscale urban model. Third, we have seen that over time the closely entangled reciprocal dynamics of the six MURMO subsystems led to a variety of substantial changes in the urban structure. These may be indicators for the following processes: attracting new landscape and urban design projects, anchoring a range of additional businesses (or supersede existing ones) or altering real-estate prices along with their local socioeconomic profiles.

Still, further testing is urgently needed in collaboration with a variety of urban stakeholders to increase the applicability of each step for real-world challenges. Moreover, it would be interesting to coordinate this methodology with other urban design strategies such as that of public participation (Innes and Booher 2000) in urban planning. This would help us to better understand the needs and preferences of inhabitants on micro-scale and therefore to come up with even more meaningful future scenarios, as well as to increase the acceptance of our resulting recommendations and interventions.

Acknowledgements The New Emscher Mobility project (NEMO) funded by the Mercator Stiftung provided partial support for this research. In the frameworks of the NEMO project, we would like to acknowledge and thank the MatSIM-team (Kaddoura 2019) for making their accessibility data available for MURMO-modelling purposes under the Smart City Scenario. J. L. would like to thank the strategic partners (Regionalverband Ruhr and Emschergenossenschaft) of the NEMO project as well as the city authorities of the Ruhr Area for providing invaluable data. J.F. acknowledges funding from the Alexander von Humboldt foundation within a Feodor-Lynen scholarship.

References

Agueda BF (2014) Urban restructuring in former industrial cities: urban planning strategies. *Territoire en mouvement Revue de géographie et aménagement* [Online], 23–24| 2014, Online since 06 September 2017, connection on 27 January 2020. http://journals.openedition.org/tem/2527; https://doi.org/10.4000/tem.2527

Batty M (2013) Big data, smart cities and city planning. Dial Hum Geograp 3(3):274–279

Bisnode Firmendatenbank. http://www.firmendatenbank.de. Accessed 19 Dec 2017

Blank M, Herrmann S (2009) Atlas der Metropole Ruhr, Vielfalt und Wandel des Ruhrgebiets im Kartenbild. ch. Technologiezentren, Keimzellen des Wandels, pp 146–147

Brenner N, Schmid C (2014) The 'urban age' in question. Int J Urban Reg Res 38(3):731–755

Copernicus. https://land.copernicus.eu/local/urban-atlas. Accessed 21 Sept 2018

Cuthbert A (2011) Urban design and spatial political economy. In: Companion to urban design. Routledge, pp 103–115

Der Emscher Umbau. https://www.eglv.de/emscher/der-umbau/. Accessed 12 Feb 2020

Flächennutzungskartierung RVR. https://www.rvr.ruhr/daten-digitales/geodaten/flaechennutzung skartierung/. Accessed 01 Aug 2020

Friedrich R, Peinke J (1997) Description of a turbulent cascade by a Fokker-Planck equation. Phys Rev Lett 78(5):863

Haag G, Weidlich W (1984) A stochastic theory of interregional migration. Geograph Anal 16(4):331–357

IGA.   https://www.rvr.ruhr/themen/oekologie-umwelt/internationale-gartenausstellung-2027/. Accessed 12 Feb 2020

Innes JE, Booher DE (2000) Public participation in planning: new strategies for the 21st century

Kaddoura I, Laudan J, Ziemke D (2019) Verkehrsmodellierung für das Ruhrgebiet: Simulationsbasierte Szenariountersuchung und Wirkungsanalyse einer verbesserten regionalen Fahrradinfrastruktur. To appear in Proceedings of Wissenschaftsforum 2019. University of Duisburg-Essen

Kohonen T (1990) The self-organizing map. Proc IEEE 78(9):1464–1480

Landesdatenbank NRW. https://www.landesdatenbank.nrw.de. Accessed 12 Sept 2017

Lengyel J, Friedrich J (2019) Multiscale urban modelling; A de-urbanization scenario in the Ruhr area. To appear in Proceedings of Wissenschaftsforum 2019. University of Duisburg-Essen

MatSIM, Multi-Agent Transport Simulation. https://www.matsim.org/. Accessed 12 Feb 2020

NEMO: https://www.nemo-ruhr.de/. Accessed 14 Feb 2020

Moosavi V, Packmann S, Vallés I (2014) SOMPY: A Python Library for Self Organizing Map (SOM). GitHub (Online). Available: https://github.com/sevamoo/SOMPY

Pope CA III, Ezzati M, Dockery DW (2009) Fine-particulate air pollution and life expectancy in the United States. N Engl J Med 360(4):376–386

Regionalverband   Ruhr.   https://www.rvr.ruhr/daten-digitales/regionalstatistik/bevoelkerung/. Accessed 11 Feb 2020

Reicher C, Kunzmann KR, Polívka J, Roost F, Wegener M (eds) (2011) Schichten einer Region: Kartenstücke zur räumlichen Struktur des Ruhrgebiets. Jovis

Roser M, Ritchie H (2020) Technological progress.. Published online at OurWorldInData.org. Retrieved from: 'https://ourworldindata.org/technological-progress [Online Resource]

Schmidt JA, Klemm S (2019) Integrative Szenarien für eine nachhaltige Mobilität in der Region Ruhr. To appear in Proceedings of Wissenschaftsforum 2019. University of Duisburg-Essen

Smart Rhino. https://www.dortmund.de/media/downloads/pdf/news_pdf/2019_6/Broschuere_S mart_Rhino.pdf. Accessed 12 Feb 2020

Stadt Dortmund: https://www.dortmund.de/en/leisure_and_culture/phoenix_see_1/index.html. Accessed 12 Feb 2020

Wehling HW (2016) Annäherungen an die Industrielle Kulturlandschaft Ruhrgebiet. Prozesse und Strukturen-Zonen, Achsen und Systeme. ICOMOS–Hefte des Deutschen Nationalkomitees 62:86–103

Wehling HW (2014). Organized and disorganized complexities and socio-economic implications in the Northern Ruhr Area. In: Understanding complex urban systems: Multidisciplinary approaches to modeling. Springer, pp 87–101

Zensus (2011) https://www.zensus2011.de. Accessed 26 July 2017

Preference-Based Planning of Urban Green Spaces: A Latent-Class Clustering Approach

Gianluca Grilli and John Curtis

Abstract Green spaces within cities are important oases of restoration and space for physical activities. Due to the health benefits provided by visits to urban green space, encouraging people to increase frequentation may be a public policy objective. To do this, individual attitudes should be considered because individuals may not respond in the same way to the increasing availability of green spaces. In this paper, we propose a latent-class analysis to probabilistically allocate respondents into homogeneous groups, each with similar attitudes towards green-space visitation. The data originated from a questionnaire survey of adult Irish citizens who resided in major Irish cities. Results identified three groups of users with varying visitation and attitudes and suggested that the conditions of the neighbourhood where people reside can influence attitudes and willingness to visit green spaces, thus raising questions about access to green spaces.

Keywords Urban planning · Environmental justice · Latent-class analysis · Smart living

1 Introduction

Green spaces (GS) within cities refer to all types of green covers, including public parks, sporting fields, riparian areas (stream and riverbanks), canal pathways, street trees, etc. GS provide several benefits to users, including opportunities for psychologically restorative experiences, physical activity, and social interactions (Langemeyer et al. 2016). The availability of GS has been linked to positive effects on physical and mental health, with such associations widely verified by empirical studies (Beil and Hanes 2013; Van Den Berg et al. 2016). The balance of evidence suggests that proximity to GS is sufficient to increase mental well-being; however, the direct use of

G. Grilli (✉) · J. Curtis
Economic and Social Research Institute, Dublin, Ireland
e-mail: gianluca.grilli@esri.ie

Trinity College Dublin, Dublin, Ireland

© The Author(s), under exclusive license to Springer Nature Switzerland AG 2021   581
A. Bisello et al. (eds.), *Smart and Sustainable Planning for Cities and Regions*,
Green Energy and Technology, https://doi.org/10.1007/978-3-030-57332-4_41

GS for recreation and physical activity allows people to gain a wider range of benefits (White et al. 2019). In addition to health-related benefits, GS provide ecosystem services that are useful to mitigate pollution and improve air quality and the general life quality of city dwellers (Shanahan et al. 2015; Tratalos et al. 2007). The use of GS is affected by individuals' attitude to consider GS as recreational sites and by their local availability (Sturiale and Scuderi 2019, 2018). It has been pointed out that GS planning at city level often lacks social justice considerations, in fact most low-income neighbourhoods have no or limited availability of GS (Wolch et al. 2014). The uneven accessibility of GS is recognized as an issue of environmental justice, and it becomes a planning priority worldwide (Tzoulas et al. 2007).

The recent scientific evidence suggests that urban and regional planning, including urban GS, is more effective when user preferences are considered, for example, using questionnaires to understand people's needs and priorities (Carlsson et al. 2015). It is often observed that preferences are not homogeneous across groups of people, but, because there are small subgroups with different expectations from public policy, a one-size-fits-all solution is difficult to achieve. An issue that is not extensively investigated so far is that respondents have different attitudes and prior knowledge about the topic that affect their preferences. Attitudinal questions are always included in surveys, but their actual use to extract information is relatively low.

In this paper, we propose a methodology to cluster urban dwellers depending on their latent attitude to using GS. The analysis is based on a probabilistic latent-class allocation that enables depicting segments of GS users, based on their current GS accessibility and their sociodemographic characteristics. The model is flexible and allows several specifications, including multivariate analysis with multiple indicators simultaneously and the inclusion of covariates to better define class membership. Our model specification includes several attitudinal-dependent variables that were collected on Likert scales, while the explanatory variables include neighbourhood quality, means of transport used to commute to GS and habits of physical activity in addition to sociodemographics. The dataset for our application originated from a questionnaire survey administered in Ireland to a sample of 1006 Irish city dwellers, stratified by age, occupation, education and place of residence.

Our results show groups with similar attitudes and orientations towards the recreational use of GS and highlight cohorts of urban dwellers with limited GS access and frequency of use that represent potential target groups for socially efficient GS planning policies. Stimulating GS visits with specific land-use planning is beneficial for individuals' well-being and for the wider society in terms of decreased social cost for healthcare and pollution and increased environmental quality.

2 Methods

2.1 Data

The dataset comprises a sample of 1006 adult Irish citizens living in the major cities in Ireland, which were interviewed face to face by a professional survey company. Respondents were contacted by telephone and invited to take part to the survey, then interviewed in their houses by professional interviewers. The survey administration was designed to be representative of the population of urban residents in Ireland and was stratified by gender, age, education and place of residence. The survey considered only city inhabitants because they were living in the policy relevant areas; the rest of Ireland is largely countryside, with plenty of green areas available already. The cities considered were all urban developments larger than 4000 inhabitants. The questionnaire was composed of 39 questions divided into five thematic sections and collected information about GS attitudes, health and physical activity habits, sociodemographics and preferences for GS attributes. For the purpose of this paper, we consider only the sections about the attitudes and sociodemographic characteristics of respondents. Part of other results were published in (Grilli et al. 2020). Respondents reported an average of 7.6 visits to GS in the prior four weeks, with a large variability (standard deviation $= 8.32$). The sample was gender-balanced (50% were male and 50% female), and the mean age was 45 years (range of 18–89). In terms of education, 44% had a secondary school education, 38% a bachelor's degree and almost 15% a master's degree or higher. Concerning occupational status, 41% was full-time employed, 13.5% had a part-time job, about 4% was self-employed, 19% were retired and 5.5% were students. Some 6% of respondents were unemployed looking for jobs, while the remaining were either unemployed not looking for work or refused to answer. The median income class was €30,000–40,000 per year and, although more than 62% of respondents refused to answer the income question, it is in line with the national average, which was €39,000 in 2019. The attitudinal questions used as dependent variables and the range of answers are reported in Fig. 1.

14. How much do you agree with the following statements? Please rate on a scale of 1-5 where
     1 is strongly disagree and 5 is strongly agree. **SHOWCARD A**
     a) Green and blue spaces are important for people's leisure
     b) Green and blue spaces are important aesthetic components of urban areas
     c) My city has a limited number of green and blue spaces
     d) I am satisfied with the number and quality of green and blue spaces of my
        hometown
     e) If green and blue spaces in my area had different features I would visit them
        more often

Strongly disagree	Disagree	Neither agree nor disagree	Agree	Strongly agree

Fig. 1 Questions used as dependent variables in the latent-class analysis

The five-point scales were converted into three points combining the answers "disagree"–"strongly disagree" and "agree"–"strongly agree". This recoding exercise is useful to reduce the number of estimated parameters, which is relatively large in latent-class analyses. The new code should not change results significantly because the order of agreement is maintained.

2.2 The Latent-Class Model

Latent-class analysis (LCA) is a powerful clustering tool that probabilistically allocates respondents into classes based on their answers to categorical questions (Lanza and Cooper 2016; Oberski2016; Pickles et al. 1995). The number of classes into which respondents are allocated should be decided a priori and the choice of the most appropriate model is usually based on goodness-of-fit measures such as Akaike Information Criterion (AIC) or Bayesian Information Criterion (BIC). An LCA model with r classes returns the probability that each respondent falls into each of the classes; if covariates are included, the LCA becomes a regression model in which the probability of belonging to the rth class depends on a set on independent variables. We denote with J the number of polytomous categorical variables answered by respondents, each with K potential outcomes. The observed values of the J variables are denoted as Y_{ijk}, which is equal to one if respondent I give the kth answer to the jth variable and zero otherwise. The probability that an individual gives a sequence of choices in class r is given by the following formula:

$$f(Y_i; \pi_r) = \prod_{j=1}^{J} \prod_{k=1}^{K} (\pi_{jrk})^{Y_{ijk}}$$

where π_{rkj} represents the class-conditional probability, estimated using a multinomial logit probability function. In LCA with polytomous variables, one class is considered as the reference level and coefficients of the other classes are interpreted with respect to the reference class. The methodology is thoroughly discussed by Linzer and Lewis (2011). The model was estimated in R using the poLCA package.

3 Results and Discussion

A set of constant-only LCA models with varying number of classes was estimated, and the most appropriate was selected using goodness-of-fit measures. Table 1 summarizes AIC and BIC calculated for models of 2–5 classes. The AIC suggested that a five-class model is appropriate, while the BIC statistic indicates a three-class model. The three-class model was chosen because the BIC statistics is more appropriate when the number of parameters varies between models (Mariel et al. 2015).

Table 1 Comparing of goodness-of-fit measure for LCA models with 2, 3, 4 and 5 classes

	2 classes	3 classes	4 classes	5 classes
N parameters	21	32	43	54
AIC	7888	7789	7746	7719
BIC	7991	7946	7958	784

Table 2 Results of the 3-class LCA model

	Class 1	Class 2	Class 3
No. visits to GS in 4 weeks	(Reference)	−0.008	−0.202***
		(0.029)	(0.042)
Has a dog		−0.013	−0.106
		(0.194)	(0.221)
Neighbourhood problems		0.575***	0.557***
		(0.088)	(0.089)
Age 18–34		0.452	0.546*
		(0.298)	(0.305)
Age 35–64		0.624**	0.531**
		(0.277)	(0.287)
Female		−0.119	−0.147
		(0.184)	(0.203)
Education		0.124	−0.079
		(0.120)	(0.138)
Constant		−1.231***	−0.627
		(0.426)	(0.450)
Class probabilities	0.300	0.478	0.222
Log-likelihood (LL)	−3821		
Respondents (N)	1006		

Source: *p < 0.10, **p < 0.05, *** p < 0.01

In addition, there is consensus that a smaller number of classes should be preferred (Hoyos et al. 2015).

The full three-class model with covariates is presented in Table 2. In an LCA regression model, the class-specific parameters are interpreted as changes with respect to the reference class, with only fixed parameters. For example, the negative coefficients associated with the number of walking days every four weeks suggest that respondents are less likely to fall in class 2 or class 3 when they make many visits or, equivalently, they are more likely to belong to class 1. Having a dog does not significantly affect class membership. This is an interesting result because it suggests that dog owners, who are more likely to frequent GS to walk the dog, do not show statistically significant differences in attitudes towards GS compared with dog non-owners (McCormack et al. 2010). The variable labelled "Neighbourhood problems" is continuous and captures the extent of problems related to antisocial behaviours (e.g., burglaries, drug abuse and public drunkenness, littering, etc.) in the respondents' neighbourhood of residence. The positive coefficient for this variable indicates that respondents with more antisocial behaviours in their neighbourhood are more likely to fall into class 2 and class 3. Age is a significant variable to explain class membership, with the reference class of respondents older than 65 more likely to belong to class 1 compared to younger respondents. Gender and education are not significant and therefore do not significantly influence class allocation.

Class probabilities are reported in the second panel of Table 2. Class 2 is the largest, with almost 48% of respondents, while the third class is the smallest with 22% of respondents.

The influence of GS visits on class allocation is interesting because it suggests three different attitudes towards GS based on users' frequency. The second class has a negative but not statistically significant coefficient, which indicates that the number of visits does not influence respondents' allocation. Class 3 has a negative coefficient, therefore respondents with larger visitation are less likely into fall in this class. Figure 2 graphically shows this result.

Respondents making one visit any four weeks are about 40% likely to fall into class 1 and around 30% to fall into either of the other two classes. The probability of class 1 membership increases with visitation up to 60% when respondents make 28 visits every four weeks, i.e. with daily visits. The probability remains the same in class 2 and decreases to negligible levels in class 3. The first class may be considered the segment of "active" users, for which positive attitudes towards increase with frequency of visitation. The second class could be labelled as the class of "indifferent" users, for which attitudes are not affected by visitation. The third class includes respondents who are not likely to make many visits to GS monthly and have negative or distant attitudes towards GS. Some people may show distant or indifferent attitudes because they spend enough time in the countryside or other natural areas, therefore

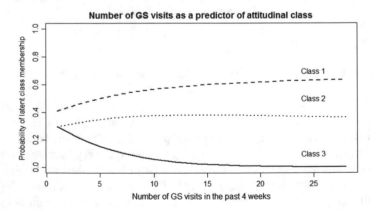

Fig. 2 Class probabilities at varying number of visits

the availability of urban GS is not important to them. Unfortunately, the data do not allow testing this hypothesis, but it represents an avenue for future research.

4 Policy Implications

A recent trend in public policy is to encourage GS visitation to improve mental and physical health. Responses to this policy may be uneven across the Irish population of citizens. The LCA analysis indicated that attitudes towards green spaces are heterogeneously distributed across respondents. The heterogeneity in preferences may be reflected in varying levels of interest and public acceptance of new GS. For this reason, the simple planning of urban GS may not be enough and should be paired with information about GS benefits. This is especially important in cases where a tight public budget requires balancing the allocation of public resources.

While GS attitudes are affected by respondents' age, other sociodemographic variables do not significantly affect attitudes. An important result is, however, related to the significant effect of neighbourhood conditions on GS attitudes. Neighbourhood quality affects frequentation due to several reasons. A first simple explanation is that residents in some areas may go out less frequently for safety reasons, especially during evening hours. Another potential explanation is the lack of locally available GS, which makes GS visits time-consuming. These findings suggest that the issue of social justice in urban GS provision should be better explored and potentially represents a planning challenge for local decision-makers.

5 Conclusion

To exploit information derived from attitudinal questions, which is often overlooked, this paper proposed a latent-class analysis to segment respondents based on their attitudes towards urban green spaces and personal characteristics, using a dataset of Irish adult citizens. Results indicated that attitudes towards green spaces are heterogeneous and three cohorts of respondents were identified. One group included respondents with a high frequency of visits, one with a low frequentation and a third group for which green-space attitudes were not affected by frequentation. In addition to visit frequency, class allocation was affected by neighbourhood conditions, which may indicate a social justice issue in green-space supply across neighbourhoods. These findings suggest that green-space planning is not a one-size-fits-all type of policy because people have different attitudes and habits. When a public policy objective is to increase visitation through creating new green areas within cities, attention should be paid on their geographical allocation, and it should be paired with information activities to raise awareness to non-users.

Acknowledgements Funding from Environmental Protection Agency is gratefully acknowledged.

References

Beil K, Hanes D (2013) The influence of urban natural and built environments on physiological and psychological measures of stress—a pilot study. 1250–1267. https://doi.org/10.3390/ijerph 10041250

Carlsson J, Eriksson LO, Öhman K, Nordström E-M (2015) Combining scientific and stakeholder knowledge in future scenario development—A forest landscape case study in northern Sweden. For Policy Econ 61:122–134. https://doi.org/10.1016/j.forpol.2015.08.008

Grilli G, Mohan G, Curtis J (2020) Public park attributes, park visits, and associated health status. Landsc Urban Plan 199:103814

Hoyos D, Mariel P, Hess S (2015) Incorporating environmental attitudes in discrete choice models: An exploration of the utility of the awareness of consequences scale. Sci Total Environ 505:1100–1111. https://doi.org/10.1016/j.scitotenv.2014.10.066

Langemeyer J, Gómez-Baggethun E, Haase D, Scheuer S, Elmqvist T (2016) Bridging the gap between ecosystem service assessments and land-use planning through Multi-Criteria Decision Analysis (MCDA). Environ Sci Policy 62:45–56

Lanza ST, Cooper BR (2016) Latent class analysis for developmental research. Child Dev Perspect 10:59–64

Linzer DA, Lewis JB (2011) poLCA: an R package for polytomous variable latent class analysis. J Stat Softw 42:1–29

Mariel P, Meyerhoff J, Hess S (2015) Heterogeneous preferences toward landscape externalities of wind turbines—combining choices and attitudes in a hybrid model. Renew Sustain Energy Rev 41:647–657. https://doi.org/10.1016/j.rser.2014.08.074

McCormack GR, Rock M, Toohey AM, Hignell D (2010) Characteristics of urban parks associated with park use and physical activity: a review of qualitative research. Heal Place 16:712–726. https://doi.org/10.1016/j.healthplace.2010.03.003

Oberski D (2016) Mixture models: latent profile and latent class analysis. In: Modern statistical methods for HCI. Springer, pp 275–287

Pickles A, Bolton P, Macdonald H, Bailey A, Le Couteur A, Sim CH, Rutter M (1995) Latent-class analysis of recurrence risks for complex phenotypes with selection and measurement error: a twin and family history study of autism. Am J Hum Genet 57:717–726

Shanahan DF, Fuller RA, Bush R, Lin BB, Gaston KJ (2015) The health benefits of urban nature: how much do we need? Bioscience 65:476–485. https://doi.org/10.1093/biosci/biv032

Sturiale L, Scuderi A (2019) The role of green infrastructures in urban planning for climate change adaptation. Climate 7:119

Sturiale L, Scuderi A (2018) The evaluation of green investments in urban areas: A proposal of an eco-social-green model of the City. Sustainability 10:4541

Tratalos J, Fuller RA, Warren PH, Davies RG, Gaston KJ (2007) Urban form, biodiversity potential and ecosystem services. Landsc Urban Plan 83:308–317. https://doi.org/10.1016/j.landurbplan. 2007.05.003

Tzoulas K, Korpela K, Venn S, Yli-Pelkonen V, Kaźmierczak A, Niemela J, James P (2007) Promoting ecosystem and human health in urban areas using Green Infrastructure: a literature review. Landsc Urban Plan 81:167–178

Van Den Berg M, Van Poppel M, Van Kamp I, Andrusaityte S, Balseviciene B, Cirach M, Danileviciute A, Ellis N, Hurst G, Masterson D, Smith G, Triguero-Mas M, Uzdanaviciute I, de Wit P, Van Mechelen W, Gidlow C, Grazuleviciene R, Nieuwenhuijsen MJ, Kruize H, Maas J (2016) Visiting green space is associated with mental health and vitality: a cross-sectional study in four european cities. Heal Place 38:8–15. https://doi.org/10.1016/j.healthplace.2016.01.003

White MP, Alcock I, Grellier J, Wheeler BW, Hartig T, Warber SL, Bone A, Depledge MH, Fleming LE (2019) Spending at least 120 minutes a week in nature is associated with good health and wellbeing. Sci. Rep 9:1–11. https://doi.org/10.1038/s41598-019-44097-3

Wolch JR, Byrne J, Newell JP (2014) Urban green space, public health, and environmental justice: The challenge of making cities just green enough. Landsc Urban Plan 125:234–244

A Smart and Open-Source Framework for Cultural Landscape Policies

Alexandru Calcatinge

Abstract This paper will emphasize that a smart and open-source framework based on meaningful and relevant open-source principles for future cultural landscapes policies is a viable option for the local policymakers. Relevant to our research will be the open-source paradigms that could be used in urban and rural development, the relationships between cultural landscapes and open-source principles, the importance of open-source innovation and the issue of open-topic argumentation in cultural landscapes policies and projects, with direct implications to the policy development in Romania. To achieve this, the paper will expose findings on some of the most important and relevant theories around concepts like cultural landscapes, smart cities, big data and open innovation. In the end, the goal is to build upon those findings and create a draft for a feasible framework for future cultural landscape and policies based on specific open-source principles.

Keywords Cultural landscapes · Open-source principles · Open innovation · Smart cities · Big data · Policy

1 Introduction

We live in a world shaped and defined by policies. Even the impact of different cultures on the [constructed] environment is generated, to some extent, by policies. Modern cultural landscapes were shaped by policymakers who based their work on the evidence of culture upon the land, and now it is the foundation of our constructed environment's evolution. This, combined with the fact that the cultural landscape, as a concept, is overly complex and ambiguous at the same time (Calcatinge 2012; 2014) is enabling a many possible circumstances to be taken into consideration when dealing with it as a working paradigm.

Nevertheless, cultural landscapes policy research can also include a more technological approach, by including topics like open-source technologies, open innovation

---

A. Calcatinge (✉)
Faculty of Urbanism, Department of Urban Planning and Regional Development, University of Architecture and Urbanism, 18-20 Academiei Street, 010014 Bucharest, Romania
e-mail: alexandru.calcatinge@gmail.com

A. Bisello et al. (eds.), *Smart and Sustainable Planning for Cities and Regions*,
Green Energy and Technology, https://doi.org/10.1007/978-3-030-57332-4_42

589

and big data (Fleishman 2020) and thus link them to the urban, rural and cultural development processes. In this respect, some of my research (Calcatinge 2016, 2017, 2018) deals with those aspects in detail, and therefore, I will only summarize some of the most important theories and technologies that will help us create a future framework for cultural landscape policies. The most important of them all and the foundation for relevant future policies are the open-source ways and the principles that defines it.

2 Method

The paper is an analytical one presenting the findings I consider relevant for creating a new kind of cultural landscape framework, based on open-source principles.

In this context, one of the most important properties of this research paper would be its ability to be seen as a system divided into layers and models (generally called modularity) in the sense that it does narrow down all the big parts that could appear, to smaller and more granular ones.

The important methodological aspects discussed earlier (modularity and layers) lead to the following *research strategy*.

The strategy for this research paper is to determine the relevance of the cultural landscape policies and then to establish a new framework for development of new policies for the protection and development of cultural landscapes. In this matter, we need to determine the minimal number of modules that could be enough to work with. The modules we consider important to the research that will also generate the research questions are: (1) understanding the cultural landscape concept; (2) determining the relevance of current policies; (3) choosing the open-source development model; (4) understanding new technologies for the policy-making process; and (5) drafting a new framework. Having those modules clearly laid down, we can determine if the working drafts of the research questions would be: (1) *Could actual cultural landscape policies be developed using smart and open-source frameworks*; and (2) *could these policies still be relevant in a constantly changing world*?

The purpose of this paper is not to give a definitive answer to all these questions, but to sketch a draft for a possible smart and open-source framework because this research is still open and will continue in the coming years, and I hope that it will generate enough interest from academia and public sector.

The main idea that should be kept in mind is that we do not want to understate the importance of the existing policies, but rather we would like to build upon them.

3 Results and Discussion

Dealing with the complex concept of cultural landscapes proved to be challenging as it involves dealing with the legislative aspect of building a relevant and critical argumentation in cultural-landscape policies (in particular) and in urban-planning processes (in general).

We have a remarkably interesting subject to deal within in this paper, and it is right at the border of different discipline studies, straddling human geography, urban studies, landscape architecture, sociology, anthropology ecology and cultural studies, to name just a few. To understand the complexities of a cultural landscape study, the professional policymakers need to understand all the modules and granularity of the specific assets of the land (Wallach 2005).

Being part of a constantly changing and increasingly complex society, all professionals involved in planning need to rethink the way they understand what is happening in the present and thus be able to plan for a better future. In this respect, I believe that anyone involved in the cultural-landscape planning process should at least see the importance of theory and modularity mainly because theory is the utmost important element of any research (Allmendinger 2009), and it is the very foundation of any practice that professionals in the fields of planning would undergo. As a result, there are some elemental concepts that need to be understood in order to cleverly link cultural landscapes with open-source paradigms, and those concepts will be analyzed in the following.

3.1 Understanding the Cultural Landscape Concept

Let us not forget that we are dealing with an overly complex concept that could lead to several ways of seeing and understanding it. In my previous research, I determined at least 21 ways to interpret and work with the concept (Calcatinge 2012: 199–200). Based on methods of interpretation defined by scholars like Ingold (1993), Cresswell (2008), Meinig (1979) and others, I "consider the concept of cultural landscape a concept of disintegrating complexities" (Calcatinge 2012: 199) and thus considered the cultural landscapes as a *worldview* by linking it to concepts of *Weltanschauung* from Martin Heidegger (Calcatinge 2012: 126–129) and Mikhail Bakhtin's *chronotope* and *chronotopy* (Calcatinge 2012: 143–186). In this paper, based on those ideas, I am linking the concept to the new paradigms of open-source development and open innovation (Calcatinge 2017, 2018).

The cultural landscape has a plethora of working definitions, but to make everything simpler, I will cite to the European Landscape Convention of the Council of Europe as the main reference, a document that Romania was amongst the first countries to ratify. Even though the Convention did not give a working definition of the concept of *cultural landscape*, I used as working definitions the ones detailed in the document for *landscape, landscape policy* and *landscape planning* (Council of Europe 2000). The landscape is a result of both human and natural actions (Council of Europe 2000: 2). One other document that defines the cultural landscape is the World Heritage Convention, which also does have a reference to the European Landscape Convention's definition of landscape right in its very definition. Thus "cultural landscapes are cultural properties and represent the "combined works of nature and of man" designated in Article 1 of the Convention" (World Heritage Centre 2008: 85). This can easily be used together with two of my own definitions, namely that the cultural landscape is a worldview/view of reality and a social dialogism/ideal chronotope (Calcatinge 2012: 200). If you want to know the chronotope's definition, you should investigate Bakhtin's writing.

Furthermore, the European Landscape Convention gives definitions for two other terms important for our research: landscape policy and landscape planning. Thus, the first one is considered "an expression by the competent public authorities of general principles, strategies and guidelines that permit the taking of specific measures aimed at the protection, management and planning of landscapes" (Council of Europe 2000: 2) and the latter "means a strong forward-looking action to enhance, restore and create landscapes" (Council of Europe 2000: 2). We would also stress that all other definitions of the concept, given by some of the most influential scholars in the field, would be equally important and could address this need for understanding the cultural landscapes with no issues. In this respect, we should mention Sauer (1963) who is considered the one who introduced the concept of cultural landscape to the Anglo-Saxon geographical academia. According to Sauer, the "cultural landscape is fashioned from a natural landscape by a culture group. Culture is the agent, the natural area is the medium, the cultural landscape is the result" (Sauer 1963: 343). While this is widely accepted, some consider the German geographers Otto Schlueter (James and Martin 1981: 177) and Friedriech Ratzel (Jones 2003: 21) to be the ones that first used the concept in academic writings. Nonetheless, ever since the beginning of the last century, the concept has seen constant changes and several definitions have been offered, thus making it a "chaotic concept" (Jones 2003: 24). Presenting all the definitions here is beyond of the scope of this paper, but for further reading on this subject, I would like you to consider my academic literature analysis from eight years ago (Calcatinge 2012: 30–106).

3.2 Determine the Relevance of Current Policies

Now that we know the definitions we are working with, let us see how relevant current policies on cultural landscapes are. Please keep in mind that we are not referring to all the policies around the world. It would be impossible to do that in this paper. We will exclusively refer to the recent changes of the local Romanian policy landscape. It is about the discussions that the new Cultural Patrimony Code[1] is generating amongst cultural landscape professionals.

First, let us determine what a policy is. According to the Cambridge Dictionary, a policy is "a set of ideas or a plan of what to do in particular situations that has been

---

[1]The documents that will form the new Cultural Patrimony Code will form a new law that is going to regulate the entire specific cultural legislative documents. Most important to notice are the preliminary thesis HG 905/2016 and the current ongoing project SIPOCA 389/ SMIS 115,895 that are developing the reports which will be the base for the future law.

agreed to officially by a group of people, a business organization, a government or a political party". More, this could be a system of certain guiding principles that will determine a desired outcome. For short, a policy could be a declaration of intent.

While I was researching the subject of new relevant policies and thus reading the preliminary theses and the report on cultural landscape for the Cultural Patrimony Code, I came across a rather old article of Jerry R. Skees that I found very relevant for today's practices, too. He states that:

> Through the years, I have learned that simply performing research and getting it published in our professional journals is not enough if we want our work to contribute to improvements in policy. The effective policy researcher must be willing to promote general understanding of the value of his(her) research. He or she must get to know policy makers who are strategic to improved policy and he/she must learn to communicate and interpret research to the layperson. We must learn to cope with politics. (Skees 1994: 43)

Furthermore, another author, Michael Horowitz, says that there are four key concepts important to the discourse of policy relevance (Horowitz 2015). Those are policy significance, accessibility, actionability and public debate. Besides those, most important of all is the quality of the work done for the policy contents. Horowitz (2015) details each of these concepts, saying that significance is "research that has implications for the policy world", accessibility "reflects the essential readability of research [...] for the policy world", and actionability "refers to a recommendation that is possible to implement for the target of the recommendation". Public debate is probably the most known to the casual citizen. It is the key to make the work for future policy relevant to the public, which in some cases is the local community.

Now let me go back to the case study for the Cultural Patrimony Code in Romania, which is developed under an EU financed project. For the final stages of the code, numerous reports (14) were developed by twelve experts hired for the project.[2] The report that is most relevant for this research is the one on cultural landscape, and it was written and researched by only one person, one of the experts hired in the project. By having only one author, the report is most certainly biased and will only reflect one person's opinion, and I find that to be one of the major weaknesses of the report and thus of the entire new code and policy, which I find to be irrelevant and flawed right from the start.

For those wishing to increase the use of their research in the policy process, there are certain risks. The more one's research is used, the more like that the researcher will be forced into providing quick analyses on very complex topics. This makes most of us uncomfortable. Further, it is easy for our colleagues to be critical of this type of work. A certain tension will develop between those who are more academically pure and those who are working with policy decision-makers. I think that this is a healthy tension that must be channelled into a constructive review process (Skees 1994:48).

This is true and should be considered by everyone who is developing any policy. Skees is taking it further by debating what happens to the quickly developed reports

---

[2]More details on the project an on the reports can be found at the following address (in Romanian language): https://www.umpcultura.ro/monumente-istorice_doc_983_rezultatele-proiectului_pg_0.htm.

and who is responsible for the work ethics in a policy development process, by pointing out the major differences between academic researchers and busy and quick to deliver policymakers (Skees 1994: 48). This is an issue open for future debate, and out of the scope of this chapter, but it remains the main problem with current cultural landscape-policy development, nevertheless.

3.3 Choosing the Open-Source Development Model

In my opinion, open source is "the" development model of the future, and you can see this if you have even the slightest interest in technology. Open source gives professionals the opportunity to freely contribute to the source code of thousands of software programs and operating systems. Anyone can contribute to open source, either by writing code, being an advocate, writing documentation, organizing activities and so on. The power and control are given to the community. In one of the most recent and iconic books on the open-source development model, Haff (2018) wrote that:

> Perhaps the biggest change is the realization that it can be such an effective development model as opposed to just a source of user freedoms. As a result, understanding modern open source has to include understanding how it's developed. (Haff 2018: 49)

In order to understand open source, one needs to understand how a project is organized. Basically, a project is based on a minimal framework that comprises of core principles, core goals, decision making and community (Haff 2018: 52). There is also an open-source way that is based on five core principles: transparency, collaboration, release cadence, meritocracy and community.[3] Of all those principles, community has a special role because it is generated by multiple people with common goals and shared values. Big projects have an exceptionally large pool of people who contribute in various ways, from smaller commits to large contributions. A community is formed of regular contributors, users that get involved and users who become contributors and mentors. Besides community, an open-source project needs to have a good set of tools and a solid documentation.

Modularity is an important aspect because it is a good way for contributors to work on small chunks and not on the entire project all at once. It is also a good way to make it more attractive to newcomers. And modularity is best served by teams. Teams are the core component of any successful enterprise or project. Inside them, individuals can work on specific projects, and multiple teams can also join to work together. This is a no brainer that every rule that applies to open-source development is valid in any type of work, even more so in policy development.

Now let us link this open-source model to cultural landscape-policy development. The most important property of an open-source solution is that it is open and, most

---

[3] According to the opensource.com website "the open source way is a set of principles derived from open source software development models and applied more broadly to additional industries and domains": https://opensource.com/open-source-way.

of the time, free. The free part comes either as a no-cost solution or as a freedom enabling one. Being free to use something and to tackle and hack it is the biggest asset of open source. With respect to the cultural landscape-policy development, the freedom should translate into an ability to freely participate into creating policies, into validating results or conducting research.

Some particularly important "rules" of open-source development are being discussed by Benson (2016), who analyses three of the most important books in open-source history. I will only extract from his paper the ideas that I found to be most relevant to our task of determining a smart and open-source framework for cultural-landscape policies. Thus, he argues that "to solve an interesting problem, start by finding a problem that is interesting to you" (Benson 2016: 489); therefore "if you have the right attitude, interesting problems will find you" (Benson 2016: 489).

Furthermore, another remarkably interesting aspect would be that "treating your users as co-developers is your least-hassle route to rapid development" (Benson 2016: 489). What does this really mean for our task at hand? Well, first it means that we will have to be open and start thinking of what kind of problems we want to solve with the policy that we want to develop, by having the right attitude and also by listening to the "users", which is the public (the community). This way, we will be able to look at the problems like a local citizen would. We will need to be open to the public because "the next best thing to having good ideas is recognizing good ideas from your users" (Benson 2016: 489).

To link this open-source model to cultural-landscape development, I argued, a few years ago, that there are four paradigms that apply and are of direct interest to us: the open-source ecology paradigm, the open-source urbanism paradigm, the open-source hardware paradigm and the open-source and DIY paradigm. (Calcatinge 2017: 323–327) These paradigms are directly connected to new technologies that we could use in our cultural landscape projects to help us create cultural-landscape policies. This, combined with the open-source development model, is the key for future proof smart policymaking for cultural landscapes.

3.4   Understanding New Technologies for the Policymaking Process

New technologies are mostly overlooked in cultural landscapes policies because those are mostly related to the cultural past. But let us not forget that the past was once present, and people that were developing during that time were innovators. In this respect, by using new technologies now, we will ensure that future researchers will see the cultural landscapes they will be studying as an imprint of knowledge for us, their past followers. In this respect, the open-innovation paradigm's importance for the developing process is of utmost importance (Calcatinge 2018: 193–194) and thus open-innovation and open-source paradigms both are the most suitable paradigms for future cultural-landscape development. For short, here is what I believe to be the relation between those two.

The relations between the cultural landscapes, open-source hardware/software and open innovation are clearly generated by the efficient use of resources and the production of knowledge. The way and scale those two are leveraged will determine the level of innovation achieved. In this respect, I constantly advice my students enrolled in my Cultural Landscape and Development course to see cultural landscapes as infinite sources of knowledge production and (re)generation, and that all the answers for a local development lay inside the area to be developed, starting from local materials, workforce and, ultimately, knowledge. (Calcatinge 2018: 197)

Knowing and understanding open-source hardware and software represents only one side of this new technology trend as the new buzzwords of the twenty-first century are big data, blockchain and smart cities. And those are, in my opinion, the future for any policy on cultural landscapes and urban development. In a nutshell, it goes like this: "The purpose of the blockchain is to achieve and maintain integrity in distributed systems"; (Drescher 2017: 17) and "The excitement about the blockchain is based on its ability to serve as a tool for achieving and maintaining integrity in purely distributed peer-to-peer systems that have the potential to change whole industries due to disintermediation" (Drescher 2017: 24), and "The problem to be solved by the blockchain is achieving and maintaining integrity in a purely distributed peer-to-peer system that consists of an unknown number of peers with unknown reliability and trustworthiness" (Drescher 2017: 32). Additionally, studies show that blockchain is successfully used, for example, in digital protection of intangible cultural heritage (Huang and Dai 2019), for developing tourism and e-governance in countries like Moldova (Pilkington et al. 2017), in arts, artefacts conservation and trusted libraries of source materials (Whitaker 2019), and in architecture, engineering and construction (Belle 2017).

Big data is another important asset of our contemporary everyday lives. According to Oxford English Dictionary, an authoritative definition of the term big data would be "data of a very large size, typically to the extent that its manipulation and management present significant logistical challenges" (Oxford English Dictionary), and it was first used by NASA researchers in 1997 (Cox and Ellsworth 1997). We use smartphones or tablets to accomplish almost everything, we use computers in almost every activity, so everything is smart, and all data are gathered for various purposes like marketing, science, studies, medicine, surveillance, etc. (Bryant et al. 2008). This is changing our lives entirely and, judging by the speed of technology adoption, every aspect of our lives will be different in the next ten to 20 years. This is where big data plays a major role because it is generated by our activities and digital footprints.

Human interactions facilitated by social media, collaborative platforms and the blogosphere generate an unprecedented volume of electronic trace data every day. These traces of human behaviour online are a unique source for understanding contemporary life behaviours, beliefs, interactions and knowledge flows. The social connections we make online, which reveal multiple types of human connection, are also recorded on a scale and to a level of granularity previously unimaginable, except possibly by science fiction writers (Matei et al. 2017: 1).

In this respect, we could conclude that our contemporary behaviour is quite different from the one of our ancestors, and this means our traits will be different from the ones we are now studying. Hence, the need for a new approach, by understanding new technologies and accepting them as tools for creating new policies. Our cultural landscapes policies cannot escape new technologies, and our future policies will have to consider new technologies, like open-source appropriate technology (OSAT) (Pearce 2012), blockchain, big data and open-source technological tools like WikiHouse, Global Village Construction Set, FarmBot, FarmHack and farmOS (Calcatinge 2018: 196).

3.5 Draft for a New Framework

Let us remember the research questions that were drafted at the beginning of the paper: (1) *Could actual cultural-landscape policies be developed using smart and open-source frameworks*; and (2) would these policies still be relevant in a constantly changing world? These questions can be answered in two ways. The easy way would be to give a direct yes/no answer, which is totally unprofessional, and the other way would be to explain the findings of the research. First let us determine what a framework is and how can it be used for our goal. According to James Tiller "a framework highlights each phase, drawing relationships between them to make sure you're on track with the objectives" (Tiller 2004: 37) and according to the Cambridge Dictionary, a framework is "a system of rules, ideas, or beliefs that is used to plan or decide something".

Creating a policy framework involves several stages, like identifying the need, gathering information, drafting, consulting, final review and approval, implementation and constant monitoring, reviewing and revising. But, by creating new frameworks and policies, a major concern arises as policy capacity of local/regional policymakers/governments could be in deficit. Policy capacity is known as "the set of skills and resources—or competences and capabilities—necessary to perform policy functions" (Wu et al. 2015: 166). As local realities are unique to each place, and policies can only partially benefit from other examples, this means that all the effort should be made by using local talent (Peters 2015; Tiernan 2015; Howlett 2015; Stone et al. 2020). And those local talented policymakers need to be trained in specific policy change techniques (Cerna 2013) and open-source principles. By applying those principles and techniques, the local communities will benefit from a collaborative knowledge construction, in which learning about their own needs, finding specific solutions and implementing them to their benefit are the greatest assets. Creating a new framework implies knowing what a "good" policy design is. According to some studies, a good policy design is based on a set of interacting activities like researching and analysing, design and recommendation, clarification of values and arguments, providing strategic advice, democratization and mediation (Mayer et al. 2004) and is generally based on three major policy models, the linear model, the stage model and the streams model (Hardee et al. 2004: 3–4). All these models have shared components. The linear model includes consecutive steps for policymakers, from prediction and choice to implementation and outcome (Hardee et al. 2004: 3); the stage model has multiple phases (stages) like "an agenda phase, a decision phase and an implementation phase" (Hardee et al. 2004: 3); and the streams model invokes a focus on timing and flow of actions, as three different streams need to connect (problems, politics and policies) (Hardee et al. 2004: 4). All three models have components that can be used in a new policy framework. Another important aspect of the framework and policy design is the effectiveness of the process (Bali et al. 2019), as problem solving should be the utmost important goal, and this should be achieved in a more dynamic way by preparing for the unpredictable (Bali et al. 2019: 3). There is also the issue of research data availability and openness (Hrynaszkiewicz et al. 2020) which should be freely available, in the true spirit of open source.

Thus, a draft for a proposed framework would consider the way problems are identified and how they are addressed, prioritized and managed. The framework will be based on a policy design that emphasizes on creativity and innovation. Modularity is also an important aspect and is used to break down complex problems into smaller, more manageable pieces. Policies should respond to most complex and wicked problems, by using instruments and data that is openly available, thus the entire process of policymaking should be open (Head 2019: 192).

4 Conclusion

As shown in the previous chapters, the framework is the structure, the skeleton of the future policy and is based on the findings of some relevant research. This skeleton has a "core" that is based upon a common set of criteria that can be applied to policies on cultural landscapes and on any other domain, and a "add-on" part that comprises of specific requirements for a specific set of policies. All these can and will be done using the open-source model, because it offers the amount of openness and transparency that a policy need. Cultural-landscape policies need to be relevant and to address a critical argumentation that will benefit the urban and regional studies altogether because the urban and regional planning practice is heavily based on a series of core principles that are similar to the ones known from the open-source development model, and among them are transparency, partnership, public debate, sustainability etc. Answering those principles will ensure that the policies will stay relevant during their intended lifetime.

References

Allmendinger Ph (2009) Planning theory, 2nd edn. Palgrave Macmillan, Hampshire

Bali AZ, Capano G, Ramesh M (2019) Anticipating and designing for policy effectiveness. In: Policy and society, vol. 38, no 1. Taylor & Francis Group, UK, pp 1–13

Belle I (2017) The architecture, engineering and construction industry and blockchain technology. In: JI G, Tong Z (eds) Digital culture, proceedings of 2017 national conference on digital technologies in architectural education and DADA 2017 international conference on digital architecture. China Architecture Industry Publishers, Nanjing, pp 279–284

Benson T (2016) Open source paradigm: a synopsis of The Cathedral and the Bazaar for health and social care. In: Journal of innovation in health informatics, vol. 23, no 2. BCS, The Chartered Institute for IT, pp 488–492

Bryant RE, Katz RH, Lazowska ED (2008) Big data computing: creating revolutionary breakthroughs in commerce, science, and society: a whitepaper prepared for the computing community consortium committee of the computing research association. http://cra.org/ccc/resources/ccc-led-whitepapers/. Accessed 06 June 2020

Calcatinge Al (2012) The need for a cultural landscape theory: An architect's approach. LIT Verlag, Wien

Calcatinge Al (2014) The cryptic path of cultural landscape studies. In: Calcatinge Al (ed) Critical spaces: contemporary perspectives in urban, spatial and landscape studies. LIT Verlag, Wien, pp 207–215

Calcatinge Al (2016) Sustainable living in the Suburban Areas. 5th international conference of thermal equipment, renewable energy and rural development TE-RE-RD 2016 proceedings, 2–4 June 2016, Bulgaria. Politehnica Press, Bucharest, pp 187–192

Calcatinge Al (2017) Open Source paradigms in urban and rural development. 6th international conference of thermal equipment, renewable energy and rural development TE-RE-RD 2017 proceedings, 8–10 June 2017, Romania. Politehnica Press, Bucharest, pp 323–328

Calcatinge Al (2018) The relations between cultural landscapes, open source hardware and open innovation in rural development. 7th international conference of thermal equipment, renewable energy and rural development TE-RE-RD 2018 proceedings, 31 May–2 June 2018, Romania. Politehnica Press, Bucharest, pp 193–198

Cambridge Dictionary (policy). https://dictionary.cambridge.org/dictionary/english/policy. Accessed 28 Dec 2019

Cambridge Dictionary (framework). https://dictionary.cambridge.org/dictionary/english/framework. Accessed 05 June 2020

Cerna L (2013) The nature of policy change and implementation: A review of different theoretical approaches. OECD Publishing, Paris

Council of Europe (2000) European landscape convention. European treaty series—No. 176. Council of Europe Publishing, Strasbourg. Online: https://rm.coe.int/1680080621. Accessed 02 Dec 2019

Council of Europe (2012) Landscape facets: reflections and proposals for the implementation of the European Landscape Convention. Council of Europe Publishing, Strasbourg. Online: rm.coe.int/09000016802f299b. Accessed 02 Dec 2019

Cox M, Ellsworth D (1997) Application-controlled demand paging for out-of-core visualization. In: Proceedings of the 8th VIS97: IEEE visualization '97 conference, Phoenix Arizona. IEEE Computer Society Press, Washington DC, pp. 235–ff

Cresswell T (2008) Landscape and the obliteration of Practice (2003). In: Nuala CJ (ed) Culture and society. critical essays in human geography. Ashgate Publishing Ltd, Aldershot

Drescher D (2017) Blockchain basics: a non-technical introduction in 25 Steps. Apress, New York

Fleishman H (2020). It's 2020. Let's stop saying Iot (Part II). https://www.forbes.com/sites/hodfleishman/2020/01/15/its-2020-lets-stop-saying-iot-part-ii/. Published on January 15, 2020. Accessed 18 Jan 2020

Haff G (2018) How open source ate software. Apress, Lancaster

Hardee K, Feranil I, Boezwinkle J, Clark B (2004) The policy circle: a framework for analyzing the components for family planning, reproductive health, maternal health and HIV/AIDS policies. POLICY Working Papers Series No. 11, June 2004. USAID: The POLICY Project

Head BD, (2019) Forty years of wicked problems literature: forging closer links to policy studies. In: Policy and society, vol. 38 no. 2. Taylor & Francis Group, UK, pp 180–197

Huang W, Dai F (2019) Research on digital protection of intangible cultural heritage based on blockchain technology. Information management and computer science (IMCS) Volume 2 number 2. Zibeline International, Malaysia, pp 14–18

Horowitz M (2015) What is policy relevance? in texas national security review. special series—the schoolhouse. Published online June 17, 2015. Online https://warontherocks.com/2015/06/what-is-policy-relevance/?singlepage=1. Accessed 12 Jan 2020

Howlett M (2015) Policy analytical capacity: The supply and demand for policy analysis in government. In: Policy and society, vol. 34, no. 3–4. Taylor & Francis Group, UK, pp 173–182

Hrynaszkiewicz I, Simons N, Hussain A, Grant, R, Goudie S (2020). Developing a research data policy framework for all journals and publishers. Data Sci J 19(5):1–15. CODATA Ubiquity Press, Paris

Ingold T (1993) The temporality of landscape. World Archaeol 25(2):152–174

James PE, Martin G (1981) All possible worlds: a history of geographical ideas. Wiley, New York

Jones M (2003) The concept of Cultural Landscape: discourse and narratives. In: Palang H, Fry G (eds) Landscape interfaces. Cultural heritage in changing landscapes. Kluwer Academic Publishers, Dordrecht, pp 21–51

Matei SA, Jullien N, Goggins SP (2017) Big data factories: collaborative approaches. Computational social series. Springer, Cham, Switzerland

Meinig DW, (1979) The beholding eye. In: Meinig, DW (ed) The interpretation of ordinary landscapes. Oxford University Press, New York, pp 33–50

Mayer I, Els van Daalen C, Bots PWG (2004) Perspectives on policy analyses: a framework for understanding and design. International journal of technology, policy and management, vol. 4 no 2. Wiley Interscience Publishers, UK, pp 169–191

Oxford English Dictionary. https://www.oed.com/view/Entry/18833#eid301162177. Accessed 05 June 2020

Pearce JM (2012) The case of open source appropriate technology. In: Environment, development and sustainability. A multidisciplinary approach to the theory and practice of sustainable development, vol. 14, no. 3. Springer, Cham, Switzerland, pp 425–431

Peters GB, (2015) Policy capacity in public administration. In: Policy and society, vol. 34, no 3–4. Taylor & Francis Group, UK, pp 219228

Pilkington M, Crudu R, Grant LG (2017) Blockchain and bitcoin as a way to lift a country out of poverty—tourism 2.0 and e-governance in the Republic of Moldova. Int J Int Technol Secur Trans 7(2):115–143

Sauer CO (1963) The morphology of landscape. In: John L (ed) Land and life. A selection from the writing of Carl Ortwin Sauer. University of California Press, Berkeley, pp 315–350

Skees JR (1994) Relevance of policy analysis: needs for design, implementation and packaging J Agricul Appl Econ 26(1):43–52. Southern Agricultural Economics Association

Stone D, de Oliveira OP, Pal LA (2020) Transnational policy transfer: the circulation of ideas, power and development models. Policy and society, vol. 39, no 1. UK, Taylor & Francis Group, pp 1–18

Tiernan A (2015) The dilemmas of organisational capacity. Policy and society, vol. 34, no 3–4. UK, Taylor & Francis Group, pp 209–217

Tiller JS (2004) The ethical hack: a framework for business value penetration testing. Auerbach Publications, Washington

Wallach B (2005) Understanding the cultural landscape. Guilford Press, New York

Whitaker A (2019) Art and Blockchain: a primer, history, and taxonomy of blockchain use cases in the arts. In: Artivate, vol. 8, no. 2. University of Arkansas Press, Arkansas, pp 21–46

World Heritage Centre (2008) Operational guidelines for the implementation of the world heritage convention. Guidelines on the inscription of specific types of properties on the World Heritage List Annex 3. UNESCO World Heritage Centre, Paris

Wu X, Ramesh M, Howlett M (2015) Policy capacity: a conceptual framework for understanding policy competences and capabilities. Policy and Society, vol. 34, no. 3-4. UK, Taylor & Francis Group, pp 165–171

Printed in the United States
by Baker & Taylor Publisher Services